Applied Tools and Techniques in Drilling Engineering

Applied Tools and Techniques in Drilling Engineering

Edited by **Alexis Federer**

SYRAWOOD
PUBLISHING HOUSE

New York

Published by Syrawood Publishing House,
750 Third Avenue, 9th Floor,
New York, NY 10017, USA
www.syrawoodpublishinghouse.com

Applied Tools and Techniques in Drilling Engineering
Edited by Alexis Federer

© 2016 Syrawood Publishing House

International Standard Book Number: 978-1-68286-164-6 (Hardback)

Printed in the United States of America.

Contents

Preface

Drilling engineering has a diverse range of applications such as sedimentary rock dating, study of the earth's structure, processes happening in inner earth, etc. This book is compiled in such a manner, that it will provide in-depth knowledge about the theories and techniques of drilling. It discusses in detail the varied aspects of drilling such as seismology, types of drilling, methods of drilling, tectonic processes, geochemistry, etc. This book will serve as a reference to a broad spectrum of readers.

Various studies have approached the subject by analyzing it with a single perspective, but the present book provides diverse methodologies and techniques to address this field. This book contains theories and applications needed for understanding the subject from different perspectives. The aim is to keep the readers informed about the progress in the field; therefore, the contributions were carefully examined to compile novel researches by specialists from across the globe.

Indeed, the job of the editor is the most crucial and challenging in compiling all chapters into a single book. In the end, I would extend my sincere thanks to the chapter authors for their profound work. I am also thankful for the support provided by my family and colleagues during the compilation of this book.

Editor

IODP Expedition 310 Reconstructs Sea Level, Climatic, and Environmental Changes in the South Pacific during the Last Deglaciation

by Gilbert F. Camoin, Yasufumi Iryu, Dave B. McInroy and
the IODP Expedition 310 Scientists

Introduction and Goals

The timing and course of the last deglaciation (19,000–6,000 years BP) are essential components for understanding the dynamics of large ice sheets (Lindstrom and MacAyeal, 1993) and their effects on Earth's isostasy (Nakada and Lambeck, 1989; Lambeck, 1993; Peltier, 1994), as well as the complex relationship between freshwater fluxes to the ocean, thermohaline circulation, and, hence, global climate during the Late Pleistocene and the Holocene. Moreover, the last deglaciation is generally seen as a possible analogue for the environmental changes and increased sea level that Earth may experience because of the greenhouse effect, related thermal expansion of oceans, and the melting of polar ice sheets.

Corals are excellent sea level indicators and can be accurately dated; therefore, studying them helps in the determination of the timing of deglaciation events and the understanding of the mechanisms driving the glacial-interglacial cycles. Coral reefs are also sensitive recorders of past climatic and environmental changes. The skeletal geochemistry of annually-banded massive corals can provide a record of sea surface temperatures (SSTs) and salinities (SSSs). Because the amplitude of the last deglacial sea level rise was at least 120 m (Barbados: Fairbanks, 1989; Bonaparte Basin: Yokoyama et al., 2001; and review in Lambeck et al., 2002), the relevant reef and sediment archives are mostly found on modern fore-reef slopes where they can be investigated by dredging, submersible sampling, and coring. However, the scarcity of such cores and related data hinders our ability to unravel the rate and timing of the last deglacial sea level rise and prevents us from understanding the role of the Pacific Ocean as a climate modulator during the course of postglacial climate change.

Sea level changes and reef development during the last deglaciation: The magnitude and rates of eustatic changes constrain the volumes of ice that accumulated on the continents during the last glacial period, including the Last Glacial Maximum (LGM), and provide direct evidence of the progress of melting of large ice sheets during the last deglaciation.

So far, only three deglaciation curves based on coral reef records have been accurately dated for times reaching the Pleistocene-Holocene boundary—in Barbados between 19,000 and 8,000 yr BP (Fairbanks, 1989; Bard et al., 1990); in Papua New Guinea between 13,000 and 6,000 yr BP (Chappell and Polach, 1991; Edwards et al., 1993), and in Tahiti between 13,850 and 3,000 yr BP (Bard et al., 1996). Of these, the only coral reef record that encompasses the whole last deglaciation is from the Barbados, where it was suggested that this period was characterized by two brief intervals of accelerated melting (meltwater pulses MWP-1A and MWP-1B at ~14,000 and 11,300 yr BP, respectively) superimposed on a smooth and continuous rise of sea level with no reversals (Fairbanks, 1989; Bard et al., 1990). These events would correspond to massive inputs of continental ice to the oceans (i.e., ~50–40 mm y^{-1}, roughly equivalent to an annual discharge of 16,000 km^3 for MWP-1A), and they are thought to have induced reef drowning events (Blanchon and Shaw, 1995). However, the abrupt and significant environmental changes that accompanied the deglacial rise in sea level have barely been investigated, leaving the accurate reconstruction of the event obscured.

The timing, the amplitude, and even the realities of those periods of accelerated sea level rise have been actively debated (Bard et al., 1996; Okuno and Nakada, 1999; Lambeck et al., 2002; Clark et al., 2004). Uncertainties con-

Figure 1. View of Tahiti from the *DP Hunter* during IODP Expedition 310. The picture was taken in the Maraa area in Southern Tahiti while stationed at Site M0007. Photo by Rolf Warthmann.

cerning the general pattern of sea level rise during this time period remain because the apparent sea level record may not be free of tectonic or isostatic complications. The Barbados sea level curve was derived from three separate drowned reefs, and each of these segments is offset from the next. Moreover, these offsets constitute the two rapid meltwater surges; however, because these sites are located close to subduction zones, sea level records derived from them may be impacted by unknown tectonic movements. It is therefore preferable to obtain sea level data from tectonically stable regions truly distant from the former ice sheets ("far-field" sites), sites that are minimally affected by isostatic rebound in relation to changes in ice loading of the lithosphere (Mix et al., 2001; Clark and Mix, 2002).

Tahiti is a volcanic island characterized by slow and regular subsidence rates (~0.25 mm yr⁻¹; Bard et al., 1996) and located at a considerable distance from the major former ice sheets. Therefore, it provides an ideal setting to reconstruct the deglacial rise in sea level and to constrain short-term events that are thought to have punctuated the period between the LGM and the present days.

Climatic variability during the last deglaciation: The tropical Pacific Ocean is known to play an important role in driving and modulating global climate variability and change on a wide range of timescales (Bjerknes, 1969; Pickard and Emery, 1990). Furthermore, climate modeling of global warming implies that the tropical Pacific is pivotal in modulating the timing, regional expression, and magnitude of climate change.

The tropical Pacific may have played an important role in glacial-interglacial timescale climate change, but currently there is debate over exactly what this role was. Additional information is therefore required for a better knowledge on climatic conditions in tropical regions during the last deglaciation, including the changes in the seasonal cycle (amplitude and structure) that are likely to yield important insights into the mechanisms and drivers of climate variability and change.

Scientific objectives of the expedition were:

A) To establish the course of postglacial sea level rise during the last deglaciation in order to 1) establish the minimum sea level during the LGM; 2) assess the validity, timing, and amplitude of meltwater pulses and thereby identify the exact sources of the ice responsible for these rapid sea level steps; 3) prove or disprove saw-tooth pattern of sea level rise; and 4) test predictions based on different ice and rheological models.

B) To define variations in SSTs and SSSs during the last deglaciation when solar insolation, sea level, and atmospheric CO_2 levels were different from today. The objectives are to 1) reconstruct interannual-decadal climate variability and seasonality (amplitude and structure) in the South Pacific; 2) reconstruct variability and change in interannual including

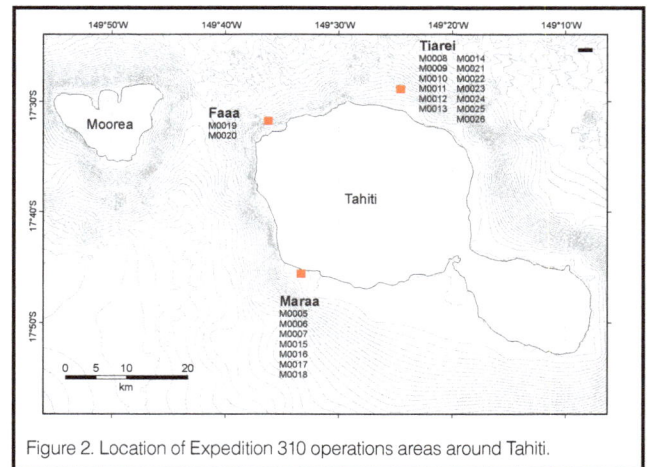
Figure 2. Location of Expedition 310 operations areas around Tahiti.

El Nino Southern Oscillation (ENSO) and decadal-interdecadal including Pacific Decadal Oscillation (PDO)/ Interdecadal Pacific Oscillation (IPO) variability; 3) compare the global variation and relative timing of postglacial warming between the tropical Pacific and the mid- and high-, northern and southern latitudes; and 4) determine major changes in tropical sea surface salinity.

C) To analyze the impact of sea level and environmental changes on reef development during the last deglaciation, with a special emphasis on the comprehensive reconstruction of environmental changes (e.g., nutrient concentrations, variations in pH and alkalinity, paleoproductivity, terrigenous and freshwater fluxes), the evolution of the geometry, biological composition, and growth mode of reef frameworks.

D) To investigate the geomicrobiology processes on the Tahiti fore-reef slopes to study potential modern counterparts of the microbialites that characterize the last deglacial reef sequence drilled in Papeete (Camoin and Montaggioni, 1994; Camoin et al., 1999). The objectives are to 1) identify the microbial communities that are involved in their formation and 2) to have a better understanding of the environmental significance of those microbial fabrics.

Drilling Sites and Operational Strategy

At Tahiti, recovery of the last deglacial reef sequence required drilling successive reef terraces of various lateral extent that occur at various depths (100, 90, 60, and 40–50 m) seaward of the living barrier reef (Camoin et al., 2003). Preliminary sedimentological and chronological results on the dredged samples have confirmed the significance of these features as unique archives of abrupt global sea level rise and climate change (Camoin et al., 2006). Based on the results of previous drillings on the Papeete reef (Bard et al., 1996; Montaggioni et al., 1997; Camoin et al., 1999 ; Cabioch et al., 1999a) and of the SISMITA cruise (Camoin et al., 2003), we drilled a transect of holes in three areas around Tahiti—offshore Faaa, Tiarei, and Maraa (Figs. 1 and 2). The final and exact locations of the drill holes was determined during the drilling cruise by mapping the nature and morphology of the seafloor with a through-pipe underwater

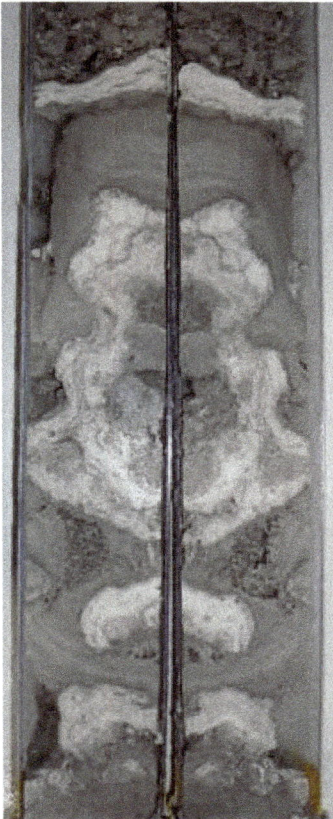

Figure 3. Core displaying coralgal frameworks heavily encrusted with microbialites (laminated fabrics overlain by dendritic growth forms).

camera. More than 600 meters of reef cores with an exceptional recovery (>90% of the rocks) and quality (Fig. 3) were recovered from thirty-seven holes at twenty-two sites (M0005–M0026) ranging from 41.6 m to 117.5 m water depth. Cores were recovered from 41.6 m to 161.8 m below sea level (mbsl). The recovery was therefore much higher than that obtained during previous Ocean Drilling Program Legs 143 and 144 (Sager et al., 1993; Premoli Silva et al., 1993) related to shallow-water carbonates.

Cored material shows that the fossil reef systems around Tahiti are composed of two major lithological units: a last deglacial sequence (Unit I) and an older Pleistocene sequence (Unit II). Those two major units have been divided into subunits based on coral assemblages and internal structure.

The set of deployed borehole geophysical instruments was constrained by the scientific objectives and the geological setting of the expedition. A suite of downhole geophysical methods was chosen to obtain high resolution images of the borehole wall (OBI40 and ABI40 televiewer tools; Fig. 4), to characterize the fluid nature in the borehole (IDRONAUT tool), to measure borehole size (CAL tool), and to measure or derive petrophysical and geochemical properties of the reef units such as porosity, electrical resistivity (DIL 45 tool), acoustic velocities (SONIC tool), and natural gamma radioactivity (ASGR tool). A total of ten boreholes were prepared for downhole geophysical measurements, which were performed under open borehole conditions (no casing) with the exception of a few spectral gamma ray logs. Nearly complete downhole coverage of the postglacial reef sequence has been obtained from 72 mbsl to 122 mbsl and from 41.65 mbsl to 102 mbsl at the Tiarei sites and the Maraa sites, respectively. Partial downhole coverage of the underlying older Pleistocene carbonate sequence has been acquired at both sites.

Samples from Tahiti drill cores were analyzed for evidence of microbial activity, possibly related to the formation of microbialites. Onboard measurements have shown a certain degree of microbial activity directly attached to rock surfaces; cultivation and microscopic observations were also carried out onboard. According to the adenosine triphosphate (ATP) activity measurements along the drill cores, the uppermost part (0–4 mbsf of the Tahiti reef slopes) is the most active zone. Pure microbiological activity was only observed in reef cavities where prokaryotic biofilms have appropriate conditions to develop. Northwestern Faaa Hole M0020A and southwestern Maraa Holes M0005C, M0007B, M0007C, M0015B, and M0018A were more active than the northeastern Tiarei sites where usually no living biofilm could be detected in cavities along the cores.

Last Deglacial Sequence

Composition: The last deglacial sequence is mostly composed of coralgal frameworks heavily encrusted with microbialites (Fig. 3), locally associated or interlayered with skeletal limestone and/or loose skeletal sediments (rubble, sand, and silt) rich in fragments of corals, coralline and green (_Halimeda_) algae, and, to a lesser extent, bryozoans, echinoids, mollusks, and foraminifers (mostly _Amphistegina_ and _Heterostegina_). The amounts of volcaniclastic sediments (e.g., silt- to cobble-sized lithic volcanic clasts, mineral fragments, clays) are highly variable, from mere sand and silt impurities in the carbonate rock units to minor components (<50 vol-%) in carbonate sand units to major components (>50 vol-%) in sand/silt (or sandstone/siltstone) interbedded with carbonate beds. The last deglacial sequence at Tiarei has a greater volcaniclastic component than the ones at Maraa and Faaa; this difference is observed on digital image scans and quantified by diffuse color reflectance spectrophotometry and magnetic susceptibility core logs.

Corals are well preserved, forming seven distinctive coral assemblages characterized by various morphologies (branching, robust branching, massive, tabular, foliaceous, and encrusting) that determine distinctive framework internal structure. Several coral assemblages are intergradational vertically and laterally. Robust branching (e.g., _Pocillopora_ and _Acropora_) and, to a lesser extent, tabular (e.g., _Acropora_) corals are usually thickly encrusted with nongeniculate coralline algae and microbialites to form dense and compact frameworks; vermetid gastropods and serpulids are locally associated with coralline algae. Foliaceous and encrusting coral colonies (e.g., _Montipora_, _Porites_, _Pavona_, _Leptastrea_, _Psammocora_, _Astreopora_, agariciids and faviids) are thinly coated with coralline algae and microbialites to form loose frameworks. Large primary cavities in coralgal frameworks are open or partially filled with skeletal sands and gravels locally mixed with volcanic elements. The open framework and centimeter- to decimeter-sized pores result in a highly variable system in which physical properties change on a centimeter scale and may range from low porosity, high density, and velocity to 100% open pore space.

Abundant microbialites represent the major structural and volumetric component of the last deglacial reef sequence (Figs. 3 and 4). The presence of microbialites at all depths in the sequence indicates that they formed in various parts of the reef tracts (Camoin et al., 2007). They developed within the primary cavities of the reef framework, where they generally overlie coralline algal crusts to form dark gray massive crusts as thick as 20 cm; they also develop in bioerosion cavities. Microbialites correspond to a late stage of encrustation of the dead parts of coral colonies, or more commonly, of related encrusting organisms (coralline algae and crustose foraminifers), implying that some time elapsed prior to the formation of the microbialites and that there was generally no direct inter-community competition for space between coralgal and microbial communities. This is also reflected by the age differences between corals, encrusting organisms, and overlying microbialite crusts (Camoin et al., 2007). Microbialites generally comprise a suite of fabrics, including two end-members represented by laminated fabrics and thrombolitic to dendritic accretions; thrombolites usually represent the last stage of encrustation. The thickness and morphology of the microbial crusts are closely related to the morphology and size of the cavities in which they are developed. In bindstone formed by encrusting coral assemblages, microbialites are dominated by thrombolitic fabrics, whereas in frameworks made of branching and massive coral colonies, they are characterized by the development of compound crusts, up to 15 cm thick, formed by a succession of laminated and thrombolitic fabrics. Preliminary results obtained on lipid biomarkers imply that carbonate precipitation processes were related to the activity of heterotrophic bacteria and were in agreement with the stable isotope data (Heindel et al., 2007).

At all sites, the top of the last deglacial carbonate sequence is characterized by the widespread development of thin coralline algal crusts indicating deep water environments. Extensive bioerosion, dark yellow-reddish to brown staining (manganese and iron) of the rock surface, and hardgrounds are common within the top 2–3 m of the sequence.

Distribution of the last deglacial reef sequence: At all sites, the last deglacial sequence displays similar trends, although it displays specific characters in each area.

The Tiarei area is characterized by the occurrence of two successive ridges seaward of the living barrier reef (Camoin et al., 2003; Camoin et al., 2006). The outer ridge coincides with a marked break in slope, and its top is located at 90–100 mbsl, whereas the inner ridge is located on an extensive terrace, and its top occurs at 60 mbsl. On the outer ridge, the last deglacial sequence was recovered from 81.7 mbsl to 122.12 mbsl at sites 9, 21, 24, 25 and 26; the thickest continuous sequence, 29.98 m, was recovered from Holes M0021A and M0021B (Fig. 6). On the inner ridge, the last deglacial sequence (recovered from 68 mbsl to 92–98 mbsl at site 23) is 24–30 m thick. On both sides of the outer ridge (Holes

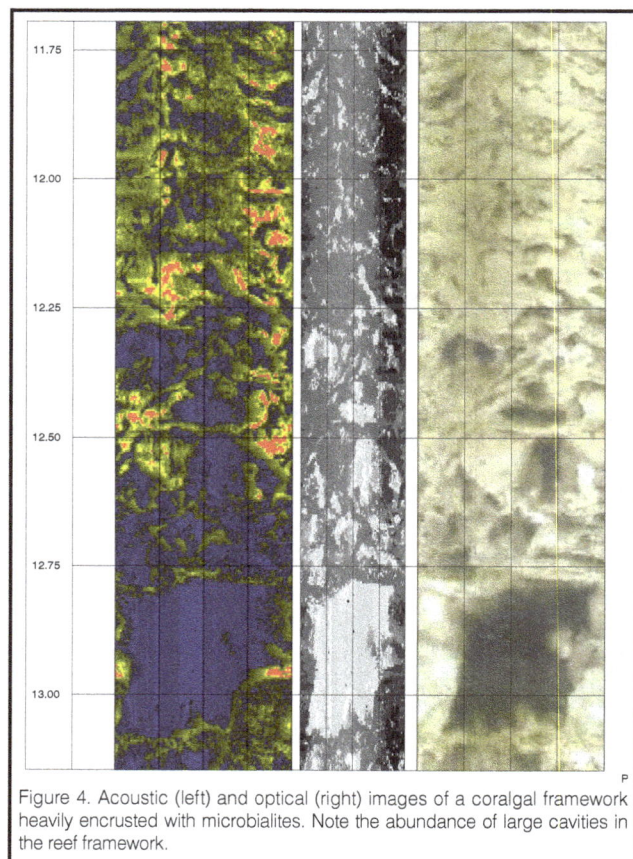

Figure 4. Acoustic (left) and optical (right) images of a coralgal framework heavily encrusted with microbialites. Note the abundance of large cavities in the reef framework.

M0010A–M0014A), the last deglacial sequence (recovered from 78.85 mbsl to 117.98 mbsl) includes reef units overlain by volcaniclastic sediments. At Hole M0008A, which is located on the flat area extending between the two ridges, the 38.7-m-thick sequence (64.15–102.85 mbsl) consists only of volcaniclastic sediments.

Two transects were drilled in the Maraa area. The western transect includes Sites M0007, M0005, and M0006 (landward to oceanward) in water depths ranging from 41.65 m to 81.58 m, and the thickness of the last deglacial sequence ranges from 33.22 m to 44.56 m at sites M0005 and M0007, respectively (Fig. 7). Sites M0017, M0015, M0018, and M0016 (landward to oceanward) in water depths ranging from 56.45 m to 81.8 m define the eastern transect; the thickness of the last deglacial sequence ranges from 35.44 m in Hole M0015A at 72.15 mbsl to 39.05 m in Hole M0018A at 81.8 mbsl (Fig. 8).

In the Faaa area, the last deglacial sequence is 62.5 m and 36.5 m thick at Sites M0019 (58.5 mbsl) and M0020 (83.3 mbsl), respectively.

Sea level issues: Coralgal assemblages identified in the Tahiti cores can be considered as reliable depth indicators for the reconstruction of sea level changes. Dominant species of coralline algal assemblages include *Hydrolithon onkodes*, *H. reinboldii*, *Lithophyllum insipidum*, *L. pygmaeum*, *Neogoniolithon brassica-florida,* and *Mesophyllum erubescens*;

Figure 5. View of corals from the surface of the Tahiti reefs. Photo by Gilbert Camoin

they provide paleowater depth estimates based on their comparison with modern counterparts. Coral assemblages are dominated by *Porites*, *Pocillopora*, *Acropora*, and *Montipora* genera, which form distinctive assemblages that are indicative of a range of modern reef environments, from the upper to middle reef slope to deep reef slope (0–30 m deep; Sugihara et al., 2006), in agreement with earlier studies on Tahiti (Montaggioni et al., 1997; Camoin et al., 1999; Cabioch et al., 1999a) and other Indo-Pacific reef sites (Camoin et al., 1997, 2004; Montaggioni and Faure, 1997; Cabioch et al., 1999b; Sagawa et al., 2001).

Corals are well preserved, as the great majority of the specimens exhibit less than 1% calcite in their skeleton, indicating that they were not subjected to diagenetic alteration. Preliminary U-series dating results of selected corals indicate that the cored last deglacial sequence covers at least the 16,000–8,000 yrs BP time span, and suggest a non-monotonous sea level rise during that period. Additional data will be necessary to constrain it accurately (Deschamps et al., 2006, 2007). Tahiti is therefore the second region, after Barbados, where coral reef records encompass the MWP-1A and MWP-1B events centered at ~14,000 yr BP and 11,300 yr BP, respectively.

Paleomagnetic and rock magnetic testing are being carried out on all of the studied cores. The natural remanent magnetization (NRM) data obtained on microbialites show

that directions have an average magnetic inclination that is very close to that expected at the site latitude from a simple dipole field. Moreover, the stratigraphic variations in inclination can be correlated between local sites, and they may also provide an independent chronostratigraphy for estimating the timing of microbialite growth within the reef primary cavities. Stratigraphic variations in several rock magnetic parameters can also be correlated between local sites, which may provide an independent record of environmental variability within the coral reef over time.

Reef development issues: At each drill site, the cored reef sequences are continuous, implying that there was no major break in reef development during the 16,000–8,000 yrs BP time span, and thus raising the question of the occurrence of reef drowning events as described in the Barbados record (Blanchon and Shaw, 1995). This possibility suggests that environmental conditions in Tahiti were optimal for reef development, and no long-term environmental changes occurred during that period, although changes in coralgal assemblages may reflect variations in environmental parameters (e.g., water depth and energy, light conditions, terrigenous fluxes, nutrient concentrations, etc.). The deeper water facies that form the top of the last deglacial sequence occur gradually shallower towards the modern reefs along the drilled transects, thus indicating a general backstepping of the reef complex as a response to sea level rise during the last deglaciation.

Paleoclimate issues: About thirty meters of the reef cores consist of massive coral colonies, mostly of the genus *Porites*, and concern successive time windows covering most of the 16,000–8,000 yrs BP period. For paleoclimatic objectives, preliminary sub-seasonal records have been already obtained on selected time windows (Felis et al., 2007).

Older Pleistocene Carbonate Sequence

An older Pleistocene carbonate sequence has been recovered below the last deglacial sequence at most sites (Figs. 6, 7, and 8). The contact between the last deglacial sequence and the older Pleistocene sequence is characterized by an irregular unconformity. The change in physical properties at this contact is sharp and abrupt—density, velocity, and magnetic susceptibility increase, and porosity decreases. The depth below present-day sea level of the top of the older Pleistocene sequence is highly variable, ranging from about 90 mbsl on the inner ridge of Tiarei and proximal Maraa sites to 122 mbsl on the outer ridge of Tiarei and distal Maraa sites. This variability indicates a rugged morphology of the Pleistocene carbonate substrate prior to the development of the last deglacial sequence.

Composition of the older Pleistocene carbonate sequence: The older Pleistocene carbonate sequence is mostly comprised of reef deposits and has been detailed in holes M0009D (Tiarei area; 122–147 mbsl; Fig. 6) and M0005D (Maraa area

92–162 mbsl; Fig. 7). Three major distinctive lithological subunits are usually closely associated:

• well lithified skeletal packstone/grainstone to floatstone/rudstone, rich in rhodoliths, and coral and algal fragments,
• well lithified coralgal frameworks exhibiting microbialite fabrics and associated with skeletal packstone/grainstone to floatstone, and
• rubbles and gravels primarily composed of coral clasts, limestone clasts, volcanic pebbles, and reworked coral colonies.

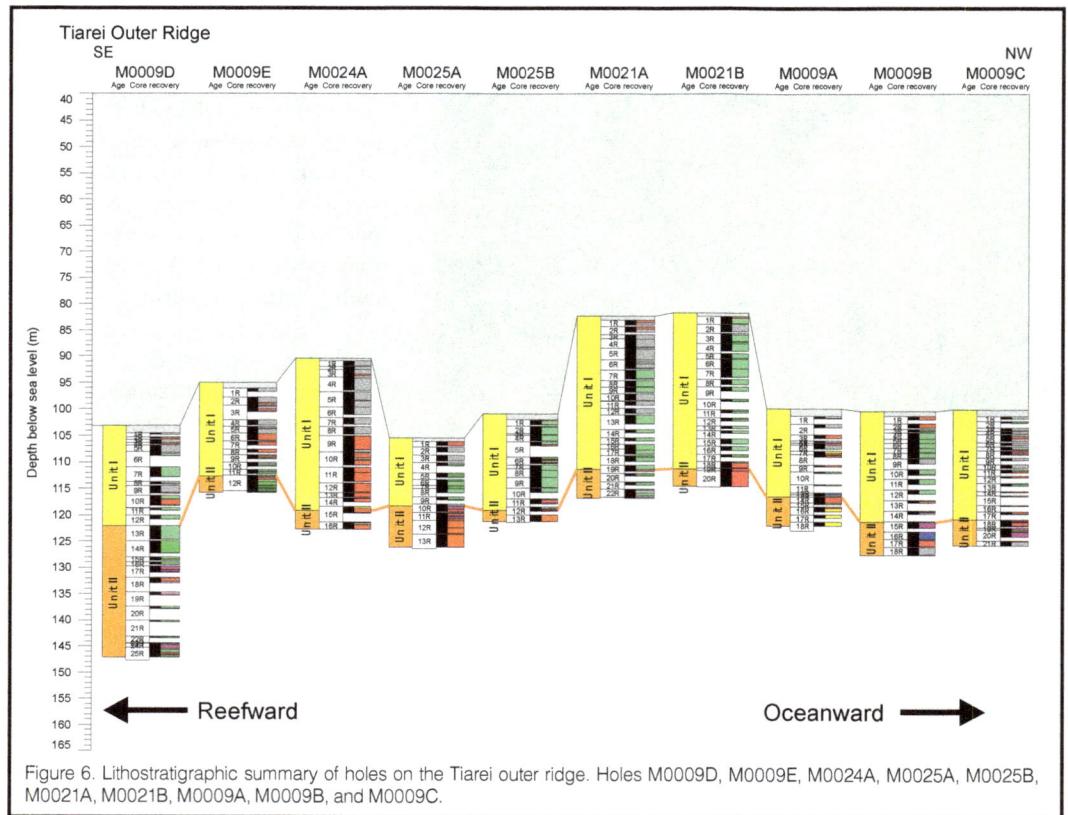

Figure 6. Lithostratigraphic summary of holes on the Tiarei outer ridge. Holes M0009D, M0009E, M0024A, M0025A, M0025B, M0021A, M0021B, M0009A, M0009B, and M0009C.

Ungraded unlithified volcaniclastic silt to sand, including scattered skeletal grains and sandy skeletal grainstone rich in volcanic grains, large coral clasts, and skeletal fragments, are clearly subordinate.

Several sedimentological features indicate that the older Pleistocene carbonate sequence has undergone several phases of diagenetic alteration resulting in tight, low-porosity, high-velocity limestones with much less variation in physical properties than observed for the last deglacial sequence. Several successive unconformities occur in the upper part of the older Pleistocene carbonate sequence and suggest that this sequence is composed of successive chronological units. Those unconformities are associated with several lines of evidence of subaerial diagenetic alteration, including the alteration of coral skeletons and skeletal grains, and the occurrence of abundant solution cavities rimmed with multiple generations of cement crusts and/or displaying several generations of sediment infillings and yellow and brown to red-brown staining. Local multiple bored and encrusted surfaces (hardgrounds) occur at the top of that sequence.

Chronology: The dating potential of the corals occurring in that sequence is generally limited by the amount of diagenesis they have suffered, but the preservation of some specimens was good enough to obtain reliable ages by U-series dating (Thomas et al., 2006).

Paleomagnetic studies provide chronological constraints on reef unit succession by identifying well dated paleointensity lows and geomagnetic excursions. The NRM is mostly carried by a mixture of titano-magnetite grains in the pseudo-single domain range. Based on the occurrence of excursional instabilities in two depth intervals, reef units have been attributed to highstand (isotopic stages 5 and 7) and lowstand reef units (isotopic stages 6 and 8) (Ménabréaz, 2007).

Geomicrobiology of Modern Reef Slopes

Biofilms recovered during the IODP Expedition 310 are diverse in structure and color. Additionally, finely laminated, 1-mm-diameter vertically upward growing columns referred to as endostromatolites were found within some reef cavities (McKenzie et al., 2007). Biofilms appear to have high diversity in macroscale observations, and they are equally diverse and heterogeneous in microscale resolution, as observed by scanning electron microscopy (SEM). In some samples, it was possible to define spherical assemblages of calcium carbonate minerals embedded in microbial exopolymeric substances (EPS).

Some evidence for heterotrophic metabolic activities is shown by exoenzyme measurements, which vary in the different biofilm samples. For instance, samples from holes M0020A (4.51 mbsf) and M0009D (3.64 mbsf) showed high phosphatase activity, suggesting the occurrence of a heterotrophic community that preferentially degrades organic-

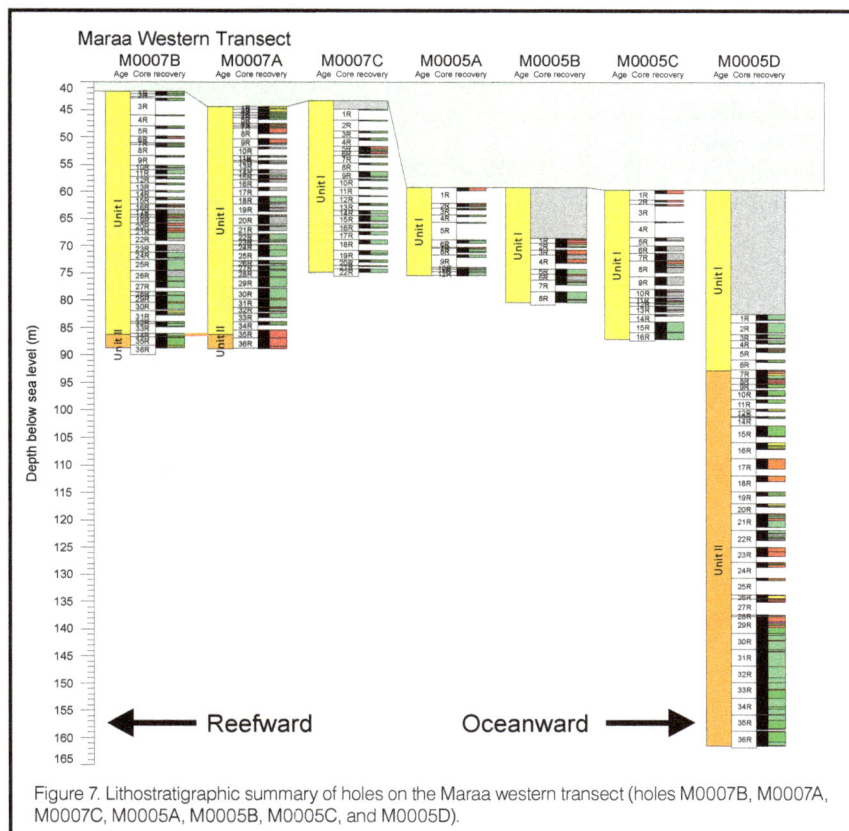

Figure 7. Lithostratigraphic summary of holes on the Maraa western transect (holes M0007B, M0007A, M0007C, M0005A, M0005B, M0005C, and M0005D).

The preliminary dating results indicate that the cored last deglacial sequence should cover at least the 16,000–8,000 yrs BP time span. Tahiti is therefore the second region, after Barbados, where coral reef records encompass the meltwater pulse events centered at ~14,000 yr BP and 11,300 yr BP (MWP-1A and MWP-1B, respectively). Huge coral colonies, chiefly *Porites* genus, were found in thirty meters of reef cores; these cores cover successive time windows in most of the 16,000–8,000 years period. The study of those colonies will address interannual/decadal-interdecadal climate variability and seasonality (amplitude and structure) in the south Pacific during the last deglaciation, and will present data that might aid modelling of future global warming. In particular, the high recovery combined with the high resolution downhole measurements provides potential for resolving in unprecedented detail the abrupt and significant environmental changes that accompanied the deglacial sea level rise.

The investigation of the 200 meters of cores retrieved from the older Pleistocene carbonate sequences might provide fragmentary information on sea level and reef growth for time windows concerning several isotopic stages.

The geomicrobiological pilot study of living biofilms bearing carbonate precipitates that were sampled *in situ* within cavities underpins a new field of research regarding the development of microbial communities on modern reef slope environments. It will be useful to the nature and environmental significance of microbialite fabrics that form an essential component of the last deglacial reef sequence.

Plans for supplementary drilling operations at four sites offshore the Australian Great Barrier Reef are currently being prepared for possible implementation during 2008–2009.

bound phosphate compounds such as phospholipids or nucleic acids. In contrast, hole M0007B (6.28 mbsf) showed only glucosidase and aminopeptidase activity, which indicates degradation and metabolization of polysaccharides and proteins.

Isolation of microorganisms from biofilm samples was performed on agar plates using a medium that is selective for heterotrophic bacteria. After two weeks incubation time, ten different heterotrophic colonies could be isolated. From experiments intended to locate anaerobes, only one isolate was found. Distinct groups of microorganisms are associated in the biofilm that could range from aerobic to anaerobic. The occurrence of pyrite well distributed in the sediment supports the prevalence of a certain degree of anoxia in the environment.

Concluding Remarks and Future Plans

The 600 meters of reef cores with a very high recovery (>90% of the carbonates) were retrieved from thirty-seven holes in 40–120 m water depth and situated on transects in three different areas around Tahiti during the IODP Expedition 310. They provided an exceptional and high resolution record of sea level, climatic, and environmental changes during part of the last deglaciation (about 400 m of cores recovered at 40–122 m below current sea level) and several time windows from the Pleistocene (about 200 m of cores recovered at 85–160 m below sea level).

Acknowledgements

We would like to thank all the people and institutions including the ODP and IODP scientific advisory structure that helped us to pursue this ambitious and challenging project and make it into a very successful expedition with unprecedented recovery of shallow water carbonate rocks.

The ECORD Science Operator (ESO) and its subcontractors made very strong efforts to make this program happen in the best conditions during all stages from planning and drilling through final core description and sampling at the premises of the IODP Bremen Core Repository.

The talented drillers from Seacore Ltd. did an incredible job by delivering exceptional cores, always with humor and grace. We cannot end without mentioning the crew of the *DP Hunter* and Captain William Roger who made our life very pleasant during the expedition.

IODP Expedition 310 Scientists

G. Camoin (Co-Chief Scientist), Y. Iryu (Co-Chief Scientist), D. McInroy (Staff Scientist), R. Asami, H. Braaksma, G. Cabioch, P. Castillo, A. Cohen, J.E. Cole, P. Deschamps, R.G. Fairbanks, T. Felis, K. Fujita, E. Hathorne, S. Lund, H. Machiyama, H. Matsuda, T.M. Quinn, K. Sugihara, A. Thomas, C. Vasconcelos, K. Verwer, R. Warthmann, J.M. Webster, H. Westphal, K.S. Woo, T. Yamada, and Y. Yokoyama.

References

Bard, E., Hamelin, B., and Fairbanks, R.G., 1990. U-Th ages obtained by mass spectrometry in corals from Barbados: sea level during the past 130,000 years. *Nature*, 346:456–458, doi:10.1038/346456a0.

Bard, E., Hamelin, B., Arnold, M., Montaggioni, L.F., Cabioch, G., Faure, G., and Rougerie, F., 1996. Deglacial sea level record from Tahiti corals and the timing of global meltwater discharge. *Nature*, 382:241–244, doi:10.1038/382241a0.

Bjerknes, J., 1969. Atmospheric teleconnections from the equatorial Pacific. *Mon. Wea. Rev.*, 97:163–172.

Blanchon, P., and Shaw, J., 1995. Reef drowning during the last deglaciation: evidence for catastrophic sea level rise and ice-sheet collapse. *Geology*, 23:4–8, doi:10.1130/0091-7613(1995)023<0004:RDDTLD>2.3.CO;2.

Cabioch, G., Camoin, G.F., and Montaggioni, L.F., 1999a. Postglacial growth history of a French Polynesian barrier reef (Tahiti, central Pacific). *Sedimentology*, 46:985–1000, doi:10.1046/j.1365-3091.1999.00254.x.

Cabioch, G., Montaggioni, L.F., Faure, G., and Ribaud-Laurenti, A., 1999b. Reef coralgal assemblages as recorders of paleobathymetry and sea level changes in the Indo-Pacific province. *Quat. Sci. Rev.*, 18:1681–1695.

Camoin, G.F., and Montaggioni, L.F., 1994. High energy coralgal-stromatolite frameworks from Holocene reefs (Tahiti, French Polynesia). *Sedimentiol*, 41:655–676, doi:10.1111/j.1365-3091.1994.tb01416.x.

Camoin, G.F., Colonna, M., Montaggioni, L.F., Casanova, J., Faure, G., and Thomassin, B.A., 1997. Holocene sea level changes and reef development in southwestern Indian Ocean. *Coral Reefs*, 16:247–259.

Camoin, G.F., Gautret, P., Montaggioni, L.F., and Cabioch, G., 1999. Nature and environmental significance of microbialites in Quaternary reefs: the Tahiti paradox. *Sediment. Geol.*, 126:271–304, doi:10.1016/S0037-0738(99)00045-7.

Camoin, G.F., Cabioch, G., Hamelin, B., and Lericolais, G., 2003. Rapport de mission SISMITA. Institut de recherche pour le développement, Papeete, Polynesia Francaise.

Camoin, G.F., Montaggioni, L.F., and Braithwaite, C.J.R., 2004. Late glacial to postglacial sea levels in the Western Indian Ocean. *Mar. Geol.*, 206:119–146.

Camoin, G.F., Cabioch, G., Eisenhauer, A., Braga, J.C., Hamelin, B., and Lericolais, G., 2006. Environmental significance of microbialites in reef environments during the last deglaciation. *Sediment. Geol.*, 185:277–295, doi:10.1016/j.sedgeo.2005.12.018.

Camoin, G.F., Westphal, H., Séard, C., Heindel, K., Yokoyama, Y., Matsuzaki, M., Vasconcelos, C., Warthmann, R., Webster, J., and IODP Expedition 310 Scientists, 2007. Microbialites: a major component of the last deglacial reef sequence from Tahiti (IODP Expedition #310). Environmental significance and sedimentological roles. Poster presentation. EGU Conference, Vienna, 15–20 April 2007.

Chappell, J., and Polach, H.A., 1991. Postglacial sea level rise from a coral record at Huon Peninsula, Papua New Guinea. *Nature*, 349:147–149, doi:10.1016/S0277-3791(01)00118-4.

Clark, P.U., and Mix, A.C., 2002. Ice sheets and the sea level of the Last Glacial Maximum. *Quat. Sci. Rev.*, 21:1–7, doi:10.1016/S0277-3791(01)00118-4.

Clark, P.U., McCabe, A.M., Mix, A.C., and Weaver, A.J., 2004. Rapid rise of sea level 19,000 years ago and its global implications. *Science*, 304:1141–1144, doi:10.1126/science.1094449.

Deschamps, P., Durand, N., Bard, E., Hamelin, B., Camoin, G.F., Thomas, A.L., Henderson, G.M., Yokoyama, Y., and IODP Expedition 310 Scientists, 2006. Extending the Tahiti postglacial sea-level record with offshore drilled corals. First results from IODP Expedition 310. SEALAIX International Symposium, Giens, France, 25–29 September 2006.

Deschamps, P., Durand, N., Bard, E., Hamelin, B., Camoin, G.F., Thomas, A.L., Henderson, G.M., Yokoyama, Y., and IODP Expedition 310 Scientists., 2007. New evidence for the exis-

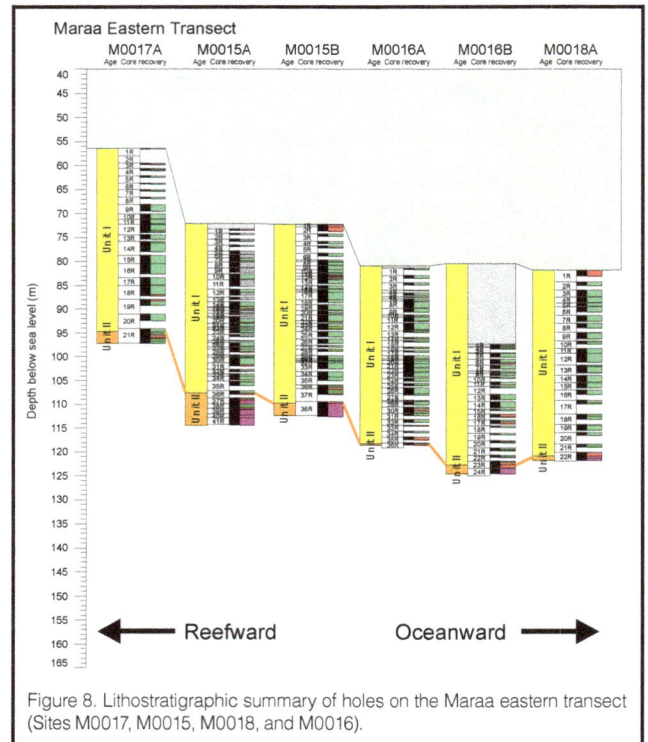

Figure 8. Lithostratigraphic summary of holes on the Maraa eastern transect (Sites M0017, M0015, M0018, and M0016).

tence of the MWP-1A from a "far-field" site. Preliminary results from the Tahiti IODP Expedition 310. EGU Conference, Vienna, 15–20 April 2007.

Edwards, R.L., Beck, W.J., Burr, G.S., Donahue, D.J., Chappell, J.M.A., Bloom, A.L., Druffel, E.R.M., and Taylor, F.W., 1993. A large drop in atmospheric $^{14}C/^{12}C$ reduced melting in the Younger Dryas, documented with ^{230}Th ages of corals. *Science*, 260:962–968, doi:10.1038/342637a0.

Fairbanks, R.G., 1989. A 17,000-year glacio-eustatic sea-level record: Influence of glacial melting rates on the Younger Dryas event and deep-ocean circulation. *Nature*, 342:637–642.

Felis, T., Asami, R., Deschamps, P., Kölling, M., Durand, N., Bard, E., and IODP Expedition 310 Scientists., 2007. Sub-seasonal reconstructions of South Pacific climate during the last deglaciation from Tahiti corals—Preliminary results from IODP Expedition 310. EGU Conference, Vienna, 15–20 April 2007.

Heindel, K., Westphal, H., Camoin, G.F., Séard, C., Birgel, D., Peckmann, J., and IODP Expedition 310 Scientists, 2007. Microbialite-dominated coral reefs as response to abrupt environmental changes during the last deglacial sea level rise. IODP Expedition #310, Tahiti. Poster presentation. EGU Conference, Vienna, 15–20 April 2007.

Lambeck, K., 1993. Glacial rebound and sea level change: an example of a relationship between mantle and surface processes. *Tectonophysics*, 223:15–37, doi: 10.1016/0040-1951(93)90155-D.

Lambeck, K., Yokoyama, Y., and Purcell, A., 2002. Into and out of the Last Glacial Maximum: sea level change during oxygen isotope stages 3 and 2. *Quat. Sci. Rev.*, 21:343–360, doi:10.1016/S0277-3791(01)00071-3.

Lindstrom, D.R., and MacAyeal, D.R., 1993. Death of an ice sheet. *Nature*, 365:214–215, doi:10.1038/365214a0.

McKenzie, J., Vasconcelos, C., Warthmann, R., and Camoin, G.F., 2007. Microbialite structures as a component of last deglacial reef sequence in Tahiti drill cores, IODP Expedition 310: initial results from geomicrobiological study. IAS Regional Meeting, Patras, Greece, 4–7 September 2007.

Ménabréaz, L., 2007. Caractérisation des faciès et magnétostratigraphie de la séquence récifale pléistocène de Tahiti (Expedition IODP 310). *Rapp. Master 2, Ecole Doctorale Sciences de l'Environnement d'Aix-Marseille* (unpublished).

Mix, A., Bard, E., and Schneider, R., 2001. Environmental processes of the ice age: land, oceans, glaciers (EPILOG). *Quat. Sci. Rev.*, 20:627–657, doi:10.1016/S0277-3791(00)00145-1.

Montaggioni, L.F., and Faure, G., 1997. Response of reef coral communities to sea level rise: a Holocene model from Mauritius (Western Indian Ocean). *Sedimentol.*, 44:1053–1070.

Montaggioni, L.F., Cabioch, G., Camoin, G.F., Bard, E., Ribaud-Laurenti, A., Faure, G., Déjardin, P., and Récy, J., 1997. Continuous record of reef growth over the past 14 k.y. on the mid-Pacific island of Tahiti. *Geology*, 25:555–558, doi:10.1130/0091-7613(1997)025<0555:CRORGO>2.3.CO;2.

Nakada, M., and Lambeck, K., 1989. Late Pleistocene and Holocene sea level change in the Australian region and mantle rheology. *Geophys. J.*, 96:497–517.

Okuno J., and Nakada, M., 1999. Total volume and temporal variation of meltwater from last glacial maximum inferred from sea level observations at Barbados and Tahiti. *Palaeogeogr.*

Palaeoclimatol. Palaeoecol., 146:283–293, doi:10.1016/S0031-0182(98)00136-9.

Peltier, W.R., 1994. Ice age paleotopography. *Science*, 265:195–201, doi:10.1126/science.265.5169.195.

Pickard, G.L., and Emery, W.J., 1990. *Descriptive Physical Oceanography, 5th Edition*, Butterworth-Heinemann (Burlington).

Premoli Silva, I., Haggerty, J., Rack, F., and Shipboard Scientific Party, 1993. Northwest Pacific Atolls and Guyots, *Proceed. ODP, Init. Repts., 144*. College Station, Texas (Ocean Drilling Program).

Sagawa, N., Nakamori, T. and Iryu, Y., 2001. Pleistocene reef development in the southwest Ryukyu Islands, Japan. *Palaeogeogr. Palaeoclimatol. Palaeoecol.*, 175:303–323.

Sager, W.W., Winterer, E.L., Firth J.V., and Shipboard Scientific Party, 1993. Northwest Pacific Atolls and Guyots, *Proceed. ODP, Init. Repts., 143*, College Station, Texas (Ocean Drilling Program).

Sugihara, K., Yamada, T., and Iryu, Y., 2006. Contrasts of coral zonation between Ishigaki Island (Japan, northwestern Pacific) and Tahiti Island (French Polynesia, central Pacific), and its significance in Quaternary reef growth histories. SEALAIX International Symposium, Giens, France, 25–29 September 2006.

Thomas, A.L., Henderson, G.M., Deschamps, P., Yokoyama, Y., Bard, E., Durand, N., Hamelin, B., Camoin, G.F., and IODP Expedition 310 Scientists, 2006. Preliminary results from the IODP Expedition 310 "Tahiti Sea Level": U-Th dating of Pleistocene reef material. SEALAIX International Symposium, Giens, France, 25–29 September 2006.

Yokoyama, Y., De Deckker, P., Lambeck, K., Johnston, P., and Fifield, L.K., 2001. Sea level at the Last Glacial Maximum: evidence from northwestern Australia to constrain ice volumes for oxygen isotope stage 2. *Palaeogeogr. Palaeoclimatol. Palaeoecol.*, 165: 281–297, doi:10.1016/S0031-0182(00)00164-4.

Authors

Gilbert F. Camoin, Centre Européen de Recherche et d'Enseignement des Géosciences de l'Environnement (CEREGE) UMR 6635 CNRS, Europôle Méditerranéen de l'Arbois, BP 80, F-13545 Aix-en-Provence cedex 4, France, e-mail: gcamoin@cerege.fr.

Yasufumi Iryu, Institute of Geology and Paleontology, Graduate School of Science, Tohoku University, Aobayama, Sendai 980-8578, Japan, email: iryu@dges.tohoku.ac.jp.

Dave B. McInroy, British Geological Survey, Murchison House, West Mains Road, Edinburgh EH9 3LA, U.K., email: dbm@bgs.ac.uk.

and **the IODP Expedition 310 Scientists**.

Related Web Links

http://www.eso.ecord.org/expeditions/310/310.htm
http://www.rcom.marum.de/English/Tahiti_Sea-Level_Expedition_2005.html

2

COBBOOM: The Continental Breakup and Birth of Oceans Mission

by Dale S. Sawyer, Millard F. Coffin, Timothy J. Reston, Joann M. Stock, and John R. Hopper

Introduction

The rupture of continents and creation of new oceans is a fundamental yet primitively understood aspect of the plate tectonic cycle. Building upon past achievements by ocean drilling and geophysical and geologic studies, we propose "The Continental Breakup and Birth of Oceans Mission (COBBOOM)" as the next major phase of discovery, for which sampling by drilling will be essential.

In September 2006, fifty-one scientists from six continents gathered in Pontresina, Switzerland to discuss current knowledge of continental breakup and sedimentary basin formation and how the Integrated Ocean Drilling Program (IODP) can deepen that knowledge (Coffin et al., 2006). Workshop participants discussed a global array of rifted margins (Fig. 1), formulated the critical problems to be addressed by future drilling and related investigations, and identified key rift systems poised for IODP investigations.

Past Achievements

Scientific ocean drilling has played an essential role in the exploration of rifted continental margins. The North Atlantic Rifted Margins (NARM) endeavor of the Ocean Drilling Program (ODP) addressed conjugate margin pair rift systems ranging from "magma-dominated" (Norway/British Isles-Greenland margins of the offshore-onshore early Tertiary North Atlantic large igneous province (LIP)) to "magma-starved" (Iberia-Newfoundland margins of Late Triassic to Early Jurassic age, Fig. 2.) Geophysical studies

and drilling results from these two conjugate pair rift systems have profoundly changed our view of the processes responsible for such margins.

The drilling of 'seaward dipping reflector' (SDR) wedges of the North Atlantic LIP off the British Isles (Roberts et al., 1984), Norway (Eldholm et al., 1987, 1989) and SE Greenland (Duncan et al., 1996; Larsen et al., 1994, 1999) confirmed them to be a thick series of subaerial lava flows covering large areas. Lavas on the landward side of the SDRs show geochemical evidence of contamination by continental crust, implying that they rose through continental crust during early rifting, whereas oceanward SDR lavas appear to have formed at a seafloor spreading center resembling Iceland. Drilling results from these margins document extreme magmatic productivity over a distance of at least 2000 km during continental rifting and breakup with temporal and spatial influence of the Iceland plume during rifting, breakup, and early seafloor spreading (Saunders et al., 1998).

Other margins such as Iberia-Newfoundland (Tucholke et al., 2007) appear magma-starved and have been hyper-extended by progressive rifting (Lavier and Manatschal, 2006) in stages, with distinct tectonic characteristics controlled by the rheological effects of the gradual thinning of continental crust and uplift of the underlying mantle (Fig. 3). When the crust has been thinned by normal faulting to less than ~10 km, it becomes entirely brittle (Pérez-Gussinyé and Reston, 2001), and tectonism then transitions to spatially focused, closely spaced normal faults that sole into a serpentine detachment at the crust mantle boundary that eventually unroofs upper mantle rocks along a detachment fault exposed at the seafloor. Exhumation of upper mantle rocks continues over a potentially wide region, until mantle uplift generates sufficient magma to initiate seafloor spreading (Tucholke et al., 2007). These results suggest that depth-dependent stretching (DDS) and detachment faulting are major controls on continental rupture and ocean formation.

The ODP also addressed active rifting along a low-angle normal fault in the Woodlark Basin that is propagating into continental lithosphere (Huchon et al., 2002; Taylor et al., 1999a, b).

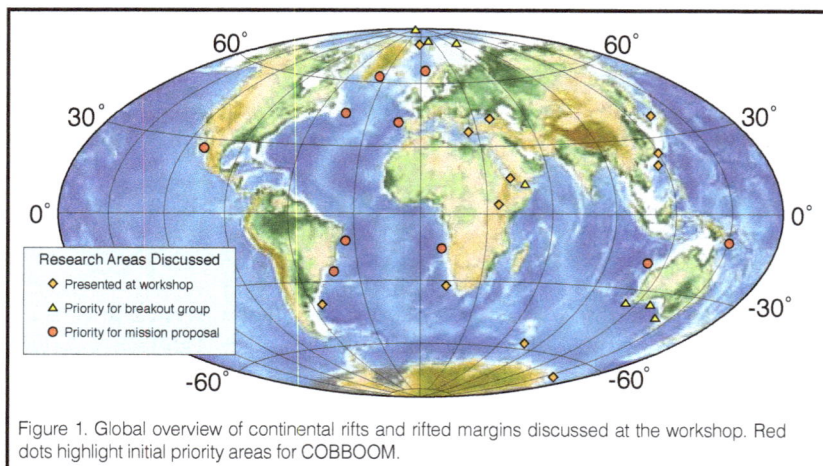

Figure 1. Global overview of continental rifts and rifted margins discussed at the workshop. Red dots highlight initial priority areas for COBBOOM.

Research Areas Discussed
◇ Presented at workshop
△ Priority for breakout group
○ Priority for mission proposal

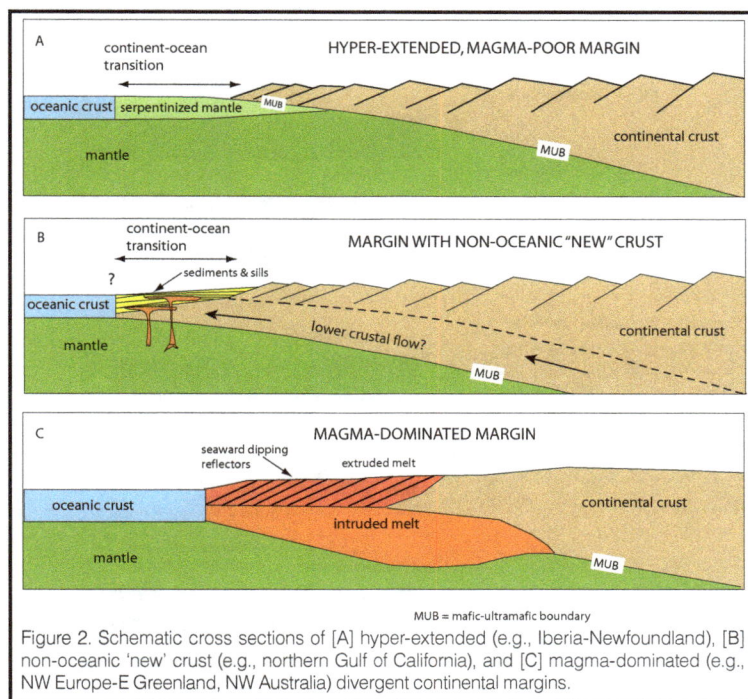

Figure 2. Schematic cross sections of [A] hyper-extended (e.g., Iberia-Newfoundland), [B] non-oceanic 'new' crust (e.g., northern Gulf of California), and [C] magma-dominated (e.g., NW Europe-E Greenland, NW Australia) divergent continental margins.

Scientific Objectives Associated with Continental Rifting

Variations in the importance and, in particular, the volume of magmatism have led to the classification of margins as "volcanic" or "non-volcanic" (Mutter et al., 1988); however, this binary dichotomy fails to adequately reflect that rifted margins form a spectrum from magma-rich to magma-poor. The key distinction is whether magmatism is more or less than expected from the degree of lithospheric thinning and passive asthenospheric upwelling of normal temperature mantle. Equally important are the timing of magmatism and the strain distribution across margins, i.e., hyper-extended versus a more abrupt transition between continental and oceanic lithosphere. Because a continuum between possible end-members may exist, the focus should be on understanding the fundamental processes causing such variations. Key aspects of continental breakup can only be addressed by drilling and associated studies (Table 1).

More specifically, we need to determine the following at multiple, carefully selected rifted margins: 1) uplift and sub-

sidence history; 2) ages and facies of synrift and syn-faulting sediment; 3) timing, volume, chemistry, and style of magmatism; 4) orientation of deformation fabrics, including faults; and 5) ages and facies of postrift sediment. Such information can be used to infer distribution of strain in space and time; deformation mechanics and dynamics; processes within the mantle, including depth and degree of melting, melt migration, and infiltration; and mantle composition, heterogeneity, and dynamics.

We propose drilling programs on well characterized and representative examples, conjugate where possible, of both active and mature rifted margins ranging from magmatic to amagmatic and abrupt to hyperextended. The rift systems described below constitute an initial focus of investigations for COBBOOM.

Gulf of California

Key Aspects and Problems to be Addressed: The active Gulf of California rift system (Fig. 4) varies along strike in crustal thickness, synrift sediment facies, amount of magmatism, structural style, and width of new seafloor (Lizarralde et al., 2007). The northern basins host an enigmatic type of crust that is 15–20 km thick, characterized by gravity anomalies and seismic velocities suggestive of silicic as opposed to basaltic material (González-Fernández et al., 2005). Low-angle normal faults are also accessible to both onshore sampling and drilling. In the central basins, magma-sediment interactions and fluid/geochemical fluxes, including methanogenesis, will be studied (Fig. 5). In southern segments, the processes and timing of the synrift to postrift transition (breakup unconformity, basin evolution, margin uplift or subsidence) will be examined.

Regional Setting and Background: The system formed from a major reorganization of the Farallon-North American plate boundary during Neogene time (Lonsdale, 1989). Narrow perched basins adjacent to seafloor spreading centers characterize the southernmost segment, whereas diffuse deformation in an apparent continental setting dominates the northern Gulf (Persaud et al., 2003). In the central Gulf, two segments of the Guaymas basin are narrow and

Table 1. Key aspects of continental breakup.

Rift initiation	Driving forces, rift localization, lithospheric strength, thermal structure.
Tectonics of rifting	Distribution of strain, rheological evolution, mechanisms of crustal thinning, strength of faults, 3-D rift geometry, mantle exhumation, transition to seafloor spreading.
Magmatism during rifting	Melt-rift interactions, mantle heterogeneities, melt production into seafloor spreading stage, controls on melt production.
Initiation of seafloor spreading	When and where, development of seafloor spreading magnetic anomalies, mantle thermal structure, mantle sources.
Sedimentary processes and basin evolution	Stratigraphic responses to rifting and breakup, stratigraphy-strain rate relationships, fault patterns and evolution, interactions among erosion, sedimentation, and tectonism.
Environmental consequences & impact	Magma interactions with sediment, hydrosphere, and the atmosphere; tectonic and magmatic controls on ocean gateways.

have slightly thicker crust, suggesting more magmatic input, whereas four segments of the south-central domain are wide, magma-poor rifts (Lizarralde et al., 2007). Simple plate kinematics cannot explain these changes in style because of constant total strain along the rift axis. Rather, along strike variation in pre-existing lithospheric and mantle structure, thermal state, and sediment inputs must have controlled this development.

Proposed Drill Sites: Several thousand kilometers of multichannel seismic lines (Persaud et al., 2003; Lizarralde et al., 2007; González-Fernández et al., 2005; Aragón-Arreola and Martín-Barajas, 2007) image key structures of the rift basins well suited for addressing the following topics: 1) the possible role of lower crustal flow to fill the gap created by rifting; 2) the role of detachment faults in early rifting and/or in delaminating continental crust; 3) differences in magmatism in adjacent rift segments along strike; and 4) the relationships between magmatism and global environmental changes (Svensen et al., 2004; Dickens, 2004).

In the northern Gulf (Fig. 4) high heat flow and thick sediment may have caused lower crustal flow and a diffuse rift with enigmatic transitional crust (González-Fernández et al., 2005). It is feasible that sites can sample possible low-angle normal faults and a complete sedimentary section constraining the age of rifting. Igneous intru-

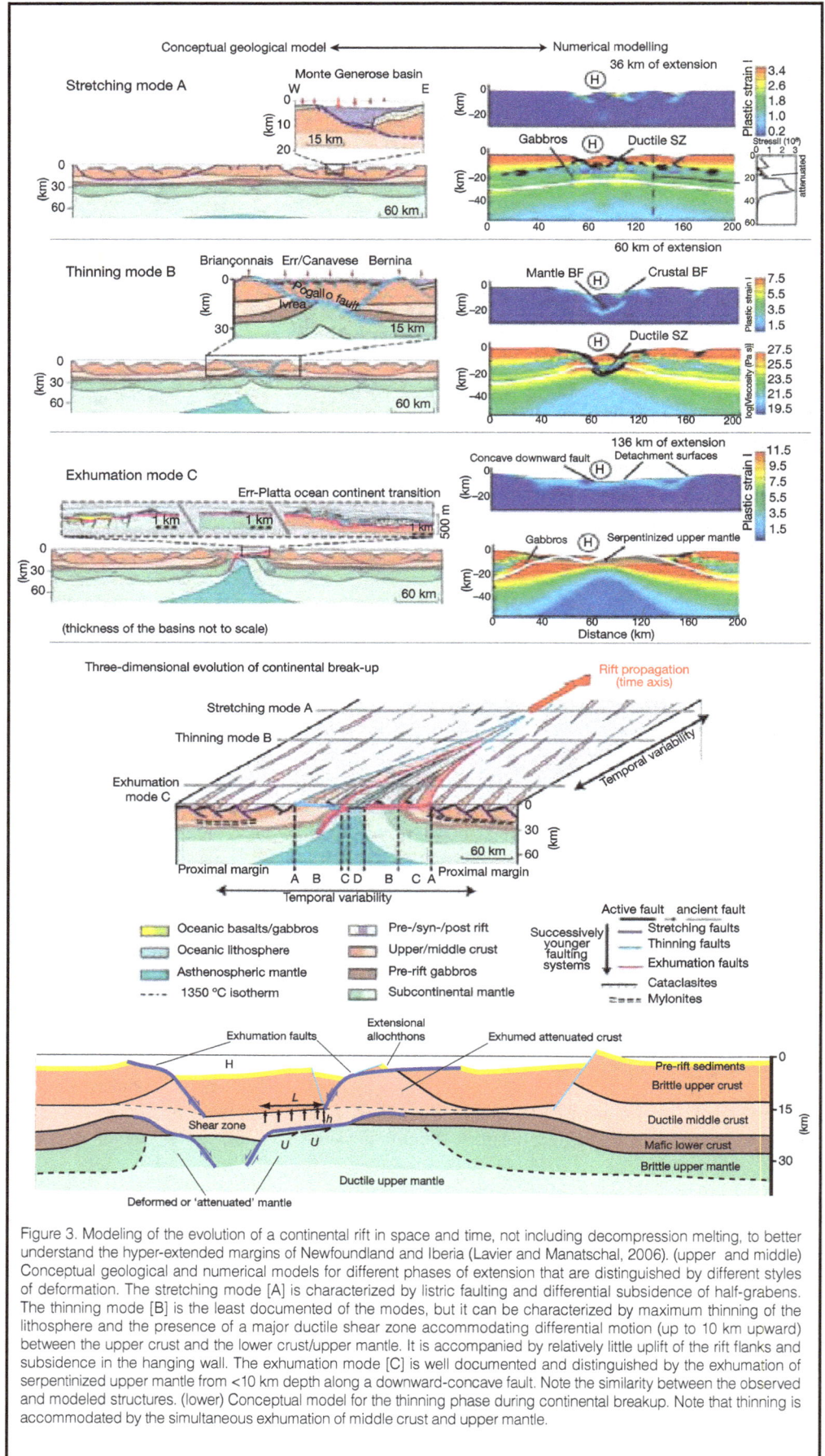

Figure 3. Modeling of the evolution of a continental rift in space and time, not including decompression melting, to better understand the hyper-extended margins of Newfoundland and Iberia (Lavier and Manatschal, 2006). (upper and middle) Conceptual geological and numerical models for different phases of extension that are distinguished by different styles of deformation. The stretching mode [A] is characterized by listric faulting and differential subsidence of half-grabens. The thinning mode [B] is the least documented of the modes, but it can be characterized by maximum thinning of the lithosphere and the presence of a major ductile shear zone accommodating differential motion (up to 10 km upward) between the upper crust and the lower crust/upper mantle. It is accompanied by relatively little uplift of the rift flanks and subsidence in the hanging wall. The exhumation mode [C] is well documented and distinguished by the exhumation of serpentinized upper mantle from <10 km depth along a downward-concave fault. Note the similarity between the observed and modeled structures. (lower) Conceptual model for the thinning phase during continental breakup. Note that thinning is accommodated by the simultaneous exhumation of middle crust and upper mantle.

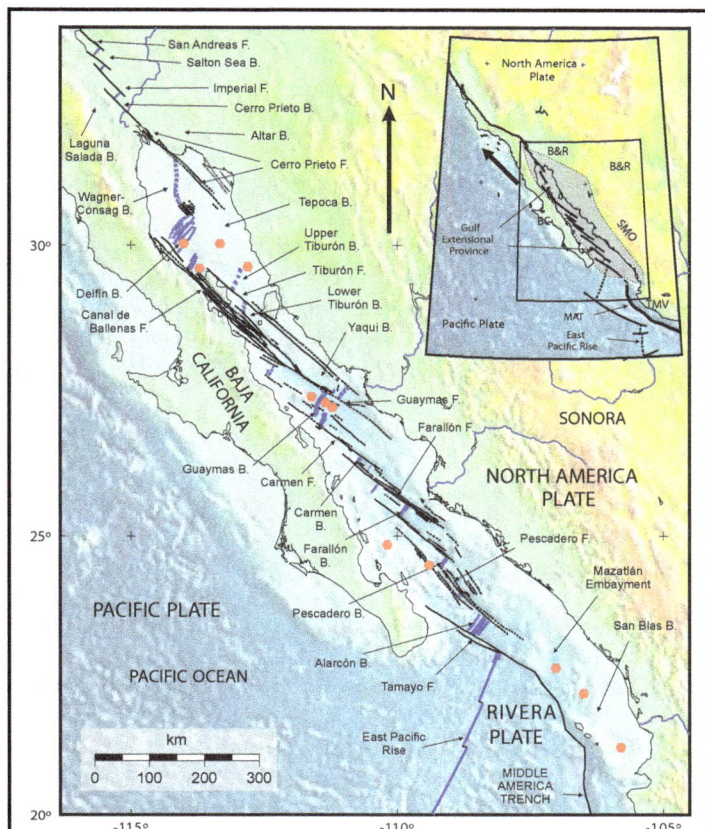

Figure 4. Tectonic map with digital elevations of the Gulf of California region, modified from Aragón-Arreola (2006). This region hosts the modern transtensional boundary between the Pacific and North America plates, comprising spreading centers or extensional basins (blue lines) and faults (black lines). Modern and abandoned basins of the plate boundary are labeled. Red dots show proposed IODP sites. Abbreviations: F=fault, B=basin. Inset: regional map of plate boundary system. Inset abbreviations: SMO=Sierra Madre Occidental; B&R = Basin and Range Province; BC=Baja California; TMV=Trans-Mexican Volcanic Belt; MAT=Middle America Trench.

sions may constrain interactions among mantle and lower crustal melts and sediment. Ignimbrites erupted during rifting will provide a detailed volcanic history and chronology for the sedimentary section. Sites in the central Guaymas basin (Fig. 4) will address methanogenesis related to igneous intrusion into organic-rich sediment, a potentially important process in the global carbon cycle that is likely to operate in most sedimented rift systems. Sites in the southern Gulf target magma-poor, tectonic extension during rifting, and the transition from rifting to seafloor spreading. Sites along the Alarcon segment, combined with observations from its landward extension (Umhoefer et al., 2007), will address the extension and subsidence history of this segment as it transitioned from magma-poor, tectonic extension to seafloor spreading.

Complementary to drilling, land studies can constrain the age and composition of rift-related volcanism, and provide additional constraints on basin history.

Woodlark Basin

Key Aspects and Problems to be Addressed: In the Woodlark Basin, westward mid-ocean ridge propagation into continen-

tal crust (Fig. 6) allows detailed investigations of continental lithosphere before, during, and after seafloor spreading commences.

Regional Setting and Background: The Woodlark rift is continuous along strike with a seafloor spreading system (Fig. 6). Since Late Miocene time, continental lithosphere of the Papuan Peninsula thickened during Australia-Pacific plate convergence (Davies and Smith, 1971) and has subsequently rifted at some of the highest known rates (Abers, 2001; Wallace et al., 2004). Seafloor spreading initiated after ~200 km of continental extension in the eastern part of the rift basin prior to 6 Ma, and propagated ~800 km to the modern rift-drift transition adjacent to Moresby Seamount (Taylor et al., 1999a). Adjacent to the westernmost spreading segment is an active, north-dipping low-angle (~30°) normal fault that is currently being dissected by igneous intrusions (Goodliffe and Taylor, 2007). Farther west, seismically active rifts (Ferris et al., 2006) continue toward the active metamorphic core complexes of the D'Entrecasteaux Islands, where exposures of high and ultra-high pressure metamorphic rocks suggest exhumation rates of ~10 km Myr^{-1} (Baldwin et al., 2004; Monteleone et al., 2007). The upper plates, separated from the lower plates by shear zones and detachment faults (Hill et al., 1992; Little et al., 2007), consist largely of undeformed mafic and ultramafic rocks. Mylonitic lineations and corrugation surfaces parallel Plio-Pleistocene plate motion vectors (Little et al., 2007).

Proposed Drill Sites: Drilling will address two fundamental issues related to the rift-drift transition in easternmost Papua New Guinea.

1) Rift-drift transition processes. A drilling transect across a nascent spreading segment will address a) the origin of the first magmas at a new spreading center; b) the relationship between magma supply rate and the development of magnetic anomalies associated with seafloor spreading; c) magma-sediment interactions; d) the state of stress at the rift-drift transition; and e) how plate motion accommodation transitions from a low-angle fault to crustal accretion. Drilling will sample the center of the spreading segment and intruded synrift sediment directly to the north and west.

2) Fault patterns and mechanisms responsible for exhuming high and ultrahigh pressure metamorphic rocks from mantle depths ahead of the westward propagating seafloor spreading rift tip, and nature of the rocks above the detachment faults associated with the active D'Entrecasteaux Islands core complexes. Two drilling transects will penetrate sediment and upper plate rocks above the detachment faults, one north of the Prevost Range core complex on Normanby Island (<30 km from the active seafloor spreading tip), and the other north of the Mailolo core complex on northwest Fergusson Island. The active submarine sections of these

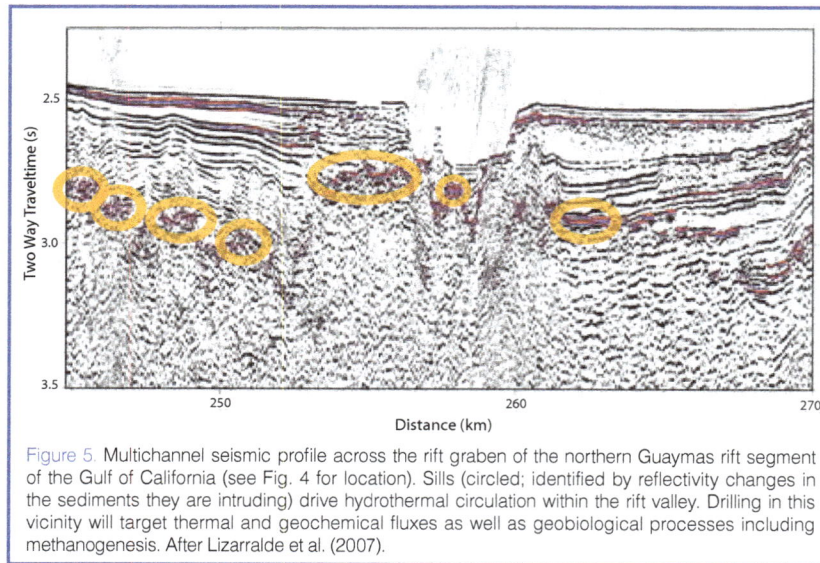

Figure 5. Multichannel seismic profile across the rift graben of the northern Guaymas rift segment of the Gulf of California (see Fig. 4 for location). Sills (circled; identified by reflectivity changes in the sediments they are intruding) drive hydrothermal circulation within the rift valley. Drilling in this vicinity will target thermal and geochemical fluxes as well as geobiological processes including methanogenesis. After Lizarralde et al. (2007).

faults extend below marine sediment that was deposited before and during motion of the hanging wall fault blocks. Drilling will constrain the timing and amount of exhumation directly west of the active seafloor spreading rift and test various models for core complex formation (Abers et al., 2002; Martinez et al., 2001) and for exhumation of high and ultrahigh pressure metamorphic rocks.

North Atlantic Magma-Dominated Margins

Key Aspects and Problems to be Addressed: The conjugate northwest Europe and east Greenland margins (Fig. 7) are characterized by voluminous magmatism associated with the Iceland plume, and their formation may have had significant environmental impact (Eldholm et al., 1989; Eldholm and Grue, 1994; Larsen and Saunders, 1998; Svensen et al., 2004; Storey et al., 2007). The magmatic productivity cannot be explained by simple decompression melting of normal temperature, sub-lithospheric mantle. Three primary competing hypotheses for excessive magmatism are 1) mantle plume with elevated temperatures (White and McKenzie, 1989); 2) small-scale convection at the base of the lithosphere (Mutter et al., 1988; King and Anderson, 1995); and 3) heterogeneities in mantle source composition (Korenaga, 2004).

Regional Setting and Background: The northeast Atlantic conjugate rifted margins show evidence for extensive magmatism including SDRs, igneous intrusions, and high seismic velocity bodies at the base of the crust attributed to magmatic underplating (Fig. 7). The conjugate margins are segmented along strike by the northwest-trending Jan Mayen Fracture Zone and the Bivrost Lineament, which separate the Møre, Vøring, and Lofoten margins and their conjugates at the Jan

Mayen microcontinent and off northeastern Greenland (Eldholm et al., 2002). The margin segments are characterized by different tectono-magmatic styles and sediment distributions, with magmatism decreasing from the Vøring segment to the north and south.

Proposed Drill Sites: The overall plan is for two shallow basement penetration transects (eleven holes) located north and south of the Jan Mayen Fracture Zone (Fig. 7), respectively, and one deep sub-basalt hole within the southern transect (Fig. 7). A dip transect (six holes; Fig. 7) to examine temporal variability of magmatism extends across the conjugate margin segment pair of central Møre/Jan Mayen Ridge. Each hole is well characterized by high quality seismic reflection data. A strike transect to sample breakup-related volcanic rocks in different margin segments as well as facies units extends along the Norwegian margin. The main segments to be drilled include 1) the central Møre margin (i.e., the location of the dip transect); 2) the southern Vøring margin (transform margin related volcanism); 3) the northern Vøring margin (voluminous volcanic complex); and 4) the southern Lofoten margin (small volcanic complexes). The deep hole—to examine temporal variability of magmatism and the nature, environment, and implications of the rift and early breakup magmatism—will be a reoccupation of a landward site on the Møre-Jan Mayen Ridge conjugate margin transect. Specific issues to be addressed by drilling are 1) melt sources and melting conditions, 2) timing of magmatism, 3) spatial and temporal variations of volcanism, 4) eruption environment and vertical movements, 5) along-axis variations in melt production, and 6) consequences of excessive magmatism for environmental change.

Overall, geophysical (including two-dimensional, three-dimensional, and wide-angle seismic) and geological (including DSDP, ODP, and commercial drilling) data sets for the North Atlantic LIP are comprehensive and of high quality.

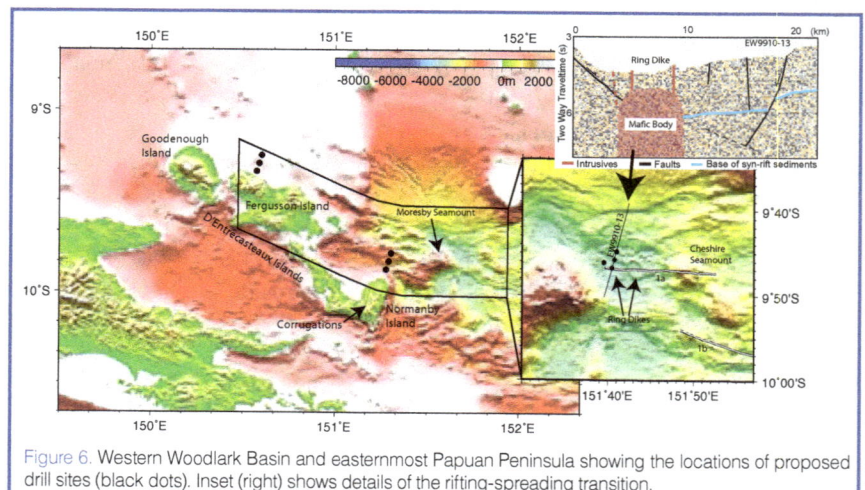

Figure 6. Western Woodlark Basin and easternmost Papuan Peninsula showing the locations of proposed drill sites (black dots). Inset (right) shows details of the rifting-spreading transition.

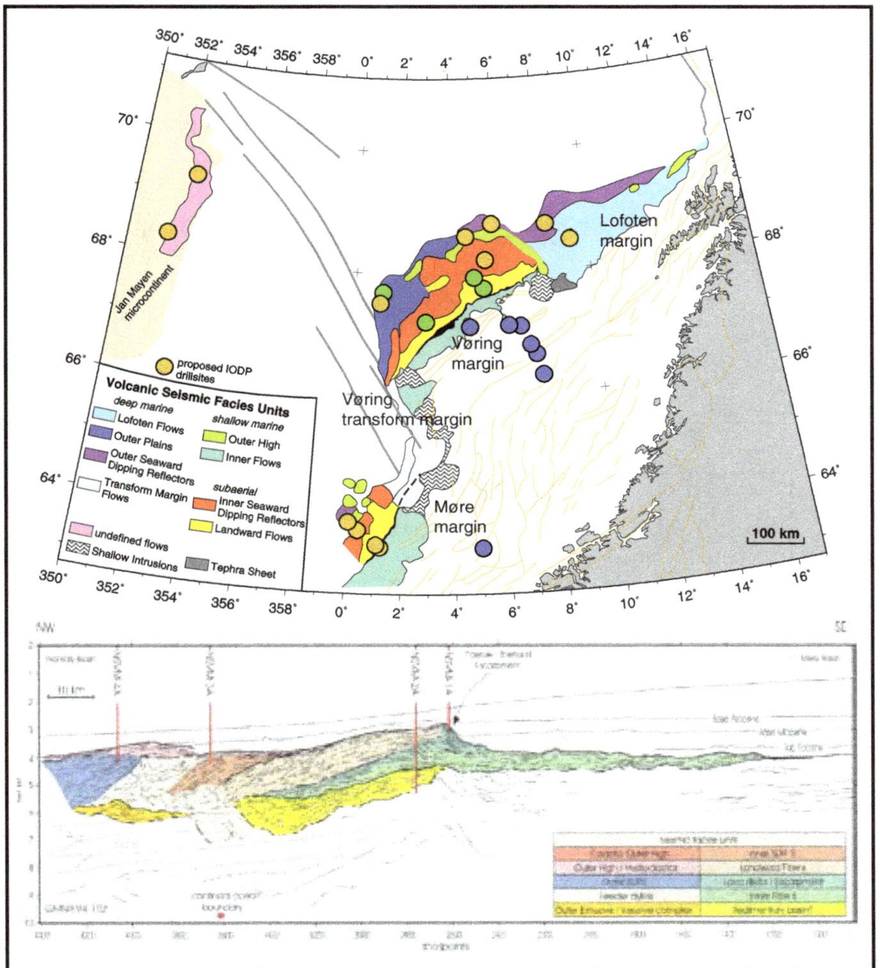

Figure 7. Upper: Distribution of volcanic seismic facies units on the mid-Norwegian margin (from Berndt et al., 2001), proposed IODP sites, and selected boreholes (DSDP/ODP drill sites in green circles; commercial boreholes in blue circles). Lower: Interpreted seismic profile across the Møre Margin (location above), revealing several distinct volcanic facies units (Planke and Alvestad, 1999; Planke et al., 2000).

identified (Srivastava et al., 2000; Russell and Whitmarsh, 2003), whereas in the north, M0 (124.5–125 Ma) appears to be the oldest anomaly. However, some evidence indicates that pre-spreading rifting continued into late Aptian time (~112 Ma; Boillot et al., 1987; Reston, 2005), supporting the idea that the earliest 'seafloor spreading' magnetic anomalies may have originated from unroofed mantle rather than igneous crust formed at a focused spreading center.

Such uncertainties in timing of key events preclude a thorough understanding of dynamic processes because neither rates nor spreading mechanisms are yet accurately known. Another outstanding problem is how the crust was thinned to only a few kilometers, challenging many tenets of lithospheric rheology and isostasy. Some combination of polyphase faulting and DDS seems likely, but the relative importance of the two is controversial (Davis and Kusznir, 2004; Reston, 2007). Similar problems characterize other rifted margins—the South Atlantic (Moulin et al., 2005), Northwest Australian (Driscoll and Karner, 1998), the Labrador Sea (Chian et al., 1995), and the Parentis basin (Pinet et al., 1987).

Proposed Drill Sites: A key objective is to determine the timing of events, which is needed for a quantitative understanding of the rates of processes associated with final thinning, crustal separation, lower crust and mantle exhumation, the onset of mantle melting, and seafloor spreading. Complete sedimentary sections on both margins are required to achieve this objective (Fig. 8). Well defined synrift wedges above a probable detachment (S) on the Iberia (Galicia Bank) margin and relatively thin sedimentary cover on the conjugate Newfoundland (Flemish Cap) margin provide unique opportunities to establish the timing of events. Another major objective is to test competing ideas on how lithosphere deforms during the final thinning phase of extension leading to exhumation of lower crust and mantle. Thinned continental crust must be sampled at key locations to achieve this objective (Fig. 8).

Flemish Cap. Drilling will help determine 1) the role of hypothesized concave-down faults in exhumation of lower crust and upper mantle during late breakup; 2) the interplay among tectonic, magmatic, and serpentinization processes in hyper-extended rift environments; 3) whether continental crust was removed completely amagmatically; 4) if initial melt products were distributed asymmetrically, with more

Newfoundland-Iberia Rift

Key Aspects and Problems to be Addressed: The Iberia and Newfoundland margins (Fig. 8) lack extensive magmatism; they are hyper-extended, characterized by polyphase and diachronous rifting, detachment faulting, and mantle serpentinization and unroofing (Pérez-Gussinyé and Reston, 2001; Reston, 2005; Tucholke et al., 2007). Thin sedimentary cover makes tectonic targets uniquely accessible to drilling. Key problems concern the timing of rifting (along and across the margins), breakup, and the onset of seafloor spreading; the mechanism(s) of extreme crustal thinning; the role of detachment faulting in mantle unroofing; and the nature of basement within the continent-ocean transition.

Regional Setting and Background: During Late Jurassic and Early Cretaceous periods, rifting localized between Newfoundland and Iberia, and the two separated. Breakup propagated from the Central Atlantic northward (Srivastava et al., 2000), but critical details remain controversial. In the south of the Newfoundland-Iberia region, seafloor spreading anomaly M3 (128–130 Ma; Gradstein et al., 2004) has been melt on the Newfoundland side; 5) when and how rifting transitioned to seafloor spreading; 6) controls on the localization of deformation into serpentinized shear zones; and 7) when asymmetries between Galicia Bank and Flemish Cap developed. The most landward sites lie on the continental slope where continental crust is <5 km thick, and will provide key information on rock types where extreme thinning with little to no upper crustal extension is observed in seismic sections. The sites will sample as much stratigraphy as possible to constrain timing as well as penetrate into basement rocks. The most seaward sites will test competing hypotheses for the formation of transition zone crust, mantle exhumation, and formation of anomalously thin oceanic crust.

Galicia Bank. Drilling will recover complete sedimentary sections at two sites that will help establish the timing and geometry of fault block movements associated with the formation of the 'S' reflection and emplacement of the peridotite ridge just prior to seafloor spreading (Fig. 8). One site (GBB-7A) will penetrate the 'S' reflection, hypothesized to represent a regional detachment surface. Data from overlying strata will constrain the timing of any motion along the surface and dip angles during motion. Coring the hypothesized detachment will reveal deformation above and below the fault while it was active. Another site (GBB-8A) will penetrate a basin and the eastern flank of the peridotite ridge. While the top of the ridge has been sampled, little is known about its internal structure and mode of formation, and when the ridge was exposed at the seafloor.

South Atlantic Margins

Key Aspects and Problems to be Addressed: Large data sets from the conjugate margins of the South Atlantic suggest that the crust here has been thinned more than can be explained by the observed faults. Drilling of well imaged synrift and early postrift sediment infill of marginal sag basins will provide critical information on timing and facies

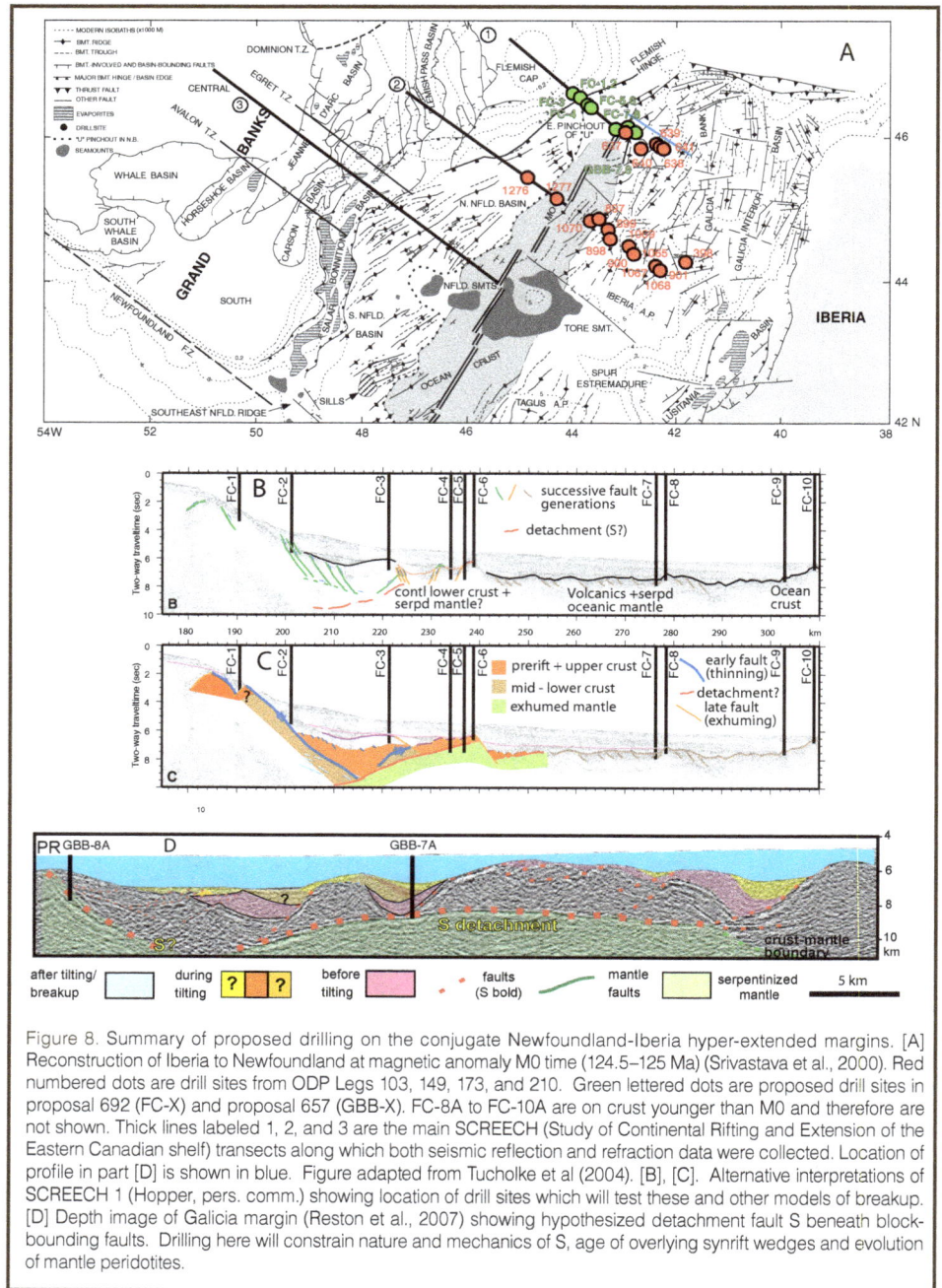

Figure 8. Summary of proposed drilling on the conjugate Newfoundland-Iberia hyper-extended margins. [A] Reconstruction of Iberia to Newfoundland at magnetic anomaly M0 time (124.5–125 Ma) (Srivastava et al., 2000). Red numbered dots are drill sites from ODP Legs 103, 149, 173, and 210. Green lettered dots are proposed drill sites in proposal 692 (FC-X) and proposal 657 (GBB-X). FC-8A to FC-10A are on crust younger than M0 and therefore are not shown. Thick lines labeled 1, 2, and 3 are the main SCREECH (Study of Continental Rifting and Extension of the Eastern Canadian shelf) transects along which both seismic reflection and refraction data were collected. Location of profile in part [D] is shown in blue. Figure adapted from Tucholke et al (2004). [B], [C]. Alternative interpretations of SCREECH 1 (Hopper, pers. comm.) showing location of drill sites which will test these and other models of breakup. [D] Depth image of Galicia margin (Reston et al., 2007) showing hypothesized detachment fault S beneath block-bounding faults. Drilling here will constrain nature and mechanics of S, age of overlying synrift wedges and evolution of mantle peridotites.

for understanding margin evolution, including the cause of the extension discrepancy.

Regional Setting and Background: Rupture between South America and Africa propagated from south to north in the Late Jurassic and Early Cretaceous time as the South Atlantic Ocean formed. The resulting South Atlantic passive margins can be divided broadly into three provinces. The first province, the South of Walvis, voluminous magmatism led to ~100-km-wide SDRs within the crust along both the Argentine and Namibian conjugate margins. The second province, the North of Walvis Ridge—the rifted margin of eastern Brazil and its conjugate Angola, Congo, and Gabon margins (Fig. 9)—also experienced volcanism during breakup, but

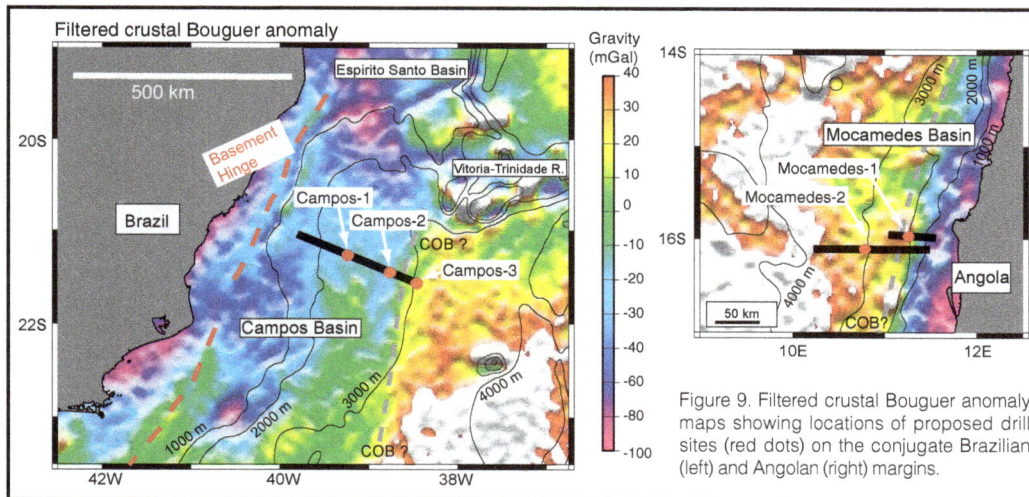

Figure 9. Filtered crustal Bouguer anomaly maps showing locations of proposed drill sites (red dots) on the conjugate Brazilian (left) and Angolan (right) margins.

not sufficiently voluminous to form SDRs. Synrift and postrift sedimentation was dominated by Aptian salt, carbonate platforms, and clastic sediment. The third province, the conjugate rifted margins of the equatorial Atlantic, are narrow compared to the hyper-extended margins to the south. Synextensional sag basins, where salt was deposited and mobilized, are found in both the Campos Basin offshore Brazil and the Kwanza Basin offshore Angola. The relative timing of Early Cretaceous synrift volcanism, evaporite deposition, and onset of seafloor spreading are controversial. The seaward edge of the Aptian salt basin may lie on thick SDR wedges (Jackson et al., 2000), implying that salt was deposited after final continental breakup; volcanic edifices in the continent-ocean transition zone probably acted as barriers between the episodically dry marginal basin and the open ocean where new oceanic crust was forming. However, Aptian salt is also abundant in the vicinity of the Congo Fan, where no thick igneous crust or SDRs are imaged in the continent-ocean transition zone. Even in the absence of a potentially bounding basement high, the seaward limit of autochthonous salt appears to lie close to the inferred landward limit of oceanic crust (Marton et al., 2000). Evaporites may even have been deposited on serpentinized mantle in the continent-ocean transition zone (Moulin et al., 2005). Mantle exhumation prior to continental breakup is well characterized from the Newfoundland-Iberia rift, and conceptual models show how continental mantle may be exhumed by low-angle normal faults (Lavier and Manatschal, 2006). However, whereas mantle was exhumed well below sea level along the Newfoundland rift (Tucholke et al., 2007), the continent-ocean transition zone of the central South Atlantic appears to have been close to sea level prior to breakup.

Proposed Drill Sites: Drill sites in the Campos Basin offshore Brazil and the Moçamedes Basin offshore Angola will establish the tectonic setting of marginal basin formation in the central South Atlantic (Fig. 9). Two sites in the Campos Basin lie at the seaward end of a sag basin; drilling will establish the timing and environment of Aptian salt basin formation. A third site on the Brazilian margin, just seaward of the

continent-ocean transition, will sample the oldest oceanic crust. In the Moçamedes Basin on the conjugate Angolan margin, basement will be penetrated where a low-angle fault appears to have ex-humed continental mantle. In total, combined site survey characterization and drilling of the sites will investigate 1) the applicability of conceptual models developed from Newfoundland-Iberia rift characteristics to South Atlantic margins; 2) the age of the first oceanic crust, and relative timing of both continental breakup and the deposition of pre-salt sag sequences; 3) the nature and composition of the crust on which pre-salt sequences were deposited; 4) interpreted exhumed continental mantle in the continent-ocean transition zone; 5) the possible existence of top-basement detachment faults; and 6) synrift and early postrift subsidence along a geophysical transect of the conjugate margins.

NW Australian Magma-Dominated Margin

Key Aspects and Problems to be Addressed: The northwest Australian magma-dominated (rifted) margin (Fig. 10) is segmented, and igneous rock volumes vary considerably along strike, without clear evidence for a related mantle plume (Mutter et al., 1988; Hopper et al., 1992; Symonds et al., 1998; Planke et al., 2000). This makes the margin a strong candidate to test the edge-driven/small-scale convection hypothesis for generating excessive magmatism. Temporal and along-strike variations in melt production and temporal and spatial relationship(s) between rifting and magmatism are secondary objectives.

Regional Setting and Background: The western Australian margin can be divided into four main segments separated by major fracture zones: Argo, Gascoyne, Cuvier, and Perth (Fig. 10). The entire margin exhibits breakup magmatism in the form of sills, SDRs, hyaloclastic buildups, and magmatic underplating that formed during Callovian (~163 Ma) Argo margin breakup and subsequent Valanginian (~138 Ma) Gascoyne, Cuvier, and Perth margin breakup (Planke et al., 2000; Symonds et al., 1998). The conjugate margins have been subducted or obducted. Two main hypotheses—mantle plume (White and McKenzie, 1989) and edge-driven/small-scale convection (Mutter et al., 1988; King and Anderson, 1998; Korenaga, 2004)—have been proposed for the formation of these massive igneous constructions (Coffin et al., 2002; Ingle et al., 2002; Müller et al., 2002). Because it cannot be convincingly tied to a well established hotspot track, the western Australian margin is a highly promising candidate for testing alternative hypotheses for magmatic margin formation (Planke et al., 2000).

If a plume was involved in the formation of the western Australian margin, then the Wallaby Plateau-Zenith Seamount province probably represents the post-breakup track of the plume. This province extends some 1200 km from the continent, across a continent-ocean transition zone likely associated with composite continental and magmatic plateaus, and into a normal ocean basin, providing an ideal opportunity to examine variations in continental contamination in time and space. Documenting the presence or absence of a geochemical plume signature within igneous basement of the Cuvier margin will therefore be a critical complement to previous and ongoing studies of the North Atlantic igneous province.

Proposed Drill Sites: The objectives are as follows: 1) to distinguish between an edge- vs. plume-driven cause for magmatism along a rifted margin; 2) to examine rift and breakup duration, and subsidence history; 3) to investigate the formation and crustal nature of marginal plateaus; 4) to understand the temporal development of multiple SDR wedges; and 5) to determine age, volume, duration, and environment of volcanism. To address these objectives,

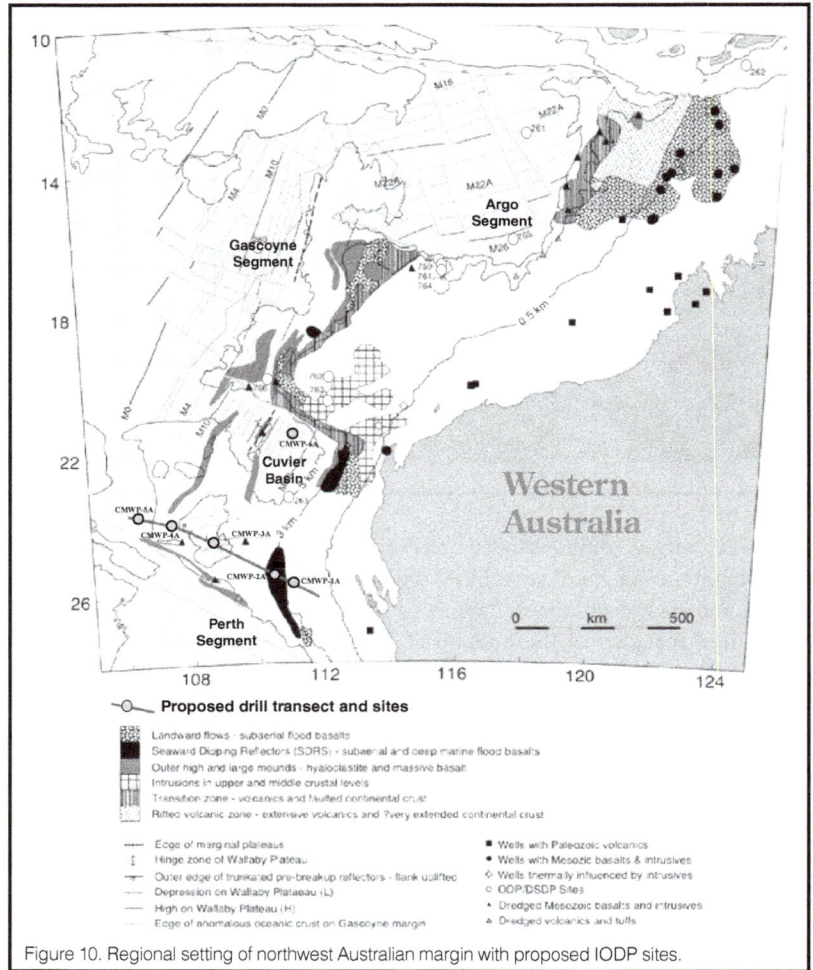

Figure 10. Regional setting of northwest Australian margin with proposed IODP sites.

a five-hole transect across the southern Cuvier margin and the Wallaby Plateau as well as a reference hole in similar age oceanic crust of the nearby Cuvier Abyssal Plain are envisioned (Figs. 10 and 11). The holes will sample i) multiple SDRs across the margin; ii) the Wallaby Plateau, and iii) oceanic crust. The petrology and geochemistry of the recovered rocks will be used to determine melting conditions, magma reservoir type (asthenosphere, lithosphere, plume), and contamination by continental lithosphere. Precision age determinations will constrain the temporal and spatial evolution of the magma source. Regionally, the west Australian margin is relatively well investigated by seismic surveys, dredging, and commercial drilling on the continental shelf.

Essential Complements to Drilling

To date, seismic studies of rifted continental margins by the academic community have been almost entirely 2-D, comprising widely spaced profiles relative to the lateral scale of faults and stratigraphic variations.

We now recognize that 2-D seismic technology does not meet the challenge of contemporary research in rifting and breakup, which are fundamentally 3-D processes. As part of its strategy in investigating rifting and breakup, the IODP should integrate acquisition and interpretation of 3-D seismic reflection data with overall program planning.

Land-based geological research has been a valuable complement to ocean drilling and marine geophysics at several rifted margin systems. The Gulf of California is a good example of a rift system that is only partly submerged. Similar opportunities are offered by the Woodlark Basin, where metamorphic core complexes and other rifting components are exposed on islands, and in the North Atlantic LIP, where flood basalts and related rocks crop out in the British Isles

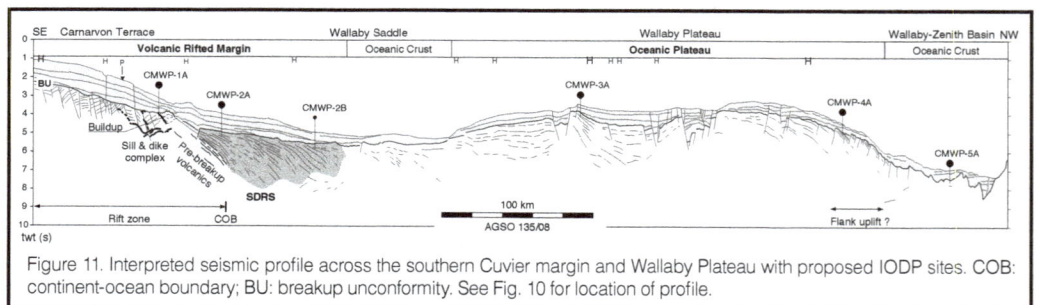

Figure 11. Interpreted seismic profile across the southern Cuvier margin and Wallaby Plateau with proposed IODP sites. COB: continent-ocean boundary; BU: breakup unconformity. See Fig. 10 for location of profile.

Figure 12. Early Tertiary flood basalt sequence in eastern Greenland associated with rifting and breakup between eastern Greenland and northwestern Europe, highlighting opportunities along some margins for synergetic onshore-offshore studies.

and Greenland (Fig. 12). The study of rift systems now exposed on land in mountain belts is another way that the advantages of field geology can contribute to the study of rift systems, as exemplified by comparisons between the Alpine Tethys rift system exposed in the Alps (Fig. 13) and the Iberia/Newfoundland rift system (Manatschal, 2004; Lavier and Manatschal, 2006).

Modeling the processes of continental rifting is increasingly important for making predictions that can be tested by drilling. Forward dynamic models now provide insights into magmatism accompanying breakup (Boutilier and Keen, 1999; Nielsen and Hopper, 2004) as well as primary controls on passive margin width, symmetries/asymmetries, and evolution from wide, diffuse rifting into narrow, localized rifting resulting in the formation of passive margins (Huismans and Beaumont, 2003, 2007; Lavier and Manatschal, 2006).

The study of continental breakup and sedimentary basin formation offers opportunities for significant collaboration between the hydrocarbon industry and the IODP. During the workshop, interest in collaboration was highest for the conjugate rifted margins of the South Atlantic.

Technical Requirements

COBBOOM will require a combination of existing IODP technology and the development of new technology. Proposed deep and challenging drilling will require the use of both riser and non-riser drills. Where water depths are less than 2500 m, we anticipate using riser capability, casing, and mud circulation to increase hole stability and improve the likelihood of deep penetration. However, some holes have the goal of sampling highly stretched continental crust and upper mantle beneath relatively thin sediment cover and in much deeper water. For such holes, ultra-long drill strings deployed in a non-riser mode and supported by extensive casing programs will be required.

Improved methods of core orientation in sedimentary and crystalline rocks will improve our ability to relate microstructures and faults in the core to the strain distribution in the rifting system. In active rift environments, borehole observatories will be used to monitor the presence and pressure of fluids in faults active during rifting. Technology currently being developed to monitor microseismicity near the boreholes will assist in understanding its relationship to fluid pressure variations.

Acknowledgements

The authors appreciate input to this paper by participants in the IODP Workshop on Continental Breakup and Sedimentary Basin Formation as well as contributors who did not attend the workshop. We thank the IODP for providing the principal funding for the workshop held in Pontresina, Switzerland on 15–18 September 2006. We thank InterMARGINS for making it possible for scientists from non-IODP member countries to participate in the workshop. Gianreto Manatschal handled the logistics of the workshop and led an incredible field trip to a rifted margin crust and exhumed upper mantle in the Alps. Finally, we acknowledge the essential contributions of Kelly Kryc of IODP-MI who organized the workshop.

References

Abers, G.A., 2001. Evidence for seismogenic normal faults at shallow dips in continental rifts. *In* Wilson, R.C.L., Whitmarsh, R.B., Taylor, B. and Froitzheim, N. (Eds.), *Non-volcanic Rifting of Continental Margins: A Comparison of Evidence from Land and Sea, Geological Society of London Special Publication* 187, 305–318.

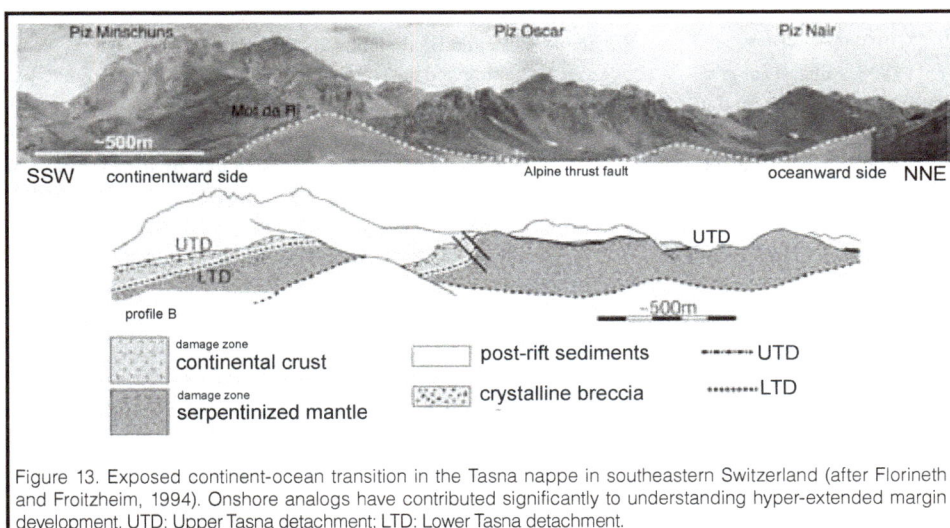

Figure 13. Exposed continent-ocean transition in the Tasna nappe in southeastern Switzerland (after Florineth and Froitzheim, 1994). Onshore analogs have contributed significantly to understanding hyper-extended margin development. UTD: Upper Tasna detachment; LTD: Lower Tasna detachment.

Abers, G.A., Ferris, A., Craig, M., Davies, H., Lerner Lam, A.L., Mutter, J.C., and Taylor, B., 2002. Mantle compensation of active metamorphic core complexes at Woodlark rift in Papua New Guinea. *Nature*, 416:862–866, doi:10.1038/nature00990.

Aragón-Arreola, M., and Martín-Barajas, A., 2007. Westward migration of extension in the northern Gulf of California, Mexico. *Geology*, 35:571–574, doi:10.1130/G23360A.1.

Baldwin, S.L., Montelone, B.D., Webb, L.E., Fitzgerald, P.G., Grove, M., and Hill, J., 2004. Pliocene eclogite exhumation at plate tectonic rates in eastern Papua New Guinea. *Nature*, 431:263–267, doi:10.1038/nature02846.

Boillot, G., Winterer, E.L., Meyer, A.W., and Shipboard Scientific Party, 1987. Introduction, objectives, and principal results: Ocean Drilling Program Leg 103, West Galicia Margin. In Boillot, G., Winterer, E.L., et al., *Proceedings of the Ocean Drilling Program, Initial Reports 103*, College Station, Texas (Ocean Drilling Program), 3–17.

Boutilier, R.R., and Keen, C.E., 1999. Small-scale convection and divergent plate boundaries. *J. Geophys. Res.*, 104:7389–7403.

Chian, D., Louden, K.E., and Reid, I., 1995. Crustal structure of the Labrador Sea conjugate margin and implications for the formation of nonvolcanic continental margins. *J. Geophys. Res.*, 100:24239–24254, doi:10.1029/95JB02162.

Coffin, M.F., Pringle, M.S., Duncan, R.A., Gladczenko, T.P., Storey, M., Müller, R.D., and Gahagan, L.A., 2002. Kerguelen hotspot magma output since 130 Ma. *J. Petrol.*, 43:1121–1139.

Coffin, M.F., Sawyer, D.S., Reston, T.J., and Stock, J.M., 2006. Continental breakup and sedimentary basin formation. *Eos Trans.* AGU, 87:528, doi:10.1029/2006EO470006.

Davies, H.L., and Smith, I.E., 1971. Geology of eastern Papua. *Geol. Soc. Amer. Bull.*, 82, 12:3299–3312, doi:10.1130/0016-7606(1971)82[3299:GOEP]2.0.CO;2.

Davis, M., and Kusznir, N.J., 2004. Depth-dependent lithospheric stretching at rifted continental margins. *In* Karner, G.D. (Ed.), *Proceedings of NSF Rifted Margins Theoretical Institute.* New York, N.Y. (Columbia University Press), 92–136.

Dickens, R.G., 2004. Hydrocarbon-driven warming. Nature, 429:513–515.

Driscoll, N.W., and Karner, G.D., 1998. Lower crustal extension across the Northern Carnarvon Basin, Australia: evidence for an eastward dipping detachment. *J. Geophys. Res.,103*:4975–4992, doi:10.1029/97JB03295.

Duncan, R.A., Larsen, H.C., Allan, J.F., et al., 1996. *Proceedings of the Ocean Drilling Program, Initial Reports, 163.* College Station, Texas (Ocean Drilling Program), 279 pp.

Eldholm, O., and Grue, K., 1994. North Atlantic volcanic margins: dimensions and production rates. J Geophys. Res., 99:2955–2968, doi:10.1029/93JB02879.

Eldholm, O., Thiede, J., and Taylor, E., 1989. Evolution of the Vøring volcanic margin. In Eldholm, O., Thiede, J., Taylor, E., et al., (Eds.), *Proceedings of the Ocean Drilling Program, Scientific Results 104*, College Station, Texas (Ocean Drilling Program), 1033–1065.

Eldholm, O., Thiede, J., Taylor, E., et al., 1987. *Proceedings of the Ocean Drilling Program, Initial Reports, 104.* College Station, Texas (Ocean Drilling Program), 783 pp.

Eldholm, O., Tsikalas, F., and Faleide, J.I., 2002. Continental margin

of Norway 62–75°N: Paleogene tectono-magmatic segmentation and sedimentation. *In* Jolley, D.W., and Bell, B.R. (Eds.), *North Atlantic Igneous Province: Stratigraphy, Tectonics, Volcanic and Magmatic Processes, Geological Society of London Special Publication* 197, 39–68.

Ferris, A., Abers, G.A., Zelt, B., Taylor, B., and Roecker, S., 2006. Crustal structure across the transition from rifting to spreading: the Woodlark rift system of Papua New Guinea. *Geophys. J. Intl.*, 166:622–634, doi:10.1111/j.1365-246X.2006.02970.x.

Florineth, D., and Froitzheim, N., 1994. Transition from continental to oceanic basement in the Tasna nappe (Engadine window, Graubunden, Switzerland): evidence for early Cretaceous opening of the Valais Ocean. *Schweiz. Mineral. Petrogr. Mitt.*, 74:437–448.

González-Fernández, A., Dañobeitia, J.J., Delgado-Argote, L.A., Michaud, F., Córdoba, D., and Bartolomé, R., 2005. Mode of extension and rifting history of upper Tiburon and upper Delfin basins, northern Gulf of California: *J. Geophys. Res.*, 110:1–17.

Goodliffe, A.M., and Taylor, B., 2007. The boundary between continental rifting and seafloor spreading in the Woodlark Basin, Papua New Guinea. *In* Karner, G.D., Manatschal, G., and Pinhiero, L.M. (Eds.), *Imaging, Mapping and Modelling Continental Lithosphere Extension and Breakup. Geological Society of London, Special Publication* 282, 213–234.

Gradstein, F.M., Ogg, J.G., and Smith, A.G., 2004. *A Geologic Time Scale 2004*, Victoria, Australia (Cambridge University Press), 589 pp.

Hill, E.J., Baldwin, S.L., and Lister, G.S., 1992. Unroofing of active metamorphic core complexes in the D'Entrecasteaux Islands, Papua New Guinea. *Geology*, 20:907–910.

Hopper, J.R., Mutter, J.D., Larson, R.L., Mutter, C.Z., and Northwest Australia Study Group, 1992. Magmatism and rift margin evolution: Evidence from northwest Australia. *Geology*, 20:853–857.

Huchon, P., Taylor, B., and Klaus, A., 2002. *Proceedings of the Ocean Drilling Program, Scientific Results, 180*, College Station, Texas (Ocean Drilling Program), 47 pp.

Huismans, R.S., and Beaumont, C., 2003. Symmetric and asymmetric lithospheric extension: Relative effects of frictional-plastic and viscous strain softening. *J. Geophys. Res.*, 108(B10), 2496, doi:10.1029/2002JB002026.

Huismans, R.S., and Beaumont, C., 2007. Roles of lithospheric strain softening and heterogeneity in determining the geometry of rifts and continental margins. *In* Karner, G., Manatschal, G., and Pinheiro, L. (Eds.), *Imaging, Mapping, and Modeling Continental Lithospheric Extension and Breakup, Geological Society of London Special Publication* 282, 107–134.

Ingle, S., Weis, D., Scoates, J.S., and Frey, F.A., 2002. Relationship between the early Kerguelen plume and continental flood basalts of the paleo-Eastern Gondwanan margins. Earth Planet. Sci. Lett., 197:35–50, doi:10.1016/S0012-821X(02)00473-9.

Jackson, M.P.A., Cramez, C., and Fonck, J.-M., 2000. Role of subaerial volcanic rocks and mantle plumes in creation of South Atlantic margins: implications for salt tectonics and source rocks. *Mar. Petr. Geol.*, 17:477–498, doi:10.1016/S0264-8172(00)00006-4.

King, S.D., and Anderson, D.L., 1995. An alternative mechanism of flood basalt formation. *Earth Planet. Sci. Lett.*, 136:269–279.

King, S.D., and Anderson, D.L., 1998. Edge-driven convection. *Earth Planet. Sci. Lett.*, 160:289–296.

Korenaga, J., 2004. Mantle mixing and continental breakup magmatism. Earth Planet. *Sci. Lett.*, 218:463–473, doi:10.1016/S0012-821X(03)00674-5.

Larsen, H.C., and Saunders, A.D., 1998. Tectonism and volcanism at the southeast Greenland rifted margin: a record of plume impact and later continental rupture. *In* Saunders, A.D., Larsen, H.C., and Wise, S. (Eds.), *Proceedings of the Ocean Drilling Program, Scientific Results 152*, College Station, Texas (Ocean Drilling Program), 503–534.

Larsen, H.C., Duncan, R.A., Allan, J.F., and Brooks, K., 1999. *Proceedings of the Ocean Drilling Program, Scientific Results 163*, College Station, Texas (Ocean Drilling Program), 173 pp.

Larsen, H.C., Saunders, A.D., Clift, P.D., et al., 1994. *Proceedings of the Ocean Drilling Program, Initial Reports 152*, College Station, Texas (Ocean Drilling Program), 977 pp.

Lavier, L.L., and Manatschal, G., 2006. A mechanism to thin the continental lithosphere at magma-poor margins. *Nature*, 440:324–328, doi:10.1038/nature04608.

Little, T.A., Baldwin, S.L., Fitzgerald, P.G., and Montelone, B., 2007. Continental rifting and metamorphic core complex formation ahead of the Woodlark spreading ridge, D'Entrecasteaux Islands, Papua New Guinea. *Tectonics*, 26:TC1002, doi:10.1029/2005TC001911.

Lizarralde, D., Axen, G.J., Brown, H.E., Fletcher, J.M., González-Fernández, A., Harding, A.J., Holbrook, W.S., Kent, G.M., Paramo, P., Sutherland, F.H., and Umhoefer, P.J., 2007. Variation in styles of rifting in the Gulf of California. *Nature*, v.448 (7152), P. 466–469, doi:10.1038/nature06035.

Lonsdale, P., 1989. Geology and tectonic history of the Gulf of California. *In* Winterer, E. L., Hussong, D.M., and Decker, R.W. (Eds.), *The Geology of North America Volume N. The Eastern Pacific Ocean and Hawaii*. Boulder, Colo. (Geological Society of America), 499–521.

Manatschal, G., 2004. New models for evolution of magma-poor rifted margins based on a review of data and concepts from West Iberia and the Alps. *Int. J. Earth Sci.*, 93:432–466.

Martinez, F., Goodliffe, A.M., and Taylor, B., 2001. Metamorphic core complex formation by density inversion and lower crust extension. *Nature*, 411:930–934, doi:10.1038/35082042.

Marton, L.G., Tari, G.C., and Lehmann, C.T., 2000. Evolution of the Angolan passive margin, west Africa, with emphasis on post-salt structural styles. *In* Mohriak, W., and Talwani, M. (Eds.), *AGU Geophysical Monograph 115*, Washington, DC (American Geophysical Union), 129–149.

Monteleone, B.D., Baldwin, S.L., Webb, L.E., Fitzgerald, P.G., Grove, M., and Schmitt, A.K., 2007. Late Miocene-Pliocene eclogite facies metamorphism, D'Entrecasteaux Island, SE Papua New Guinea. J. Metamorph. Geol., 25:245–265, doi:10.1111/j.1525-1314.2006.00685.x.

Moulin, M., Aslanian, D., Olivet, J.-L., Contrucci, I., Matias, L., Géli, L., Klingelhoefer, F., Nouzé, H., Réhault, J.-P., and Unternehr, P., 2005. Geological constraints on the evolution of the Angolan margin based on reflection and refraction seismic

data (ZaïAngo project). *Geophys. J. Intl.*, 162:793–810, doi:10.1111/j.1365-246X.2005.02668.x.

Müller, R.D., Mihut, D., Heine, C., O'Neill, C., and Russell, I., 2002. Tectonic and volcanic history of the Carnarvon Terrace: constraints from seismic interpretation and geodynamic modelling. *In* Gorter, J. (Ed.), *The Sedimentary Basins of Western Australia 3*, Perth, Australia (Petroleum Exploration Society of Australia), 719–740.

Mutter, J.C., 1993. Margins declassified. *Nature*, 364:393–394, doi:10.1038/364393a0.

Mutter, J.C., Buck, W.R., and Zehnder, C.M., 1988. Convective partial melting: 1. a model for the formation of thick basaltic sequences during the initiation of spreading. *J. Geophys. Res.*, 93:1031–1048.

Nielsen, T.K., and Hopper, J.R., 2004. From rift to drift: mantle melting during continental breakup. *Geochem. Geophys. Geosyst.*, 5:Q07003, doi:10.1029/2003GC000662.

Pérez-Gussinyé, M., and Reston, T.J., 2001. Rheological evolution during extension at passive non-volcanic margins: onset of serpentinization and development of detachments to continental break-up. *J. Geophys. Res.*, 106:3691–3975.

Persaud, P., Stock, J.M., Steckler, M.S., Martín-Barajas, A., Diebold, J.B., González-Fernández, A., and Mountain, G.S., 2003. Active deformation and shallow structure of the Wagner, Consag, and Delfín Basins, northern Gulf of California, Mexico. *J. Geophys. Res.*, 108:2355, doi: 10.1029/2002JB001937.

Pinet, B., Montadert, L., Curnelle, R., Cazes, M., Marillier, F., Rolet, J., Tomassino, A., Galdeano, A., Patriat, P., Brunet, M.F., Olivet, J.L., Schaming, M., Lefort, J.-P., Arrieta, A., and Riaza, C., 1987. Crustal thinning on the Aquitaine shelf, Bay of Biscay, from deep seismic data. *Nature*, 325:513–516, doi:10.1038/325513a0.

Planke, S., Symonds, P.A., Alvestad, E., and Skogseid, J., 2000. Seismic volcanostratigraphy of large-volume basaltic extrusive complexes on rifted margins. *J. Geophys. Res.*, 105:19335–19351, doi:10.1029/1999JB900005.

Reston, T.J., 2005. Polyphase faulting during development of the west Galicia rifted margin. *Earth Planet. Sci. Lett.*, 237:561–576, doi:10.1016/j.epsl.2005.06.019.

Reston, T.J., 2007. Extension discrepancy at North Atlantic nonvolcanic rifted margins: depth-dependent stretching or unrecognized faulting? *Geology*, 35:367–370, doi:10.1130/G23213A.1.

Reston, T.J., Leythaeuser, T., Booth-Rea, G., Sawyer, D., Klaeschen, D., and Long, C., 2007. Movement along a low-angle normal fault: The S reflector west of Spain. *Geochem. Geophys. Geosyst.*, 8:Q06002, doi:10.1029/2006GC001437.

Roberts, D.G., Schnitker, D., et al., 1984. *Initial Reports of the Deep Sea Drilling Project, Leg 81*, Washington, DC (U.S. Government Printing Office), 923 pp.

Russell, S.M., and Whitmarsh, R.B., 2003. Magmatic processes at the west Iberia non-volcanic rifted continental margin: evidence from analyses of magnetic anomalies. *Geophys. J. Intl.*, 154:706–730, doi:10.1046/j.1365-246X.2003.01999.x.

Saunders, A.D., Larsen, H.C., and Wise, S.W., Jr., 1998. *Proceedings of the Ocean Drilling Program, Scientific Results 152*, College Station, Texas (Ocean Drilling Program), 554 pp.

Srivastava, S.P., Sibuet, J.-C., Cande, S., Roest, W.R., and Reid, I.D.,

2000. Magnetic evidence for slow seafloor spreading during the formation of the Newfoundland and Iberian margins. *Earth Planet. Sci. Lett.*, 182:61–76.

Storey, M., Duncan, R.A., and Swisher, C.C., III, 2007. Paleocene-Eocene thermal maximum and the opening of the Northeast Atlantic. *Science*, 316:587–589, doi:10.1126/science.1135274.

Svensen, H., Planke, S., Malthe-Serenssen, A., Jamtveit, B., Myklebust, R., Eidem, T.R., and Rey, S.S., 2004. Release of methane from a volcanic basin as mechanism for initial Eocene global warming. *Nature*, 429:542–545, doi:10.1038/nature02566.

Symonds, P. A., Planke, S., Frey, Ø., and Skogseid, J., 1998. Volcanic evolution of the Western Australian continental margin and its implications for basin development. *In* Purcell, P.G.R.R. (Ed.), *The Sedimentary Basins of Western Australia 2: Proceedings of the PESA Symposium*, Perth, Australia (Petroleum Exploration Society), 33–54.

Taylor, B., Goodliffe, A.M., and Martinez, F., 1999a. How continents break up: insights from Papua New Guinea. *J. Geophys. Res.*, 104:7497–7512.

Taylor, B., Huchon, P., Klaus, A., et al., 1999b. *Proceedings of the Ocean Drilling Program, Initial Reports 180*, College Station, Texas (Ocean Drilling Program), doi:10.1029/1998JB900115.

Tucholke, B.E., Sawyer, D.S., and Sibuet, J.-C., 2007. Breakup of the Newfoundland-Iberia rift. *In* Karner, G.D., Manatschal, G., and Pinheiro, L.M. (Eds.), *Imaging, Mapping and Modeling Continental Lithosphere Extension and Breakup*, Geol. Soc., Spec. Publ., 282:9–46.

Tucholke, B.E., Sibuet, J.-C., Klaus, A., et al., 2004. *Proceedings of the Ocean Drilling Program, Initial Reports 210*, College Station, Texas (Ocean Drilling Program), 78 pp.

Umhoefer, P.J., Schwennicke, T., Del Margo, M.T., Ruiz-Gerald, G., Ingle, J.C., Jr., and McIntosh, W., 2007. Transtensional fault-termination basins: an important basin type illustrated by the Pliocene San Jose Island basin and related basins in the southern Gulf of California, Mexico. *Basin Res.,* 19:297–322, doi:10.1111/j.1365-2117.2007.00323.x.

Wallace, L.M., Stevens, C., Silver, E.A., McCaffrey, R., Loratung, W., Hasiasta, S., Stanaway, R., Curley, R., Rosa, R., and Taugaloidi, J., 2004. GPS and seismological constraints on active tectonics and arc-continent collision in Papua New Guinea: implications for mechanics of micro-plate rotations in a plate boundary zone. *J. Geophys. Res.*, 109:B05404, doi:10.1029/2003JB002481.

White, R.S., and McKenzie, D.P., 1989. Magmatism at rift zones: the generation of volcanic continental margins and flood basalts. *J. Geophys. Res.*, 94:7685–7729.

Authors

Dale S. Sawyer, Department of Earth Science, Rice University, MS 126, P.O. Box 1892, Houston, Texas 77251-1892, U.S.A., e-mail: dale@rice.edu.

Millard F. Coffin, Ocean Research Institute, University of Tokyo, 1-15-1 Minamidai, Nakano-ku, Tokyo 164-8639, Japan.

Timothy J. Reston, Subsurface Group, Earth Sciences, School of Geography, Earth and Environmental Sciences, University of Birmingham, B15 2TT, U.K.

Joann M. Stock, Seismological Laboratory 252-21, California Institute of Technology, 1200 East California Boulevard, Pasadena, Calif. 91125, U.S.A.

John R. Hopper, Department of Geology and Geophysics, Texas A&M University, College Station, Texas 77843-3115, U.S.A.

Related Web Links

IODP: www.iodp.org
InterMARGINS: www.intermargins.org

Figure and Photo Credits

Figure 12. Photo courtesy of M. Storey (University of Leicester, U.K.).
Figure 13. Photo courtesy of G. Manatschal (CGS-EOST, Strasbourg, France).

3

Exploring Subseafloor Life with the Integrated Ocean Drilling Program

by Steven D'Hondt, Fumio Inagaki, Timothy Ferdelman, Bo Barker Jørgensen, Kenji Kato, Paul Kemp, Patricia Sobecky, Mitchell Sogin and Ken Takai

Introduction

Deep drilling of marine sediments and igneous crust offers a unique opportunity to explore how life persists and evolves in the Earth's deepest subsurface ecosystems. Resource availability deep beneath the seafloor may impose constraints on microbial growth and dispersal patterns that differ greatly from those in the surface world. Processes that mediate microbial evolution and diversity may also be very different in these habitats, which approach and probably pass the extreme limits of life. Communities in parts of the deep subsurface may resemble primordial microbial ecosystems, and may serve as analogues of life on other planetary bodies, such as Mars or Europa, that have or once had water.

Cell concentration estimates suggest remarkably abundant microbial populations in subseafloor sediments (Whitman et al., 1998; Parkes et al., 2000), but current models do not account for their possible impact on global biogeochemical cycling. Furthermore, we lack fundamental knowledge of microbial community composition, diversity, distribution, and metabolism in subseafloor environments. Exploration of this system presents a rich opportunity to understand microbial communities at Earth's extremes. Microbes in different subseafloor environments often encounter and must survive conditions of high pressure, high temperature, and extreme starvation. The limits of subsurface life are not yet known in terms of any environmental properties, including depth, temperature, energy availability, and geologic age; however, it is known that subseafloor microbes play a significant role in chemical reactions that were previously thought to have been abiotic, including iron and sulfur cycling as well as ethane and propane generation (Hinrichs et al., 2006). Some subseafloor communities probably derive food (electron donors) without sunlight (Jørgensen and D'Hondt, 2006). In addition, while molecular analyses and cultivation experiments demonstrate a surprisingly high diversity of microbial life in the subseafloor (Reed et al., 2002; Inagaki et al., 2003, 2006; Teske, 2006; Webster et al., 2006; Batzke et al., 2007), the relative abundances and roles of *Archaea*, *Bacteria*, *Eukarya*, and viruses remain largely unknown. In short, subseafloor life constitutes one of the least explored biomes on Earth and deserves intense exploration.

The Integrated Ocean Drilling Program (IODP) provides researchers with tremendous opportunities to better under-stand the abundance, activity, diversity, and limits of deep subseafloor microbial communities through drilling expeditions. A workshop titled "Exploration of Subseafloor Life with IODP" was convened in October 2006 to solicit recommendations and guidance from a broad community of scientists and to address scientific issues and technical challenges for exploring microbial life in the deep subseafloor. The ninety diverse workshop participants included molecular biologists, microbiologists, microbial ecologists, geologists, biogeochemists, drilling experts, and engineers.

Breakout sessions were focused on four key scientific areas: (1) biogeography, (2) genes, cells, populations, and communities, (3) habitability, and (4) technology. The principal conclusions of these breakout groups were as follows:

Biogeography: Four aspects of biogeography were considered particularly important. The first is a thorough characterization of the full range of deep subsurface habitats and their microbiota including deep subseafloor communities that are controlled by surface inputs (for example, via sediment accumulation) and their changes with age and depth. They also include microbial communities that derive their energy from non-photosynthetic processes (e.g., thermogenesis of organic compounds, subduction zone processes, serpentinization, and radiolysis of water) and basaltic environments deep beneath the seafloor. The last is potentially Earth's largest habitat by volume, but its significance as a habitat for microbial life has yet to be confirmed. Second, spatial and temporal controls on diversity need to be explored through (1) detailed fine-scale analysis of appropriately stored cores from previously drilled holes, (2) new drilling expeditions that target contrasting sedimentary environments, and (3) institution of deep subseafloor microbial observatories. Such studies will identify evolutionary controls on diversity and go beyond current surveys of biodiversity changes with depth and habitat. Third, we also need to know the mechanisms and rates of evolution under potentially slow growth, low predation, and severe energy limitation conditions. Do these situations select for novel metabolic and life history strategies? Finally, the extent to which the deep biosphere is generally connected or isolated from the surface biosphere is unclear. Is there a subsurface community unique to the deep biosphere? Does horizontal spreading of microbes cause the same biogeochemical zones to contain the same communities in different oceans, or are there barriers to dispersal?

Genes, cells, populations, and communities: The workshop participants recommended that microscopic observations of sediment and rocks should be expanded to include modern cell staining procedures that maximize biological information. In combination with studies of nucleic acids and organic biomarkers, such direct counts can provide information on the abundance, distribution, and extent of microorganisms in the subseafloor biosphere. Relatively detailed information on microbial populations can be gathered by employing specific microscopic techniques that provide fundamental information about the phylogenetic status and activity of individual cells. Because viral populations have the potential to play important roles in cell death and in horizontal transfer of genes between microorganisms, efforts should also be made to assess the abundance of viruses in both sediment porewater and crustal fluids. Studies of microbial abundance will answer major questions regarding the extent of the subseafloor biosphere, per-cell rates of microbial activities, and the roles of specific populations in major biogeochemical cycles. In addition, with rapid progress in molecular microbial ecology methods, it is now possible to determine the overall genetic potential of microbial communities and to link this potential to specific gene function and expression. This combination of approaches will provide rich information about subseafloor microbial genes, cells, populations, and communities.

Habitability: Subseafloor sediments and crust comprise two of the largest habitats on Earth. Exploring these habitats poses three major challenges for the biogeochemical and geomicrobiological scientific community. The first of these challenges, defining and mapping the limits to habitability in deep subseafloor sediments and crust, was deemed by the breakout group participants to be of highest priority. These limits are set by a variety of physical and chemical properties, such as temperature, availability of energy and nutrients, pH, pressure, water availability, and salinity. Mapping global distributions of physical properties, concentrations, and fluxes of both reductants and oxidants that provide a hospitable environment for microbial activity in the subseafloor ocean should be an obvious product of a concerted scientific drilling effort. The second critical challenge is the detection of life and the consequences of microbial activity, especially in low energy systems. The third challenge is to determine the role of deep subseafloor habitats in biosphere-geosphere coupling (e.g., how the subseafloor biosphere affects global biogeochemical cycles). Thorough and quantitative descriptions of key environments, microbial activities, and mass fluxes are required. The geological and biological processes that control transitions in states of habitability and that fuel growth and survival of microbial communities in deep subseafloor environments remain to be determined, but they can be realized through appropriate IODP projects. The opportunity to characterize these transitions with high

resolution sampling across a variety of deep subseafloor environments should be pursued by using a broad array of potential microbial signatures—pore water metabolic products, alterations of elemental isotope distributions, enzymatic activities, RNA, and intact polar lipids.

Technology: A number of issues related to technology were discussed. The primary objective was to assure that specific procedures of coring, sample handling, archiving, and routine measurements would be maintained to facilitate microbiological characterization of the deep biosphere. The participants recognized on-going efforts that have been undertaken to facilitate coring and sample handling for study of subseafloor life. Workshop recommendations were intended to build on this existing framework. They included the establishment of an IODP microbiology legacy sampling protocol that would be essential to further studies of the deep biosphere and to facilitate interdisciplinary investigations among IODP scientists. All data generated as part of this legacy sampling should be integrated with the existing IODP database structure, and more microbiological analyses must be added. The IODP is clearly the best vehicle to gain an understanding of the microbial life of the deep biosphere, though appropriate modifications of existing sampling and analytical protocols may be necessary.

Efforts must be made to develop and refine new shipboard and *in situ* technologies, with one major focus being on CORKs (circulation obviation retrofit kit). CORK technology can provide a platform for culturing active microbial members of the population *in situ* as well as continual access to subsurface porewater for continuous biogeochemical monitoring of microbial activity within the subseafloor.

Scientific Objectives

Several major scientific objectives were described by multiple working groups as ideally addressed by IODP capabilities. These include (1) discovering the limits of life and habitability on Earth; (2) determining the extent and nature of deep subseafloor life; (3) identifying the fundamental processes that control the dispersal and evolution of subsurface life; and (4) quantifying the rates and consequences of interactions between the deep subseafloor biosphere and the surface world.

Limits of Life and Habitability on Earth

There are very few natural environments on Earth's surface where life is absent; however, limits to life are expected in the subsurface world. Consequently, deep drilling of the ocean crust and sediments is uniquely positioned to access and explore the physical and chemical limits to life.

Transitions between states of habitability, as shaped by transitions in environmental properties, are laid out over extensive spatial and temporal scales in the subsurface

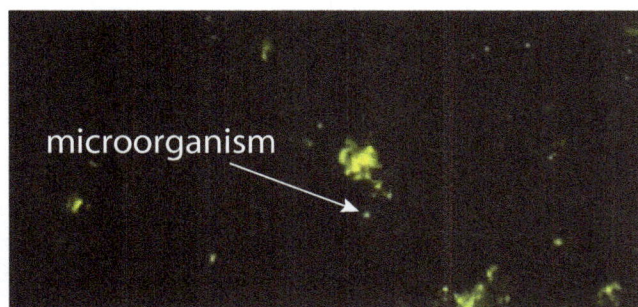

Figure 1. Microorganism from sediment deep beneath the Arctic seafloor (IODP Leg 302). The microbe glows because it has been stained with a fluorescent dye (SYBR Green) that preferentially binds to DNA. Photomicrograph from David C. Smith..

realm. The opportunity to characterize these transitions with high resolution sampling, across a variety of subseafloor environments, is unparalleled for characterizing the fundamental constraints on life. Deep subseafloor drilling is uniquely capable of realizing this potential.

The surface and subsurface realms differ fundamentally in their potential to refine our concept of habitability. The surface world is extensively inhabited. It is characterized by abundant flows of energy. Its uninhabitable spaces are rare. They principally result from dramatic extremes of temperature or aridity, and they typically begin at sharp spatial and temporal transitions from abundantly inhabited spaces. The abrupt aspect of these transitions is incongruent with the nature and resolution of sampling required to refine our sense of habitability and the limits to life. The deep subsurface environment contrasts these characteristics in several respects that offer significant promise for characterizing fundamental constraints on life and habitability. The current state of the art in ocean drilling virtually ensures that the limits of habitability and life will be breached with respect to physical, chemical, and energetic constraints, both individually and in combination. For example, energy flow in subseafloor settings is often orders of magnitude lower than in the surface world (D'Hondt et al., 2002). Subseafloor gradients in environmental conditions and biological states occur over distances that permit highly resolved sampling.

High temperature provides an obvious example of a limit to life and habitability that can be effectively explored through deep-ocean drilling. The presently accepted upper temperature limit for life is 121°C (Kashefi and Lovley, 2003), although higher temperature limits have been inferred by a small number of studies. Drilling into subseafloor sediments, igneous rocks, and hydrothermal deposits with temperatures that span the range of 100°C–250°C can be used to provide data that may determine upper temperature limits for life.

An additional parameter that undoubtedly limits the occurrence of life is the availability of energy (electron donors and electron acceptors). Subseafloor environments include environments where energy availability is probably too low to maintain life, and environments where life-sus-

taining fluxes of electron donors (food) may be independent of photosynthesis and the surface world. In order to characterize these environments, their potential inhabitants, and their control on the occurrence of life, these chemical constraints should be studied in a broad range of environments, including, but not limited to, sediments with very low organic content, deep crustal rocks, and subseafloor regions of active serpentinization.

Environmental limitations on cell density and community complexity may also include the mineralogy of sediments and sub-sediment rocks, interaction between cells and minerals, surface area, porosity and permeability, minimum pore size, and the availability of water.

Extent and Nature of Life in the Deep Subseafloor

The discovery of a diverse and active subseafloor microbial community in deep sediments and crustal rocks has fundamentally changed our perception of life on Earth, yet, the extent and nature of this subsurface life remains largely unknown. To address this issue, information on the abundance, diversity, and activities of *Archaea*, *Bacteria*, *Eukarya*, and viral populations needs to be elucidated.

Abundance—Microbial cells are widespread and often abundant in subseafloor sediment (Parkes et al., 2000) (Fig. 1), but the geographic distribution, composition, and total biomass of subseafloor life is largely unknown, because major provinces of ocean sediment and crust have never been examined for life, and many categories of organisms (archaea, eukaryotes, viruses) have never been sought in subseafloor samples. Routine microscopic observations of sediment and rocks should be expanded to consistently include direct cell counts. Further information on subseafloor populations can be gathered with microscopic techniques that provide phylogenetic and activity information (such as fluorescent *in situ* hybridization [FISH], catalyzed reporter deposition [CARD]-FISH, 5-cyano-2,3-ditolyl tetrazolium chloride [CTC], and live/dead staining). To examine viral populations, which have the potential to play important roles in cell death and horizontal gene transfer, efforts should be made to assess their presence in the porewater of sediments (as in Breitbart et al., 2004) and in crustal fluids. Detectable viral populations should be sequenced or tested for an effect on sediment organisms. In addition, efforts to explore the existence and abundance of eukaryotes in sediment and oceanic crust need to be made. The examination of microbial abundance will help answer major questions regarding the extent of the subseafloor and the rate of activity per cell in biogeochemical cycles. These results will serve as a major guide for microbiological investigations of subseafloor sediment and basalt.

Diversity—Microorganisms are the most abundant and diverse life forms on the planet and are the fundamental drivers of global biogeochemical cycles. Despite their ubiquity, surprisingly little is known about their diversity in the sub-seafloor, where sediments and basalt are key environments for studying microbial populations and community structures. Many important questions can only be solved by first understanding what microorganisms are present and what functional roles they serve in their communities. Given the combinations of extreme physical and chemical conditions, many of which are only found in the subseafloor (high pressure, low carbon, etc.), this environment is likely to contain assemblages of endemic microorganisms well-adapted to these conditions. Therefore, the subseafloor represents a rich and potentially novel source of genetic material for study and for mining for novel enzymes and enzymes adapted to extreme conditions that can be applied to a wide variety of biotechnological applications.

To understand subseafloor diversity, it will be necessary to conduct surveys of appropriate phylogenetic (16S rRNA genes (Fig. 2) and 18S rRNA genes) and functional gene markers. A variety of methods exist for such microbial diver-sity analyses (denaturing gradient gel electrophoresis (DGGE), terminal restriction fragment length polymor-phism (T-RFLP), clone libraries, and tag sequencing); indi-vidual investigators should choose the method that is most appropriate based on their specific questions, equipment, and available resources. Unusual and rare microorganisms are to be expected in the deep subsurface, and they may require novel 16S rRNA sequencing approaches for their detection (Sogin et al., 2006). In addition, other markers for phylogenetic diversity should also be considered (e.g., *recA*, *gyrase*). While phylogenetic approaches are useful for under-standing diversity, they do not provide a direct understanding of function. To better investigate the genetic potential of the subseafloor and the physiological roles of its inhabitants, a suite of investigations would need to be integrated. Molecular approaches include laser capture microdissection (a method for extracting a single cell), which can be coupled to whole genome amplification to map the entire genome of individual cells (Podar et al., 2007). In addition, mRNA amplification for gene expression profiles may be of particular importance for use in low biomass environments.

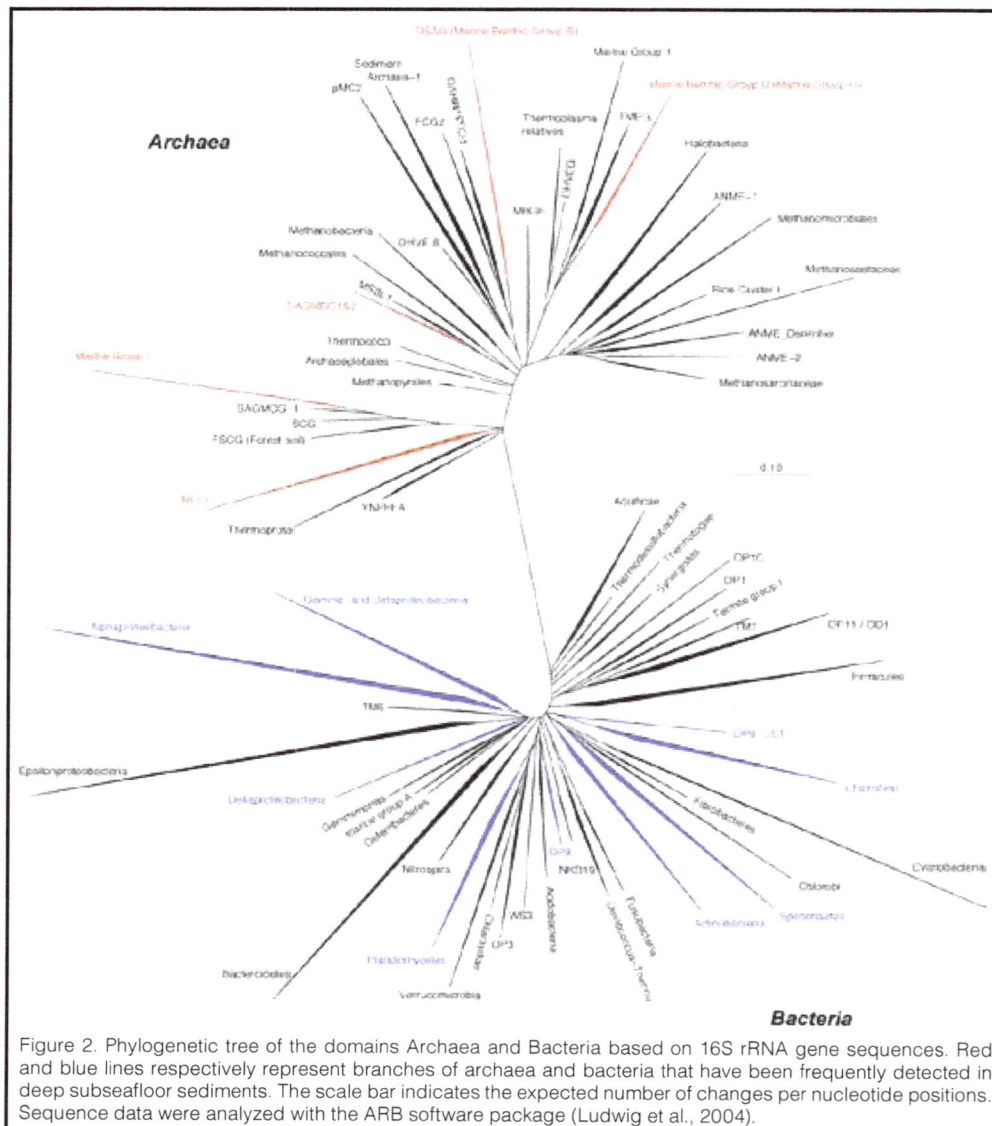

Figure 2. Phylogenetic tree of the domains Archaea and Bacteria based on 16S rRNA gene sequences. Red and blue lines respectively represent branches of archaea and bacteria that have been frequently detected in deep subseafloor sediments. The scale bar indicates the expected number of changes per nucleotide positions. Sequence data were analyzed with the ARB software package (Ludwig et al., 2004).

The application of genomic methods to the study of microbial communities directly from the environment (e.g., metagenomics) also represents a range of powerful new approaches. Initial forays into the generation of metagenomic data might best be performed on low diversity environments of at least moderate biomass. This would increase the ability to obtain a larger contiguous sequence from shotgun sequencing and assembly. The creation of genomic libraries also represents an important form of archiving, since these libraries represent a snapshot of the environment from which they were constructed and can be mined by multiple investigators over time.

Metagenomic data, par-ticularly a deep coverage reference metagenome, would create a critical resource for furthering the application of post-genomic methods designed to inves-tigate gene presence or absence (DNA), transcription (mRNA), and translation (protein). For instance, micro-array approaches can be applied to determine the presence or absence of genes from a particular region of the subsea-floor using comparative genomic hybridizations or to examine gene expression profiles.

Novel culturing methods include those that rely on dilution to extinction in very-low-nutrient media that have been used to obtain bacterial isolates from oligotrophic freshwater (e.g., Crater Lake, Oregon) and marine (e.g., the Sargasso Sea; Rappé et al., 2002) systems and are adaptable to use in studies of subseafloor life. A critical need in environmental microbiology is to elucidate the ecological roles of the most abundant habitat-specific phylogenetic clades. Therefore, cultivation of representative organisms from dominant phylogenetic clades is an important step in initiating additional studies linking diversity with function. Cultured microorganisms are also excellent targets for whole genome sequencing and comparative analyses to aid the creation and data mining efforts of metagenomic and post-metagenomic approaches as well as studies of population genetics.

Activity—IODP can provide access to many biogeochemical regimes within the subseafloor from which a polyphasic approach that incorporates geochemical and molecular techniques should be used to characterize microbial metabolic activities. By extending current shipboard protocols, geochemical analyses of cored sediments can determine concentrations of key terminal electron acceptors (e.g., NO_3^-, Fe(III), Mn(IV), H_2, CO_2, SO_4^{2+}) and electron donors (CH_4, Fe(II), H_2S, low molecular weight fatty acids), as well as additional soluble geochemical constituents (isotopic ratios). These data can then serve as guides to specific metabolic processes within the sediment (D'Hondt et al., 2002, 2004; Parkes et al., 2005; Hinrichs et al., 2006) and basalt (Cowen et al., 2003). Molecular assays that target domain level and lineage specific fractions of the population using SSU rRNA genes (DNA-based) and SSU rRNA (RNA-based) can identify the structure of total communities (Inagaki et al., 2006) and metabolically active communities (Sørensen and Teske, 2006), respectively (Fig. 3). Fluorescent in situ hybridization assays that target SSU rRNA molecules within active microbial cells can detect and quantify subsurface microbial populations (Schippers et al., 2005; Biddle et al., 2006). The abundance of functional genes and quantification of enzymes (hydrogenase, phosphatase) can then be utilized to determine the relative activity of specific metabolic pathways that may contribute to local geochemistry. Stable isotope probing (SIP) can also be incorporated to identify carbon utilization, addressing a number of questions including the contribution of heterotrophic versus autotrophic communities within the subsurface. Specific metabolic pathways previously identified within IODP core sediments can therefore be characterized using these and other molecular techniques to identify the lineages responsible for these pathways and begin to address unresolved issues, including but not limited to spa-tial heterogeneity, isolation of subseafloor communities, and horizontal gene flow.

A majority of this microbiological work would be accomplished postcruise. However, efforts will be made to develop and refine new shipboard and in situ technologies. The outcome of both geochemical and molecular measurements will provide a better understanding of the interaction of microbes and their biogeochemical environment, build better biogeochemical models in the oceans, and link surface life and the contribution of subseafloor life to global carbon and other biogeochemical cycles.

Dispersal and Evolution of Subseafloor Life

Exploration of subseafloor life offers an unmatched opportunity to explore the fundamental processes and mechanisms that have determined and continue to drive the evolution and dispersal of life on Earth.

The fundamental processes that influence the geographic distribution of all life include evolutionary processes that shape the diversity of microbial assemblages through their effects on the growth and death of individuals and populations; dispersal processes that serve to isolate or connect microbial assemblages; and environmental factors that favor winners and losers among the microbes present in a given environment.

Almost nothing is known about how these processes and their consequences play out in the subsurface world. The extent to which subseafloor life is an archive for organisms from the surface world, or a source of genomic, evolutionary, and ecological innovation for the surface world, can be debated. Do evolutionary pressures and mechanisms of microbial dispersal operate differently in the subseafloor communities than in surface communities? Is subseafloor diversity a subset of the diversity in the upper world, or have subseafloor processes of organismal dispersal and environmental selection led to identifiable endemic assemblages? Have novel groups of organisms evolved to take advantage of energy sources that are not commonly considered important in the surface world, such as radiolysis of water, serpentinization reactions, and other subsurface energy sources? Is there selection for or against major metabolic and life history strategies, e.g., metabolic specialization versus a generalism? Can one identify signatures of surface life in subseafloor microbial assemblages, and conversely, can one find evidence that subseafloor life is exported to the surface world?

Geosphere-Biosphere Interactions

Interactions between the deep subseafloor ecosystem and the geosphere occur on scales from very local to global. These interactions may ultimately affect climate, evolution of the global biosphere, and structure and function of the

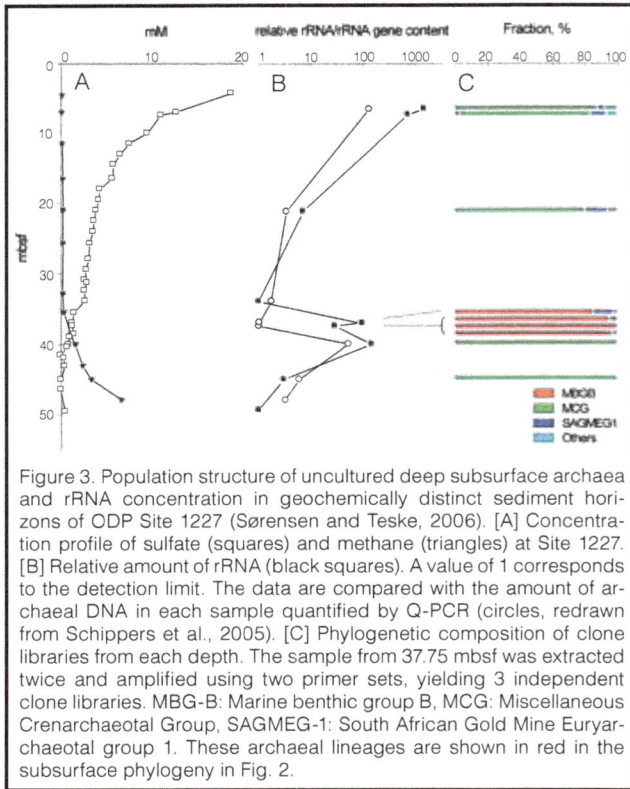

Figure 3. Population structure of uncultured deep subsurface archaea and rRNA concentration in geochemically distinct sediment horizons of ODP Site 1227 (Sørensen and Teske, 2006). [A] Concentration profile of sulfate (squares) and methane (triangles) at Site 1227. [B] Relative amount of rRNA (black squares). A value of 1 corresponds to the detection limit. The data are compared with the amount of archaeal DNA in each sample quantified by Q-PCR (circles, redrawn from Schippers et al., 2005). [C] Phylogenetic composition of clone libraries from each depth. The sample from 37.75 mbsf was extracted twice and amplified using two primer sets, yielding 3 independent clone libraries. MBG-B: Marine benthic group B, MCG: Miscellaneous Crenarchaeotal Group, SAGMEG-1: South African Gold Mine Euryarchaeotal group 1. These archaeal lineages are shown in red in the subsurface phylogeny in Fig. 2.

lithosphere. Scientific drilling can address the question of how subseafloor microbial processes and products are related to processes and properties of the overlying surface world (the ocean and the atmosphere). These include, but are not limited to 1) the global balance of oxidants and reductants, 2) the acid-base system of the world ocean, 3) primary productivity uncoupled from the photosynthetic surface world, 4) physical and chemical alteration of deep sediment, and 5) crustal geologic structure.

Scientific drilling of continental margins has demonstrated that massive levels of methane exist along continental margins and may principally be derived from the deep subseafloor biosphere (Fig. 4). Paleoclimatic studies suggest that this energy-rich compound enters the overlying ocean water and the atmosphere and dramatically alters climate. Rates of methanogenesis in deep marine sediments along continental margins are poorly constrained, and recent evidence strongly implicates the important role of microbes in the production of higher hydrocarbons such as ethane and propane (Hinrichs et al., 2006). Another climate driver includes the production and consumption of nutrient compounds. For example, we have no clear idea of the biological fixation, removal, and/or production of nitrogen in deep subseafloor environments. Profiles of dissolved and solid phase N compounds suggest that these compounds are being cycled in deep subseafloor sediment. The acid-base system of the deep ocean is also subject to deep subseafloor processes. Subseafloor metabolism produces and consumes alkalinity and may have a significant effect on ocean pH and thus climate. Understanding the overall effect of the subseafloor

biosphere on the oceanic acid/base balance (and climate) requires quantification of the net rates of metabolic redox reactions and the factors that control these rates. The redox state of buried carbon and its modification with burial and age as it enters subduction zones have clear implications for global carbon and redox budgets. Exchange of oxidants and reductants between the mantle, crust, and ocean may be significantly altered by deep subsurface processes (Hayes and Waldbauer, 2006). These are all global aspects of coupling of the geosphere and deep subseafloor processes that cause us to ask: what would life in the surface biosphere look like if the deep subsurface biosphere did not exist?

There are numerous potential mechanisms for connecting material and energy flow between deep geosphere processes and life in deep subseafloor environments. On Earth's surface, sunlight overwhelmingly provides the energy for primary productivity; in the subsurface however, the nature of primary productivity is not yet constrained. For example, processes that sustain subsurface primary productivity may be potentially more diverse than those that sustain surface primary productivity, and they may ultimately feed into the ocean-climate system. Because of the high likelihood for abiotic production of hydrogen and simple organic compounds, places where water interacts with ultramafic rock are obvious drilling targets. This type of interaction takes place in a variety of tectonic settings, including rifted continental margins, lava-starved mid-ocean ridge segments, and suprasubduction zone forearcs. The lithosphere community (e.g., 21st Mohole Mission) has a large interest in exploring geodynamic and petrogenetic aspects of these settings. Radiolysis of water may provide a cryptic source of energy for primary production in colder subseafloor environments (Jørgensen and D'Hondt, 2006).

Moreover, the impact of deep sub-surface processes on lithologic properties in crustal and sedimentary environments is largely unknown. Key properties of porosity, permeability, surface area, and the viscosity of interstitial fluids may be significantly altered by sub-surface life processes.

To appreciate the extent and importance of geosphere-biosphere interactions in the deep subseafloor, scientific ocean drilling must focus on (1) understanding the processes, pathways, and products of interaction between microbial activity in the subseafloor and the ocean/atmosphere system; (2) elucidating the magnitude and rates of these processes and their ability to influence the environment and climate of the ocean/atmosphere system; and (3) determining the evolution of these interactions over geological time.

Expeditionary Strategies

The IODP provides a unique facility for studying deep subseafloor life. No other organization provides direct access to deep subsurface environments throughout the world ocean. Nowhere else has a comparable sampling capability been developed. Microbiological samples need to be evaluated for contamination by seawater or drilling fluid, as a single cell or DNA molecule might cause erroneous results, and IODP offers essential contamination assessment unlike any other scientific operation. IODP offers a controlled framework for sample handling, archiving, and analysis in microbiology, biogeochemistry, and related disciplines that allows for data integration and cross-sample comparisons. A large set of chemical and physical metadata generated on board is easily integrated into biological experiments. Furthermore, the program provides sites appropriate for long-term *in situ* experiments on and in the subseafloor, on a broad range of spatial and temporal scales.

Multiple expeditionary strategies will be necessary to explore subseafloor life and habitats. These include (1) single expeditions dedicated to specific subseafloor life objectives; (2) incorporation of dedicated holes or sites into legs scheduled for other purposes (e.g., through an Ancillary Program Letter); (3) participation as a shipboard or shore-based scientist on expeditions scheduled for other purposes; (4) routine measurements and routine archiving of appropriate samples on many or all IODP expeditions; and (5) in special cases, such as required by installation of subseafloor microbial observatories, multi-expedition (multi-platform) projects. In every case, including routine legacy sampling, consideration of microbiological sampling should be an integral part of expedition planning.

Technology Recommendations

The Ocean Drilling Program (ODP) and the IODP have built a solid framework for occasional study of subseafloor life; however, a number of issues must be addressed to improve studies of deep subseafloor microbial communities and to render such studies routine. The primary objective is to assure that, whenever and wherever possible, coring and sampling handling will facilitate characterization of subseafloor life.

Core recovery (quality and quantity): Coring technology should be optimized to increase quality and, when possible, to increase quantity of material. Advances from industry and from ODP experiments can be leveraged to achieve this goal. Approaches to minimize drilling disturbance and fluid circulation include alteration of existing drilling practices (e.g., use of mud motor, "drilling over" to extend use of hydraulic piston corers to great depth); monitoring activity at the drill bit (to improve quality); bit/shoe redesign; and alteration(s) of core barrel and liners. These changes will greatly enhance the microbiological and biogeochemical value of cored material. For example, they will provide samples of crucial niches at geologic interfaces (e.g., lithologic contacts and fractured materials), and they will improve the rigor and quality of downstream analyses.

Figure 4. Methane hydrate (white patches) in freshly drilled, organic-rich, methane-supersaturated deep subsurface sediment (Peru Trench, ODP Site 1230, 5086 m water depth, 96 m below seafloor).

Diagnostic monitoring of core quality: Quality control issues have been previously addressed with methods to quantify the intrusion of drilling fluid (Smith et al., 2000; House et al., 2003; Lever et al., 2006). Contamination is particularly problematic with igneous samples, and this issue needs to be carefully addressed.

Use of Contamination Tracers—Quantitative contamination tracing should be done on all cores to be used for microbiological study. Existing tracers for monitoring and quantifying contamination [a perfluorocarbon tracer (PFT) and fluorescent beads] have been key components of recent advances in study of subsurface life (Smith et al., 2000; House et al., 2003; Lever et al., 2006), but contamination tracing can and should be greatly improved. Detection limits should be enhanced by further improvement of techniques and by use of a chemical tracer less volatile than the current PFT. Other categories of tracers (e.g., quantum dots) and tools (e.g., 3-D imaging techniques) should be considered to enhance the quality, rigor, and simplicity of analyses. A standardized nucleic acid based method should be developed for monitoring cross-contamination during drilling. The geochemistry and microbiology programs should be directly coupled to monitor gases and contamination tracers.

Onboard handling protocol—Microbiological samples should be aseptically handled and, when appropriate, anaerobically maintained. An appropriate temperature-controlled handling method should be established. Rigorous monitoring of reagents should be conducted.

Personnel—A dedicated, full-time microbiologically trained technician is essential for diagnostic monitoring, archival sampling, and routine measurements. In addition, one or more microbiologists and one or more biogeochemists are needed to maximize scientific return on all cruises.

Maintenance of in situ *conditions*: As study of subsurface life grows within IODP, the program will need to consider methods for maintaining *in situ* conditions during coring and sample handling. Many analyses are compromised by depressurization and uncontrolled temperature fluctuations during core retrieval and handling. Core processing should be optimized to minimize changes in temperature, e.g., by recovering cores as quickly as possible and sampling them in a temperature-stabilized environment. IODP should also explore methods for insulating the core after removal from the core barrel.

To date, all ocean drilling microbiological samples have undergone depressurization prior to subsampling. Development of sub-sampling tools, or experimental chambers, that can be mated to the pressure coring systems and used to acquire samples or to initiate experiments, is possible. Sensors can be adapted into the design of pressure core barrels to provide time series measurements of important parameters. There may be opportunities for microbiologically relevant spectral measurements using sapphire windows built into the barrels or by changing barrel composition (e.g., aluminum rather than steel) to allow volumetric imaging (X-ray CT, NMR, or PET) for microbial subsampling. However, microbiologists and biogeochemists must identify specific needs and objectives for these experiments before engineering design can be successfully integrated with science requirements.

Sample collection and archiving: Routine collection and preservation of microbiological samples and routine measurements of biological parameters have not been a component of drilling operations. We recommend that a routine sampling program be implemented on all legs and on all three drilling platforms. Microbiological sampling requires specific procedures for core handling that is best accomplished by a dedicated hole.

As an initial step toward this routine program, we recommend an experiment in integrating biodiversity sampling, archiving, and measurements as part of every drilling leg. The development of this database will establish baseline information that could be screened for parameters of interest by individual investigators (porewater chemistry, ocean basin, subseafloor depth, cell density variation, etc.), who could then request samples for the investigator's analysis of choice. A two-year experiment within the IODP that includes a well advertised, clear and straightforward sample request plan is recommended, with the following guidelines.

Requirements for sample handling and sample storage depend on the downstream analyses (see below). The following considerations are pertinent to samples that will be used for shipboard analyses and those that will be shipped to shore-based laboratories or repositories for postcruise studies. Although it is anticipated that most samples will be collected from anaerobic environments, low productivity and/or low carbon environments may be oxygenated, and thus samples should be stored appropriately. In all cases, avoidance and quantification of sample contamination with non-indigenous microbes during drilling and sample handling is of paramount importance.

Biomarker samples—Frozen samples are used for analyzes of nucleic acids, lipid biomarkers, amino acids, etc. These analyses are the central component of biodiversity studies. These samples should be collected as soon as possible and immediately frozen in liquid nitrogen or a -80°C freezer to ensure the safe delivery of samples from the sampling site to the repositories. Samples stored in ultra-low temperature freezers can be maintained in an anaerobic environment by adapting the method of Cragg et al. (1992).

Samples for cultures and activity experiments—Samples that will be used for culture-based isolations and microbial activity measurements should be stored at 4°C until analyzed. Samples that are to be used for microbial culturing must be protected from temperature and/or oxygen fluctuations (e.g., samples from an anaerobic environment should be maintained under anaerobic conditions). Samples for anoxic culturing work should be transferred to gas-tight trilaminate bags containing an oxygen scrubber.

Microscopy samples—Samples used for microscopy (e.g., direct cell counts, FISH, microautoradiography) are fixed with aldehydes such as formaldehyde or glutaraldehyde, washed with ethanol:PBS (1:1) solution, and stored at -20°C. For FISH-based sample storage it is recommended that samples be preserved according to established protocols (Pernthaler et al., 2002). For these samples, the particular assay dictates the details necessary in the fixation process.

Sample shipping—Because maintaining proper temperature for each category of analysis is essential, samples must be shipped under appropriate temperature conditions (e.g., frozen samples should be shipped in dry ice, and culturing samples should be shipped under 4°C refrigeration). A temperature logger should be included in each microbiological or biogeochemical shipping container to provide the thermal history of the samples during transit.

Sample archiving—IODP should establish a repository for routinely collected samples that are collected and stored for subsequent microbiological analysis. The subsamples should be collected as soon as possible after removal from the core in sterile syringes and stored appropriately as described below. This legacy sample should be taken from the middle of the core in near proximity to samples taken for biogeochemical, contamination tracing, and other microbiological analyses.

Standardization of sample-handling protocols—An IODP laboratory protocol book is suggested to help in standardizing procedures and techniques for microbiological sampling, shipping and archiving. This book can be made available electronically on the IODP web site.

Routine Microbiological Measurements: The openly accessible DSDP/ODP/IODP database of routine measurements is a tremendous strength of the scientific drilling program. This allows for continued analysis of the data whether it is using new techniques or global syntheses of data (Parkes et al., 2000; D'Hondt et al., 2002). Therefore, it is necessary to institute routine measurements that can be realistically obtained during IODP drilling projects and which provide useful data to assist in the study of subsurface microbiology. All of these data should be made available to the shipboard party via the standard IODP database.

Metabolic products and reactants—Concentrations of some electron acceptors (e.g., dissolved SO_4^{2-}) and some electron donors (e.g., CH_4) are already measured routinely as part of the shipboard geochemistry program, which should be expanded to include a much larger range of metabolites. These include, for example, dissolved iron and dissolved manganese in anoxic formation water, and dissolved oxygen and nitrate in the upper sediment column of all sites and above the sediment/basalt interface at open-ocean sites.

Biomass—Biomass quantification should be instituted as a routine measurement.

Total cell counts. Counting of total microbial cells is essential for quantifying subsurface biomass. Each legacy sample for cell counts should be divided in half, with one half counted on ship, and the other preserved and placed at -20°C for comparative counts or more specific quantification on shore. While most ODP/IODP cell counts have been generated using acridine orange direct counts (Cragg et al., 1990), we recommend phasing in the use of a nucleic acid stain with lower background and more stable fluorescence (e.g., SYBR Green) for cell enumeration (Lunau et al., 2005 for sediment analysis; Santelli et al., 2006 for basalt). Total cell counts tend to result in maximum estimates of biomass; however, the detection limit can be reduced by orders of magnitude by separating the cells from the sediment prior to enumeration (Kallmeyer et al., 2006). To standardize cell counts, we suggest that IODP consider adding a flow cytometry instrument to the shipboard laboratory.

As soon as possible, additional methods for assessing biomass should be compared by study of subsamples from multiple sites and cores. Candidate methods include:

(a) Phospholipids. Intact phospholipids can be used to estimate the total microbial biomass and broad community composition in sediment samples (White et al., 1979; Zink et al., 2003). This can be achieved with an HPLC-MS system by quantitative and qualitative analysis of the intact polar membrane lipids, which are diagnostic of live cells (Sturt et al., 2004; Biddle et al., 2006).

(b) ATP. Quantification to estimate active biomass has been used successfully in cores (Egeberg, 2000).

People qualified to undertake these techniques are rare. Members of the IODP community should work to build a pool of expertise sufficient to undertake one or more of these assays during or immediately after most expeditions.

Community composition—Further information on microbial populations can be gathered by using microscopic techniques such as FISH, CARD-FISH, CTC, and live/dead staining, which distinguish different phylogenetic groups and distinguish between active cells and dead or inactive cells. Samples can also be monitored for viruses.

Standardization of laboratory protocols—An IODP handbook is recommended to describe standard microbiological procedures (routine surface decontamination prior to sample handling, biological waste decontamination, etc). The Explanatory Notes from ODP Expedition 201 may serve as a guide for what may be achieved with such a handbook (D'Hondt et al., 2003). As with the suggested microbiology sampling handbook, this book can be made available electronically on the IODP web site.

Technology transfer and data dissemination: Because microbiologists generate some types of samples and data that are unique to their field, some additional issues need to be addressed.

Sequence data—Sequencing of nucleic acids has become the accepted standard method for identifying microorganisms. The usefulness of the data resides in the ability to compare sequences. This is accomplished by submission of sequences to internationally recognized, publicly accessible databases.

Molecular biology offers a suite of tools that provide a powerful strategy for gaining new insights into the diversity of microbial life of the subseafloor. This strategy is particularly powerful when complemented with culture-based methods. The coordinated structure of IODP offers a unique opportunity to explore microbial life in the subseafloor using these integrated approaches. To maximize this opportunity requires capture of genetic data (including sequences of 16S rRNA and 18S rRNA genes) from samples collected throughout the program and to make the genetic data available to a wider scientific audience. For this approach to succeed, the provenance of all the genetic information must be catalogued and integrated with the metadata obtained by other IODP efforts (e.g., lithology, geochemistry) in a relational database. To best complete this task, we recommend that the IODP consider building a relational database suitable to the data types collected in this program, as well as exploring interfacing with other repositories of genetic and metadata where feasible and appropriate. Examples of these databases include GenBank, the Ribosomal Database Project (RDP), and CAMERA.

Culture isolates—A common goal for many microbiologists is to obtain pure cultures of microorganisms to perform detailed studies on their physiological capabilities, produce specific enzymes or metabolic byproducts, and so on. It is common practice to place subsamples of the cultures in publicly accessible culture collections. In keeping with the open, international cooperation established during the previous decades of scientific ocean drilling, IODP should require that cultures of microorganisms isolated from cores be deposited in a publicly accessible culture collection (Takai et al., 2005). As the program grows and more microbial cultures become available, we recommend that a deep subseafloor culture collection be established. A good example and possible leverage would be to consider a repository in the U.S. Department of Energy's Subsurface Microbial Culture Collection (SMCC).

Microbiological Observatories: In situ experimentation is ultimately the best mechanism for determining *in situ* processes, including microbial colonization, mineral-microbe interactions and effects of tectonic/volcanic events on subseafloor communities. It is necessary to conduct time-series and manipulative experiments to constrain the roles, if any, that microorganisms may play in rock alteration and secondary mineral formation. The power of *in situ* experimentation for studying colonization and mineral alteration has been demonstrated at the seafloor (Edwards et al., 2003, 2004; Edwards, 2004) and in terrestrial systems (Bennett et al., 1996; Edwards et al., 1998; Hiebert and Bennett, 1992). Subseafloor microbiological experiments are now being conducted for the first time as part of the Juan de Fuca hydro-geology experiments made possible by IODP Leg 301 (Fisher et al., 2005; Nakagawa et al., 2006). Similar experiments are proposed for some subsurface life proposals active in the IODP system. Support for CORK operations and technology is imperative for the success of such projects. This includes necessary support for existing technologies, such as the CORK hardware and associated sensors (pressure, temperature, strain and tilt sensors) and samplers (OSMO). Additional *in situ* capabilities will be necessary to meet future microbiology needs. These needs include samplers and incubators for microbiological analyses, fluid transfer systems, pumping and power systems for seafloor sampling from bore holes, and *in situ* sensors for key metabolic species, such as hydrogen and methane. Study of active hydrothermal systems will require development of high temperature sampling and technology for *in situ* measurements. These developments must be supported and encouraged by IODP and governmental funding agencies in IODP countries.

Inclusion of microbiological observatories in IODP proposals requires early identification of critical design specifications. Experimental modules that could be deployed in boreholes will require iterative design efforts by scientists and engineers.

Partnership with Other Organizations

Study of subseafloor life may provide significant opportunities to partner with industry and government agencies. For example, the U.S. Department of Energy is interested in advanced drilling technology for sampling in high-temperature/high-pressure environments. Such partnerships may

require alignment of interests and new models for IODP collaboration and funding. Existing industry tools—such as the modular dynamic (formation) tester (MDT) with a flow control module for fluid sampling across the borehole wall with quality assurance methodology to reduce contamination—may greatly facilitate downhole integration of microbiological and formation fluid sampling. The challenge will be to find ways to deploy these tools in IODP boreholes and work out economic models for their deployment. IODP applications of existing tools may require alternative methods of tool deployment (e.g., wireline reentry, ROV guidance of tools into boreholes, or use of larger-diameter drill pipe) to make the desired measurements.

Acknowledgements

We thank all of the workshop participants for helping to build a solid community-based foundation for future IODP studies of subsurface life. We particularly thank Kelly Kryc, Holly Given, and the breakout group chairs (Wolfgang Bach, Heribert Cypionka, Katrina Edwards, Philippe Gaillot, Julie Huber, John Parkes and Andreas Teske). Without their efforts, the workshop would not have been a success.

References

Batzke, A., Engelen, B., Sass, H., and Cypionka, H., 2007. Phylogenetic and physiological diversity of cultured deep-biosphere bacteria from Equatorial Pacific Ocean and Peru Margin sediments. *Geomicrobiology J.*, (in press), doi:10.1016/S0009-2541(96)00040-X.

Bennett, P.C., Hiebert, F.K., and Choi, W.J., 1996. Microbial colonization and weathering of silicates in a petroleum-contaminated aquifer. *Chem. Geol.*, 132:45–53.

Biddle, J.F., Lipp, J.S., Lever, M.A., Lloyd, K.G., Sorensen, K.B., Anderson, R., Fredricks, H.F., Elvert, M., Kelly, T.J., Schrag, D.P., Sogin, M.L., Brenchley, J.E., Teske, A., House, C.H., and Hinrichs, K.-U., 2006. Heterotrophic Archaea dominate sedimentary subsurface ecosystems off Peru. *Proc. Natl. Acad. Sci. U.S.A.*, 103:3846–3851.

Breitbart, M., Felts, B., Kelley, S., Mahaffy, J.M., Nulton, J., Salamon, P., and Rohwer, F., 2004. Diversity and population structure of a near-shore marine-sediment viral community. *Proc. R. Soc. Lond. B*, 271(1539):565–574, doi:10.1098/rspb.2003.2628.

Cowen, J.P., Giovannoni, S.J., Kenig, F., Johnson, H.P., Butterfield, D., Rappe, M.S., Hutnak, M., and Lam, P., 2003. Fluids from aging ocean crust that support microbial life. *Science*, 299:120–123, doi:10.1126/science.1075653.

Cragg, B.A., Parkes, R.J., Fry, J.C., Herbert, R.A., Wimpenny, J.W.T., and Getliff, J.M., 1990. Bacterial biomass and activity profiles within deep sediment layers. *In* Suess, E., von Huene, R., et al. (Eds.), *Proc. ODP, Sci. Res. 112*, College Station, Texas (Ocean Drilling Program), 607–619.

Cragg, B.A., Harvey, S.M., Fry, J.C., Herbert, R.A., and Parkes, J.R., 1992. Bacterial biomass and activity in the deep sediment layers of the Japan Sea, Hole 798B. *In* Pisciotto, K.A., Ingle,

J.C., Jr., von Breymann, M.T., Barron, J., et al. *Proc. ODP, Sci. Res. 127/128 Pt 1*, College Station, Texas (Ocean Drilling Program), 761–773.

D'Hondt, S., Jørgensen, B.B., Miller, D.J., Aiello, I.W., Bekins, B., Blake, R., Cragg, B.A., Cypionka, H., Dickens, G.R., Hinnrichs, K.-U., Holm, N., House, C., Inagaki, F., Meister, P., Mitterer, R.M., Naehr, T., Niitsuma, S., Parkes, J., Schippers, A., Skilbeck, C.G., Smith, D.C., Spivack, A.J., Teske, A., Wiegel, J., 2003. Controls on microbial communities in deeply buried sediments, eastern equatorial Pacific and Peru Margin Sites 1225–1231. *Proc. ODP, Init. Rep. 201* [CD-ROM]. Available from Ocean Drilling Program, Texas A & M University, College Station Texas 77845–9547, USA. Web site: http://www-odp.tamu.edu/publications/201_IR/201ir.htm.

D'Hondt, S., Jorgensen, B.B., Miller, D.J., Batzke, A., Blake, R., Cragg, B.A., Cypionka, H., Dickens, G.R., Ferdelman, T., Hinrichs, K.U., Holm, N.G., Mitterer, R., Spivack, A., Wang, G.Z., Bekins, B., Engelen, B., Ford, K., Gettemy, G., Rutherford, S.D., Sass, H., Skilbeck, C.G., Aiello, I.W., Guerin, G., House, C.H., Inagaki, F., Meister, P., Naehr, T., Niitsuma, S., Parkes, R.J., Schippers, A., Smith, D.C., Teske, A., Wiegel, J., Padilla, C.N., and Acosta, J.L.S., 2004. Distributions of microbial activities in deep subseafloor sediments. *Science*, 306:2216–2221, doi :10.1126/science.1101155.

D'Hondt, S., Rutherford, S., and Spivack, A.J., 2002. Metabolic activity of subsurface life in deep-sea sediments. *Science*, 295:2067–2070, doi:10.1126/science.1064878.

Edwards, K.J., 2004. Formation and degradation of seafloor hydrothermal sulfide deposits. *In* Amend, J.A., Edwards, K.J., and Lyons, T. (Eds.), *Sulfur Biogeochemistry – Past & Present*, Boulder, Colo. (Geological Society of America), 83–96.

Edwards, K.J., Bach, W., McCollom, T.M., and Rogers, D.R., 2004. Neutrophilic iron-oxidizing bacteria in the ocean: habitats, diversity, and their roles in mineral deposition, rock alteration, and biomass production in the deep-sea. *Geomicrobiology J.*, 21(6):393–404, doi:10.1080/01490450490485863.

Edwards, K.J., McCollom, T.M., Konishi, H., and Buseck, P.R., 2003. Seafloor bio-alteration of sulfide minerals: results from *in situ* incubation studies. *Geochim. Cosmochim. Acta*, 67(15):2843–2856, doi:10.1016/S0016-7037(03)00089-9.

Edwards, K.J., Schrenk, M.O., Hamers, R.J., and Banfield, J.F., 1998. Microbial oxidation of pyrite: Experiments using microorganisms from an extreme acidic environment. *Amer. Mineral.*, 83(12):1444–1453.

Egeberg, P.K., 2000. Adenosine 5'-Triphosphate (ATP) as a proxy for bacteria numbers in deep-sea sediments and correlation with geochemical parameters (Site 994). *In* Paull, C.K., Matsumoto, R., Wallace, P.J., and Dillon, W.P. (Eds.), *Proc. ODP, Sci. Res. 164*, College Station, Texas (Ocean Drilling Program), 393–398.

Fisher, A.T., Urabe, T., Klaus, A., and the Expedition 301 Scientists, 2005. *Proc. IODP,* 301: College Station, Texas (Integrated Ocean Drillig Program Management International, Inc.), doi:10.2204/iodp.proc.301.2005.

Hayes, J.M., and Waldbauer, J.R., 2006. The carbon cycle and associated redox processes through time. *Philos. Trans. R. Soc. Lond. B Biol. Sci.*, 361(1470):931–50, doi:10.1098/rstb.2006.1840.

Hiebert, F.K., and Bennett, P.C., 1992. Microbial control of silicate weathering in organic-rich ground water. *Science*, 258:278–281, doi:10.1126/science.258.5080.278.

Hinrichs, K.-U., Hayes, J.M., Bach, W., Spivack, A.J., Hmelo, L.R., Holm, N.G., Johnson, C.G., and Sylva, S.P., 2006. Biological formation of ethane and propane in the deep marine subsurface. *Proc. Natl. Acad. Sci. U.S.A.*, 103(40):14684–14689, doi:10.1073/pnas.0606535103.

House, C., Cragg, B., Teske, A., and Party, S.S., 2003. Drilling contamination tests during ODP Leg 201 using chemical and particulate tracers. *Proc. ODP Init. Rep..201*, College Station, Texas (Ocean Drilling Program), 1–19.

Inagaki, F., Nunoura, T., Nakagawa, S., Teske, A., Lever, M., Lauer, A., Suzuki, M., Takai, K., Delwiche, M., Colwell, F.S., Nealson, K.H., Horikoshi, K., D'Hondt, S., and Jørgensen, B.B., 2006. Biogeographical distribution and diversity of microbes in methane hydrate-bearing deep marine sediments, on the Pacific Ocean Margin. *Proc. Natl. Acad. Sci. U.S.A.*, 103:2815–2820.

Inagaki, F., Suzuki, M., Takai, K., Oida, H., Sakamoto, T., Aoki, K., Nealson, K.H., and Horikoshi, K., 2003. Microbial communities associated with geological horizons in coastal subseafloor sediments from the Sea of Okhotsk. *Appl. Environ. Microbiol.*, 69:7224–7235, doi:10.1073/pnas.0511033103.

Jørgensen, B.B., and D'Hondt, S., 2006. A starving majority deep beneath the seafloor. *Science*, 314:932–943, doi:10.1126/science.1133796.

Kallmeyer, J., Anderson, R., Smith, D.C., Spivack, A.J., and D'Hondt, S., 2006. Separation of microbial cells from deep sediments, NASA Astrobiology Institute Biennial Meeting 2005. *Astrobiology*, 6(1):271.

Kashefi, K., and Lovley, D.R., 2003. Extending the upper temperature limit for life. *Science*, 301:934, doi:10.1126/science.1086823.

Lever, M.A., Alperin, M., Engelen, B., Inagaki, F., Nakagawa, S., Steinsbu, B.O., Teske, A., and IODP Expedition 301 Shipboard Scientific Party, 2006. Trends in basalt and sediment core contamination during IODP Expedition 301. *Geomicrobiology J.*, 23:517–530, doi:10.1080/01490450600897245.

Ludwig, W., Strunk, O., Westram, R., Richter, L., Meier, H., Yadhukumar, Buchner, A., Lai, T., Steppi, S., Jobb, G., Förster, W., Brettske, I., Gerber, S., Ginhart, A.W., Gross, O., Grumann, S., Hermann, S., Jost, R., König, A., Liss, T., Lüßmann, R., May, M., Nonhoff, B., Reichel, B., Strehlow, R., Stamatakis, A., Stuckmann, N., Vilbig, A., Lenke, M., Ludwig, T., Bode, A., and Schleifer, K.-H., 2004. ARB: a software environment for sequence data. *Nucleic Acids Res.* 32:1363–1371, doi: 10.1093/nar/gkh293.

Lunau, M., Lemke, A., Walther, K., Martens-Habbena, W., and Simon, M., 2005. An improved method for counting bacteria from sediments and turbid environments by epifluorescence microscopy. *Environ. Microbiol.*, 7:961–968, doi:10.1111/j.1462-2920.2005.00767.x.

Nakagawa, S., Inagaki, F., Suzuki, Y., Steinsbu, B.O., Lever, M.A., Takai, K., Engelen, B., Sako, Y., Wheat, C.G., Horikoshi, K., and Integrated Ocean Drilling Program Expedition 301 Scientists, 2006. Microbial communities in black rust exposed to hot ridge-flank crustal fluids. *Appl. Environ. Microbiol.*, 72:6789–6799, doi:10.1128/AEM.01238-06.

Parkes, R.J., Cragg, B.A., and Wellsbury, P., 2000. Recent studies on bacterial populations and processes in subseafloor sediments: *A review. Hydrogeol. J.*, 8:11–28, doi:10.1007/PL00010971.

Parkes, R.J., Webster, G., Cragg, B.A., Weightman, A.J., Newberry, C.J., Ferdelman, T.G., Kallmeyer, J., Jørgensen, B.B., Aiello, I.W., and Fry, J.C., 2005. Deep sub-seafloor prokaryotes stimulated at interfaces over geological time. *Nature*, 436:390–394, doi:10.1038/nature03796.

Pernthaler, A., Pernthaler, J., and Amann, R., 2002. Fluorescence *in situ* hybridization and catalyzed reporter deposition (CARD) for the identification of marine bacteria. *Appl. Environ. Microbiol.*, 68:3094–3101, doi:10.1128/AEM.68.6.3094-3101.2002.

Podar, M., Abulencia, C.B., Walcher, M., Hutchison, D., Zengler, K., Garcia, J.A., Holland, T., Cotton, D., Hauser, L., and Keller, M., 2007. Targeted access to the genomes of low-abundance organisms in complex microbial communities. *Appl. Environ. Microbiol.*, 73:3205–3214, doi:10.1128/AEM.02985-06.

Rappé, M.S., Connon, S.A., Vergin, K.L., and Giovannoni, S.J., 2002. Cultivation of the ubiquitous SAR11 marine bacterioplankton clade. *Nature*, 418:630–633, doi:10.1038/nature00917.

Reed, D.W., Fujita, Y., Delwiche, M.E., Blackwelder, D.B., Sheridan, P.P., Uchida, T., and Colwell, F.S., 2002. Microbial communities from methane hydrate-bearing deep marine sediment in a forearc basin. *Appl. Environ. Microbiol.*, 68:3759–3770, doi:10.1128/AEM.68.8.3759-3770.2002.

Santelli, C.M., Edgcomb, V., Bach, W., and Edwards, K.J., 2006. Diversity of endolithic bacteria in seafloor basalt. *European Geosciences Union (EGU) Meeting*, Vienna, Austria, 3 April 2006 (Poster Presentation).

Schippers, A., Neretin, L.N., Kallmeyer, J., Ferdelman, T.G., Cragg, B.A., Parkes, J.R., and Jørgensen, B.B., 2005. Prokaryotic cells of the deep sub-seafloor biosphere identified as living bacteria. *Nature*, 433:861–864, doi:10.1038/nature03302.

Smith, D.C., Spivack, A.J., Fisk, M.R., Haveman, S.A., Staudigel, H., and Shipboard Scientific Party, 2000. Tracer-based estimates of drilling-induced microbial contamination of deep sea crust. *Geomicrobiology J.* 17:207–219. doi:10.1080/01490450050121170.

Sogin, M.L., Morrison, H.G., Huber, J.A., Welch, D.M., Huse, S.M., Neal, P.R., Arrieta, J.M., and Herndl, G.J., 2006. Microbial diversity in the deep sea and the under explored "rare biosphere". *Proc. Natl. Acad. Sci. U.S.A.*, 103:12115–12120, doi:10.1073/pnas.0605127103.

Sorensen, K.B., and Teske, A., 2006. Stratified communities of active archaea in deep marine subsurface sediments. *Appl. Environ. Microbiol.*, 72:4596–4603, doi:10.1128/AEM.00562-06.

Sturt, H.F., Summons, R.E, Smith, K., Elvert, M., and Hinrichs, K.U., 2004. Intact polar membrane lipids in prokaryotes and sediments deciphered by high-performance liquid chromatography/electrospray ionization multistage mass spectrometry - new biomarkers for biogeochemistry and microbial ecology. *Rap. Comm. Mass Spec.*, 18:617–628, doi:10.1002/rcm.1378.

Takai, K, Moyer, C.L., Miyazaki, M., Nogi, Y., Hirayama, H., Nealson, K.H., and Horikoshi, K., 2005. *Marinobacter alkaliphilus* sp. nov., a novel alkaliphilic bacteriumm isolated from subseafloor alkaline serpentine mud from Ocean Drilling Program (ODP) Site 1200 at South Chamorro Seamount, Mariana Forearc. *Extremophiles*, 9:17–27, doi:10.1007/s00792-004-0416-1.

Teske, A., 2006. Microbial communities of deep marine subsurface sediments: molecular and cultivation surveys. *Geomicrobiology J.*, 23:357–368, doi:10.1080/01490450600875613.

Webster, G., Parkes, R.J., Cragg, B.A., Newberry, C.J., Weightman, A.J., and Fry, J.C., 2006. Prokaryotic community composition and biogeochemical processes in deep subseafloor sediments from the Peru Margin. *FEMS Microbiol. Ecol.*, 58:65–85, doi:10.1111/j.1574-6941.2006.00147.x.

White, D.C., Davies, W.M., Nickels, J.S., King, J.D., and Bobbie, R.J., 1979. Determination of the sedimentary microbial biomass by extractable lipid phosphate. *Oecologia*, 40:51–62, doi:10.1007/BF00388810.

Whitman, W.B., Coleman, D.C., and Wiebe, W.J., 1998. Prokaryotes: The unseen majority. *Proc. Natl. Acad. Sci. U.S.A.*, 95:6578–6583, doi:10.1073/pnas.95.12.6578.

Zink, K.G., Wilkes, H., Disko, U., Elvert, M., and Horsfield, B., 2003. Intact phospholipids — microbial "life markers" in marine deep subsurface sediments. *Organic Geochem.*, 34:755–769, doi:10.1016/S0146-6380(03)00041-X.

Authors

Steven D'Hondt, Graduate School of Oceanography, University of Rhode Island, Narragansett Bay Campus, South Ferry Road, Narragansett, R.I. 02882, U.S.A., e-mail: dhondt@gso.uri.edu.

Fumio Inagaki, Kochi Institute for Core Sample Research, Japan Agency for Marine-Earth Science and Technology (JAMSTEC), B200 Monobe, Nankoku, Kochi, 783-8502, Japan.

Timothy Ferdelman, Max Planck Institute (MPI) for Marine Microbiology, Celsiusstr. 1, D-28359, Bremen, Germany.

Bo Barker Jørgensen, Max Planck Institute (MPI) for Marine Microbiology, Celsiusstr. 1, D-28359, Bremen, Germany.

Kenji Kato, Department of Geosciences, Faculty of Science, Shizuoka University, Shizuoka, 422-8529, Japan.

Paul Kemp, Center for Microbial Oceanography: Research and Education, 1000 Pope Road, Marine Sciences Builiding, Honolulu, Hawaii 96822, U.S.A.

Patricia Sobecky, School of Biology, Georgia Institute of Technology, Atlanta, Ga. 30332, U.S.A.

Mitchell Sogin, Marine Biological Laboratory, 7 MBL Street, Woods Hole, Mass., 02513-1015, U.S.A.

Ken Takai, Subground Animalcule Retrieval Program, Extremobiosphere Research Center, Japan Agency for Marine-Earth Science and Technology (JAMSTEC), 2-15 Natsushima-cho, Yokosuka, Kanagawa, 237-0061, Japan.

4

Scientific Drilling in a Central Italian Volcanic District

by M. Teresa Mariucci, Simona Pierdominici, and Paola Montone

Introduction and Goals

The Colli Albani Volcanic District, located 15 km SE of Rome (Fig. 1), is part of the Roman Magmatic Province, a belt of potassic to ultra-potassic volcanic districts that developed along the Tyrrhenian Sea margin since Middle Pleistocene time (Conticelli and Peccerillo, 1992; Marra et al., 2004; Giordano et al., 2006 and references therein). Eruption centers are aligned along NW-SE oriented major extensional structures guiding the dislocation of Meso-Cenozoic siliceous-carbonate sedimentary successions at the rear of the Apennine belt. Volcanic districts developed in structural sectors with most favorable conditions for magma uprise. In particular, the Colli Albani volcanism is located in a N-S shear zone where it intersects the extensional NW- and NE-trending fault systems. In the last decade, geochronological measurements allowed for reconstructions of the eruptive history and led to the classification as "dormant" volcano. The volcanic history may be roughly subdivided into three main phases marked by different eruptive mechanisms and magma volumes. The early Tuscolano-Artemisio Phase (ca. 561–351 ky), the most explosive and voluminous one, is characterized by five large pyroclastic flow-forming eruptions. After a ~40-ky-long dormancy, a lesser energetic phase of activity took place (Faete Phase; ca. 308–250 ky), which started with peripheral effusive eruptions coupled with subordinate hydromagmatic activity. A new ~50-ky-long dormancy preceded the start of the late hydromagmatic phase (ca. 200–36 ky), which was dominated by pyroclastic-surge eruptions, with formation of several monogenetic or multiple maars and/or tuff rings.

Periodic unrest episodes have been directly observed in the area of most recent volcanism, posing a threat not only to this densely populated area but also to the nearby city of Rome (Amato and Chiarabba, 1995; Chiarabba et al., 1997; Pizzino et al., 2002). Unrest activities include i) intermittent seismic swarms of shallow depth and small-moderate magnitude earthquakes, ii) episodes of CO_2, radon, and H_2S emissions, and iii) pulses of surface uplift. In the framework of two multidisciplinary projects (DPC and FIRB, see details below), the Colli Albani 1 borehole (CA1, 350 m) was drilled to directly investigate the volcano at depth (Mariucci et al., 2007).

In this project, detailed analyses of borehole data will be coupled to geological, geophysical, and geochemical studies. The main goals are to better characterize the shallow crust structure beneath the volcanic complex, to define the present-day stress field, and to understand the formation and emanation of hazardous gases. Finally, a seismometer will be installed at depth in order to acquire outstanding seismological records without anthropogenic noise.

Drilling and On-site Measurements

The borehole is on public land about 10 km south of Rome close to Santa Maria delle Mole village, adjacent to the west rim of the Tuscolano-Artemisio caldera. Considering the high gas concentrations in the aquifers (mainly CO_2 and H_2S), which have caused illness and casualties among habitants and animals, a blowout preventer was installed at the wellhead. Moreover, for safety reasons H_2S concentrations and combustible gases were constantly monitored.

Drilling was conducted in one stage during April and May 2006 by an Italian company (SO.RI.GE. srl) using a Casagrande C8 machine and wire-line technique. A vertical borehole and continuous coring down to 350 m were planned

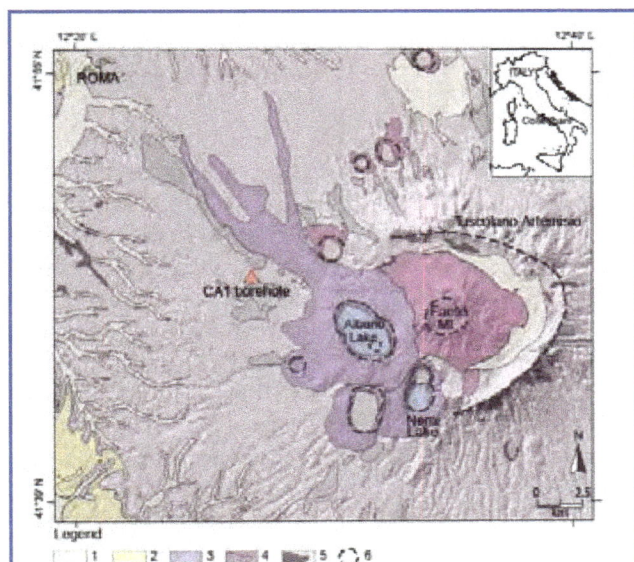

Figure 1. Geological map of Colli Albani area and CA1 borehole location. Color legend: 1) Holocene alluvial deposits; 2) Plio-Quaternary sedimentary deposits; 3) "Hydromagmatic phase" (200-36 ky); 4) Deposits of "Faete phase" and related lava flows (308-250 ky); 5) "Tuscolano Artemisio phase" and related lava flows (561–351 ky); 6) Crater rims; (modified after Marra et al., 2004).

to sample volcanic and sedimentary sequences. Wireline coring allowed for a very good core recovery (about 99%) that enabled reconstruction of the detailed stratigraphy of the volcanic units down to the Plio-Pleistocene sedimentary basement. From top to bottom the borehole stratigraphy consists of the typical volcanic succession pertaining to the Tuscolano-Artemisio Phase of activity, spanning the interval 561–351 ky and a Plio-Pleistocene sedimentary sequence characterized by ~150 meters of consolidated clays underlying a ~20-m-thick sand layer followed by sand at the hole bottom (the last ~5 m).

The borehole was drilled with standard rotary technique down to 25.5 m, followed by wire-line diamond coring with downscaling diameters (178 mm for the first 18 m, 152 mm up to 25.5 m, 122.7 mm up to 204 m, all with 85 mm core;

Figure 2. Blowout event.

96.1 mm from 204 m to 300 m, with 63.5 mm cores and 75.6 mm up to 350 m, with cores of 47.6 mm). Each 3-m-long core section was cut in pieces of 1-m length and packed in wooden boxes, now curated at Istituto Nazionale di Geofisica e Vulcanologia (INGV)-Rome. Some quick geotechnical analyses were performed on site, such as the Schmidt Hammer and the Barton Comb, to compute the joint compressive strength (JCS) and the joint roughness coefficient (JRC), respectively. A quick analysis of fractures and faults was also performed on the cores in the field.

Three hydraulic fracturing tests had been planned between 320 m and 350 m for a first evaluation of the local stress field in the area, in particular to better constrain the minimum horizontal stress. Unfortunately, a blow-out occurred (Fig. 2) during the positioning of the hydraulic fracturing probe in the well, causing a collapse of the deepest part of the borehole. Overpressured fluids (mainly in the form of gases) leaked out of the well for about one hour with a wellhead pressure of ~30 bar, sputtering up to about 5 m above ground level for one day with pressure stabilizing around 15 bar. This phenomenon, probably caused by the presence of a sandy unit at 345–350 m, allowed collecting fluid samples for quick geochemical analyses on-site and in the laboratory. Fluids are an aqueous mixture of CO_2, H_2S, helium, aromatic hydrocarbons, and other gases. Although hampering some geophysical investigations, this blow-out gave direct access to the deepest fluids ever sampled in the area up to now. After fluid samples were collected, the flux was stopped by pumping water into the borehole and restoring the initial conditions.

In order to evaluate the *in situ* physical properties of volcanic rocks and to estimate the state of stress, the following geophysical logs were performed using the International Continental Scientific Drilling Program (ICDP) - Operational Support Group (OSG) slim-hole probes: spectrum gamma ray, sonic, magnetic susceptibility, electric resistivity (dual-laterolog), four-arm dipmeter, and borehole televiewer. Spectral gamma logs were recorded down to 270 m, whereas other tools were only

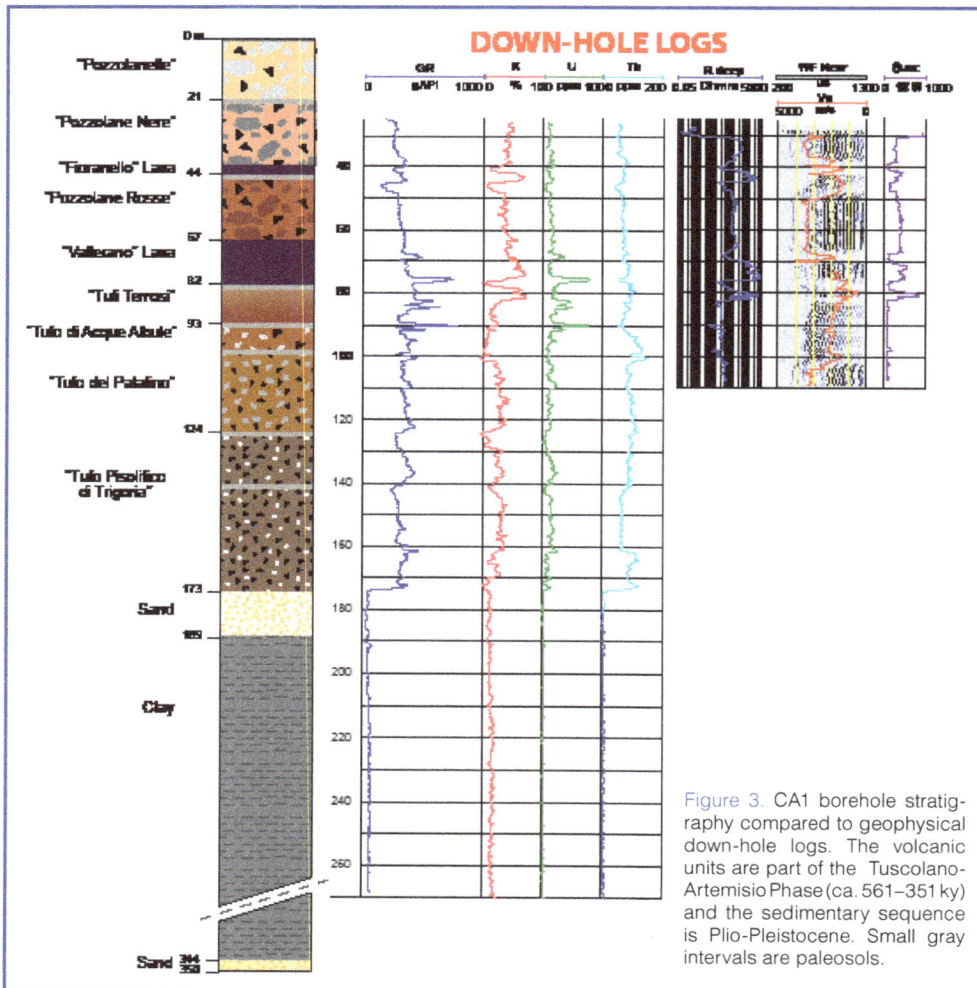
Figure 3. CA1 borehole stratigraphy compared to geophysical down-hole logs. The volcanic units are part of the Tuscolano-Artemisio Phase (ca. 561–351 ky) and the sedimentary sequence is Plio-Pleistocene. Small gray intervals are paleosols.

deployed down to 110 m as a consequence of the blow-out (Fig. 3).

The borehole was cased down to 200 m and plugged with cement in the lower part hosting a broad-band seismometer. Now it serves as a seismic station for the Italian national centralized seismic network managed by INGV.

Ongoing Research

Some data analyses and laboratory measurements on samples are still ongoing. We analyzed downhole logs and compared them to the detailed stratigraphy obtained from cores, defining the main physical characters of the main volcanic units. Moreover, selected samples were taken on the most representative volcanic units to determine physical properties such as elastic wave velocities under different pressure conditions. Breakout analyses from both caliper and televiewer data are performed to get the horizontal minimum stress direction and to analyze fractures from televiewer images. All cores from the logged interval were scanned using an optical DMT Core Scanner at GeoForschungsZentrum in Potsdam (Germany) for digital documentation and structural analysis. A lot of fault planes with striae along the sedimentary cores were recognized, and these will be analyzed using an innovative high resolution thermal field emission scanning electron microscope to highlight some characteristics of the microstructures. Ar-Ar dating will be applied to define the age of a drilled lava flow that is poorly known so far, and mineralogical and geochemical analyses will unravel the characteristics of the volcanic units. Biostratigraphic analysis of nannoplankton and foraminifera and detailed measurements of magnetic susceptibility of Pliocene clays will allow us to define the depositional environment and former positions of shorelines. Dynamic tests on the clays will provide data useful to define the local seismic response of the consolidated clays that form the basement of the volcanic sequence below the City of Rome. Measurements of natural gamma radiation on the cores will be compared with downhole measurements. The goal is to integrate core data with new geodetic and seismological data for physical and numerical modeling to understand the behavior of the whole volcanic complex.

Acknowledgements

This drilling project was funded by the Italian Department of Civil Protection (Project INGV-DPC V3.1 "Colli Albani"), the Italian Ministry of University and Research (Project FIRB "Research and Development of New Technologies for Protection and Defense of Territory from Natural Risks", W.P. C3 "Crustal Imaging in Italy"), and INGV Departments "National Earthquake Center" and "Seismology and Tectonophysics", Rome, Italy. We thank the ICDP Operational Support Group (GFZ Potsdam, Germany), in particular J. Kück, C. Carnein, and R. Conze.

Research Unit 9 of Project INGV-DPC V3.1

M.T. Mariucci, F. Marra, P. Montone, S. Pierdominici, F. Florindo (Instituto Nazionale di Geofisica e Vulcanologia, Rome, Italy).

References

Amato, A., and Chiarabba, C., 1995. Recent uplift of the Alban Hills volcano (Italy): evidence for magmatic inflation? *Geophys. Res. Lett.*, 22(15):1985–1988, doi:10.1029/95GL01803.

Chiarabba, C., Amato, A., and Delaney, P.T., 1997. Crustal structure, evolution, and volcanic unrest of the Alban Hills, Central Italy. *Bull. Volcanol.*, 59(3):161–170, doi:10.1007/s004450050183.

Conticelli, S., and Peccerillo, A., 1992. Petrology and geochemistry of potassic and ultrapotassic volcanism in central Italy - petrogenesis and inferences on the evolution of the mantle source. *Lithos,* 28(3/6):221–240.

Giordano, G., De Benedetti, A.A., Diana, A., Diano, G., Gaudioso, F., Marasco, F., Miceli, M., Mollo, S., Cas, R.A.F., and Funiciello, R., 2006. The Colli Albani mafic caldera (Roma, Italy): Stratigraphy, structure and petrology. *J. Volcanol. Geotherm. Res.*, 155:49–80, doi:10.1016/j.jvolgeores.2006.02.009.

Mariucci, M.T., Pierdominici, S., Florindo, F., Marra, F., and Montone, P., 2007. How a borehole can help volcanology: the scientific drilling in the Colli Albani volcanic area (Italy). Workshop on "Volcano Tectonics", *Second Cuban Earth Sciences Convention*, Havana, Cuba, 20–23 March 2007, Abstract. Poster available on http://www.earth-prints.org.

Marra, F., Taddeucci, J., Freda, C., Marzocchi, W., and Scarlato, P., 2004. Recurrence of volcanic activity along the Roman Comagmatic Province (Tyrrhenian margin of Italy) and its tectonic significance. *Tectonics*, 23:TC4013, doi:10.1029/2003TC001600.

Pizzino, L., Galli, G., Mancini, C., Quattrocchi, F., and Scarlato, P., 2002. Natural gas hazard (CO_2, ^{222}Rn) within a quiescent volcanic region and its relations with tectonics: the case of the Ciampino-Marino area, Alban Hills volcano, Italy. *Natural Hazards*, 27:257–287.

Authors

M. Teresa Mariucci, Simona Pierdominici, Paola Montone, Istituto Nazionale di Geofisica e Vulcanologia sezione di Sismologia e Tettonofisica, Via di Vigna Murata 605, 00143, Rome, Italy, e-mail: mariucci@ingv.it.

Photo Credits

Fig. 2: Photo by S. Pierdominici (INGV)

Related Web Link

http://hdl.handle.net/2122/2063

5

The Second Deep Ice Coring Project at Dome Fuji, Antarctica

by Hideaki Motoyama

Introduction

Throughout the history of the polar icecaps, dust and aerosols have been transported through the atmosphere to the poles, to be preserved within the annually freezing ice of the growing ice shields. Therefore, the Antarctic ice sheet is a "time capsule" for environmental data, containing information of ancient periods of Earth's history. To unravel this history and decode cycles in glaciations and global change is among the major goals of the Dome Fuji Ice Coring Project.

With an elevation of 3810 m, Dome Fuji (39°42'E, 77°19'S) is the second highest dome summit on the Antarctic ice sheet and might be one of the locations holding the oldest ice in

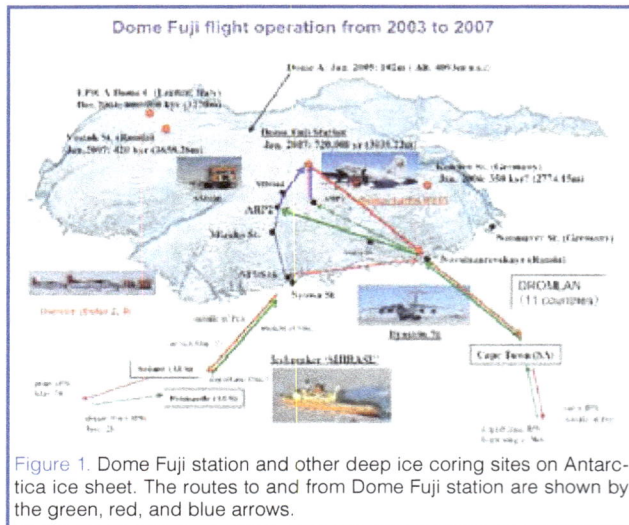

Figure 1. Dome Fuji station and other deep ice coring sites on Antarctica ice sheet. The routes to and from Dome Fuji station are shown by the green, red, and blue arrows.

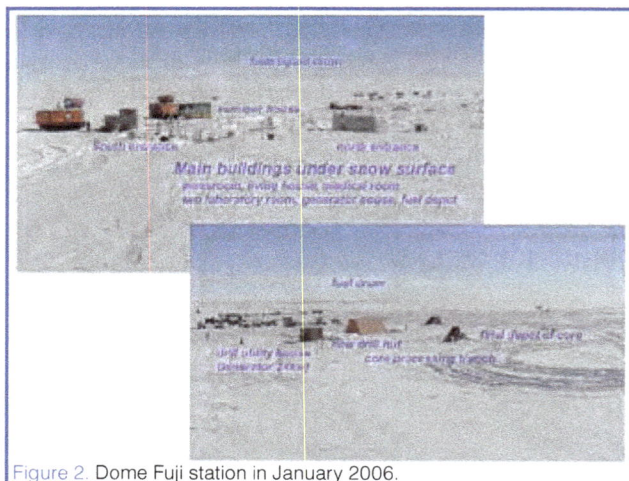

Figure 2. Dome Fuji station in January 2006.

Antarctica. The base of the ice underneath Dome Fuji is estimated to have formed at the beginning of the glacial cycle in the Quaternary era. Analysis of this ice can shed light on the mechanism of Quaternary glacial cycles. The second deep ice coring project at Dome Fuji, Antarctica reached a depth of 3028.52 m (3810 m above sea level) during the austral summer season in 2005/2006 (Figs. 1 and 2). The recovered ice cores contain records of global environmental changes going back about 720,000 years. During the recent 2006/2007 season the Japanese Antarctic Research Expedition (JARE) team finally reached a depth of 3035.22 m at the Dome Fuji station on 26 January 2007.

First Deep Ice Core Drilling at Dome Fuji

In December 1996, during the first deep ice core drilling campaign at Dome Fuji, the 37th JARE team had succeeded coring the ice to a depth of 2503 m. Due to a shortage of antifreeze supply, the drilling team temporarily stopped drilling and waited to receive supplies to arrive with the next drilling team (JARE 38). However, while reaming and chip recovery were repeated using the ice core drill in order to keep the hole open, the drill became stuck at the end of December at a depth of approximately 2300 m. The subsequent 39th and 40th JARE teams attempted to recover the drill by injecting high density liquid (hydrochlorofluorocarbon) 200 m below the surface of the ice sheet (100 m below the liquid level), but this attempt failed, and the borehole had to be abandoned.

A second deep ice coring project was commenced at Dome Fuji during 2001. It was decided to draw on new drilling technology developments and to establish a new borehole about 43 m north of the abandoned one with the aim of full penetration to bedrock. A pilot hole was drilled to a depth of 122 m, and a casing was installed by the 42nd JARE wintering team in 2001. In 2002 the 43rd JARE team constructed a new deep ice core drilling house, updated the station equipment, and began repairing the logistic equipment. The 44th Dome Fuji wintering team completed and prepared the new deep ice core drilling system in 2003—a completely new deep ice core drill of 4.0 m core barrel length. It subsequently proved to be of true world class quality, when it was able to recover 3.84 m of the ice core in each operation with almost no unexpected difficulties caused by ice chip clogging. Due to logistical considerations, only summer operations were performed.

Deep Ice Core Drilling Technology

The deep ice coring drill used during this project is an electro-mechanical liquid-filled type. A schematic diagram and photos are shown in Fig. 3, and the main specifications of the drills used at Dome Fuji are summarized in Table 1. The ice drill consists of a core barrel, a chip chamber, a pressure tight section, and an anti-torque section (Takahashi et al., 2001). Three cutters are attached to cut an ice core of 94 mm diameter, leaving a borehole of 135 mm diameter. To prevent borehole closure during drilling, the borehole is filled with an anti-freezing fluid, n-butyl acetate. Its density is about equal to the ice, and the viscosity at temperatures below −50°C is low. Since the second deep ice coring project drilled only during the austral summer season, the design of the drill could be improved to increase the productivity under this premise. The equipment was able to penetrate up to 3.84 m for each core, as opposed to the 2.3 m cored during the first deep ice coring project. Effective transportation and storage of the cutting chips generated by the drill turned out to be one of the biggest problems. Technicians experimented with various pumps to solve this problem, and finally an archimedean screw pump was used, which is operated by rotating a core barrel (Fig. 3C) through a spiral spring located within a double tube. A propeller-like booster attached to the driving shaft of the core barrel provides momentum for the transportation of the borehole liquid and cutting chips to a chip chamber (Fig. 3B).

A special pipe perforated with many small holes was manufactured for storing the cutting chips, while the liquid could easily pass through the perforations (Fig. 3A). However, the cutting chips create a countercurrent in the chip chamber during drill ascent, leading to leakage of the chips from the chip chamber. A current prevention system, including a new check valve and direct current (DC) drill motor, was adopted to prevent this from happening.

Difficulties and Progress

In the first season, 2003/2004, the final drilling depth achieved was 362 m despite significant logistics problems with weather and transportation of equipment. However, with the considerable experience gained in the 2003/2004 season, it was possible to drill ice cores smoothly during the summer season of 2004/2005. The hole was deepened by approximately 1500 m, reaching a total drilling depth of 1850.35 m.

To reach the bedrock of Antarctica under the ice sheet, 1200 meters had to be drilled in the last projected summer season 2005/2006. Hence, it was necessary to arrive at the Dome Fuji station at the earliest possibility. The team arrived at the station on 17 November 2005. The drilling resulted in a record high 133 m of drill core per week without encountering problems, and a drilling depth of 3000 m was achieved on 12 January 2006. Through most drilling runs, a 3.7-m ice core of excellent quality was obtained. When the drilling

depth exceeded 3000 m near the bedrock, the ice temperature was close to the pressure melting point. The cutting chips immediately froze to become ice, which made chip transportation within the corer very difficult. At this depth, nearly four hours were required for each single ice coring operation, with performance rapidly decreasing. Finally, only ten centimeters of the ice core could be drilled on average with each core. Because it had been expected that the "warm" ice would cause problems, the normal drill was replaced with a special short teflon-coated drill in an attempt to determine the most suitable drilling method. The final drilling depth was 3028.52 m on 23 January 2006, when drilling had to stop to provide sufficient time for demobilization of the operation and crew.

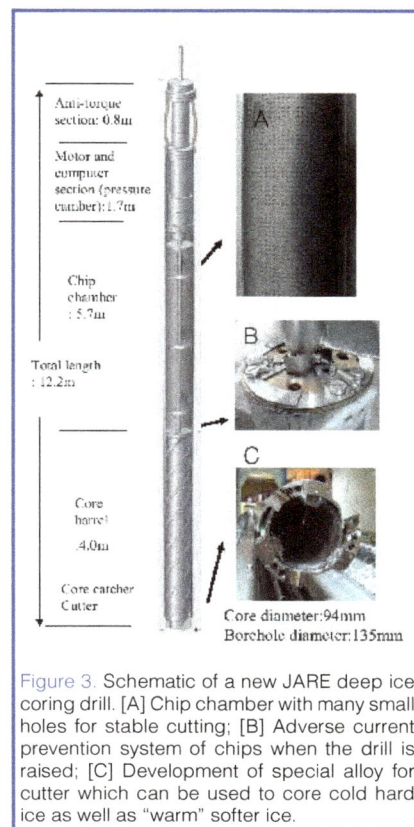

Figure 3. Schematic of a new JARE deep ice coring drill. [A] Chip chamber with many small holes for stable cutting; [B] Adverse current prevention system of chips when the drill is raised; [C] Development of special alloy for cutter which can be used to core cold hard ice as well as "warm" softer ice.

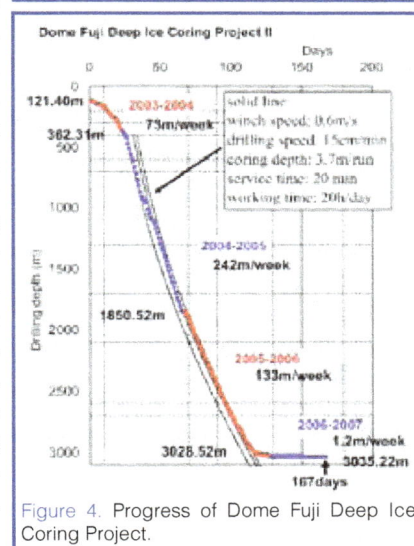

Figure 4. Progress of Dome Fuji Deep Ice Coring Project.

Ultimately, to reach the bedrock, the deep ice core drilling was extended for another year. In the fourth drilling season, 2006/2007, the total drilling period was 39 days. The total drilling length was 6.70 m, and the final drilling depth reached was 3035.22 m. The average core length was approximately 10 cm, which was half the length expected. The overall progress of deep ice core drilling throughout the seasons 2003 to 2007 is summarized in Fig. 4.

When a drilling depth of 3034.34 m was reached, a special type of small ice pieces appeared to be abundant in the chip chamber and in the frozen water chip accumulating on the

Table 1: Specs of the JARE phase 1 drill and the improved model used for normal and "warm" ice during Phase 2.

Item	Phase 1 Model	Phase 2 Model	Phase 2 Model (for warm ice)
Type	Electro-Mechanical Drill	Same as Phase 1	no change
Core ØxL	94 mm x 2,200 mm	94 mm x 3,840 mm	94 mm x 2,000 mm
Cutting Speed	15-20 cm/min	Same as Phase 1	no change
Static Pressure	30MPa	Same as Phase 1	no change
Drill Size ØxL	122 mm x 8,593 mm	122 mm x 12,200 mm	122 mm x 8,106 mm
Cutter	3 x Block Type	Same as Phase 1	Special
Core Barrel ØxL	101.6 mm x 2,321 mm	101.6 mm x 4,000 mm	101.6 mm x 2,256 mm
Chip Chamber ØxL & Density	112 mm x 3,260 mm ρ = 500 kg m^{-3}	112 mm x 5,533 mm ρ = 550 kg m^{-3} Hole: 1.2 mm Ø x 45,000	112 mm x 3,160 mm ρ = 550 kg m^{-3} Hole: 1.2 mm Ø
Chip Pump	Archimedean Pump & 1 Turn Screw Booster	Archimedean Pump & 1 or 0.75 Turn Screw Booster x 2-3	Archimedean Pump & 1 or 0.75 Turn Screw Booster x 2
Motor Output Power	AC Brushless Motor 600 W for 15 min at 12,000 rpm	DC Permanent Magnet Motor with Brushes, 600W for 15 min. at 4,000 rpm.	no change
Reduction Gear Type & Ratio	4 Stage Planetart Gear 1/170	Harmonic Drive Type: CSF17, 1/100, 1/80 (, 1/50)	no change
Electronics	Monitoring Computer (10 Parameters)	Same as Phase 1 (version 2)	no change
Pressure Chamber ØxL & Pressure	122 mm x 1,700 mm 30 MPa	Same as Phase 1	no change
Anti-Torque	3 x Leaf Spring	Same as Phase 1	no change
Cable ØxL	7-H-314K, 7.72 mm x 3,500m	Same as Phase 1	no change
Hole Liquid	n- butyl acetate	Same as Phase 1	no change
Special Items		1. System to Prevent Adverse Current of Chips 2. Super Banger	1. Special Cutter Mount 2. Teflon Coated Drive Shaft, Screw Booster, Cutter, Core Catcher, Outer Tube, Core Barrel

ice core (Fig. 5). The crystal structure of these strange ice pieces differed from that of the cutting ice chips. The conclusion was that water beneath the ice sheet had probably leaked into the borehole and had frozen in the drill. In addition, the ice core was found to be contaminated with small rocks. Hence, since liquid water existed near the bedrock, the drilling machine was covered with ice when it was positioned in the ground to drill through the ice sheet, which had a temperature of −55°C or lower. The shape of the ice underneath the drill resembled frozen drops of water. Drilling was carefully continued for the next days, and finally, the last ice core was recovered topped with mysterious white frozen water from a depth of 3035.22 m below the surface.

Preliminary Analysis of the Ice Core

The oxygen isotope ratio of the ice core was measured to determine its age. This ratio fluctuates depending on the paleotemperatures, and it can be used to study the past glacial-interglacial cycles in great detail. The ages of the ice cores were estimated by comparing the determined age with the Dome C ice core data from the European Project for Ice Coring in Antarctica (EPICA). As a result, the deepest ice cores at Dome Fuji were estimated to be approximately 720,000 years old. Traces of atmospheric

gases were trapped as air bubbles in the ice sheet and will be analyzed.

The ice cores recovered from the Dome Fuji station confirmed that the history of global environmental changes could be continuously recorded from 720,000 years in the past. More analysis will be conducted to clarify the Earth's climate, microorganisms present in ice, and space climate. Currently, ice core studies are being conducted in cooperation with the National Institute of Polar Research in Tokyo, Japan, other universities, and other institutes.

For more information about the Dome Fuji Deep Ice Coring Project see the Web link below.

References

Fujii, Y., Azuma, N., Tanaka, Y., Nakayama, M., Kameda, T., Shinbori, K., Katagiri, K., Fujita, S., Takahashi, A., Kawada, K., Motoyama, H., Narita, H., Kamiyama, K., Furukawa, T., Takahashi, S., Shoji, H., Enomoto, H., Saitoh, T., Miyahara, M., Naruse, R., Hondoh, T., Shiraiwa, T., Yokoyama, K., Ageta, Y., Saito, T., and Watanabe, O., 2001. Deep ice core drilling to 2503 m depth at Dome Fuji, Antarctica. *Natl. Inst. Polar Res., Spec. Issue*, 56:103–116.

Takahashi, A., Fujii, Y., Azuma, N., Motoyama, H., Shinbori, K., Tanaka, Y., Watanabe, O., Narita, H., Nakayama, Y., Kameda, T., Fujita, S., Furukawa, T., Takata, M., and Miyahara, M., 2001. Improvements to the JARE deep ice core drill. *Natl. Inst. Polar Res., Spec. Issue*, 56:117–125.

Author

Hideaki Motoyama, National Institute of Polar Research, Kaga 1-9-10, Itabashi-ku, Tokyo 173-8515, Japan, email: motoyama@pmg.nipr.ac.jp.

Related Web Link

http://polaris.nipr.ac.jp/~domef/home/eng/index-e.html

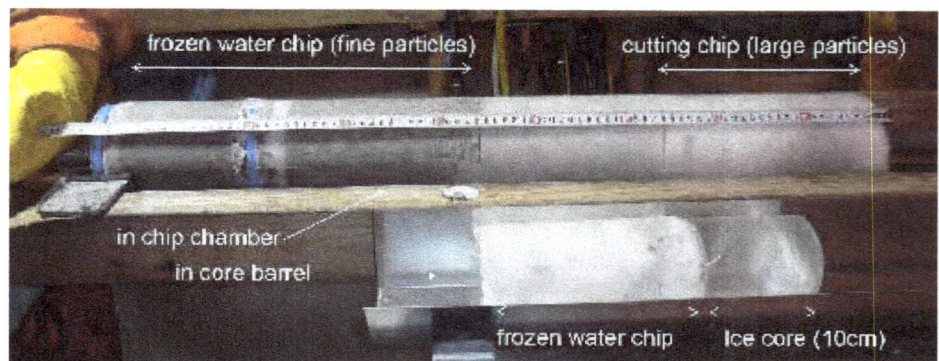

Figure 5. Cutting chips of ice core and a lot of frozen water chips.

6

Tenaghi Philippon (Greece) Revisited: Drilling a Continuous Lower-Latitude Terrestrial Climate Archive of the Last 250,000 Years

by Jörg Pross, Polychronis Tzedakis, Gerhard Schmiedl, Kimon Christanis, Henry Hooghiemstra, Ulrich C. Müller, Ulrich Kotthoff, Stavros Kalaitzidis, and Alice Milner

Introduction and Goals

With the dramatically increasing manifestation of anthropogenic forcing on the Earth's climate, understanding the mechanisms and effects of abrupt climate change is crucial to extend the lead time for mitigation and adaptation. In this context, the climate variability during the Quaternary represents the closest analogy to present-day climate change. Unprecedented insights into both short-term (i.e., decadal-to centennial-scale) and long-term (i.e., orbital-scale) climate variability over the last 740 kyr have been derived from ice cores from polar regions (Dansgaard et al., 1993; EPICA community members, 2004). These records show that the higher latitudes repeatedly witnessed temperature changes of more than 10°C within human time scales (Severinghaus et al., 1998). Considerably less information is available on the characteristics of abrupt climate change in the middle and lower latitudes and on their imprint on terrestrial environments. These regions are, however, home to the majority of the Earth's population, and consequently they will witness the greatest impact of future climate change on people's lives.

Located in a strategic position between the higher-latitude (i.e., North Atlantic Oscillation-influenced) and lower-latitude (i.e., monsoonally-influenced) climate systems, the Mediterranean region is particularly sensitive in recording abrupt climate change and its imprint on terrestrial ecosystems. Moreover, terrestrial climate archives from the Mediterranean borderlands yield rich, diverse biotic signals also during colder intervals because the region's climate was relatively mild even under fully glacial boundary conditions. In contrast to higher-latitude records, this warrants the detailed analysis of short-term climate variability in terrestrial environments throughout the full range of climatic boundary conditions of the Quaternary.

In light of the above, the climate archive of Tenaghi Philippon (site within the Drama Basin, Eastern Macedonia, Greece; Fig. 1) plays an exceptional role. Since the initiation of pollen-based vegetation analyses from drill cores in the late 1960s (Wijmstra, 1969), it has been increasingly recognized as one of the best terrestrial archives of Quaternary climate history in Europe. This prominent position is due to (1) its temporal length, spanning the last 1.35 million years and comprising at least nineteen consecutive glacial-interglacial cycles (Tzedakis et al., 2006); (2) its completeness, as evidenced by the close climato-stratigraphical correspondence with global deep-sea records; and (3) its proximal position with regard to glacial refugia of thermophilous plants in SE Europe, which reduces the time lag between atmospheric forcing and vegetation response as documented in pollen data.

Because previous investigations of the Tenaghi Philippon climate archive were restricted to a temporal resolution within the Milankovitch time band, and the core material used in these investigations has deteriorated, the potential of this site for the analysis of abrupt climate change has remained virtually untapped. Therefore, a campaign has been initiated to re-drill this archive; it is funded by the German Research Foundation, the Wilhelm Schuler Foundation, and the Royal Society (U.K.). The aim of this initiative is an interdisciplinary analysis of short-term climate variability under interglacial, "semi-glacial", and glacial boundary conditions of the Quaternary. Given the close proximity of the Tenaghi Philippon site to the Aegean Sea, special emphasis is placed on the identification of short-term environmental perturbations during intervals coeval with sapropel formation in the Eastern Mediterranean Sea. Disciplines include palynology, sedimentology, stable isotope geochemistry, coal petrology, photogrammetric and magnetic susceptibility core logging, magnetostratigraphy, and radiometric (^{14}C, ^{40}Ar/^{39}Ar) dating.

Drilling at Tenaghi Philippon

The Philippi peatland, which includes the Tenaghi Philippon site, is situated in the intramontane Drama Basin, Eastern Macedonia, Greece (Fig. 1). Owing to rapid subsidence that may have started in Late Miocene times, the

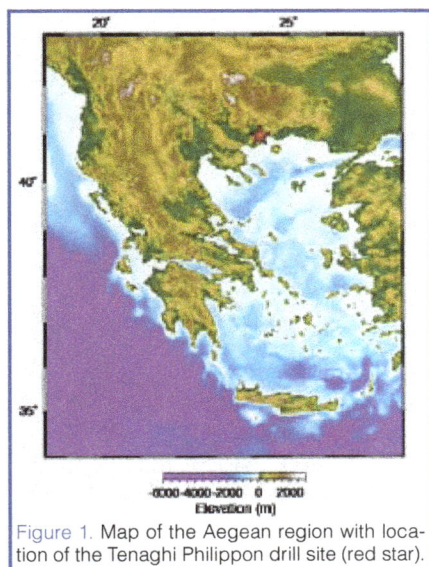

Figure 1. Map of the Aegean region with location of the Tenaghi Philippon drill site (red star).

Drama Basin constituted a limnic to telmatic setting throughout the Middle and Late Quaternary. Much of the basin fill that accumulated during this time consists of peat, resulting in the largest peat and lignite deposit of SE Europe. Based on the results of previous scientific drilling during the late 1960s to mid-1970s, the sedimentary succession at Tenaghi Philippon is known to comprise peat, mud, lake marls, and clays until 198 m depth; further downhole, and until the maximum depth drilled (280 m), clastic input increases such that sediments are palynologically non-productive (Wijmstra and Smit, 1976a, b; Van der Wiel and Wijmstra, 1987).

A 60-m-long core from Tenaghi Philippon (40°58.40'N, 24°13.42'E; 40 m above sea level) was drilled over three weeks in April 2005 using a WIRTH Eco1 drilling rig (Fig. 2) and a special, non-rotating probe driven by a pneumatic hammer system ("Dystel hammer"). The core, which has excellent recovery (97.8%), is now stored in the core repository of the Institute of Geosciences, University of Tübingen, Germany. Magnetic susceptibility measurements were performed on the entire core. After splitting the core into an archive half and a working half, it was photogrammetrically scanned and lithologically described. To facilitate non-destructive magnetostratigraphic measurements (e.g., to identify the Blake and Lachamps reverse-polarity events), core segments that may comprise these events according to the preliminary age model (see below) have been left intact.

Initial Results

A first, preliminary age model for the entire core has been established based on the analysis of pollen samples from the bottom of each 1-m-long core segment. The overview record presented in Fig. 3 indicates a succession of pollen zones reflecting open, steppe-like vegetation intercalated with zones reflecting the prevalence of Mediterranean forests (Fig. 4). Figure 3 also shows a preliminary correlation with the SPECMAP chronology (Imbrie et al., 1984) and insolation values (Berger and Loutre, 1991). We interpret the interval from 0 m to 5 m to represent the Holocene and Late Glacial, with Late Glacial climate fluctuations being yet unresolved because of low resolution. The interval from 5 m to 19 m is approximately correlative to Marine Isotope Stages (MIS) 2 to 4. The intervals from 19 m to 34 m and 34 m to 42 m represent MIS 5 and MIS 6, respectively. The stratigraphic interpretation below 42 m is less straightforward due to the low-resolution age model available at present. According to our current interpretation, the interval from 42 m to 57 m correlates with MIS 7, and the

Figure 2. Drilling rig WIRTH Eco1 used for drilling in the Drama Basin. The Phalakron mountain range (2232 m a.s.l.), which borders the Drama Basin to the northeast, is visible in the background.

interval down-core corresponds to MIS 8. Although absolute age control is not yet available, our data indicate that the core comprises at least the last 250 kyr, potentially even the last 290 kyr.

Tephra layers in the core, which have the potential for $^{40}Ar/^{39}Ar$ dating, provide a unique opportunity for age control independent from both palyno-stratigraphic and radiocarbon dating. Moreover, they allow the direct correlation with other terrestrial and marine climate archives from the Eastern Mediterranean region. As depicted in Fig. 3, the magnetic susceptibility curve exhibits prominent spikes that point to the positions of tephra layers. The inspection of layers with high magnetic susceptibility within the already split part of the core resulted in the identification of tephra layers with glass and pumice shards. The layer at 7.59 m represents the PhT2 tephra derived from the Cape Riva eruption of Santorini at ~22 kyr BP and correlates with Y-2 in the Mediterranean Sea. The PhT3 tephra at 12.80 m is

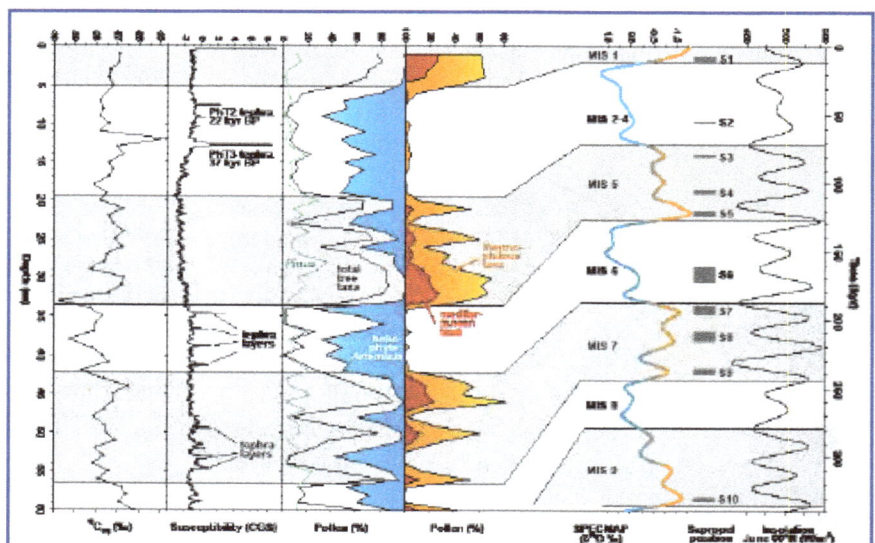

Figure 3. First results from the new core, comprising palynological, magnetic susceptibility and carbon isotope data. The tentative correlation of the pollen data with the SPECMAP chronology (Imbrie et al., 1984) and insolation values (Berger and Loutre, 1991) suggests that the record comprises at least the last 250 kyr. The magnetic susceptibility curve shows positions of PhT2 and PhT3 tephra layers and also indicates positions of older tephra layers.

Figure 4. Typical glacial pollen spectrum from the new core (Sample is from a core depth of 10.0 m, corresponding to Marine Isotope Stage 3).

correlative to the Campanian Ignimbrite and Y-5 in the Mediterranean Sea, having a $^{40}Ar/^{39}Ar$ age of ~37 kyr BP (S. Wulf, 2006, pers. comm.).

High resolution studies of the core are currently underway, including palynology and sedimentology (Frankfurt, Leeds), stable isotope geochemistry, radiocarbon and tephra dating (Frankfurt), and coal petrology (Patras). The analytical phase of this multi-disciplinary project is scheduled for the next three years. Following the compilation of data by each group/discipline, a synthesis study is planned for the final phase of the project. The integration of the resulting data will allow new insights into the characteristics of abrupt (decadal- to centennial-scale) climatic change and their consequences for terrestrial environments in the Mediterranean region.

Acknowledgements

Theodoros Kalliontzis, Andreas Balikas, Constantinos Tsompanoglou, and Nikos Nikolaidis provided invaluable support during field work. Ferdinand Stölben and his team (www.stoelbenbohr.de) did an excellent job in recovering high quality core material. Funding by the German Research Foundation (project Pr 651/3), the Wilhelm Schuler Foundation, and The Royal Society (United Kingdom) is gratefully acknowledged.

References

Berger, A.L., and Loutre, M.F., 1991. Insolation values for the climate of the last 10 million years. *Quat. Sci. Rev.*, 10:297–317, doi:10.1016/0277-3791(91)90033-Q.

Dansgaard, W., Johnsen, S.J., Clausen, H.B., Dahljensen, D., Gundestrup, N.S., Hammer, C.U., Hvidberg, C.S., Steffensen, J.P., Sveinbjornsdottir, A.E., Jouzel, J., and Bond, G., 1993. Evidence for general instability of past climate from a 250-kyr ice-core record. *Nature*, 364:218–220, doi:10.1038/364218a0.

EPICA community members, 2004. Eight glacial cycles from an Antarctic ice core. *Nature*, 429:623–628.

Imbrie, J., Hays, J.D., Martinson, D.G., McIntyre, A., and Mix, A.C., 1984. The orbital theory of Pleistocene climate: support from a revised chronology of the marine $\delta^{18}O$ record. *In* Berger, A.L., Imbrie, J., Hays, J., Kukla, G., and Saltzman, B. (Eds.), *Milankovitch and Climate Part I*. Dordrecht, The Netherlands (Kluwer), 269–305.

IOC, IHO, and BODC, 2003. *General Bathymetric Chart of the Oceans.* Centenary edition of the GEBCO Digital Atlas. CD-ROM. The Intergovernmental Oceanographic Commission and the International Hydrographic Organization, as part of the. British Oceanographic Data Centre (Liverpool, U.K.). http://www.bodc.ac.uk/products/bodc_products/gebco/

Severinghaus, J.P., Sowers, T., Brook, E.J., Alley, R.B., and Bender, M.L., 1998. Timing of abrupt climate change at the end of the Younger Dryas interval from thermally fractionated gases in polar ice. *Nature*, 391:141–146, doi:10.1038/34346.

Tzedakis, P.C., Hooghiemstra, H., and Pälike, H., 2006. The last 1.35 million years at Tenaghi Philippon: revised chronostratigraphy and long-term vegetation trends. *Quat. Sci. Rev.*, 25:3416–3430, doi:10.1016/j.quascirev.2006.09.002.

Van der Wiel, A.M., and Wijmstra, T.A., 1987a. Palynology of the lower part (78–120 m) of the core Tenaghi Philippon II, Middle Pleistocene of Macedonia, Greece. *Rev. Palaeobot. Palynol.*, 52:73–88, doi:10.1016/0034-6667(87)90047-9.

Van der Wiel, A.M., and Wijmstra, T.A., 1987b. Palynology of the 112.8–197.8 m interval of the core Tenaghi Philippon III, Middle Pleistocene of Macedonia, Greece. *Rev. Palaeobot. Palynol.*, 52:89–117, doi:10.1016/0034-6667(87)90048-0.

Wijmstra, T.A., 1969. Palynology of the first 30 m of a 120 m deep section in northern Greece. Acta Bot. Neerland., 18:511–527.

Wijmstra, T.A., and Smit, A., 1976. Palynology of the middle part (30–78 meters) of a 120 m deep section in northern Greece (Macedonia). *Acta Bot. Neerland.*, 25:297–312.

Authors

Jörg Pross, Ulrich C. Müller, Ulrich Kotthoff, Institute of Geosciences, University of Frankfurt, Altenhöfer Allee 1, D-60438 Frankfurt, Germany, e-mail: joerg.pross@em.uni-frankfurt.de.

Polychronis Tzedakis, Alice Milner, School of Geography, University of Leeds, West Yorkshire LS2 9JT, U.K.

Gerhard Schmiedl, Geological-Paleontological Institute, University of Hamburg, Bundesstraße 55, D-20146 Hamburg, Germany.

Kimon Christanis, Stavros Kalaitzidis, Department of Geology, University of Patras, GR-265.00 Rio-Patras, Greece.

Henry Hooghiemstra, Institute for Biodiversity and Ecosystem Dynamics, University of Amsterdam, Kruislaan 318, NL-1098 SM Amsterdam, The Netherlands.

Related Web Link

www.stoelbenbohr.de

Photo and Figure Credits

Fig. 1. Bathymetry and elevation taken from the GEBCO digital atlas (IOC, IHO, and BODC, 2003).

Fig. 2. Photo by Ulrich C. Müller.

Fig. 4. Photo by Ulrich Kotthoff.

7

Directional Drilling and Stimulation of a Deep Sedimentary Geothermal Reservoir

by Ernst Huenges, Inga Moeck, and the Geothermal Project Group

Introduction

Strata of Lower Permian sandstones and volcanics are widespread throughout Central Europe, forming deeply buried (on average, 4000-m) aquifers in the North German Basin with formation temperatures of up to 150°C. Stimulation methods to increase their permeability by enhancing or creating secondary porosity and flow paths are investigated by deep drilling. The goal is to map the potential for the generation of geothermal electricity from such deep sedimentary reservoirs using a doublet of boreholes—one to produce deep natural hot water and the other to re-inject the water after use. For these purposes, an *in situ* downhole laboratory was established in Gross Schönebeck, north of Berlin, Germany (Fig. 1).

At present, two 4.3-km-deep boreholes have been drilled. The first well (GrSk 3/90), originally completed in 1990 as a gas exploration well abandoned due to non-productivity, was reopened in 2000 and hydraulically stimulated in several treatments between 2002 and 2005. In 2006, the second well (GrSk 4/05), planned for extraction of thermal waters, was drilled to form a doublet system of hydraulically connected boreholes. In this second well the Lower Permian sandstones and the underlying volcanic rock are targeted for stimulation by hydrofracturing. The resulting reservoir should have an increased productivity with a minimal requirement for auxiliary energy to drive the thermal water loop (reservoir-surface-reservoir) and with minimal risk of a temperature short circuit of the system during the planned 30-year utilization period. The current experiment is designed to demonstrate sustainable hot water production from the reservoir between the two wells.

Background

Increasing demands for renewable energy are leading to utilization of geothermal resources from areas with typical (low) continental thermal gradients, as found in western and central Europe. For the exploitation of such low-enthalpy reservoirs, it is necessary to enhance the geothermal system. Two basic technologies based on hydraulic fracturing of the reservoir by

variations in fluid pressure (Economides and Nolte, 1989; Huenges and Kohl, 2007) can be applied:

• creation of an artificial heat exchanger at depth and using surface water for heat extraction from mostly dry rocks, e.g., Soultz-sous-Forêts (Baumgärtner et al., 2004)
• creation of artificial pathways at depth to enhance the water flow from water-bearing reservoir rocks, e.g., Gross Schönebeck (Huenges et al., 2004).

Lower Permian strata comprising upward fining siliciclastic rocks underlain by volcanic rocks (Fig. 1) are well-known from extensive gas exploration and production in NE Germany. Suitable framing for the study includes (1) formation temperatures above 120°C, in rocks at >3000 m depths, (2) large and regional extent of representative reservoir rocks, and (3) a variety of lithologies available for investigation. An abandoned gas exploration well (GrSk 3/90) at Gross Schönebeck completely meets these requirements and gives access to hot, water-bearing Lower Permian successions. It was therefore selected from a suite of existing wells, reopened in December 2000, and deepened from 4264 m to 4309 m to serve as a geothermal *in situ* laboratory.

Hydraulic Stimulation

Nine months after the well was reopened, a re-equilibrated temperature of 149°C was measured at 4285 m depth. The formation pressure was determined from long-term pressure

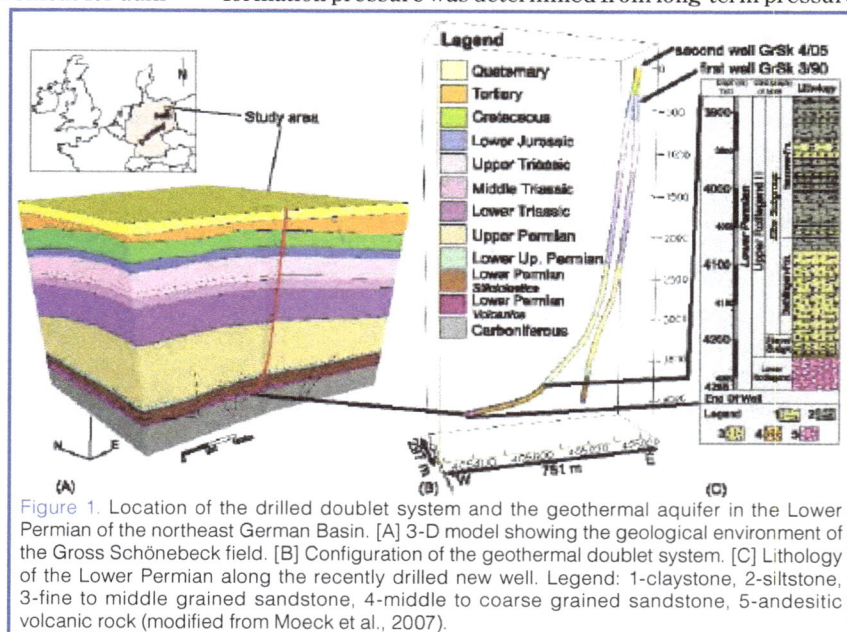

Figure 1. Location of the drilled doublet system and the geothermal aquifer in the Lower Permian of the northeast German Basin. [A] 3-D model showing the geological environment of the Gross Schönebeck field. [B] Configuration of the geothermal doublet system. [C] Lithology of the Lower Permian along the recently drilled new well. Legend: 1-claystone, 2-siltstone, 3-fine to middle grained sandstone, 4-middle to coarse grained sandstone, 5-andesitic volcanic rock (modified from Moeck et al., 2007).

logs showing equilibrium conditions close to 44.9 ± 0.3 MPa at 4220 m depth. A series of stimulation experiments was performed. First, open hole hydraulic gel-proppant fracturing treatments were conducted in two pre-selected sedimentary reservoir zones in the Lower Permian sandstones at a depth of ~4 km. The main inflow zones could clearly be identified. In a second step, massive water fracturing treatments were applied over the entire open hole interval from 3874 m to 4309 m depth. Pressure response analyses and well logs indicated the creation of vertical fractures and a bilinear flow regime in the reservoir, implying that an enhanced geothermal system suitable for geothermal power production had formed (Zimmermann et al., 2005; Huenges et al., 2006).

Directional Drilling of a Second Well

Hydraulic-thermal modeling based on data from the first well, along with regional structural analyses, identified the best possible well path geometry for the second well (Zimmermann et al., 2007). The borehole was designed parallel to the minimum horizontal stress direction and perpendicular to potentially hydraulic fractures to decrease auxiliary energy requirements for the thermal water loop in the planned doublet. Furthermore, this setup provides a low risk of a temperature short circuit of the system within the projected thirty years of utilization. Due to infrastructural requirements, the new well (GrSk 4/05) was located at the same drill site as GrSk 3/90 (27 m distance) but with a bottom hole some 500 m apart due to the reservoir requirements. Therefore, the new drilling operations required (1) a large hole diameter due to the deep static water table of the reservoir and the respective withdrawal during production (housing for the submersible pump), (2) directional drilling to intersect the target horizon at the derived offset from the existing hole and to increase the inflow conditions through well inclination in addition to later multiple fracturing, and (3) a special drilling mud concept to avoid formation damage of the reservoir as much as possible.

Initially, the large hole diameter (23") drilling experienced difficulties in clay-dominated sections requiring pumping capabilities beyond 4000 L min^{-1}. Complete casing cementation was necessary, because thermally induced stress from hot water might have caused casing damage on the non-cemented pipes. Total fluid loss and uncontrolled hydrofracturing occurred during the bottom up cementation of the 16" crossover 13-3/8" casing, conducted with a mean slurry density of 1450 kg m^{-3}. Therefore, squeeze cementation was performed from top of the well to the former cement infiltration zone. The successful placement of the cement was controlled by thermal logging.

Following drilling of a 1600-m-thick Upper Permian evaporate section counterbalanced with a mud density of 2000 kg m^{-3}, a 9-5/8" liner was installed, which (despite a strength with a safety factor of 1.8) collapsed in the bottom region

after reduction of the mud density was reduced to 1060 kg m^{-3}. Presumably, this failure was caused by additional stress components from anisotropic stress from the well inclination of ~20° in connection with the presence of highly ductile rocksalt (temperatures of 110°C in 3800 m depth). Stress concentration in interbedded anhydritic layers might have increased anisotropic stresses. The collapsed 9-5/8" liner was replaced with a combined 7" x 7-5/8" liner after sidetracking. The latter caused further challenges, as the setting of the mechanical anchor of the whipstock required its modification for reliable operation in mud with 40% barite content. Furthermore, the borehole design needed to be adjusted due to the loss of one casing dimension. Therefore, the borehole was deepened with 5-7/8" diameter drilling into the geothermal reservoir of the Lower Permian section.

In order to avoid drilling mud which would invade the formation and reduce its permeability, the reservoir below 3900 m was drilled with a near-balanced mud density of 1030 kg m^{-3}. Borehole wall breakouts at 3940 m forced a cleaning run and an elevation of mud pressure to 1100 kg m^{-3}. This specific mud pressure was the result of a geomechanical study investigating the initiation of borehole breakouts in reservoir successions under low mud pressures (Moeck et al., 2007). Another reason for increasing the mud weight was the occurrence of H$_2$S below the 7-5/8" casing shoe and within the fissured lowermost Upper Permian formation. To prevent gas inflow, the mud density was partly increased up to 1200 kg m^{-3}, and a specially designed marble flour-based mud was used to minimize fluid losses into the coarse sandstone formation. Due to the danger of differential sticking and formation damage, the mud weight was lowered in subsequent drilling operations. No significant fluid losses were observed during the final drilling and casing operations.

Accessing the Geothermal Reservoir

The well reached the target along the planned borehole track (Fig. 2). A 5" liner combined with a non-cemented section of pre-perforated pipes at the bottom was installed in the lowermost section at 4400 m depth. The presence of Lower Permian middle to fine grained sandstones of the Dethlingen Formation at this depth was confirmed by cutting and well log analyses. In the well, the Lower Permian sediments reached a thickness of 340 m at the flank of structural high of the sandstones (Fig. 3). Reservoir sandstone layers with permeabilities up to 160 mD lay within the succession and have

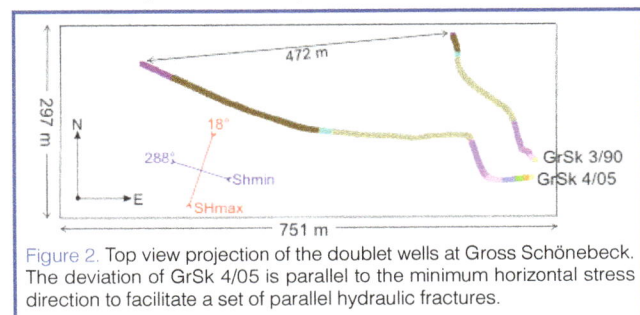

Figure 2. Top view projection of the doublet wells at Gross Schönebeck. The deviation of GrSk 4/05 is parallel to the minimum horizontal stress direction to facilitate a set of parallel hydraulic fractures.

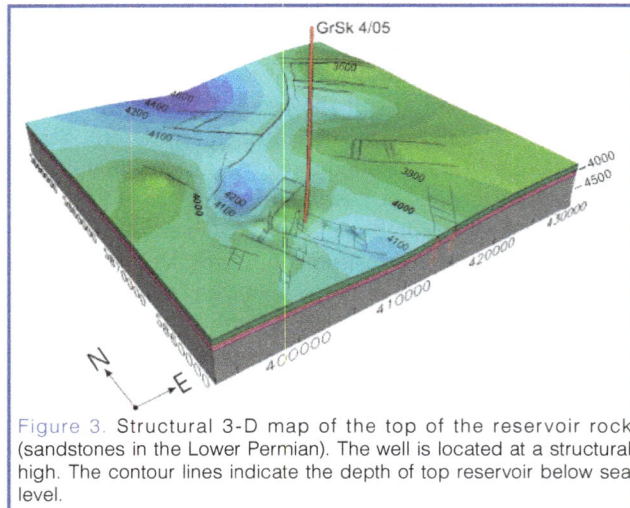

Figure 3. Structural 3-D map of the top of the reservoir rock (sandstones in the Lower Permian). The well is located at a structural high. The contour lines indicate the depth of top reservoir below sea level.

a vertical thickness of >80 m. The well inclination of 45° implies that up to 150 m of the well is within this permeable sandstone. The well deviation is oriented at 288° to optimize the hydraulic fracturing design (Fig. 2 and Holl et al., 2005). Hydrofracs are planned in the volcanic rock and some in the sandstones. Based on the previously mentioned hydraulic-thermal modeling, a distance of no less than 450 m between the bottoms of the two wells was realized to avoid a thermal breakthrough of the injected cold water directly into the production well.

Conclusions

In the Northeast German Basin, 4000-m-deep Lower Permian sandstones and volcanic rocks have been explored for geothermal energy production near Gross Schönebeck. The research strategy we applied consists of (i) re-using a former gas exploration well for logging and hydraulic stimulation campaigns, (ii) understanding the reservoir behavior based on data recovery from hydraulic treatments, (iii) optimizing the planned reservoir exploitation by analyzing the performance variances of well paths, (iv) completing the geothermal doublet system by drilling a second well, (v) future stimulating and testing the new well and installing a thermal water loop using a doublet system, and (vi) installing a binary geothermal power plant if sufficient reservoir conditions are continued. The experiences gained, especially in (iv), show that drilling a large hole diameter (23") is feasible but challenging especially in clay dominated layers; that directional drilling can be applied as a standard operation; and that a variable mud concept needs to be applied in order to react to unforeseen operational requirements such as formation damage, breakouts, or inflows. In this project, technical and scientific challenges were successfully met, and the lessons that were learned provided essential knowledge for developing future drilling strategies in deep sedimentary geothermal systems, especially in the Central European Basin System.

Acknowledgement

The authors want to thank the German Federal Ministry for the Environment for funding (BMU ZIP 0327508, BMU 0329951B, BMU 0329991).

Geothermal Project Group:

Ernst Huenges, Inga Moeck, Ali Saadat, Wulf Brandt, Axel Schulz, Heinz-Gerd Holl, David Bruhn, Günter Zimmermann, Guido Blöcher, and Lothar Wohlgemuth.

References

Baumgärtner, J., Jung, R., Hettkamp, T., and Teza, D., 2004. The status of the Hot Dry Rock Scientific Power Plant at Soultz-sous-Forêts. Z. Angew. Geol., 2:12–17.

Economides, M.J., and Nolte, K.G., 1989. Reservoir Stimulation. Houston, Texas (Schlumberger Educational Services).

Holl, H.-G., Moeck, I., and Schandelmeier, H., 2005. Characterisation of the tectono-sedimentary evolution of a geothermal reservoir - implications for exploitation (Southern Permian Basin, NE Germany). Proceedings World Geothermal Congress 2005, Antalya, Turkey, 24–29 April 2005, 1–5.

Huenges, E., and Kohl, T., 2007. Stimulation of reservoir and micro-seismicity - summary of the Ittingen workshop June 2006, ENGINE – Enhanced Geothermal Innovative Network for Europe, Mid-Term Conference Potsdam, Germany, 9–12 January 2007.

Huenges, E., Holl, H.-G., Legarth, B., Zimmermann, G., Saadat, A., and Tischner, T., 2004. The stimulation of a sedimentary geothermal reservoir in the North German Basin: case study Groß Schönebeck. Z. Angew. Geol., 2:24–27.

Huenges, E., Trautwein, U., Legarth, B., and Zimmermann, G., 2006. Fluid pressure variation in a sedimentary geothermal reservoir in the North German Basin: case study Groß Schönebeck. Pure Appl. Geophys., 163(10):1–12.

Moeck, I., Backers, T., and Schandelmeier, H., 2007. Assessment of mechanical wellbore stability by numerical analysis of fracture growth. Proc. EAGE 69th Conference & Exhibition, London, 11–14 June 2007, D047: 1-5.

Zimmermann, G., Reinike, A., Blöcher, G., Milsch, H., Gehrke, D., Holl, H.-G., Moeck, I., Brandt, W., Saadat, A., and Huenges, E., 2007. Well path design and stimulation treatments at the geothermal research well GT GRSK 4/05 in Groß Schönebeck. Proc. 32nd Workshop on Geothermal Reservoir Engineering, Stanford University, Stanford, Calif., 22–24 January 2007, SGP-TR-183.

Zimmermann, G., Reinicke, A., Holl, H.-G., Legarth, B., Saadat, A., and Huenges, E., 2005. Well test analysis after massive Waterfrac treatments in a sedimentary geothermal reservoir, Proc. World Geothermal Congress 2005, Antalya, Turkey, 24–29 April 2005, 1–5.

Authors

Ernst Huenges, Inga Moeck, GFZ Potsdam, Telegrafenberg, D-14473, Potsdam, Germany, e-mail: huenges@gfz-potsdam.de.

Related Web Link

http://www.gfz-potsdam.de/pb5/pb52/projects/Machbarkeit/ewelcome.html

8

Contribution of Borehole Digital Imagery in Core-Log-Seismic Integration

by Philippe Gaillot, Tim Brewer, Philippe Pezard, and En-Chao Yeh

Introduction

The Integrated Ocean Drilling Program (IODP) and the International Continental Drilling Program (ICDP) use many new technologies to increase the quality of core data. A problem with drilling deep oceanic or continental crust, as well as shallow fractured rock or karstic formations, is that core recovery can be low, and much of the recovered material often consists of small, disrupted core pieces that are frequently biased toward particular types of rocks (lithologies). Even when core recovery is high (in the case of sedimentary formation), recovered cores are not always oriented; as a result, detailed structural and paleomagnetic studies are impaired.

In contrast, logging provides nearly continuous records of the *in situ* chemical and physical properties of the penetrated formation, which can be used to extrapolate the various lithologies in areas of reduced core recovery (Brewer et al., 1998).

With the advent of modern imaging tools, a fundamentally new concept has been introduced (Serra, 1989; Lovell et al., 1998). Formations are no longer scanned by a single sensors creating a single scalar log, but can be sampled multiple times horizontally and at a high rate vertically to form a dense matrix of measurements being displayed as an image. Since the mid-eighties there has been an explosive development in imaging technology, principally in terms of tools but also in terms of producing the image. The progress has been linked with the availability of downhole digitization of signals and the possibility of transmitting large data volume in real time. Where the standard logs are sampled every ~15 cm (6"), image logs may be sampled every ~0.25 cm (0.1") or less; where the standard logs have one measurement per depth point, image logs may have 360 or more.

So, digital borehole images (mm-scale) can potentially bridge the scale gap between the (dm-scale) standard logs and (μm to dm-scale) core measurements. Indeed, continuous and oriented borehole wall images provide high resolution lithologic, textural, and structural information, filling information gaps where core recovery is low and allowing orientation of cores when core recovery and undoubted features are recognized on both core and logging images. In turn, continuous and quantitative information extracted from these images can help in bridging the scale gap between log and the large scale (macroscopic, >100 m) properties of the penetrated formation. While physical principles and applications of electrical, acoustic, and optical borehole imaging tools are easily schematized in Fig. 1 (see http://publications.iodp.org/sd/05/suppl/ for details), the present paper shows some examples of quantitative analyses of electrical images that illustrate various contributions of borehole images to a fully inte-

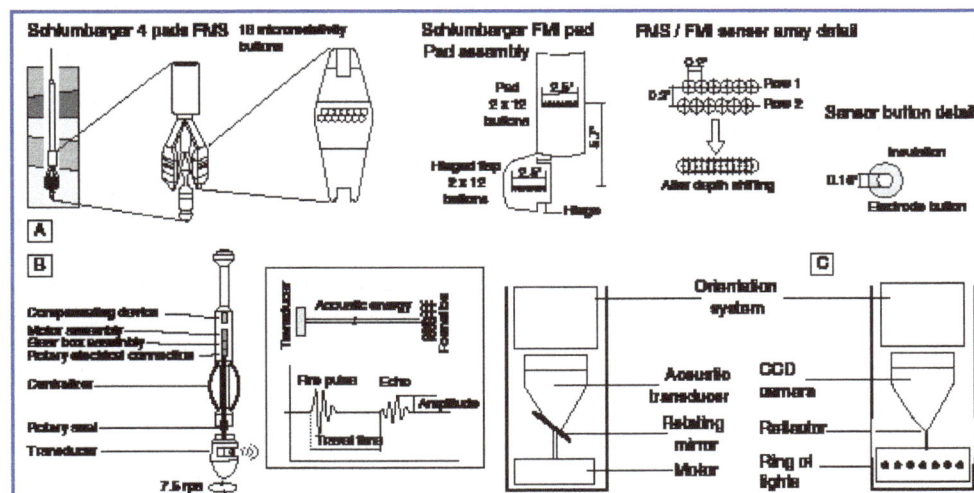

Figure 1. Sketch of electrical, acoustic and optical imaging tools. [A] Schlumberger Formation MicroScanner (FMS) with four pads, 16 buttons per pad, covering 25%–40% of hole diameter, and Schlumberger Fullbore Formation MicroImager (FMI) with 4 pads, 4 hinged flaps, and 24 buttons on each pad and flap. The hinged flap is able to increase coverage by up to 80% (modified from Ekstrom et al., 1987). The buttons are aligned in two rows; processes for depth corrections shift the recorded resistivity to one row. Each button consists of an electrode surrounded by insulation. [B] The Ultrasonic Borehole Imager features a high resolution transducer that provides acoustic images of the borehole wall. The transducer emits ultrasonic pulses at a frequency of 250 kHz or 500 kHz (low and high resolution, respectively), which are reflected by the borehole wall and then received by the same transducer. Amplitude and travel time of the reflected signal are then determined. [C] Optical televiewers generate a continuous oriented 360° image of the borehole wall unwrapped using an optical imaging system (downhole CCD camera which views a reflection of the borehole wall in a conic mirror- sketch of advanced logic technology ALT OBI40). Like electrical imaging tools and acoustic televiewers, the optical televiewers include a full orientation device consisting of a precision 3-axis magnetometer and 2 accelerometers, thus allowing for accurate borehole deviation data to be obtained during the same logging run and for accurate and precise orientation of the image.

grated core-log-seismic integration workflow.

Case Studies

Borehole images filling lithostratigraphic core gaps:
Electrical borehole wall images acquired with the Schlumberger Formation MicroScanner (FMS) in Ocean Drilling Program (ODP) Hole 896A are a good example of the contribution of borehole images to reconstruct the lithostratigraphy when core recovery is low (<30%) (ODP Leg 148 Shipboard Scientific Party, 1993; Alt et al., 1996). The alternative lithostratigraphy that is constructed by combining standard scalar logging data and FMS images contains considerably more brecciated units (~30%) than

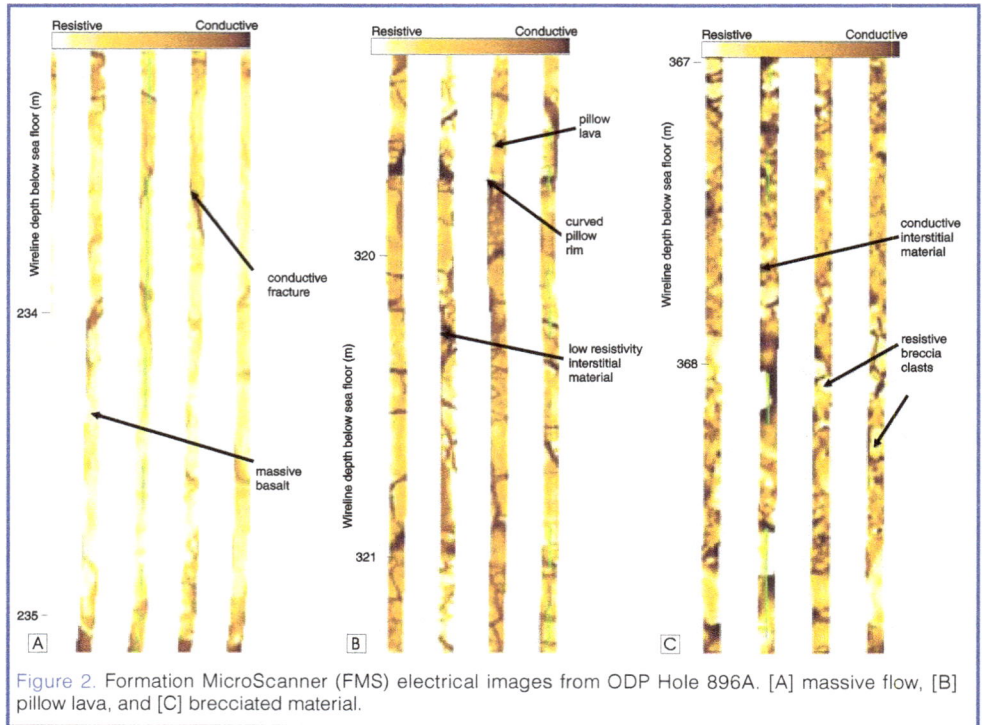

Figure 2. Formation MicroScanner (FMS) electrical images from ODP Hole 896A. [A] massive flow, [B] pillow lava, and [C] brecciated material.

suggested by the shipboard core descriptions (10%). This disparity, explained as the reflection of preferential recovery of less fractured massive flows, emphasized the necessity to fully integrate core and logging results in boreholes with reduced core recovery. In Hole 896A, different volcanic lithologies can be identified on the FMS image data by variations in electrical conductivity (Fig. 2). Massive units appear on the FMS images as extensive areas with a uniformly low conductivity and predominantly straight, branching fracture patterns (Fig. 2A). Pillow lavas show variable conductivity within a small area, but this is less variable than for brecciated units. Individual pillow lavas can often be distinguished on the FMS data owing to the curved nature of the pillow boundaries (Fig. 2B). Interstitial material usually has high conductivity. Breccias are characterized on the FMS image data by high conductivity, which is highly variable within a small area. The presence of small high-resistivity clasts can often be noted (Fig. 2C).

Orientation of cores using borehole images: In a more general manner, although cores obtained through ocean and continental drilling provide valuable information, recovered core pieces are often initially inaccurately located and unoriented in a geographical reference frame. In these situations, logging image data are essential to supplement and enhance structural data from the recovered core. One core imaging tool used by ODP during ODP Leg 176 was a digital core scan device that produces 360° images of the outside of the core (Dick et al., 1999). These digital photographs can be unwrapped and displayed as 2-D images showing the core's entire outer surface (Fig. 3). The images can then be used to locate and measure the dip and orientation of structural features (veins, fractures, boundaries between different rock

types) in the core. Assuming that the core recovery is high enough and features on core and borehole wall images are distinctive enough, such structural features on the core and logging FMS images can be matched and so allow the accurate location and orientation of individual core pieces, a key step in structural and paleomagnetic studies (MacLeod et al., 1995; Harvey and Lovell, 1998; Haggas et al., 2001). In the San Andreas Fault zone, Iturino et al (2001) used a combination Logging While Drilling (LWD) Resistivity At the Bit (RAB) image and core x-ray computed tomography images.

Formation scale information deduced from borehole images: Any paleoenvironmental reconstruction or investigation of large-scale geological bodies (through, upper oceanic crust, reservoir) requires the extension of the 1-D view of the borehole to a regional-formation scale view. The first step involves a classification of the core and downhole log responses into relatively homogeneous sub-groups (units) based on (1) a lithofacies determination relying on visual core description and measurement (core units) and (2) a visual or statistical analysis of available logging data (log units). Due to the *in situ* and continuous nature of downhole data in respect to expensive and discontinuous nature of core data, methods based on multivariate statistics of downhole log response (electro-facies-based classification; See Ravenne, 2002 for a review.) have been developed to estimate spatial distribution of heterogeneous subsurfaces (e.g., regionalized classification relying on statistical relationships between laboratory-determined hydrologic properties and field-measured geophysical properties to estimate spatial distributions of porosity, permeability, and diagenetic characteristics; Moline and Bahr, 1995). Such methods aim to predict the lithology of the penetrated formation, but more widely attempt to provide additional formation-scale hydrologic (flow unit, reservoir permeability) or geologic (structural, environmental) information. Taking into account the increase in computational power and the higher resolution and coverage of the borehole images in respect to standard logs, such methods have been generalized to include borehole images information (e.g., texture) and have been coupled to modeling tools.

Core, borehole image, log, and formation-scale integration: Log data and digital borehole images collected from the Hole-B of the ICDP Taiwan Chelungpu-fault Drilling Project have been analyzed to establish the relationships between deformation structures and *in situ* stress, and to identify the rupture zone of the 7.6 Mw 1999 Chi-Chi earthquake. Based on standard scalar logs, three log units and five subunits are recognized as consistent with lithological units defined from visual core description (Fig. 4). Analysis of the Schlumberger Fullbore Formation MicroImager (FMI) resistivity data has also been automatically performed. This analysis, relying on the local and multi-scale properties of the wavelet transform formalism, consists in measuring the number and characteristics (size and electrical contrast) of each electric feature recorded in the high resolution logs, and so provides new high resolution "quantitative and integrative" logs—namely, the density of electrical features (Npm), the average size (wavelength, W) and average electrical contrast (C) of the detected features. In turn, combined with visual interpretation of sedimentary and structural features recognized on borehole images, these logs can be correlated to large-scale formation units and fault zones independently defined from the scalar logs and core data providing extra small-scale information on the nature of the downhole variations in natural gamma radiation (NGR), electrical resistivity, and sonic velocity (Vp) (Fig. 4).

Conclusion

Digital borehole images have the same depth coverage of scalar logs, with a resolution higher than these logs and often higher than standard core logging measurements. Relying on different physical bases, imaging tools provide a palette of high resolution, continuous, and oriented 360° views of the borehole wall from which the character, relation, and orientation of lithologic and structural planar features, as well as texture, can be defined to support detailed core analyses. Analyzed with specific methods relying on multivariate statistics or automatic extraction of quantitative attributes, these images and high resolution information derived from them allow filling the scale gap between core/sample measurements and large-scale formation properties, thus opening an avenue in addressing scientific challenges related to formation characterization.

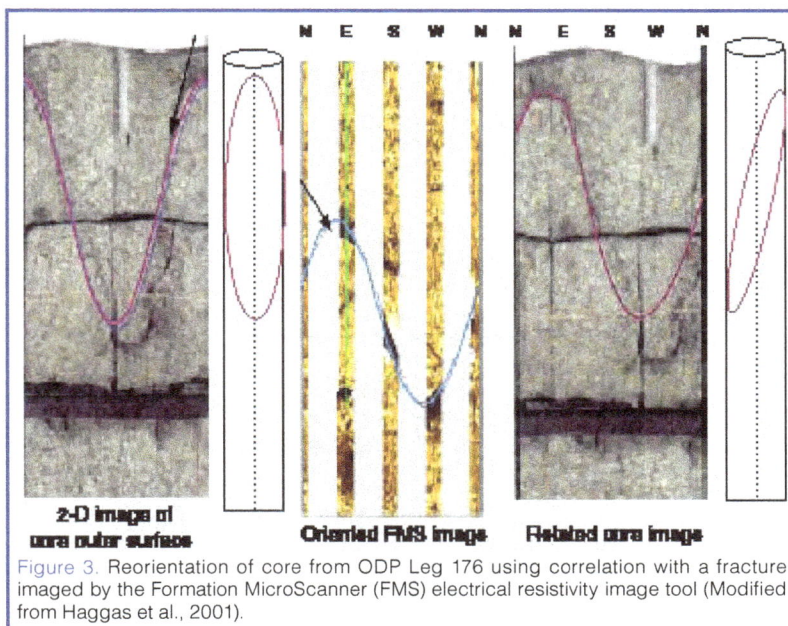

Figure 3. Reorientation of core from ODP Leg 176 using correlation with a fracture imaged by the Formation MicroScanner (FMS) electrical resistivity image tool (Modified from Haggas et al., 2001).

References

Alt, J.C., Kinoshita, H., Stokking, L.B., and Michael, P.J. (Eds.), 1996. *Proc. ODP, Sci. Results, 148*. College Station, Texas (Ocean Drilling Program).

Brewer, T.S., Harvey, P.K., Lovell, M.A., Haggas, S., Williamson, G., and Pezard, P., 1998. Ocean floor volcanism: constraints from the integration of core and downhole logging measurements. *Geol. Soc. Lond., Spec. Publ.*, 136:341–362.

Dick, H.J.B., Natland, J.H., Miller, D.J., et al., 1999. *Proc. ODP, Init. Repts., 176* [Online]. Available from http://www-odp.tamu. edu/publications/176_IR/176TOC.HTM.

Ekstrom, M. P., Dahan, C. A. Chen, M. Lloyd, P. M. and Rossi, D. J., 1987. Formation imaging with microelectrical scanning arrays. Log Analyst, 28, 294–306.Haggas, S., Brewer, T.S., Harvey, P.K., and Iturrino, G., 2001. Relocating and orienting cores by the integration of electrical and optical images: a case study from Ocean Drilling Program Hole 735B. *J. Geol. Soc. (London, U.K.)*, 158:615–623.

Haggas, S., Brewer, T.S., Harvey, P.K., and Iturrino, G., 2001. Relocating and orienting cores by the integration of electrical and optical images: a case study from Ocean Drilling Program Hole 735B. *J. Geol. Soc. (London, U.K.)*, 158:615–623.

Harvey, P.K., and Lovell, M.A., (Eds.), 1998. *Core-Log Integration. Geol. Soc. Spec. Publ., 136*, 400 pp.

Iturino, G.J., Goldberg, D., and Ketcham, R., 2001. Integration of core and downhole images in the San Andreas fault zone. *EarthScope Workshop: Making and Breaking a Continent*, Snowbird, Utah, 10–12 October 2001. Abstract online at http://www.scec.org/instanet/01news/es_abstracts/Iturrino_et_al.pdf.

Leg 148 Shipboard Scientific Party, 1993. Site 504. *In* Alt, J.C., Kinoshita, H., Stokking, L.B., et al., *Proc. ODP, Init. Repts., 148*. College Station, Texas (Ocean Drilling Program), 27–121. Available online at http://www-odp.tamu.edu/publications/148_IR/VOLUME/CHAPTERS/ir148_02.pdf.

Lovell, M.A., Harvey, P.K., Brewer, T.S., Williams, C., Jackson, P.D., and Williamson, G., 1998. Application of FMS images in the Ocean Drilling Program: an overview. *In* Cramp, A.,

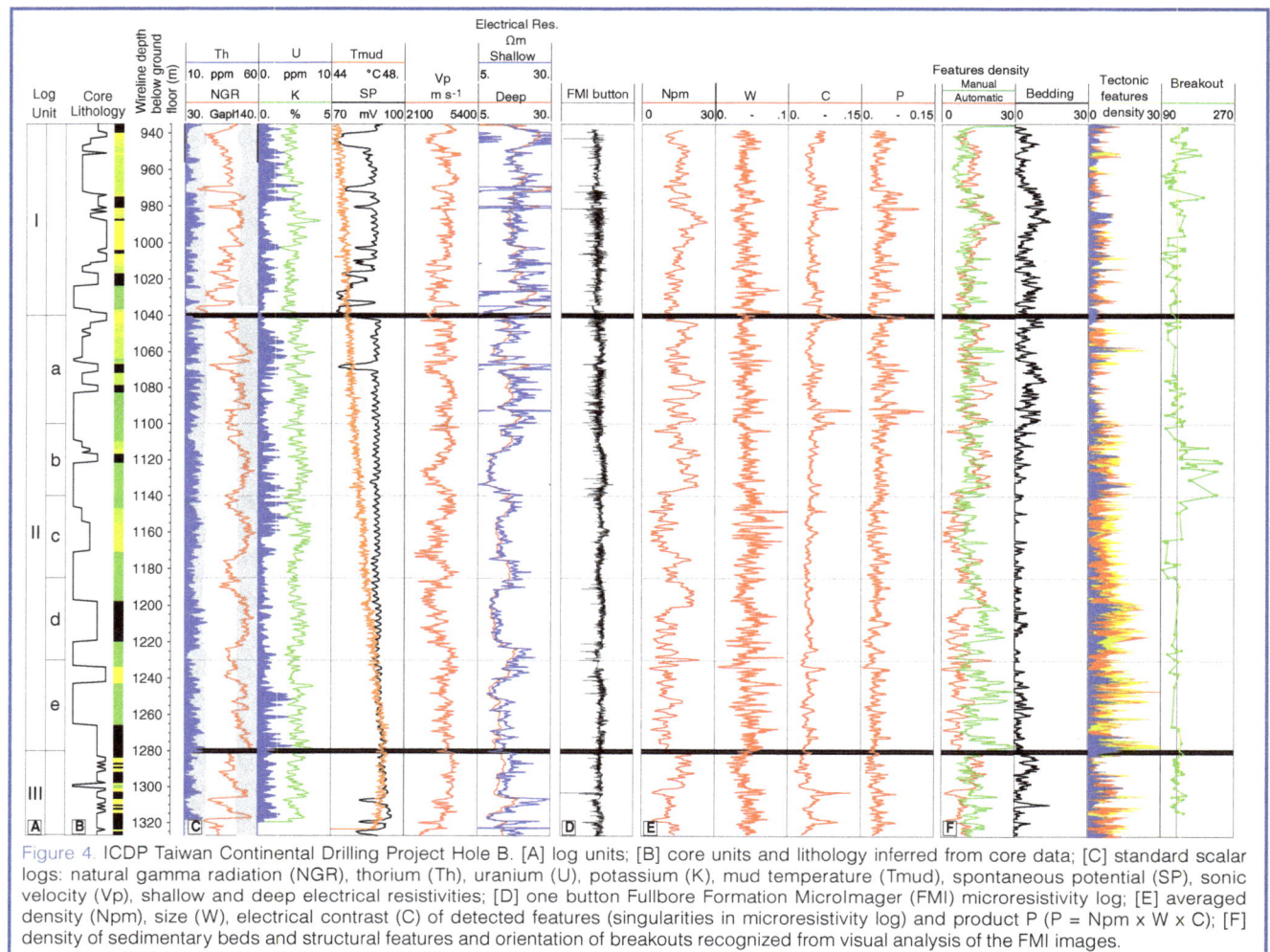

Figure 4. ICDP Taiwan Continental Drilling Project Hole B. [A] log units; [B] core units and lithology inferred from core data; [C] standard scalar logs: natural gamma radiation (NGR), thorium (Th), uranium (U), potassium (K), mud temperature (Tmud), spontaneous potential (SP), sonic velocity (Vp), shallow and deep electrical resistivities; [D] one button Fullbore Formation MicroImager (FMI) microresistivity log; [E] averaged density (Npm), size (W), electrical contrast (C) of detected features (singularities in microresistivity log) and product P (P = Npm x W x C); [F] density of sedimentary beds and structural features and orientation of breakouts recognized from visual analysis of the FMI images.

MacLeod, C.J., Lee, S.V., and Jones, E.J.W. (Eds.), *Geological Evolution of Ocean Basins: Results from the Ocean Drilling Program. Geol. Soc. Spec. Publ.*, 131:287–303.

MacLeod, C.J., Célérier, B., and Harvey, P.K., 1995. Further techniques for core reorientation by core-log integration: application to structural studies of lower oceanic crust in Hess Deep, eastern Pacific. *Sci. Drill.*, 5:77–86.

Moline, G.R., and Bahr, J-M., 1995. Estimating spatial distributions of heterogeneous subsurface characteristics by regionalized classification of electrofacies. *Mathemat. Geol.*, 27(1):3–22.

Ravenne, Ch., 2002. Stratigraphy and oil: a review, part 2 characterization of reservoirs and sequence stratigraphy: quantification and modeling. *Oil Gas Sci. Technol. Rev. IFP*, 57:(4):311–340.

Serra, O., 1989. *Formation MicroScanner Image Interpretation.* Houston, Texas (Schlumberger Educational Services), SMP-7028.

Authors

Philippe Gaillot, CDEX-IFREE, Japan Agency for Marine-Earth Science and Technology, Yokohama Institute for Earth Science, 3173-25 Showa-machi, Kanazawa-ku, Yokohama, Kanagawa, 236-0001 Japan, e-mail: gaillotp@jamstec.go.jp.
Tim Brewer, Department of Geology – Geophysics and Borehole Research, University of Leicester, University Road, Leicester, LE1 7RH, U.K.

Philippe Pezard, Laboratoire de Géophysique et d'Hydrodynamique en Forage, Geosciences Montpellier, University of Montpellier 2, France.
En-Chao Yeh, Department of Geosciences, National Taiwan University, No.1, Sec. 4, Roosevelt Road, Taipei 106, Taiwan.

Related Web Links

http://dx.doi.org/10.2204/iodp.proc.310.2007
http://dx.doi.org/10.2973/odp.proc.ir.176.1999
http://www.scec.org/instanet/01news/es_abstracts/Iturrino_et_al.pdf

Co-author **Dr. Tim Brewer** collapsed and died on Saturday morning, 14th July, 2007, while attending a conference in Barcelona. This is obviously shocking and very sad news and came as a complete surprise to everyone.

Tim was a senior member of staff in the Department of Geology at the University of Leicester but also the lead coordinator in the European Petrophysical Consortium, part of the ECORD Science Operator for the Integrated Ocean Drilling Program.

He will be sadly missed by his friends and colleagues around the world.

9

On the Fidelity of "CORK" Borehole Hydrologic Observatory Pressure Records

by Earl Davis and Keir Becker

Introduction

Long-term formation pressure monitoring in Ocean Drilling Program (ODP) and Integrated Ocean Drilling Program (IODP) boreholes using evolving Circulation Obviation Retrofit Kit (CORK) hydrologic observatory technology has led to unanticipated applications as a result of the growing duration of recording intervals and the improvement of measurement fidelity. Current capabilities provide geologically meaningful observations over a broad range of time scales from static state to 1 Hz, allowing investigations of many coupled hydrologic, geodynamic, and seismologic phenomena. In this review, we present observations that provide constraints on current limits to recording fidelity, and examples of how leakage can affect pressure observations.

Background

The capability to seal and monitor the hydrologic state of deep-ocean boreholes drilled by the ODP was developed in 1990, and since that time, a broad range of experiments using this technology has been carried out to study the hydrogeology of ridge axes, older oceanic crust, and accretionary and non-accretionary subduction prisms. Original "CORK" installations (Fig. 1) employed a seal at the top of a standard solid steel casing string to isolate a single window of interest at depth for pressure monitoring and fluid sampling (Davis et al., 1992). Subsequent refinements (Fig. 1) have allowed pressure monitoring and fluid sampling at multiple formational levels outside and below the casing (Mikada et al., 2002; Jannasch et al., 2003; Becker and Davis, 2005). In both configurations, temperature measurements can be made inside and below the casing. Monitoring in installations completed to date has led to the determination of the static formation state and the driving forces and rates of fluid flow as originally planned, as well as to unexpected new insights regarding elastic and hydrologic formation properties, and secular and episodic strain in a variety of geologic settings and over a broad range of temporal and spatial scales (Davis and Becker, 2004; Becker and Davis, 2005; Kastner et al., 2007). In this article, we present data excerpts that illustrate the current level of recording fidelity, and we review problems experienced with some of the installations. Examples are drawn primarily from ocean crustal Site 1026 recently instrumented with new high resolution instrumentation, and sedimentary Sites 1173 and 808 at the Nankai subduction zone. While significant new insight has been gained through temperature monitoring and continuous fluid sampling, we limit this discussion to the measurement of pressure.

Theoretical Limits to Resolution

Limits to data quality are imposed by the characteristics of the measuring devices employed, and by the hydraulics of the CORK systems in context of the formations in which they are installed. All CORK systems deployed to date have been equipped with absolute pressure sensors (Paroscientific, Inc.) that utilize a quartz transducer loaded by the ocean- or formation-water pressure via a Bourdon tube. The pressure-sensitive oscillation frequency of the transducer crystal is determined by comparing its output signal to a fixed reference frequency. Early electronics employed an integer cycle counter and provided a resolution of roughly 1 ppm full-scale at 1 Hz. In practice, memory and battery capacity limited sampling intervals to 1 hr early in the development history, and this was later refined to 10 minutes. The intrinsic characteristics of the transducers allow much higher resolution to be achieved, and a recently developed fractional period counter (Bennest Enterprises, Ltd.) allows pressure to be

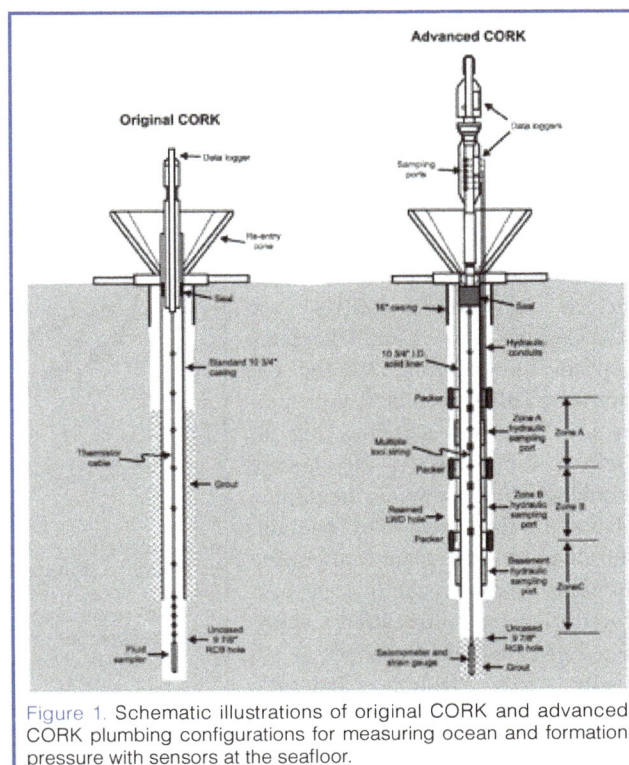

Figure 1. Schematic illustrations of original CORK and advanced CORK plumbing configurations for measuring ocean and formation pressure with sensors at the seafloor.

determined to a few tens of ppb (~2 Pa, or 0.2 mm of equivalent water head) at 1 Hz. Autonomous deployments are still restricted to measurement frequencies lower than this, but instruments equipped with Bennest fractional-period counters soon to be connected to seafloor power and communications cables will allow full use of this "high-frequency" resolution. The highest autonomous sampling rate employed to date (15 s sample period at Site 1026) shows oceanographic signal levels approaching the theoretical limit (Fig. 2A).

At the low-frequency end of the spectrum, fidelity limits are imposed by sensor drift. Seafloor records from Sites 808 and 1173 (located 13 km from one another) show rates of drift that decrease from ~5 kPa yr^{-1} in the first few months after deployment, to ~0.1 kPa yr^{-1} over longer periods of time (Fig. 2B). Other sites show a similar behavior. Detection of natural secular change of formation pressures or water depths at rates less than this requires periodic calibration checks with a mobile sensor. Intermediate-period (<1 yr) oceanographic signals (Fig. 2B) also impose limits on detecting tectonic signals, although local stable reference sites can help to overcome this problem.

Hydraulic-system limits on fidelity originate primarily from the inability of the formation to deliver or receive fluid rapidly enough to accommodate compliance in the CORK plumbing as formation pressure changes. The most influential formation and CORK system properties are hydraulic storage compressibility and permeability, and plumbing volume and fluid compressibility, respectively. Original CORKs utilized the full volume of the cased holes to transmit formation pressure to the sensors at the seafloor (Fig. 1). A significant reduction in the system fluid volume and associated compliance was gained by later CORKs (collectively referred to herein as "advanced CORKs", or ACORKs) which used small-diameter hydraulic lines running from the sensors to relatively small-volume permeable formation screens (Fig. 1; Table 1). Compliance of the all-steel plumbing components is negligible relative to the water they contain. Entrained gas can strongly affect the plumbing compressibility, but purging during (and often after) deployment has been done to reduce or eliminate this factor. Hydraulic resistance and compliance in the annulus outside the casing screens in

sedimentary sections are difficult to assess, but they are probably insignificant, since low sediment strength does not allow the formation to support its overburden, and collapse around the screens is likely to be complete and benign. A discussion of these factors and their possible influence on the frequency response of CORK systems is provided by Sawyer et al. Dependence of the system response on the primary factors of sediment compressibility, permeability, and CORK-system compliance, can be understood using a formulation of Bredehoeft and Papadopulos (1980) for well-bore response to a step-wise change in formation pressure. Parameters that cover the range of CORK installations are provided in Table 1, and the results are summarized in Fig. 3. In Fig. 3A, dimensionless time incorporates scaling for formation permeability and compressibility; when converted to real time (Fig. 3B), the orders-of-magnitude difference in sediment and basalt permeabilities result in much faster response times for a CORK in basalt compared to a CORK in sediment, despite the fact that their type curves are similar (Fig. 3A). The differences among the type curves also involves a scaling for system volume, leading to a much faster actual response time for an ACORK vs. a CORK in sediment (Fig. 3B).

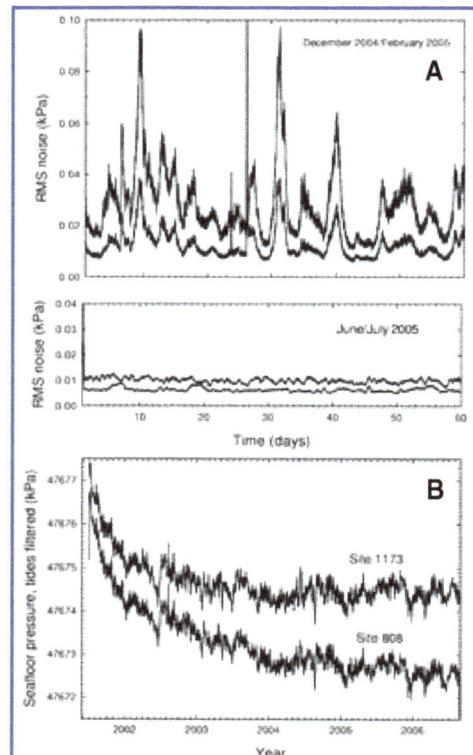

Figure 2. Indications of high recording fidelity from oceanographic signals. [A] Typical RMS high-frequency (<5 min period) seafloor- and formation-pressure signal levels at ODP Site 1026 in stormy and calm periods. Short bursts of high amplitudes in late December are teleseismic in origin (e.g., Fig. 5a). [B] Seafloor pressure records from ODP Sites 808 and 1173 (filtered to eliminate tidal and higher frequency signals) are dominated by slowly decaying sensor drift and intermediate-period oceanographic signals.

Observational Indications of High Data Quality

The best examples of current intrinsic recording capability at high frequencies are provided by a new-generation (Bennest fractional period counter) instrument installed on the Juan de Fuca Ridge flank in Hole 1026B in 2004 during IODP Expedition 301. This modified (Fisher et al., 2005) CORK samples highly permeable oceanic crust, and CORK system effects are probably small (Fig. 3B). Time-series data suggest uniform elastic formation response to oceanographic loading over a broad range of frequencies (Fig. 4). A slight flattening of the seafloor- and formation-pressure power spectra at high frequencies may indicate approach to the sensor noise floor, although this is not surprising, given that the high-frequency signal levels during oceanographically quiet times are only ~10 Pa, i.e., close to the 2 Pa measurement resolution (Fig. 2A). Other indications of recording fidelity at high frequency have been provided by signals of geologic origin. An example of pressure signals from the Sumatra earthquake of 2004 includes seismic body waves that are smaller in amplitude in the for-

Table 1. Range of parameters for formation and CORK system properties (500 m hole assumed)

Property	CORK in basalt	CORK in sediment	ACORK in sediment
Fluid compressibility	0.4×10^{-9} Pa^{-1}	0.4×10^{-9} Pa^{-1}	0.4×10^{-9} Pa^{-1}
Formation compressibility	10^{-9} Pa^{-1}	10^{-8} Pa^{-1}	10^{-8} Pa^{-1}
Permeability	10^{-10} m^2	10^{-18} m^2	10^{-18} m^2
Fluid viscosity	0.4×10^{-3} Pa s	0.4×10^{-3} Pa s	0.4×10^{-3} Pa s
System volume	30 m^3	30 m^3	0.064 m^3
Window length	50 m	50 m	7.6 m
Window radius	0.15 m	0.15 m	0.15 m
Porosity	0.1	0.4	0.4
Storage compressibility	1.04×10^{-9} Pa^{-1}	1.02×10^{-8} Pa^{-1}	1.02×10^{-8} Pa^{-1}

mation than in the water column, and surface waves that are larger in the formation (Fig. 5A). The formation/seafloor amplitude ratio for the P waves is consistent with that for ocean waves and tides (~0.3). The amplitude ratio for the surface waves (~2.5) is much larger.

As yet, no high resolution instruments have been installed in any sedimentary formations, and the limits of the old-generation instruments do not allow oceanographic signals to be used to assess the point at which CORK or ACORK hydraulic systems filter high-frequency hydrologic signals. Signals of geologic origin do provide relevant information. Observation of a static strain step recorded at Site 1173 (Fig. 5b) provides confidence that this low-system-volume ACORK provides reasonable high-frequency fidelity, despite the low formation permeability. Response to what is probably a step-wise change in formation pressure (reflecting coseis-

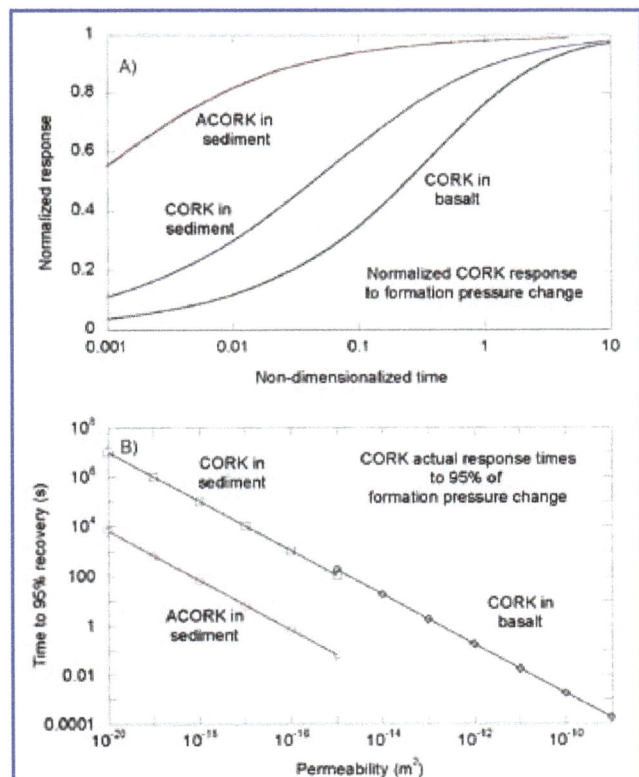

Figure 3. Simulated CORK and ACORK system response to a step change in formation pressure. [A] Characteristic Bredehoeft and Papadopulos (1980) type-curves for three example configurations (see parameter values in Table 1). [B] Dependence on formation permeability of actual CORK response times to 95% of the formation pressure change i.e., when curves reach 0.95 in [A].

mic volumetric contraction; Davis et al., 2007) is rapid (≤10 minutes, as predicted by Fig. 3). Seismic waves are also seen in this record, although the 10-min sampling interval causes the recorded signal to be highly aliased. In the future, higher resolution and higher sampling frequency will allow the full system fidelity to be properly examined.

Pitfalls

While CORK pressure monitoring to date has been generally very successful and the fidelity generally high, hydraulic leakage caused by physically damaged, hydrothermally deteriorated, or missing seals has impaired the fidelity of monitoring at several sites. Where leaks occur, measured formation pressure can be affected directly by pressure loss through the leaks, although most problems have arisen indirectly through effects of thermal buoyancy and thermal expansion associated with steady or transient flow up or down the leaking borehole along the geothermal gradient. A record from Site 1025 on the Juan de Fuca Ridge flank (Fig. 6a) illustrates a case where thermal buoyancy contributes most to a leakage-induced pressure perturbation. When a fluid sampling valve was opened, the record became noisy, the tidal signal was distorted, and the static pressure rose above the natural formation pressure as a consequence of the buoyancy of the warm discharge through the CORK casing from the slightly super-hydrostatic formation. Such perturbations arise whenever pressures are measured in high-permeability formations in the presence of leaks.

A different effect has been witnessed in two ACORKs that penetrate low-permeability sediments, namely Site 1173, where fluid sampling valves connected to some of the umbilical tubes were manipulated during monitoring, and Site 808, where pressures in multiple screens were affected by flow inside the casing from a permeable fault zone at depth. In this latter instance, a plug was to have been installed inside the main casing (Fig. 1), but drilling difficulties precluded this operation. Most of the screens appear to have been hydrologically well isolated and coupled to the formation at this site, but there was no check against flow up the inside of the casing from its bottom end. Several initially puzzling observations can be explained by thermal expansion and contraction of fluid in the lines and in low-permeability parts of the formation outside the screens caused by variable heating and cooling by the unwanted vertical flow:

Tidal loading response: The first of these observations is the occurrence of unusual formation tidal signals. Normally, formation response should reflect elastic deformation of the formation matrix under the influence of seafloor loading, with little or no phase between the loading and response. Given the compressibility of the sediments at Nankai (Bourlange et al., 2005), the amplitude of the formation pressure signal should be reduced to roughly 90% of that at the seafloor (Wang and Davis, 1996). In several instances, very unusual response is seen relative to these expectations. During part of the early history of recording at the deepest screen at Site 1173, when valves to higher screens were unintentionally left open, the amplitude of the formation tidal signal at the deepest screen actually exceeded that at the seafloor, and the sign was reversed (Fig. 6B). At Site 808, tidal signals at most of the screens are characterized by large phases and unreasonable levels of attenuation during most of the recording period (Fig. 6C). This behavior has been described in detail by Sawyer et al. who have attributed it to purely hydrologic interaction between the ACORK measurement system and the formation. Inexplicably large compliance in the ACORK plumbing and a large resistive "skin" effect at the screens are required to account for most of the behavior observed, however, and the signal amplification and sign reversal seen in Fig. 6b cannot be accounted for in this way at all.

High-frequency noise: Another noteworthy characteristic of the records from Site 808 is the persistence of high frequency pressure variations riding on the tidal signal that are highly correlated among multiple levels (Fig. 6C). The magnitude of this noise is consistently greatest for those screens that display the most anomalous tidal response. Whatever the source of this noise, it must act consistently over the full section of the hole to produce the coherence observed.

"Cross-talk": A third observation that provides key evidence for thermal expansion effects is of apparent cross-talk between certain screens at Site 1173. This is particularly well illustrated by the example at the end of the record shown in Fig. 6D, when a small-diameter geochemical sampling line was opened for testing. This line is terminated in screen 4, and as expected, pressure measured in the parallel monitoring line connected to that screen fell slightly when the sampling valve was opened. At the same time, pressures in the lines terminating at screens 1 and 3 rose by an amount roughly ten times greater than the drop registered at screen 4. A similar but seemingly opposite interaction was witnessed when the monitoring line valves were closed in 2002. Pressures at screen 1 dropped at the time of valve closure by 80 kPa to a seemingly inexplicable sub-hydrostatic level at the time the valves to other lines in the umbilical were closed.

Strain transients: Perhaps the most enigmatic observation is of impulsive pressure transients observed at Site 808

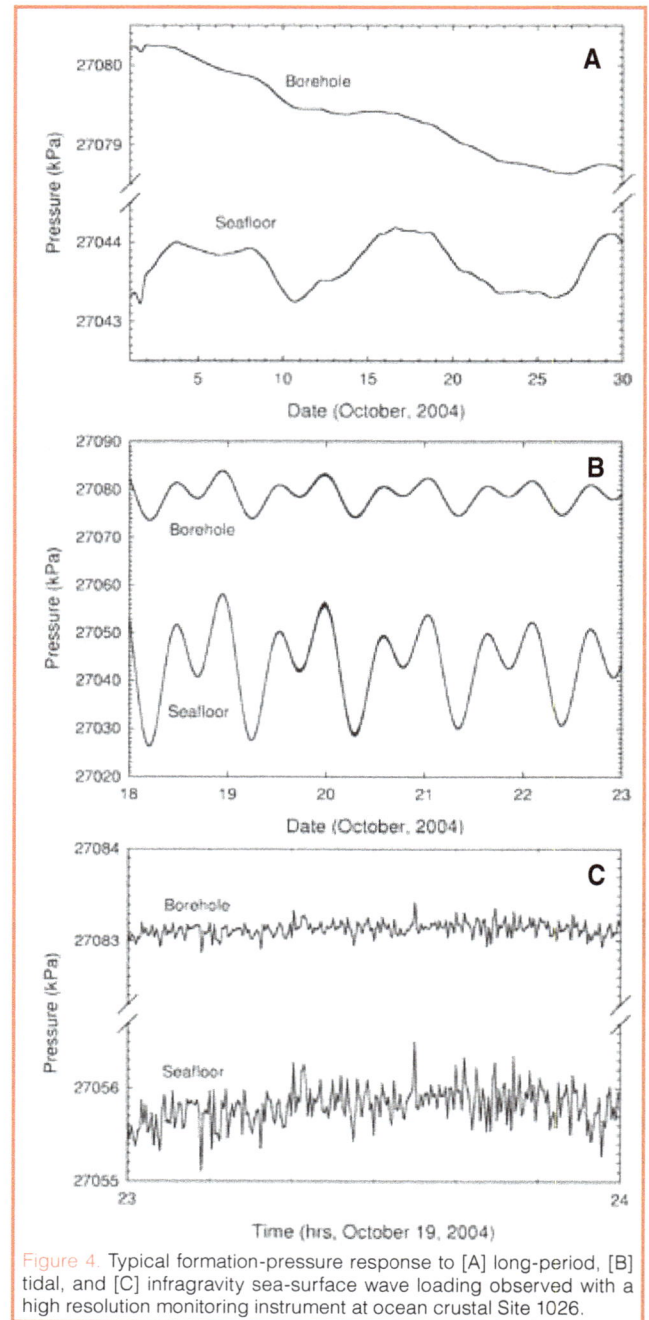

Figure 4. Typical formation-pressure response to [A] long-period, [B] tidal, and [C] infragravity sea-surface wave loading observed with a high resolution monitoring instrument at ocean crustal Site 1026.

at the times of two seismic events (Fig. 6E and 6F). The transients contrast with simultaneous step-wise changes in pressure observed at Site 1173 (e.g., Fig. 5B); the earlier transient pulse was initially interpreted to be a signature of near-field transient strain associated with slip on the decollement immediately beneath the observation screens (Davis et al., 2006). This interpretation is almost certainly wrong.

All of these unusual observations can be accounted for by a single mechanism—thermal expansion and contraction of water in the umbilical tubing and screens and in the formation surrounding the screens, caused by variable vertical fluid flow inside the large diameter casing (Site 808) and in the opened sampling/monitoring lines of the hydraulic umbilical (Site 1173). Given the large contrast between the

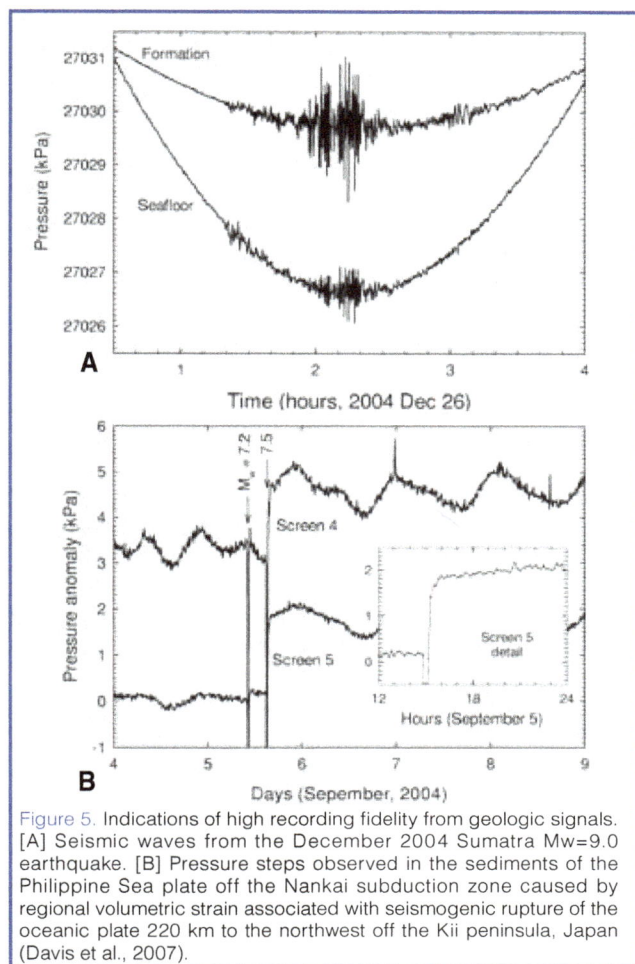

Figure 5. Indications of high recording fidelity from geologic signals. [A] Seismic waves from the December 2004 Sumatra Mw=9.0 earthquake. [B] Pressure steps observed in the sediments of the Philippine Sea plate off the Nankai subduction zone caused by regional volumetric strain associated with seismogenic rupture of the oceanic plate 220 km to the northwest off the Kii peninsula, Japan (Davis et al., 2007).

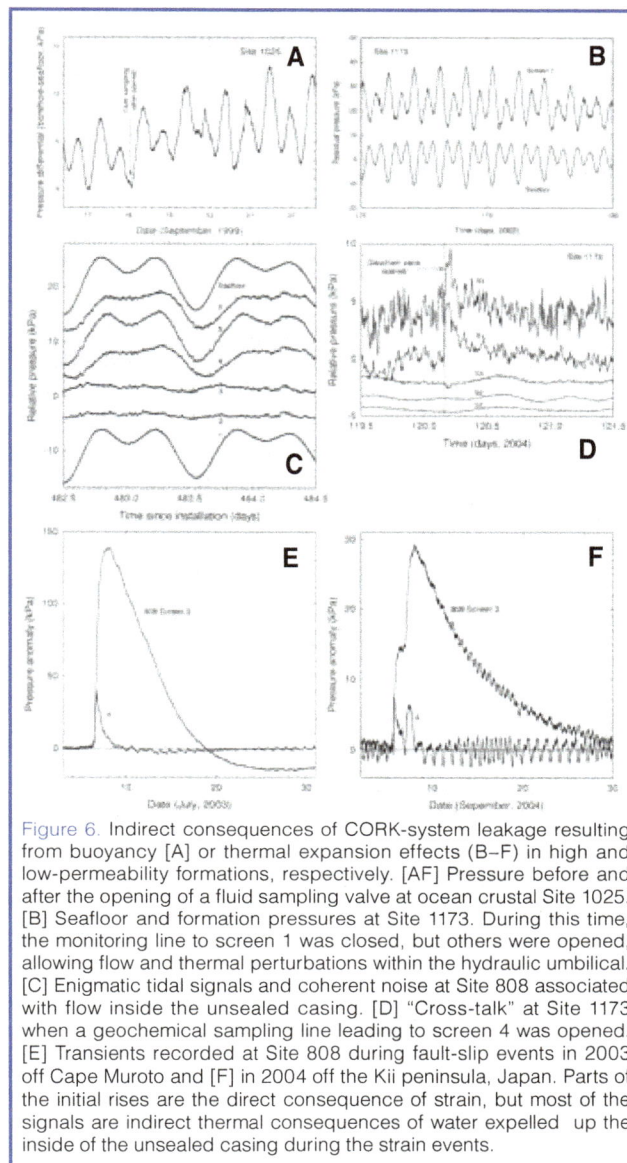

Figure 6. Indirect consequences of CORK-system leakage resulting from buoyancy [A] or thermal expansion effects (B–F) in high and low-permeability formations, respectively. [AF] Pressure before and after the opening of a fluid sampling valve at ocean crustal Site 1025. [B] Seafloor and formation pressures at Site 1173. During this time, the monitoring line to screen 1 was closed, but others were opened, allowing flow and thermal perturbations within the hydraulic umbilical. [C] Enigmatic tidal signals and coherent noise at Site 808 associated with flow inside the unsealed casing. [D] "Cross-talk" at Site 1173 when a geochemical sampling line leading to screen 4 was opened. [E] Transients recorded at Site 808 during fault-slip events in 2003 off Cape Muroto and [F] in 2004 off the Kii peninsula, Japan. Parts of the initial rises are the direct consequence of strain, but most of the signals are indirect thermal consequences of water expelled up the inside of the unsealed casing during the strain events.

thermal expansivity and volumetric compressibility of water ($\sim$0.4 x 10^{-3} K^{-1} and $\sim$0.4 x 10^{-9} Pa^{-1}, respectively), temperature sensitivity is inevitable wherever screens are well coupled to low-permeability sediments (like those of the lower Shikoku basin facies). The contrast yields a sensitivity of pressure to temperature of roughly 1 kPa mK^{-1}. Pressures may be produced with this efficiency if the temperature variation is coherent along the full distance from the screens to the sea-floor (an inevitable outcome of flow up or down the geothermal gradient), and if the sediment permeability at a given screen is sufficiently low to prohibit drainage at the rate of volumetric expansion. Whether this condition is met can be evaluated by considering the contrast between the thermal diffusivity ($\sim10^{-7}$ m^2 s^{-1}) and the hydraulic diffusivity. Thermal diffusion can indeed outpace hydrologic diffusion in sediments like those of the Lower Shikoku Basin facies, which have a hydraulic diffusivity of $\sim$2.5 × 10^{-9} m^2 s^{-1} (Bourlange et al., 2005; Gamage and Screaton, 2005; Table 1). Some inefficiency will result from thermal expansion of the tubing itself, and from the attenuation and time lag of thermal diffusion between the source of the thermal perturbation and the umbilical tube, screen, or sediment where the consequences of expansion are observed. The first factor is insignificant, given the large contrast in expansivity between water and steel. The importance of the second will vary with the frequency of the signal and with the time constant for the

thermal path (e.g., $\sim$1 hr for conduction between the inside of the 10.75" casing and the formation, screens, or umbilical outside, and $\sim$1 min for conduction between tubes within the umbilical).

Summary and Recommendations

The fidelity of CORK formation pressure records appears to be high over a broad frequency range. After pressure transducers equilibrate to ambient pressure, long-term drift is typically about 0.1 kPa yr^{-1}. Response at high measurement frequencies (a few Pa at 1 Hz) is good in high permeability formations, but CORK-system compliance may influence the fidelity of observations in very-low-permeability lithologies, even with the low-system-volume advantage offered by ACORK configurations.

Several lessons have been learned through inevitable installation errors, and these should be highlighted.

1) Consequences of leakage can degrade fidelity over a broad frequency range and they can be difficult to discriminate from real signals. A good example of this is provided by impulse transients witnessed at two strain events at the Nankai accretionary prism (Figs. 6E, 6F) which are now believed to be the result of umbilical- and formation-fluid thermal expansion caused by strain-induced flow inside the unsealed casing.

2) Leakage can be efficiently assessed by the measurement of temperature. At Nankai, installation of temperature sensors was planned, but was precluded by the same operational problem that prevented the installation of an internal casing seal. Great care must be used in the future to prevent and detect leaks.

3) Entrained CORK fluid volume should be minimized and screen area maximized to enhance the inherent system fidelity by reducing the time constant associated with the coupling between the measurement system and the formation. This will have the added benefit of reducing the geochemically perturbing effects of "foreign" water.

Acknowledgments

We thank the ODP and the captains and crew of the *JOIDES Resolution* for ongoing support for long-term borehole monitoring experiments. Appreciation is also given to T. Pettigrew, R. Meldrum, and R. Macdonald for engineering assistance. We are grateful to the pilots and crews of the submersibles *Alvin, Nautile, and Shinkai 6500* and remotely-operated vehicles *ROPOS, Jason, Kaiko 12k*, and *Kaiko 7k* for instrument installations and data downloads. We would also extend our gratitude to the captains and crews of the respective support vessels for their multi-faceted support during site visits. Financial support has been provided by the U.S. National Science Foundation (NSF), the Geological Survey of Canada (GSC), and the Japanese Center for Marine Science and Technology (JAMSTEC).

References

Bredehoeft, J.D., and Papadopulos, S.S., 1980. A method for determining the hydraulic properties of tight formations. *Water Resources Res.*, 16:223–238.

Becker, K., and Davis, E.E., 2005. A review of CORK designs and operations during the Ocean Drilling Program. *In* Fisher, A.T., Urabe, T., Klaus, A., and the Exp. 301 scientists (Eds.), *Proc. IODP, Init. Repts. 301*. College Station, Texas (IODP), 1–28 (DVD).

Bourlange, S., Jouniaux, L., and Henry, P., 2005. Data report: Permeability, compressibility, and friction coefficient measurements under confining pressure and strain, Leg 190, Nankai Trough. *In* Mikada, H., Moore, G.F., Taira, A., Becker, K., Moore, J.C., and Klaus, A. (Eds.), *Proc. ODP Sci. Results 190/196*, College Station, Texas (IODP), 1–16 (DVD).

Davis, E.E., and Becker, K., 2004. Observations of temperature and pressure: Constraints on ocean crustal hydrologic state, properties, and flow. *In* Davis, E.E., and Elderfield, H. (Eds.), *Hydrogeology of the Oceanic Lithosphere*, Australia (Cambridge University Press), 225–271.

Davis, E.E., Becker, K., Pettigrew, T., Carson, B., and MacDonald, R., 1992. CORK: A hydrologic seal and downhole observatory for deep ocean boreholes. *In* Davis, E.E., Mottl, M.J., and Fisher, A.T. (Eds.), *Proc. ODP Init. Rep. 139*, College Station, Texas (ODP), 43–53.

Davis, E.E., Becker, K., Wang, K., Obara, K., Ito, Y., and Kinoshita, M., 2006. A discrete episode of seismic and aseismic deformation of the Nankai subduction zone accretionary prism and incoming Philippine Sea plate. *Earth Planet. Sci. Lett.*, 242:73–84, doi:10.1016/j.epsl.2005.11.054.

Davis, E.E., Wang, K., Becker, K., and Kinoshita, M., 2007. Co- and post-seismic crustal contraction and fault-zone dilatation, Nankai subduction zone. *Nature*, in press.

Fisher, A.T., Wheat, G.C., Becker, K., Davis, E., Hannasch, H., Schroeder, D., Dixon, R., Pettigrew, T., Meldrum, R., Macdonald, R., Nielsen, M., Fisk, M., Cowen, J., Bach, W., and Edwards, K., 2005. Scientific and technical design and deployment of long-term subseafloor oberatories for hydrogeologic and related experiments, IODP Expedition 301, eastern flank of Juan de Fuca Ridge. *In* Fisher, A.T., Urabe, T., Klaus, A., and the Expedition 301 Scientists, Proc. IODP, 301: College Station, Texas (IODP-MI, Inc.). doi:10.2204/iodp.proc.301.103.2005.

Gamage, K., and Screaton, E., 2005. Data report: Permeabilities of Nankai accretionary prism sediments. *In* Midaka, H., Moore, G.F., Taira, A., Becker, K., Moore, J.C., and Klaus, A. (Eds.), *Proc. ODP Sci. Res. 190/196*, College Station, Texas (ODP), 1–22 (DVD).

Jannasch, H.W., Davis, E.E., Kastner, M., Morris, J.D., Pettigrew, T.L., Plant, J.N., Solomon, E.A., Villinger, H.W., and Wheat, C.G., 2003. CORK II: Long-term monitoring of fluid chemistry, fluxes, and hydrology in instrumented boreholes at the Costa Rica subduction zone. *In* J.D. Morris, H.W. Villinger, A. Klaus, et al. (Eds.), *Proc. ODP, Init. Rep. 205*, College Station, Texas (ODP).

Kastner, M., Becker, K., Davis, E.E., Fisher, A.T., Jannasch, J.W., Solomon, E.A., and Wheat, C.G., 2007. New insights into the hydrogeology of the oceanic crust through long-term monitoring. *Oceanogr.*, 19:46–57.

Mikada, H., Becker, K., Moore, J.C., Klaus, A., and the Shipboard Scientific Party, 2002. Deformation and fluid flow processes in the Nankai Trough accretionary prism: Logging while drilling and advance CORKs. *Proc. ODP, Init. Rep. 196*, College Station, Texas (ODP).

Sawyer, A.H., Flemings, P.B., Elsworth, D.E., and Kinoshita, M., in press, Response of Submarine Hydrologic Monitoring Instruments to Formation Pressure Changes: Theory and Application to Nankai ACORKs, Journal of Geophysical research.

Wang, K., and Davis, E.E., 1996. Theory for the propagation of tidally induced pore pressure variations in layered subseafloor formations. *J. Geophys. Res.*, 101:11483–11495, doi:10.1029/96JB00641.

Authors

Earl Davis, PGC, Geological Survey of Canada, 9860 West Saanich Road, North Saanich, Sidney, British Columbia V8L 4B2, Canada, e-mail: edavis@nrcan.gc.ca.

Keir Becker, RSMAS/MGG, University of Miami, 4600 Rickenbacker Causeway, Miami, Fla. 33149-1098, U.S.A.

10

Aurora Borealis – Development of a New Research Icebreaker with Drilling Capability

by Nicole Biebow and Jörn Thiede

Introduction

The International Polar Year (IPY), with its attempts to coordinate and foster cooperation on an international level in an unprecedented way, offers a unique chance for a leap of progress in our understanding of polar processes and their dynamics with their influence on the adjacent continents and the global environment. However, polar research both on land and in the sea cannot achieve the progress needed without novel and state of the art technologies and infrastructure.

There are many novel tools presently being developed for polar research. In this report we will concentrate on the planning for a new research icebreaker, *Aurora Borealis* (Fig. 1), with an all-season capability of endurance in permanently ice-covered waters and with the possibility to carry out deep-sea drilling in ice-covered basins.

Scientific Relevance of the *Aurora Borealis* Project

Polar research and, in particular, the properties of northern and southern high latitude oceans are currently a subject of intense scientific debate and investigations, because they are (in real time) and have been (over historic and geologic time scales) subject to rapid and dramatic climatic variations. Polar regions react more rapidly and intensively to global change than other regions of the Earth. Examples of these modern changes include news about shrinking of the Arctic sea-ice cover (potentially leading to an opening of sea passages to the north of North America and Eurasia, and in the long run to a "blue" Arctic Ocean) and news about the calving of giant table icebergs from the ice shelves of Antarctica. Until now it has not been clear how many of these profound shifts in all parts of the Arctic are natural fluctuations or are due to human activity. Since this is a phenomenon occurring over decades, long time data series of atmospheric and oceanic conditions are needed for its understanding and prediction of its further development.

Global climate models demonstrate the importance of the polar areas in forcing of the ocean/climate system. The presence or absence of snow and ice influences global heat distribution through its effect on the albedo, and the polar oceans are the source of dense, cold bottom waters, which influence thermohaline circulation in the world oceans. This global conveyor is a major determinant of Earth's climate.

Despite the strong seasonality of polar environmental conditions, research in the central Arctic Ocean up to now could essentially only be conducted during the summer months, when the Arctic Ocean is accessible only by the strongest research icebreakers.

In spite of the critical role of the Arctic Ocean in climate control, it is the only sub-basin of the world's oceans that has essentially not been sampled by the drill ships of the Deep-Sea Drilling Project (DSDP) or the Ocean Drilling Program (ODP), and its long-term environmental history and tectonic structure is therefore poorly understood. Exceptions are the ODP Leg 151 and the more recent Integrated Ocean Drilling Program's (IODP) Expedition 302 (Arctic Coring Expedition, ACEX, within the central Arctic; Myhre et al., 1995; Moran et al., 2006). The lack of data represents one of the largest gaps of information in modern Earth science (Nansen Arctic Drilling Program, 1992, 1997), also relevant for the field of hydrocarbon exploration. Therefore, the new research icebreaker *Aurora Borealis* (Fig. 1) should be equipped with proper drilling facilities to drill in deep, permanently ice-covered ocean basins. The icebreaker must also be powerful enough to keep station against the drifting sea-ice cover and will have to be equipped with dynamic positioning.

The *Aurora Borealis* project impacts on two scientific communities which in part overlap in interests. The first one is

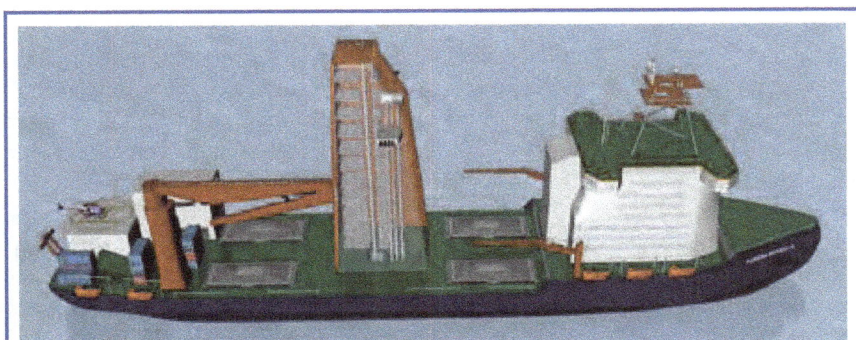

Figure 1. Initial design of the Aurora Borealis

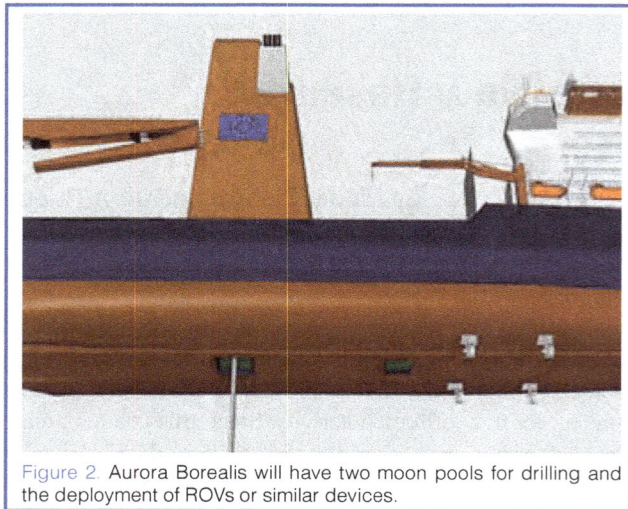

Figure 2. Aurora Borealis will have two moon pools for drilling and the deployment of ROVs or similar devices.

the general polar science community that requires a ship for conducting year-round field and marine work and has a wide spectrum of scientific perspectives. The second is the deep-sea drilling community that would use the ship mainly during the summer months with optimal ice conditions to study the structure and properties of the crust below the Arctic Ocean and to unravel the history of environmental and climate changes. While the ACEX expedition in 2004 is the only case of high Arctic drilling, substantial progress has been made around Antarctica by the drilling platforms of the DSDP and ODP during ice-free seasons. Also, deployment of small drill rigs from sea-ice very close to shore (ANDRILL, Cape-Roberts-Project) and shallow drilling from the ice-breaker *R/V Nathaniel B. Palmer* (SHALDRILL) have taken place.

The scientific objectives of the *Aurora Borealis* project are outlined in the "Science Perspective", published under the same name by the European Polar Board (EPB) of the European Science Foundation (ESF) in collaboration with ECORD (European Consortium of Ocean Research Drilling). A detailed accounting of the scientific objectives and research prospects can be found in these documents (Thiede and Egerton, 2004).

Technical Details

The research icebreaker *Aurora Borealis* will be the most advanced polar research vessel in the world with a multi-functional role of drilling in deep ocean basins and supporting climate/environmental research and decision support for stakeholder governments for the next few decades. The new technological features will include azimuth propulsion systems, satellite navigation, ice-management support, and the deployment and operation of remotely operated vehicles (ROVs) and autonomous underwater vehicles (AUVs) from the twin moon pools (Fig. 2). The most unique feature of the vessel is the deep drilling rig, which will enable sampling of the ocean floor and sub-sea in up to 4000 m of water and with 1000 m penetration at the most inhospitable places on Earth.

In the long term the drilling capability will be deployed in both polar regions, and *Aurora Borealis* will be the only vessel worldwide that could undertake this type of scientific investigation. The possibility to flexibly equip the ship with laboratory and supply containers, and the variable arrangement of other modular infrastructure (in particular, winches, cranes, etc.), free deck-space, and separate protected deck areas, will allow the planned research vessel to cover the needs of most disciplines in marine research, including the capability to carry out geophysical investigations (seismic reflection and refraction, gravity, magnetic, swath bathymetry mapping system, sediment echo sounder). The ship can be deployed as a research icebreaker in polar seas because it will meet the specifications of the highest ice-class for polar icebreakers. The vessel will be a powerful research ice-breaker with 44,000 tons displacement and a length of 196 m, with 50-megawatt azimuth propulsion systems. It will have high ice performance to penetrate autonomously (single ship operation) into the central Arctic Ocean with 2.5 meters of ice cover, during all seasons of the year. A large fuel capacity is required because of the excessive power requirements for drilling and maintaining station in the central Arctic (or other severely ice covered waters) during what are envisaged to be long expeditions. This factor is decisive for the large size of the ship. The construction of *Aurora Borealis* requires several new technical solutions and will provide an extended technical potential and knowledge for marine technologies and the ship building industry

Perspectives of the *Aurora Borealis* Project

Many northern nations have a particular interest in understanding the Arctic environment with its high potential for environmental change in response to global warning. In addition, considerable living and non-living resources are likely to be found below the Arctic Ocean and its adjacent continental margins. However, modern research vessels capable of penetrating into the central Arctic are few and mostly inadequate. Therefore, a new state-of-the-art research icebreaker is urgently required to fulfill the needs of polar research. This new icebreaker would be conceived as an optimized and multi-national science platform from the keel up and will allow long international and interdisciplinary expeditions into the central Arctic Ocean during all seasons of the year.

An efficient use of this icebreaker requires the formation of a consortium of several countries and a substantial build-up of their polar research institutions to ensure an efficient employment of the research vessel during all seasons of the year. Extensive and well-developed Arctic research programs exist in several countries, particularly in the Scandinavian countries, the U.S.A., Russia, and Germany. Each country has different organizations or working groups with rather diverse structures. The construction of *Aurora Borealis* as a joint European/international research ice-breaker would result from a considerable commitment of the participating nations to coordinate and expand their polar research programs to operate this expensive ship continuously and with the necessary efficiency. *Aurora Borealis* would contribute to meet the Arctic drilling challenge within IODP; however, in a long-term perspective the *Aurora Borealis* would also be used to address Antarctic research targets, both in its mode as a regular research vessel as well as a polar drill ship.

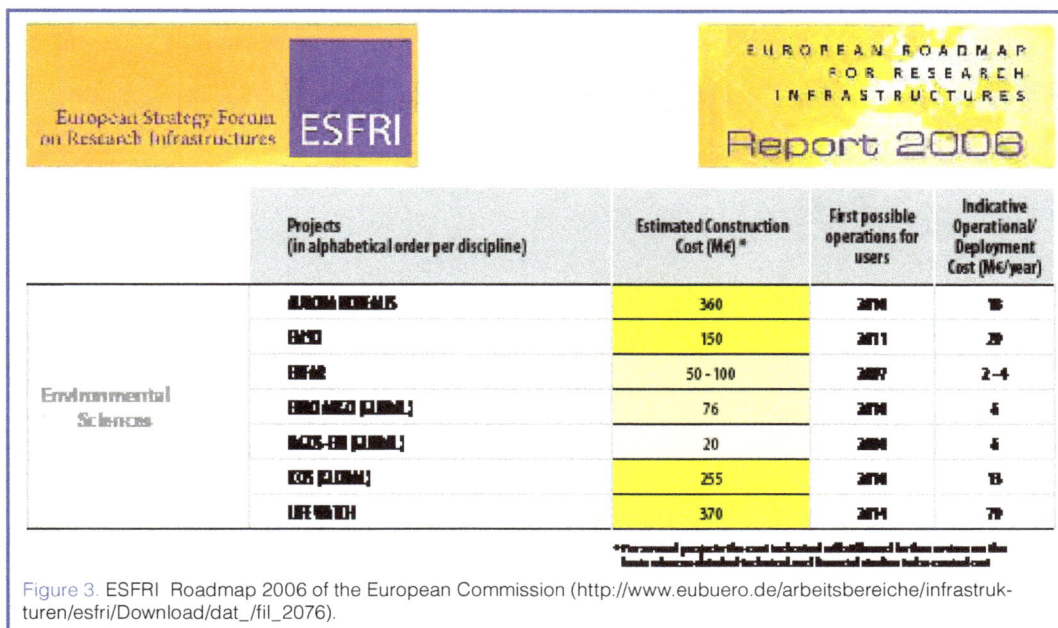

Figure 3. ESFRI Roadmap 2006 of the European Commission (http://www.eubuero.de/arbeitsbereiche/infrastrukturen/esfri/Download/dat_/fil_2076).

The German Science Council evaluated the *Aurora Borealis* project in May 2005 and recommended the construction of the research icebreaker in 2006. Since March 2007 the German Federal Ministry for Science and Education (BMBF) has been funding a portion of the preparatory work for *Aurora Borealis*. In this project the final engineering work for the development of the vessel is carried out under coordination of the Alfred-Wegener-Institute and the University of Applied Sciences in Bremen. Additionally, the engagement of the European science community will be promoted by organizing workshops in different European countries to discuss science plans and technical requirements for the *Aurora Borealis*.

The European Commission identified the project for the European Strategy Forum on Research Infrastructures (ESFRI) Roadmap (Fig. 3). The Commission found that it reached the highest scientific priority for developing this large-scale infrastructure for basic research in Europe. A European consortium of sixteen institutions, funding agencies, and companies from eleven European nations including Russia has already formed to develop management structures for this unique facility and to implement it into the European Research Area.

References

Moran, K., Backman, J., Brinkhuis, H., Clemens, S.C., Cronin, T., Dickens, G.R., Eynaud, F., Gattacceca, J., Jakobsson, M., Jordan, R.W., Kaminski, M., King, J., Koc, N., Krylov, A., Martinez, N., Matthiessen, J., McInroy, D., Moore, T.C., Onodera, J., O'Regan, A.M., Pälike, H., Rea, B., Rio, D., Sakamoto, T., Smith, D.C., Stein, R., St. John, K., Suto, I., Suzuki, N., Takahashi, K., Watanabe, M., Yamamoto, M.,

Frank, M., Jokat, W., and Kristoffersen, Y., 2006. The Cenozoic palaeoenvironment of the Arctic Ocean. *Nature*, 441:601–605, doi:10.1038/nature04800.

Myhre, A.M., Thiede, J., Firth, J.V., Ahagon, N., Black, K.S., Bloemendal, J., Brass, G.W., Bristow, J.F., Chow, N., Cremer, M., Davis, L., Flower, B.P., Fronval, T., Hood, J., Hull, D.M., Koc, N., Larsen, B., Lyle, M.W., McManus, J., O'Connell, S., Osterman, L.E., Rack, F.R., Sato, T., Scherer, R.P., Spiegler, D., Stein, R., Tadross, M., Wells, S., Williamson, D., Witte, B., Wolf-Welling, T., Marin, J.A., 1995. Underway geophysics. *Proc. Ocean Drill. Prog. Init. Repts.*, 151:47–48.

Nansen Arctic Drilling Program, 1992. The Arctic Ocean record: key to global change (Initial Science Plan). *Polarforschung* 61(1):1–102.

Nansen Arctic Drilling Program, 1997. *An implementation plan for the Nansen Arctic Drilling Program*. Washington, DC (Joint Oceanographic Institutions), 42 pp.

Thiede, J., and Egerton, P. 2004. *Aurora Borealis: a Long-term European Science Perspective for Deep Arctic Ocean Research 2006–2016*. Strasbourg (European Science Foundation), 80 pp.

Authors

Nicole Biebow, Alfred Wegener Institute (AWI), Am Handelshafen 12, D-27570, Bremerhaven, (Building E-3335), Germany, e-mail:Nicole.Biebow@awi.de.

Jörn Thiede, Alfred Wegener Institute (AWI), Am Handelshafen 12, D-27570, Bremerhaven, (Building E-3221), Germany.

Related Web Link

http://www.eubuero.de/arbeitsbereiche/infrastrukturen/esfri/Download/dat_/fil_2076

Figure Credits

Figs. 1 and 2 Graphics by Delgado / Michaelis, AWI

11

Scientific Drilling with the Sea Floor Drill Rig MeBo

by Tim Freudenthal and Gerold Wefer

Introduction

In March 2007 the sea floor drill rig MeBo (short for "**Me**eresboden-**Bo**hrgerät", 'sea floor drill rig' in German) returned from a 17-day scientific cruise with the new German research vessel *Maria S. Merian*. Four sites between 350 m and 1700 m water depth were sampled at the continental slope off Morocco by push coring and rotary drilling. Up to 41.5-m-long sediment cores were recovered from Miocene, Pliocene, and Pleistocene marls. MeBo bridges the gap between conventional sampling methods from standard multipurpose research vessels (gravity corer, piston corer, dredges) and drill ships. Most bigger research vessels will be able to support deployment of the MeBo. Since the drill system can be easily transported within 20-ft containers, worldwide operation from vessels of opportunity is possible. With the MeBo a new system is available for marine geosciences that allows the recovery of high quality samples from soft sediments and hard rock from the deep sea without relying on the services of expensive drilling vessels.

Rationale

A variety of research targets in marine sciences—including gas hydrates, mud mounds and mud volcanoes, ore formation, and paleoclimate—can be addressed by shallow drilling (30–100 m below sea floor) in the deep sea (Quinn and Mountain, 2000; Sager et al., 2003; Herzig et al., 2003). In general, standard sampling tools like gravity corers or dredges only allow recovery of fairly short cores from soft sediments or fragments of bedrock lying on the sea surface. Drill ships providing deeper penetration are expensive and typically booked far in advance, if available at all; therefore at the Marum Center for Marine Environmental Sciences (Marum) at the University of Bremen we developed the drill rig MeBo (Freudenthal and Wefer, 2006) that can be deployed from standard research vessels.

System Concept

The MeBo is deployed on the sea bed and is remotely controlled from the vessel (Fig.1). The rig is lowered to the sea bed using a steel-armored umbilical with a diameter of 32 mm. The deployment depth is currently limited to a maximum of 2000 m below sea level by the length and strength of the umbilical. Four legs are extended before landing to increase the stability of the rig (Fig. 2). Copper wires and fiber optic cables within the umbilical are used for energy supply from the vessel and for communication between the MeBo and the control unit on the deck of the vessel, respectively. The drill rig is powered by four hydraulic pumps that are driven with electric motors. A variety of sensors, video cameras, and lights are used for monitoring the drill performance.

The mast with the feeding system forms the central part of the drill rig (Fig. 2). The drill head provides the required torque and rotary speed for rock drilling; it is mounted on a guide carriage that moves up and down the mast with a maximum push force of 4 tons. A water pump provides sea water for flushing the drill string, for cooling of the drill bit, and for removing the drill cuttings. The system utilizes commercial rotary core barrels with diamond or tungsten carbide bits. In addition it can push core barrels into soft formations, and case the boreholes.

The MeBo stores drilling rods, casing tubes, and push-coring and rotary barrels on two rotating magazines that

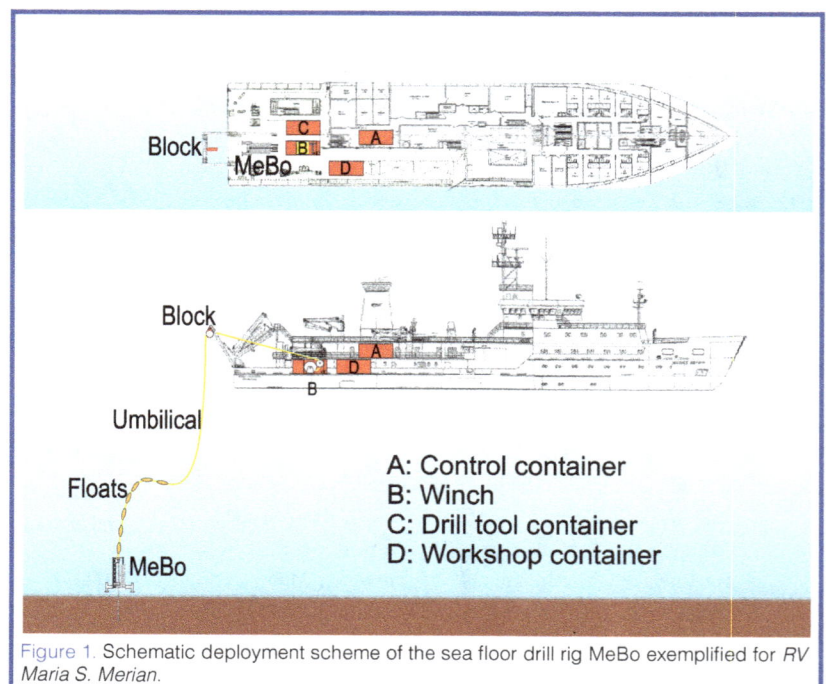

Figure 1. Schematic deployment scheme of the sea floor drill rig MeBo exemplified for *RV Maria S. Merian*.

A: Control container
B: Winch
C: Drill tool container
D: Workshop container

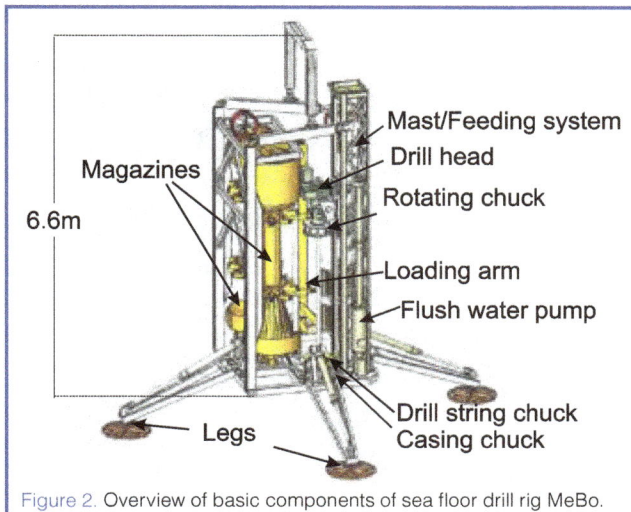

Figure 2. Overview of basic components of sea floor drill rig MeBo.

may be loaded with a mixture of tools as required for a specific task. The loading arm is used in combination with one rotating and two fixed chucks for building the drill string up and down. When drilling is started, a core barrel is taken off the magazine and loaded below the drill head. After the thread connection between barrel head and drill head is closed, the first three meters can be sampled. The barrel is then stored together with the drilled core in the magazine, and the next empty core barrel is lowered into the drilled hole. A 3-m rod is added, and the next three meters can be sampled. With a storing capacity of 17 barrels, 16 rods and 15 casing tubes, the MeBo has the capability to drill up to 50 m into the sea floor, to recover cores with 74–84 mm diameter, and to stabilize the drilled hole down to a depth of 40 m.

Easy transportation was a key requirement for the system design, since the MeBo is deployed worldwide from vessels of opportunity (Fig. 3). All parts belonging to the MeBo system, including control unit, workshop, a rack for the drill tools, and the winch, are containerized (Fig. 3). The MeBo unit fits into an open top high cube 20-ft container when the four legs are retracted. In addition, a launch and recovery system that was developed for the deployment of the MeBo from the German research vessel *Meteor* can be adapted to the deck configuration of other vessels. It, too, fits into a 20-ft open top container when dismantled. Altogether, the complete MeBo system is shipped within six 20-ft containers.

Development and Tests

Previous developments of sea bed drill rigs proved the advantages of drilling from a stable platform at the sea floor but demonstrated also the challenges of remotely controlled drilling and drill string handling (Johnson, 1991; Stuart, 2004). Only a few prototype systems of sea bed drill rigs, including the Rockdrill from the British Geological Survey, the Japanese BMS, and the Australian PROD (Portable Remotely Operated Drill), are operated worldwide, with only the PROD system capable of reaching drilling depths of more than 30 m. Since this system is dedicated to operations in the

marine geotechnical survey industry (Stuart, 2004), we decided to establish a new development—the sea floor drill rig MeBo—optimized for the requirements and facilities for scientific investigations of the sea floor.

The development was realized at the Marum with support from a variety of national and international companies. Next to overall system design, project management, and system integration, the Marum Center was responsible for the sensors as well as control hardware and software of the system. Most of the mechanical and hydraulic parts of the drill rig were developed, manufactured, and assembled by Prakla Bohrtechnik GmbH (Peine, Germany), a company specializing in the fabrication of multi-purpose land drill rigs for core drilling and well installation. The energy supply and telemetry system were designed and delivered by Schilling Robotics (Davis, California), Hogenkamp (Lilienthal, Germany), and STA GmbH (Bremerhaven, Germany). Schilling Robotics also developed the rotating magazines and the loading arm. The umbilical was designed and manufactured by Norddeutsche Seekabelwerke (Nordenham, Germany), while the winch was delivered by MacArtney (Esbjerg, Denmark).

After an engineering and construction phase of about one year, the system was tested successfully in July/August 2005 in deep water at the continental slope off Morocco with the German research vessel *Meteor* (Wefer et al., 2006). Within the following year, two further expeditions were conducted with the Irish research vessel *Celtic Explorer*: one for shallow

Figure 3. View of the work deck of *RV Meteor* during a deployment of the sea floor drill rig MeBo.

water tests of the MeBo system in the Baltic Sea, and a first scientific cruise on the Porcupine Bank west of Ireland for the Irish Shelf Petrol Studies Group (ISPSG) of the Irish Petroleum Infrastructure Programme Group 4.

During these two cruises, the MeBo was deployed twenty times between 20 m and 1700 m. Push coring for soft sediments and rotary drilling for hard rocks, as well as stabilization of the drilled hole by setting casings, were successfully accomplished. Altogether, 127 meters were drilled in sand, gravel, marl, till, conglomerates, breccia, granite, and gneiss, and 57.2 meters of core were recovered (Fig. 4). The recovery rate was especially good for hard rocks and consolidated cohesive sediments.

Figure 4. Examples of sea floor samples recovered with the sea floor drill rig MeBo. [A] consolidated pliocene marl, continental slope off Morocco; [B] granite, Porcupine Bank; [C] conglomerate, Porcupine Bank; [D] gneiss, Porcupine Bank.

Scientific Results of MSM04/4

In March 2007 the MeBo was deployed for the first time for paleoclimate research on the research vessel *Maria S. Merian* at the continental slope off Morocco (Fig. 5) with two major goals:

Figure 5. Launch of the sea floor drill rig MeBo from the *Maria S. Merian*.

1) Two sites with extremely high sedimentation rates of more than 80 cm kyr^{-1} were cored for high resolution records of abrupt climate changes during the last glacial period.

2) Two other sites were drilled at and near the former DSDP drilling site 369. A big slide at this part of the continental slope of Morocco allowed for direct sampling of sediments of Miocene age by shallow drilling.

Altogether we recovered approximately 154 meters of sediments at these four sites with a maximum drilling depth of 41.55 m and about 120 m core recovery. A highlight at the last drilling site was to recover within forty-eight hours a nearly 40-m-long sediment core of middle to late Miocene age, according to first shipboard nannofossil stratigraphy (Fig. 6). This time period is of special interest for paleoclimate research because it comprises large changes in carbon burial in deep-sea sediments as well as a major step in the formation of the East Antarctic continental ice shield (Vincent and Berger, 1985; Holbourne et al., 2005). Core recovery rate at this site (1720 m water depth) by rotary drilling with hard metal bits was about 100%. Without the services of the MeBo, it wouldn't have been possible to sample these Miocene consolidated marls from a standard research vessel like *RV Maria S. Merian*.

Outlook

With the development of the MeBo system, a substantial improvement of the sampling possibilities for the marine geosciences was achieved. For shallow drilling and coring, the MeBo provides a new and cost effective alternative to the services of drill ships. Worldwide, it is the only system available for marine geosciences that can reach drilling depths of up to 50 m from standard research vessels. MeBo also has the major advantage that the drilling operations are performed from a stable platform independent of any ship movements caused by waves, wind, or currents. Currently,

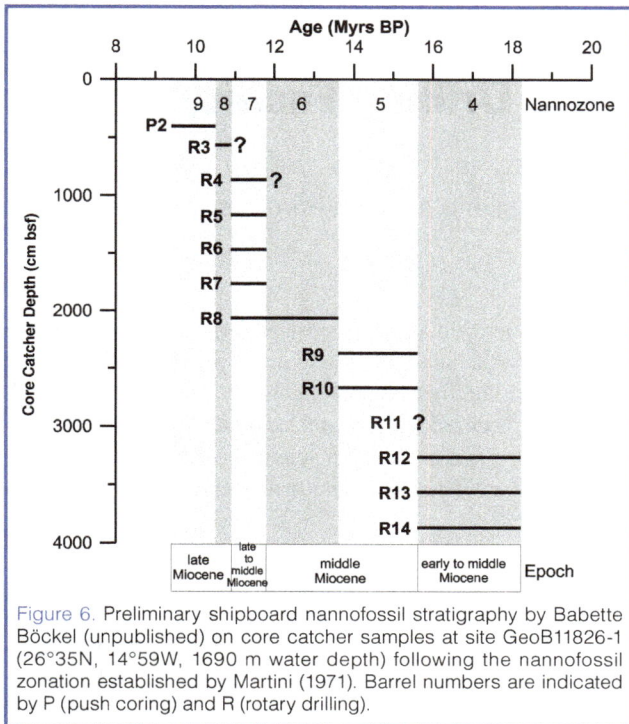

Figure 6. Preliminary shipboard nannofossil stratigraphy by Babette Böckel (unpublished) on core catcher samples at site GeoB11826-1 (26°35N, 14°59W, 1690 m water depth) following the nannofossil zonation established by Martini (1971). Barrel numbers are indicated by P (push coring) and R (rotary drilling).

we are working together with Prakla Bohrtechnik GmbH to upgrade the system to use wire-line coring system, which will accelerate the drilling procedure and allow the deployment of logging tools for in-hole data logging. Further plans include the development of a pressure core barrel for the recovery of sediments enriched in gas under *in situ* pressure conditions, as well as the development of sensor strings to be deployed with the MeBo for long-term monitoring of hydraulic processes and heat flow in the sea floor.

Acknowledgements

The development of the MeBo was funded by the German Federal Ministry of Education and Research and by the State Government of Bremen. The technicians and engineers at the Marum Center and cooperating companies accomplished an extraordinary performance in getting the MeBo-system running within a short development time. We thank the Leitstelle Meteor/Merian and the Irish Marine Institute for their expedition support and the crews of *R/V Meteor, R/V Celtic Explorer* and *R/V Maria S. Merian* for their support during the various deployments of the MeBo. The German Research Foundation (DFG) and the DFG Research Center Ocean Margins (RCOM) supported MeBo-cruises M65/3, CE0511, and MSM04/4. We thank the Irish Shelf Petrol Studies Group of the Irish Petroleum Infrastructure Programme Group 4 (comprising Chevron Upstream Europe, ENI *Irelandd* BV, Island Oil & Gas plc, Lundin Exploration BV,, Providene Resources plc, Ramco Energy plc, Shell E&P Ireland Ltd, Statoil Exploration (Ireland) Ltd, Total E&P UK plc and the Petroleum Affairs Division of the Department of Communications, Marine and Natural Resources) for their faith in the MeBo system, for being the first scientific user of the MeBo, and for financing its deployment on the Porcupine Bank (CE0619).

References

Freudenthal, T., and Wefer, G., 2006. The sea-floor drill rig "MeBo": robotic retrieval of marine sediment cores. *PAGES News*, 14(1):10.

Herzig, P.M., Petersen, S., Kuhn, T., Hannington, M.D., Gemmell, J.B., Skinner, A.C., and SO-166 Shipboard Scientific and Technical Party, 2003. Shallow drilling of seafloor hydrothermal systems using R/V Sonne and the BGS Rockdrill: Conical Seamount (New Ireland Fore-Arc) and Pacmanus (Eastern Manus Basin), Papua New Guinea. *InterRidge News*, 12(1):22–26.

Holbourne, A., Kuhnt, W., Schulz, M., and Erlenkeuser, H., 2005. Impacts of orbital forcing and atmospheric carbon dioxide on Miocene ice-sheet expansion. *Nature*, 438:483–487, doi:10.1038/nature04123.

Johnson, H.P., 1991. Next generation of seafloor samplers. *EOS*, 72(7):65–67, doi:10.1029/90EO00045.

Martini, E., 1971. Standard Tertiary and Quaternary calcareous nannoplankton zonation, *In* Farinaci, A. (Ed.), *Proceedings of the Second Planktonic Conference Roma 1970,* Rome (Edizioni Tecnoscienza), 739–785.

Quinn, T.M., and Mountain, G.S., 2000. Shallow water science and ocean drilling face challenges. *EOS*, 81(35):397–400, doi:10.1029/00EO00293.

Sager, W., Dick, H., Fryer, P., and Johnson, H.P., 2003. Report from a workshop. Requirements for robotic underwater drills in U.S. marine geologic research., Texas A&M University, College Station, *Texas, 3–4 November 2000.* http://www. usssp-iodp.org/PDFs/DrillRep51403.pdf.

Stuart, S., 2004. The remote robot alternative. *Intl. Ocean Syst.*, 8(1):23–25.

Vincent, E., and Berger, W.H., 1985. Carbon dioxide and polar cooling in the Miocene: The Monterey hypothesis. *In* Sundquist,, E.T., and Broecker, W.S. (Eds.), *The carbon cycle and atmospheric CO2: natural variations Archean to present, AGU Geophysical Monographs* 32, 455–468.

Wefer, G., Bergenthal, M., Buhmann, S., Diekamp, V., Düßmann, R., Engemann, G., Freudenthal, T., Hemsing, V., Hill, H.-G., Kalweit, H., Klar, S., Könnecker, H.-O., Lunk, T., Renken, J., Rosiak, U., Schmidt, W., Truscheit, T., and Warnke, K., 2006. Report and preliminary results of *R/V METEOR* cruise M65/3, Las Palmas – Las Palmas (Spain) 31 July–10 August 2005, *Berichte, Fachbereich Geowissenschaften, Universität Bremen*, 254: 24p.

Authors

Tim Freudenthal, Marum Center for Marine Environmental Sciences, University of Bremen, Leobener Str. D-28359 Bremen, Germany, e-mail: freuden@marum.de.

Gerold Wefer, Marum Center for Marine Environmental Sciences, University of Bremen, Leobener Str. D-28359 Bremen, Germany.

Related Web Link

http://www.rcom.marum.de/English/Sea_floor_drill_rig_MeBo.html

12

InnovaRig – The New Scientific Land Drilling Facility

by Lothar Wohlgemuth, Ulrich Harms, and Jürgen Binder

Introduction

Deep drilling is becoming an increasingly important tool to study fundamental processes at depth, such as earthquake nucleation in fault zones or volcanic structures and eruption mechanisms or other basic Earth science research topics. At the same time, there are growing demands for new sustainable energy sources (e.g., geothermal energy) and for underground storage of carbon dioxide. Drilling for such missions often takes place in unexplored, structurally or geotechnically difficult environments that require special drilling, coring, and testing capabilities. Furthermore, continuous coring, deviated drilling, and complex testing are frequently required within these kinds of research projects. However, the worldwide market for drilling devices appropriate for this is small and currently is stressed by very high hydrocarbon exploration activity. Accordingly, scientific projects are often unable to contract the right drilling rig and service, or cannot get it for the planned timeframe or at an affordable price.

Several projects in the framework of the International Continental Scientific Drilling Program (ICDP) were severely hampered by the fact that suitable rigs were not available or that they did not have the capability or the flexibility for various drilling, coring, or testing options necessary for scientific operations. For ICDP and other projects the GeoForschungsZentrum Potsdam therefore developed a new deep drilling and coring installation in cooperation with the company Herrenknecht GmbH in Schwanau (southern Germany). On 14 May 2007 the novel deep drilling and coring installation called InnovaRig was officially commissioned for a first operations test at the manufacturer's workshop (Fig. 1). From summer 2007 on, drilling of up to 5 km (~16,000 ft) depth can be realized through the derrick with a hook load of 3500 kN. One of the key features for scientific drilling in InnovaRig is that it allows for fast changes from rotary drilling to wireline diamond coring or vice versa in order to sample key lithologies and switch to faster and inexpensive rotary drilling in less important geological sections.

The Helmholtz Association of German Research Centers funded the development of the facility that is owned by the GFZ and will be made available for scientific and industrial projects through a commercial operator (Geoforschungsbohrgesellschaft). The contractual basis for this arrangement is designed to allow for industry missions during pauses between scientific drilling projects, thus avoiding costly standstill (or a need for permanent scientifically funded operations). For this reason the InnovaRig has been designed to be technologically and economically attractive for industry purposes as well. For scientific drilling projects, the day rates charged will not include depreciation of investment costs.

Technical Characteristics

In the InnovaRig, the usual standard of a rope hoist carrying the drillstring or casings is replaced by a hydraulic cylinder drawworks with 22 m stroke. Drill pipe is handled using "hand-off technologies" with semi-automated connection of two pipes to one stand in a horizontal position in a bridge magazine outside the derrick, while a new type of pipe handler transports stands into the tower from a horizontal magnetic pipe racking system on ground. The pipe handler provides practically unlimited capacity and does not compromise working on the rig floor due to setback areas. All kinds of pipe and casing in sizes between 2 7/8" and 24½" can be handled in the system with tripping speeds of up to 500 m h^{-1}. The drillstring is driven by two separate top-drive

Figure 1. InnovaRig with hydraulic pipe handling unit on lower right side. Photo by Herrenknecht Vertical.

systems with a broad range of rotary speeds and an auxiliary rotary table. Furthermore, the mud system, tanks, and pumps are constructed to be flexible for adapting to the various drilling procedures. The complete system including rig with pipe handler, pumps, mud tanks and other equipment can be skidded for easy relocation during multiple well operations.

The design and construction of the complete InnovaRig were performed to achieve a maximum potential for fast, inexpensive operations without any negative effects on efficiency, safety, and the environment. Key features (see Table 1 for details) are:

• the modular set-up and complete containerization which allow rapid conversion, mobilization, and skidding,

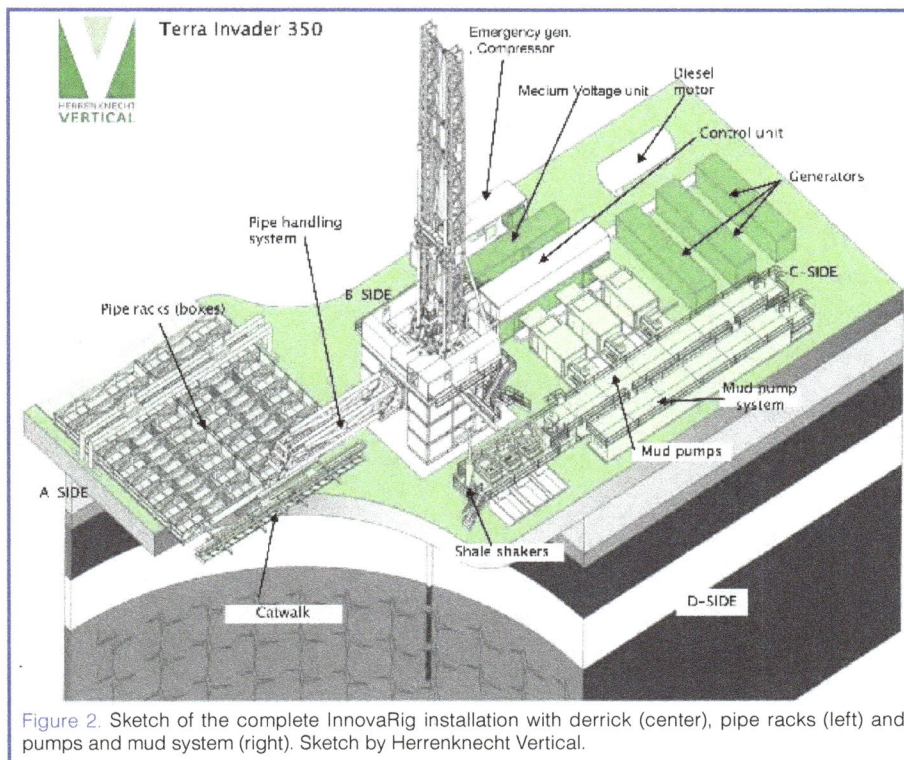

Figure 2. Sketch of the complete InnovaRig installation with derrick (center), pipe racks (left) and pumps and mud system (right). Sketch by Herrenknecht Vertical.

• the ability to rapidly switch between various drilling options including airlift drilling in large diameters (0–500 m depths), standard rotary drilling, continuous wireline diamond drilling, casing drilling as well as underbalanced drilling,

• the high degree of automation, in particular the semi-automatic pipe-handling for safe operations, condensed rig workload, and a minimum staffing,

• the high integration of devices for scientific measurements and tests to allow for a very fast switch from drilling to science operations,

• the minimization of area required for the drill site and the project-dependent use of rig modules, and

• the option to use either public or rig-installed power sources.

In terms of energy consumption and environment protection, InnovaRig can be operated through internal and/or public power supply, with biodegradable mud additives and greases. The rig is fully shielded against noise to allow deployments close to housing areas and will be extended for almost "waste-free" operations. In addition to standard equipment for rotary drilling such as 6.5-km drillstring and heavy pipes, the facility is outfitted with 5.5-km wireline pipes plus coring system and winch, 1000-m pipes for airlift drilling, as well as a 10,000 psi blowout preventer. In the case of low diameter wireline drilling in mining dimensions, the mud flow system can be easily rescaled to the volume of the circulating drilling fluid.

An important restriction is that although the rig can be partly downsized, it is a really heavy piece of equipment (Figs. 1 and 2) whose mobilization costs for more than 60 truckloads and day rates will not be suitable for shallow borings to less than ~2 km depth.

Special Installations for Science Operations

Geophysical wireline logging tools need to be deployed in research wells on short notice without lengthy backfitting. Accordingly, InnovaRig has cable guidance, sonde racks, spaces for winches, etc. integrated and in easy reach. At any time during drilling, the working platform is clear for setting up and recording wireline logs or conducting experiments such as vertical seismic profiling.

To allow a regular automatic sampling of drill chips, a sampling facility for cuttings and drill mud is provided at the mud cleaning line. This is extremely useful during non-coring phases to determine a couple of important parameters of the drilled rock column. In addition, mud samples can also be used to determine formation fluid inflow zones at depth which change the ion load in the drilling fluid. A mud gas extraction unit is also integrated to assure air contamination-free analysis of gaseous components from drilled rocks in the mud. Tracers or reference gases can be injected into the well during drilling operations through extra pipes flanged to the injection pumps. Such devices can, for example, be used to determine the lag time of the mud circulation and thereby detect the location of inflows.

Table 1. InnovaRig technical data

MAST		HOIST	
Height	51.8 m	Type	Hydraulic double-cylinder system
Hook load	3,500 kN (regular)	Stroke	22 m
		Power	2,000 kW
SUBSTRUCTRUE		**ROTARY TABLE**	
Type	box-on-box	Bore	953 mm
Height	9 m rig floor (9 x 10m dimension)	Nominal load	4,450 kN
Casing load	3,500 kN	Dynamic load	3,500 kN
BOP trolleys	2 x 250 kN	Drive	hydraulic (max. 200 min-1, 600 kW)
ROTARY TOP DRIVE		**CORING TOP DRIVE**	
Nominal load	4,450 kN	Nominal load	1,500 kN
Power	800 kW	Power	350 kW
Max. dyn. torque	48,000 Nm	Max. dyn. torque	12,000 kN
Max. RPM	220 min^{-1}	Max. RPM	500 min^{-1}
HYDRAULIC ROUGHNECK		**MUD PUMPS**	
Max. diameter	254 mm frame 1	Type	Electric (2 + 1 opt.)
Max. torque	508 mm frame 2	Power	1,300 kW
Max. load	4,540 kN	Max. pressure	350 bar
		Max. flowrate	2,200 L min^{-1}
ELEVATORS		**ROTARY TONGS**	
Max. diameter	254 mm frame 1	Type	Hydraulic clamping
Max. diameter	508 mm frame 2	Diameter range	73 mm–508 mm
Max. load	4,540 kN		
PIPE HANDLER		**MAGNETIC PIPE RACKING SYSTEM**	
Drive	hydraulic	Type	Horizontal
Max. diameter	620 mm	Drive	Electric
Min. diameter	73 mm	Nominal load	45,000 N per magnet group
Lifting capacity	45,000 N		

Science containers, such as a core and cuttings lab, a microbiology unit, or a containerized geochemistry lab, have extra space with power, water, and communication hook-up reserved within the setup of the drill site. The integrated data acquisition system records rig parameters, drilling data, and scientific data according to project needs. All these data are captured through a central bus system and stored in the systems main computer that offers direct connections to other data banks and Internet, and it is capable of handing over all data directly to the ICDP's Drilling Information System, DIS (Conze et al., 2007).

Outlook

The implementation of the idea to produce and operate a novel "science-owned" type of deep land drilling facility is a new pathway in scientific drilling, since land drilling endeavors are usually performed with contracted service rigs of opportunity. However, the lack (or at best, the very high costs) of suitable tools on the world market advanced this development. As it is anticipated that there will be no continuous science operations for this tool, neither within ICDP nor in other research-related drilling projects, InnovaRig will be operated on an as-needed basis also for industry. Scientific projects will have priorities, but commercial missions will be performed as long as no research project is being conducted.

Due to its outstanding, novel capabilities, InnovaRig can be used now in a very broad range of deep drilling missions for research and in industry projects. In terms of engineering, the facility is expected to establish a modern, inexpensive standard with hydraulic drive and pipe handling, as well as variable drilling capabilities. For science, the tool will provide the long-needed flexibility for coring, whenever necessary and useful, convenient installations for research without costly conversions. For these reasons, InnovaRig will support advancing Earth sciences with drilling projects.

The first deployment will be performed during this summer for a commercial 4-km-deep geothermal drilling project in southern Germany. Further applications are not yet contracted, but negotiations for the use in ICDP and in further geothermal research projects are currently underway. Proposals for the use of the InnovaRig are welcome.

References

Conze, R., Wallrabe-Adams, H.J., Graham, C., and Krysiak, F., 2007. Joint data management on ICDP and IODP mission specific platform expeditions. *Sci. Drill.*, 4:32–34.

Authors

Lothar Wohlgemuth and **Ulrich Harms**, Operational Support Group ICDP, GFZ Potsdam, Telegrafenberg A34, 14473 Potsdam, Germany, e-mail: wohlgem@gfz-potsdam.de. **Jürgen Binder**, Herrenknecht Vertical GmbH, 77963 Schwanau, Germany.

Related Web Links

www.icdp-online.org
www.vertical-herrenknecht.de

13

The COral-REef Front (COREF) Project

by Yasufumi Iryu, Hiroki Matsuda, Hideaki Machiyama, Werner E. Piller, Terrance M. Quinn, and Maria Mutti

Introduction

The First International Workshop on the COral-REef Front (COREF) project was held on 14–19 January 2007 in Okinawa-jima, southwestern Japan to discuss objectives, required laboratory analyses and techniques, potential drilling sites, and scientific proposals for the Integrated Ocean Drilling Program (IODP) and the International Continental Scientific Drilling Program (ICDP). This article briefly introduces the project and reports the outcome of the First International Workshop on the COREF Project.

COREF Project

The COREF Project (Iryu et al., 2006) involves ocean and land scientific drilling into Quaternary reef deposits in different settings in the Ryukyu Islands (Fig. 1). Major scientific objectives are to examine the following questions:

(1) What are the nature, magnitude, and driving mechanisms of coral-reef front migration in the Ryukyus?

(2) What is the ecosystem response of coral reefs in the Ryukyus to Quaternary climate changes?

(3) What is the role of coral reefs in the global carbon cycle?

To clarify the stratigraphic succession and lithofacies distribution, it is crucial to sample both highstand and lowstand reefs currently ranging from ~200 m above sea level on the islands to ≥150 m in elevation below sea level on the shelf and shelf slope. To obtain complete stratigraphic coverage, land drilling will need to be combined with ocean drilling.

Secondary objectives include (i) the timing and causes of coral-reef initiation in the Ryukyus, (ii) the position of the Kuroshio Current during glacial periods and its effects on coral-reef formation, and (iii) early carbonate diagenetic responses as a function of compounded variations in climate, eustacy, and depositional mineralogies (subtropical aragonitic to warm-temperate calcitic).

Workshop

Participating in the workship were a total of twenty-four scientists from seven countries/areas (Austria, French Polynesia, Germany, Japan, Korea, Taiwan, U.S.A.) representing multidisciplinary fields.

During the first day, a field excursion to an Upper Miocene to Pliocene siliciclastic slope to forearc basin deposits (Shimajiri Group), Pleistocene coral-reef carbonates (Ryukyu Group), and transitional lithofacies (Chinen Formation) illustrated the paleoceanographic transition from the "mud sea" to the "coral sea" (Fig. 2). During the following three days, the geologic setting of the Ryukyu Islands and the stratigraphic scheme of the Plio-Pleistocene carbonate succession were presented in detail, and existing datasets relevant to the COREF Project were reviewed at the Global Oceanographic Data Center of the Japan Agency for Marine-Earth Science and Technology (GODAC/JAMSTEC; http://www.godac.jp/top/en/index.html). Discussions also addressed the regional geologic framework, the availability of biota (corals and foraminifers), diagenetic features as paleoenvironmental proxies, and critical issues related to age determination of the carbonate sequences.

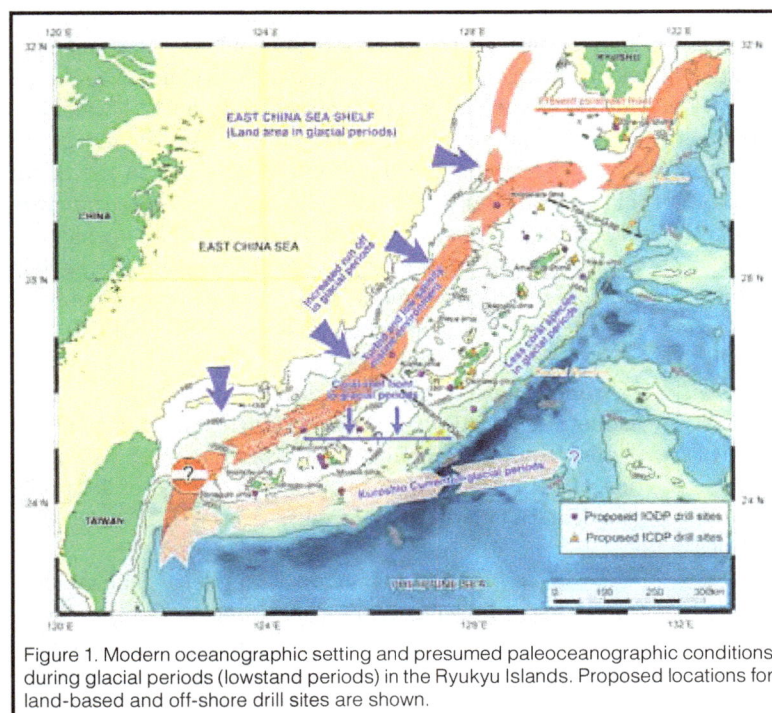

Figure 1. Modern oceanographic setting and presumed paleoceanographic conditions during glacial periods (lowstand periods) in the Ryukyu Islands. Proposed locations for land-based and off-shore drill sites are shown.

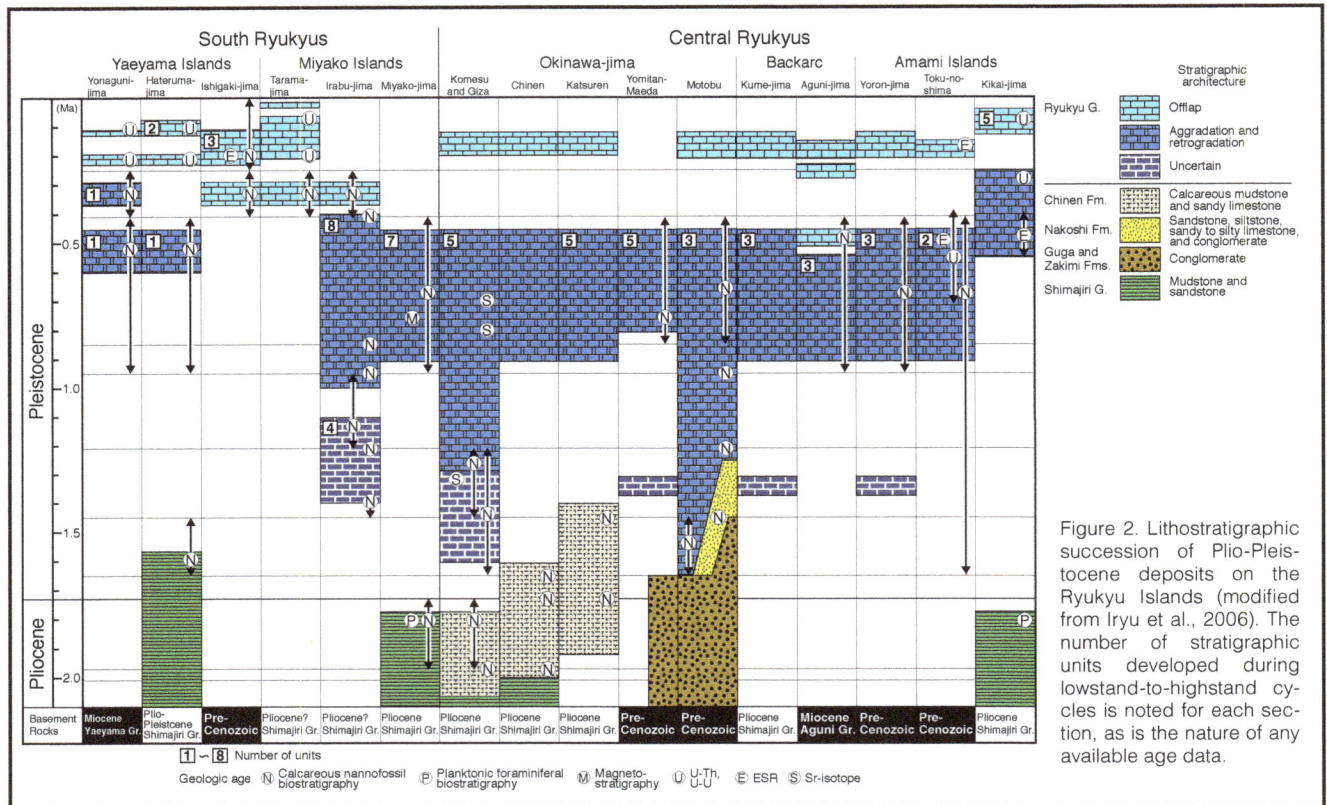

Figure 2. Lithostratigraphic succession of Plio-Pleistocene deposits on the Ryukyu Islands (modified from Iryu et al., 2006). The number of stratigraphic units developed during lowstand-to-highstand cycles is noted for each section, as is the nature of any available age data.

Finally, the workshop participants selected potential drill sites for the COREF Project (Fig. 1).

Major Issues

Drilling sites: The workshop selected drill sites on transects along and across the Ryukyu Island Arc. The northeast-southwest transect along the Ryukyu Island Arc extends from 24°N (south Iriomote-jima) to 31°N (west of Tane-ga-shima), covering islands from subtropical to warm-temperate regions (Fig. 1). The northernmost site is located on the northern limit of the modern coral-reef formation. At present, the distance between areas characterized by reefal coral communities and those by non-reefal coral communities (midway between Amami-o-shima Island and Tane-ga-shima Island; Fig. 1) is approximately 150 km (for definitions of reefal/non-reefal coral communities, see Veron, 1995). Therefore, the drilling sites on the northeast-southwest transect were designed to be located within an interval <200 km. Drilling on this transect will provide information on the nature and magnitude of coral-reef front migrations between glacial and interglacial periods. The northwest-southeast transects across the Ryukyu Island Arc are located near Amami-o-shima, Okinawa-jima, and Miyako Islands. These drilling sites are located from proximal (reef) via distal (off-reef) parts of ancient carbonate factories to shelf slopes toward the Okinawa Trough and the Ryukyu Trench. These drilling transects will recover a complete stratigraphic succession of the Quaternary carbonate deposits in the Ryukyus at different latitudes.

Coral-reef ecosystem: In the South and Central Ryukyus, corals build extensive fringing reefs dominated by a highly diverse assemblage of acroporid and poritid corals (Sugihara et al., 2003). Conversely, reefs at Tane-ga-shima, near the northern limit of coral-reef distribution, are thin, narrow, sparsely distributed, and dominated by only a few high-latitude coral species (Ikeda et al., 2006). North of the reef front, coral communities are dominated by faviid corals as well as a few other species. The latter are particularly abundant at high latitudes or even endemic to mainland Japan and commonly form large monospecific stands (Japanese Coral Reef Society and Ministry of the Environment, 2004). The compositions of larger foraminiferal assemblages in the North Ryukyus are distinguished from those found in the South and Central Ryukyus by the disappearance of *Calcarina gaudichaudii* and *C. hispida* (Sugihara et al., 2006). Nongeniculate coralline algae constitute the third taxonomic key-group emphasized during the workshop, because of their potential use as depth indicators in the fossil record (Iryu, 1992).

Workshop participants stressed the importance of establishing a schematic diagram summarizing the distribution of corals, benthic foraminifers, and coralline algae as a function of bathymetry, irradiance, water energy, and latitude (north and south of the reef front).

Age control: As the main body of the Pleistocene reef and off-reef deposits formed in the Ryukyus before 0.3 Ma (Fig. 2), they are beyond the limit of the ^{230}Th/^{234}U dating method. Thus, biostratigraphy, strontium (Sr) isotope stratigraphy, and magnetostratigraphy are the three principal techniques providing chronostratigraphic constraints to carbonate sequences to be recovered during the COREF Project.

Calcareous nannofossil biostratigraphy provides a good chronological constraint to shallow water carbonates (Yamamoto et al., 2006). However, two major problems exist– 1) although a high abundance of nannofossils within mixed siliciclastic and carbonate sediments is expected, few fossils may be found from well-indurated and diagenetically altered carbonate rocks, and 2) time resolution of calcareous nannofossil biostratigraphy (twelve datum plains in the Quaternary sequence) is hardly sufficient to resolve the effect on coral reefs of Quaternary glacio-eustatic sea level oscillations. However, precise dating by Sr isotope stratigraphy is possible for shallow water carbonates older than 1 Ma and free of siliciclastic grains/clasts. Magnetostratigraphy is a powerful tool for precise temporal correlation and accurate dating of sediments, even for recrystallized and dolomitized carbonates; however, there are few references on magneto-stratigraphic dating and rock magnetic characterization of the Pleistocene carbonate sequences in the Ryukyus. Sakai and Jige (2006) showed that bacterial magnetite minerals in the deposits carry an original depositional remanent magnetization useful for magnetostratigraphic dating.

Multiple techniques, such as electron spin resonance (ESR) and thermoluminescence dating methods, will therefore need to be used to date the samples of the COREF Project.

Towards the Achievement of the COREF Project

The workshop participants agreed that both land and off-shore drilling is required to address the scientific objectives of COREF and that proposals should be submitted to IODP and ICDP for such drilling. A thorough data mining of the literature on the Pleistocene carbonates and their basement rocks, especially Cenozoic sequences in the Ryukyus, is needed because most works were published in local journals with limited distribution. The proponent group welcomes geochemists studying thermoluminescence and ESR dating methods to join the project team.

Acknowledgements

This workshop was jointly funded by the ICDP and the Japan Drilling Earth Science Consortium (J-DESC). Meeting room and facilities were provided by GODAC/JAMSTEC.

References

Ikeda, E., Iryu, Y., Sugihara, K., Ohba, H., and Yamada, T., 2006. Bathymetry, biota, and sediments on the Hirota reef, Tane-ga-shima; the northernmost coral reef in the Ryukyu Islands. Island Arc, 15:407–419, doi:10.1111/j.1440-1738.2006.00538.x.

Iryu, Y., 1992. Fossil nonarticulated coralline algae as depth indicators for the Ryukyu Group. Transactions and Proceedings of the Palaeontological Society of Japan, New Series, 167:1165–1179.

Iryu, Y., Matsuda, H., Machiyama, H., Piller, W.E., Quinn, T.M., and Mutti, M., 2006. An introductory perspective on the COREF Project. Island Arc, 15:393–406, doi:10.1111/j.1440-1738.2006.00537.x.

Japanese Coral Reef Society and Ministry of the Environment, 2004. Coral Reefs of Japan. Tokyo (Ministry of the Environment), 356 pp.

Sakai, S., and Jige, M., 2006. Characterization of magnetic particles and magnetostratigraphic dating of shallow water carbonates in the Ryukyu Islands, northwestern Pacific. Island Arc, 15:468–475.

Sugihara, K., Masunaga, N., and Fujita, K., 2006. Latitudinal changes in larger benthic foraminiferal assemblages in shallow water reef sediments along the Ryukyu Islands, Japan. Island Arc, 15(4):437–454, doi:10.1111/j.1440-1738.2006.00540.x.

Sugihara, K., Nakamori, T., Iryu, Y., Sasaki, K., and Blanchon, P., 2003. Late Holocene sea level changes and tectonic uplift in Kikai-jima, Ryukyu Islands, Japan. Sediment. Geol., 159:5–25.

Veron, J.E.N., 1995. Corals in Time and Space. Sydney, Australia (University New South Wales Press), 321 pp.

Yamamoto, K., Iryu, Y., Sato, T., Chiyonobu, S., Sagae, K., and Abe, E., 2006. Responses of coral reefs to increased amplitude of sea level changes at the Mid-Pleistocene Climate Transition. Palaeogeogr. Palaeoclimatol. Palaeoecol., 241:160–175.

Authors

Yasufumi Iryu, Institute of Geology and Paleontology, Graduate School of Science, Tohoku University, Aobayama, Sendai 980-8578, Japan, e-mail: iryu@dges.tohoku.ac.jp

Hiroki Matsuda, Department of Earth Sciences, Faculty of Science, Kumamoto University, Kurokami 2-39-1, Kumamoto 860-8555, Japan.

Hideaki Machiyama, Kochi Institute for Core Sample Research, Japan Agency for Marine-Earth Science and Technology (JAMSTEC). Monobe-otsu 200, Nangoku, Kochi 783-8502, Japan.

Werner E. Piller, Institute of Earth Sciences (Geology and Palaeontology), University of Graz, Heinrichstrasse 26, A-8010 Graz, Austria.

Terrence M. Quinn, John A. and Katherine G. Jackson School of Geosciences, Department of Geological Sciences, The University of Texas at Austin, 1 University Station C1100, Austin, Texas 78712-0254, U.S.A.

Maria Mutti, Institut für Geowissenschaften, Universität Potsdam-Postfach 60 15 53, D-14415 Potsdam, Germany.

Related Web Links

http://www.godac.jp/top/en/index.html
http://www.dges.tohoku.ac.jp/igps/iryu/COREF/
http://coref.icdp-online.org/

Upcoming Workshops

Colorado Plateau Coring Workshop 2007

13–16 Nov. 2007, St. George, Utah, U.S.A.

The Colorado Plateau is the textbook example of layered sedimentary rocks in North America, representing the depositional history of the western Cordillera during much of the Paleozoic and Mesozoic. A focused coring program in Triassic through Lower Jurassic strata on and east of the Colorado Plateau would result in a quantum leap in our insight into issues of Pangean chronology, paleogeography, paleoclimate, and biotic evolution that also include those associated with the Triassic-Jurassic boundary. Topics include how the transition from the Paleozoic to a modern terrestrial ecosystems took place, how largely fluvial systems respond to cyclical climate, what the global or regional climate trends vs. plate position changes in "hot house" Pangea were, and how the major tectonic and eustatic events of the time were recorded in low accommodation continental settings.

The workshop is sponsored by DOSECC and NSF. More information and registration are at http://www.ldeo.columbia.edu/~polsen/cpcp/CPCP_home_page.html.

12th Ann. Continental Scientific Drilling WS

June 2008, Moab, Utah, U.S.A. Application deadline: 15 April 2008

Drilling in the Earth's continental crust allows study of otherwise inaccessible subsurface geological processes and structures. Drilling has led to many important geological discoveries on paleoclimate, impacts, volcanoes, mantle plumes, active faults, etc. The workshop, sponsored by DOSECC, will include presentations on international and multidisciplinary drilling projects and topics; a field trip and a reception are also planned to allow participants ample opportunity to exchange ideas. All geoscientists interested in using drilling as a tool are invited.

Limited funding is available for travel. Members of the scientific community who wish to contribute or participate in the workshop are invited to submit an application. The workshop details will be posted in early 2008 on DOSECC's website: http://www.dosecc.org. For more information contact David Zur (dzur@dosecc.org).

ICDP Workshop to Investigate Hominin-Paleoenvironmental History

Unconfirmed date and venue: late 2008, Nairobi, Kenya.

This workshop will consider the scientific opportunities and technical challenges of obtaining sediment cores from several of the most important fossil hominin and early Paleolithic artifact sites in the world, located in Kenya and Ethiopia. The objective will be to drill in near-continuous sedimentary sequences close to areas of critical importance for understanding hominin phylogeny, covering key time intervals for addressing questions about the role of environmental forcing in shaping human evolution. These sites are all currently on land, but consist of thick lacustrine sedimentary sequences with rapid deposition rates. Therefore, the proposed sites combine the attributes of relatively low-cost targets (in comparison with open water, deep lake sites) and the potential for highly continuous and informative paleoenvironmental records obtainable from lake beds.

Info will be available at http://magadi.icdp-online.org.

The Thrill to Drill: Continental Scientific Drilling Townhall Meeting

10 December 2007, San Francisco, California, U.S.A.

As previous years, DOSECC and ICDP will host a townhall meeting at the Fall AGU 2007 in San Francisco. Feel free to drop in to meet people and learn and chat about the drilling programs. Time and date await confirmation, so visit http://www.dosecc.org or http://www.icdp-online.org for the latest information. We look forward to welcoming you in San Francisco!

Multidisc. Observatory & Laboratory of Experiments Along a Drilling in Central Italy

Spring 2008, Central Italy.

The workshop will prepare a drilling project to investigate the shallow crust and the inner structure of normal faults in Northern Apennines to study geophysical and geochemical processes controlling normal faulting and earthquake ruptures during moderate-to-large seismic events as well as the low angle normal fault paradox. The sites in the Umbria-Marche sector of Northern Apennines offer a unique opportunity to reach a complex system of anthitetic normal faults: an active fault dipping SW at 40°–45°, which ruptured during a recent earthquake sequence in 1997 (Colfiorito fault) and a low angle normal fault dipping 15°–25° towards ENE (Alto Tiberina Fault).

More info available from http://apennines.icdp-online.org.

Testing Extensional Detachment Paradigm in the Sevier Desert Basin (Western U.S.)

Tentative date and venue: 15–18 July 2008, Salt Lake City, Utah, U.S.A.

Low-angle normal faults or detachments are widely regarded as playing an important role in crustal extension and the development of passive continental margins. No consensus exists on how to resolve the mechanical paradox implied or to account for the general absence of evidence for seismicity. Drilling to a depth of 2–4 km in the Sevier Desert basin of west-central Utah will test the extensional detachment paradigm through coring, downhole logging, biostratigraphic, isotopic and fission-track dating, magnetostratigraphy, and *in situ* measurement of pore pressure, permeability, fluid chemistry, temperature and stress orientation/magnitude.

More info at: http://www.ldeo.columbia.edu/sevier/icdp/.

DSDP and ODP Legacy Publications Online

Published volumes detailing nearly 40 years of scientific discoveries from ocean drilling research are now freely accessible online. All findings and data published in the Ocean Drilling Program (ODP) and the Deep Sea Drilling Project (DSDP) pulication series and program reports are online. ODP publications are available at http://www.odplegacy.org (click on Samples, data & publications). DSDP publications are available at http://www.deep-seadrilling.org.

The *Proceedings of the Ocean Drilling Program* includes an *Initial Reports* volume of shipboard reports for each ODP research cruise and a companion *Scientific Results* volume of peer-reviewed postcruise research results. ODP first began publishing its *Proceedings* online in 1997. Through the digitization effort, scanned versions of ODP *Proceedings* volumes originally published between 1986 and 1996 have been made Web-accessible.

The HTML tables of contents provide navigation to individual chapter files. PDF chapter files generated from more than 185,000 pages through the digitization project started as scanned images of each original page. Through the use of optical character recognition (OCR), a searchable text layer was added, allowing the user to copy and paste text from the final PDF files. The digitized volumes include links to individual core photographs scanned from original film as part of a separate ODP legacy project.

Every chapter in both the *Proceedings of the Ocean Drilling Program* and the *Initial Reports of the Deep Sea Drilling Project* has a digital object identifier (DOI) associated with it. With information about DSDP and ODP publications deposited with the DOI, publishers can now link directly online to cited papers across the DSDP and ODP series.

The digitization project was carried out at the Texas A&M University Digital Library in a joint venture with ODP sponsored by the U.S. National Science Foundation.

ODP was a 20-year international partnership of scientists and research institutions organized to explore the evolution and structure of the Earth through scientific ocean drilling. It conducted drilling operations in the world's oceans from January 1985 through September 2003. The program succeeded DSDP, which began drilling operations in 1968 and concluded its explorations in 1983. The Integrated Ocean Drilling Program (IODP) has been building upon the legacy of success of both its predecessor programs since 2004.

Conversion of the *JOIDES Resolution* Progressing

Joint Oceanographic Institutions (JOI) announced in April 2007 that Overseas Drilling Limited has signed a contract with Jurong Shipyard PTE LTD for the overhaul and enhancement of the research vessel *JOIDES Resolution*. The shipyard contract covers work through 2007, when the vessel is to be delivered for sea trials before returning to service for the Integrated Ocean Drilling Program (IODP) in 2008.

There have been great efforts by NSF and the USIO to keep the Scientific Ocean Drilling Vessel (SODV) project moving forward, despite severe cost pressures. As of July 2007 purchase of new science equipment was nearly complete, refurbishment of some existing equipment continued, and land based testing and integration had begun. Many ship and drilling service life extension projects have been completed and others are continuing. The drilling equipment, including the drilling derrick, were removed from the ship and refurbishment is well underway (see photos). To prepare the ship for new science facilities and accommodations, demolition of the existing facilities was required. To that end the lab stack, lifeboats, davits, and bridge have been removed and the accommodations have been gutted. Operations are scheduled to begin in 2008.

Since 2004 the *JOIDES Resolution* has been the U.S. platform for the IODP. For the majority of the previous two decades, the ship was employed by the

Visit IODP at Conferences & Exhibitions

Meet program scientists, get program updates, and pick up new informational resources!

GSA—28–31 Oct. 2007, Denver, Colo., U.S.A.
AGU—10–14 Dec. 2007, San Franc., Calif., U.S.A.
AAAS—14–18 Feb. 2008, Boston, Mass., U.S.A.
AAPG—20–23 Apr. 2008, San Antonio, Tex., U.S.A.
OTC—5–8 May 2008, Houston, Tex., U.S.A.
EGU—13–18 Apr. 2008, Vienna, Austria
JPGU—May 2008, Chiba, Japan
IGC—6–14 Aug. 2008, Oslo, Norway

Conference exhibitions may be added or changed. Visit http://www.iodp.org for calendar updates.

Ocean Drilling Program, predecessor to the IODP. For the latest information and more spectacular pictures about the SODV project visit http://joiscience.org/sodv/.

IODP Phase 2 Started

With the start of the first expedition of the NanTroSEIZE project in October 2007, the second phase of IODP has begun. During IODP Expedition 314 the Japanese drilling vessel *Chikyu* will drill a subduction zone off southern Japan. The first phase of IODP spanned from 2004 to 2005, comprising operation of twelve research expeditions. Ten expeditions were undertaken by the riserless U.S. operated drilling vessel *JOIDES Resolution* and two expeditions were performed by mission-specific platforms operated by the European Consortium for Ocean Research Drilling (ECORD). After a drilling hiatus of almost two years the IODP's options are now enhanced by the use of the *D/V Chikyu*, a ship providing riser drilling technology supplied to the program by the Japan Agency for Marine Earth-Science and Technology. In addition the U.S. ship *JOIDES Resolution* is being significantly remodeled, enhanced and upgraded, which will allow scientists to address new and previously inaccessible drilling targets.

In preparation for the upcoming expeditions, the first *Scientific Prospectuses* for Expeditions 314–316 have been published and made available on the Web (http://www.iodp.org/scientific-publications/). Applications for upcoming expeditions can be submitted through IODP's Web page at http://www.iodp.org/apply-to-sail/.

ICDP Training Course 2007

Each year the ICDP conducts a one-week field course to train scientists and engineers on basics in scientific drilling. The key aspects of the courses cover project planning and management, drilling technology, borehole measurements, scientific on-site analyses, and information management. The training is open for advanced students, PhD students, post-docs from ICDP member countries and is recommended

for managers and scientists from forthcoming ICDP projects.

The training will be held in Windischeschenbach, Germany, the drill site of the scientific deep drilling project KTB, from 5–9 November this year. A highlight will be an excursion to InnovaRig, the newly developed facility for scientific drilling (see Wohlgemuth et al., this issue). The detailed program can be found in "News>>Upcoming Events" on the ICDP webpage: http://www.icdp-online.org.

First Results of ICDP Bosumtwi Project Published

The first results of the impact and geophysical aspects of the ICDP drilling project have just been published in a special double issue (April–May 2007) of the international journal Meteoritics and Planetary Science. See http://www.ingentaconnect.com/content/arizona/maps for contents and abstracts. As a service to the ICDP community, these 27 articles have also been placed online at the ICDP Bosumtwi Web site under "references", see http://www.icdp-online.org/contenido/icdp/front_content.php?idcat=446. More work, especially also on the lake sediments as well as impactites, is in progress.

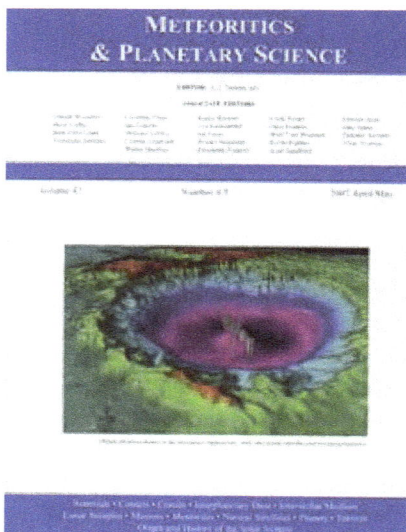

ICDP welcomes new members Italy and Spain

Italy and Spain have recently completed Memoranda of Understanding on membership and participation in the International Continental Scientific Drilling Program (ICDP). A Consortium

of the Instituto Nazionale di Geofisika e Vulcanologica (INGV) and Centro Regionale di Competenza Analisi e Monitoraggio del Rischio Ambientale CRdC (AMRA) are partners in ICDP for Italy while the Spanish Ministry of Education and Science joined in for the Spanish Earth science community.

ESSAC Office in Aix-en-Provence after 1 Oct. 2007

After two efficient and fruitful years in Cardiff, U.K., the ESSAC office rotates to Aix-en-Provence in France on 1 October 2007. It will be located at CEREGE, Europôle Méditerranéen de l'Arbois, BP80, 13545 Aix-en-Provence cedex 4, France. Gilbert Camoin (gcamoin@cerege.fr) will take over from Chris MacLeod to serve as the new ESSAC Chair. The new ESSAC Science Coordinator is Bonnie Wolff-Boenish, who was previously Program Manager of the German Priority Program (PP) International Continental Scientific Drilling Program (ICDP) at the University of Potsdam since 2005.

Scientific Drilling Special Issue on Fault Zones

Scientific Drilling will print its first special issue in October 2007 on the results of the Fault Zone Workshop held in May 2006 in Miyazaki, Japan. This issue will contain the white paper as the result of the workshop, as well as selected abstracts from the presentations given. The editorial board consists of the workshop steering committee: Harold Tobin, Steve Hickman, Jan Behrmann, Hisao Ito, and Gaku Kimura. Since only a limited number of copies will be printed, it will not be distributed like the regular issues of *Scientific Drilling*. Interested scientists or libraries are welcome to send an email with their mailing address to journal@iodp-mi-sapporo.org to request a printed copy. Requests are served as long as supplies last. The special issue on the Fault Zone Workshop will also be available online in PDF format from http://www.iodp.org/scientific-drilling/.

14

IODP Expedition 323—Pliocene and Pleistocene Paleoceanographic Changes in the Bering Sea

by Kozo Takahashi, A. Christina Ravelo, Carlos Alvarez Zarikian, and the IODP Expedition 323 Scientists

Abstract

High-resolution paleoceanography of the Plio-Pleistocene is important in understanding climate forcing mechanisms and the associated environmental changes. This is particularly true in high-latitude marginal seas such as the Bering Sea, which has been very sensitive to changes in global climate during interglacial and glacial or Milankovitch time scales. This is due to significant changes in water circulation, land-ocean interaction, and sea-ice formation. With the aim to reveal the climate and oceanographic history of the Bering Sea over the past 5 Ma, Integrated Ocean Drilling Program (IODP) Expedition 323 cored a total of 5741 meters of sediment (97.4% recovery) at seven sites covering three different areas: Umnak Plateau, Bowers Ridge, and the Bering slope region. Four deep holes range from 600 m to 745 m spanning in age from 1.9 Ma to 5 Ma. The water depths (819 m to 3173 m) allow characterization of past vertical water mass distribution such as the oxygen minimum zone (OMZ). The results highlight three key points. (1) The first is an understanding of long-term evolution of surface-water mass distribution during the past 5 Ma including past sea-ice distribution and warm and less eutrophic subarctic Pacific water mass entry into the Bering Sea. (2) We characterized relatively stagnant intermediate water mass distribution imprinted as laminated sediment intervals that have been ubiquitously encountered. Today, the OMZ impinges upon the sediments at ~700–1600 m water depths. In the past, the OMZ appears to have occurred mainly during interglacial periods. Changes in low oxygen-tolerant benthic foraminiferal faunas clearly concur with this observation. (3) We also characterized significant changes between glacial episode of terrigenous sedimentary supply and interglacial episode of diatom flux.

Introduction and Goals

The rate and regional expression of recent global warming is difficult to understand and even more difficult to predict because of the complex nature of the climate system, whose components interact nonlinearly with various time lags and on various timescales. Paleoclimatic and paleoceanographic studies provide opportunities to study the dynamics of the climate system by examining how it responds to external forcing (e.g., greenhouse gases and solar radiation changes) and how it generates internal variability due to interacting Earth-system processes. Of note is the amplified recent warming of the high latitudes in the Northern Hemisphere (Solomon et al., 2007), which is presumably related to sea-ice albedo feedback and teleconnections to other regions; both the behavior of sea ice-climate interactions and the role of large-scale atmospheric and oceanic circulation in climate change can be studied with geologic records of past climate changes in the Bering Sea.

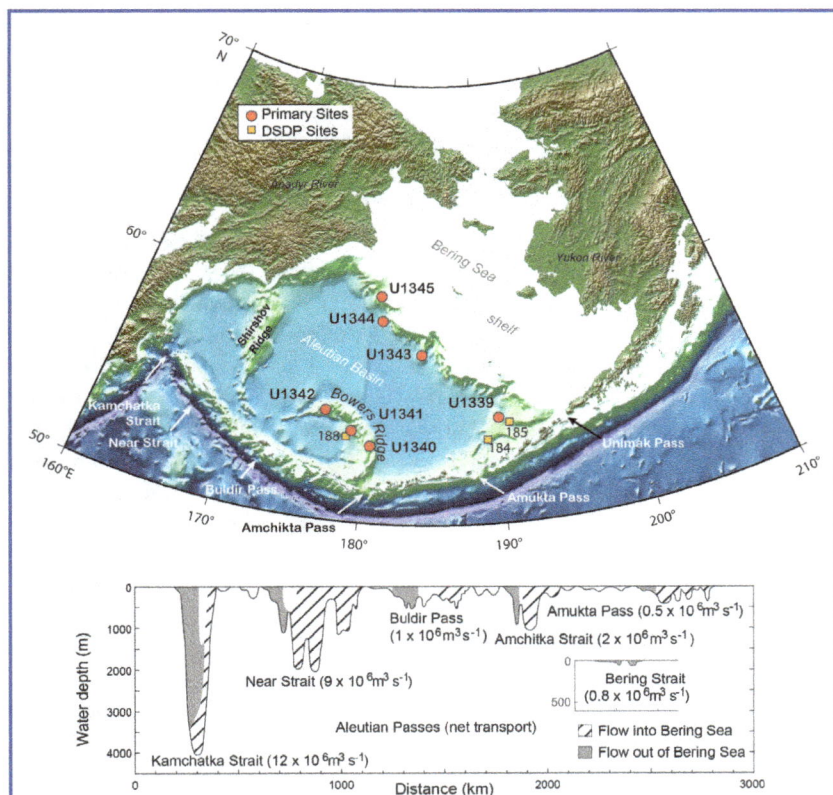

Figure 1. Map illustrating the locations of the seven sites drilled and cored during IODP Exp. 323 in the Bering Sea, along with cross-sections of the passes with volume transport (Sv) in the Aleutian Island arc and the Bering Strait. Note that the horizontal and vertical scales of the Bering Strait are twice that of the Aleutians (from Stabeno et al., 1999; Takahashi, 2005).

Prior to IODP Expedition 323 (Exp 323 hereafter), little was known about the sedimentology and climate history of the Bering Sea outside of a few piston core studies (Cook et al., 2005; Okazaki et al., 2005; Katsuki and Takahashi, 2005; Takahashi et al., 2005) and Sites 188 and 185 (Scholl and Creager, 1973), which were drilled by the Deep Sea Drilling Project (DSDP) in 1971 with old drilling technology and poor recovery. Past studies using piston cores in the Bering Sea indicated that, while current conditions in the Bering Sea promote seasonal sea-ice formation, during the Last Glacial Maximum (LGM) conditions sustained perennial or nearly perennial sea-ice cover (Tanaka and Takahashi, 2005), attesting to the potential utility of sedimentary records in the Bering Sea to examine past sea-ice distributions. In paleoceanographic studies of the North Pacific, the Bering and Okhotsk seas have been implicated as sources of dense oxygenated intermediate water that possibly impacted oceanic and climate conditions throughout the Pacific on glacial-interglacial (Gorbarenko, 1996; Matsumoto et al., 2002) and millennial (Hendy and Kennett, 2003) timescales. In addition, changes in Bering Sea environmental conditions could be related to sea-level and circulation changes, which alter flow patterns through narrow straits that connect the Bering Sea to the Arctic Ocean to the north and the Pacific Ocean to the south. The lack of pertinent Bering Sea material prevented the evaluation of these and other ideas for a long time.

The scientific objectives of Exp 323 are as follows: (1) to elucidate a detailed evolutionary history of climate and surface ocean conditions since the earliest Pliocene in the Bering Sea, where amplified high-resolution changes of climatic signals are recorded; (2) to shed light on temporal changes in the origin and intensity of North Pacific Intermediate Water (NPIW) and possibly deeper water mass formation in the Bering Sea; (3) to characterize the history of continental glaciation, river discharge, and sea ice formation in order to investigate the link between continental and oceanic conditions in the Bering Sea and on adjacent land areas; (4) to investigate linkages through comparison to pelagic records between ocean/climate processes that occur in the more sensitive marginal sea environment and processes that occur in the North Pacific and/or globally (This objective includes evaluating how the ocean/climate history of the Bering Strait gateway region may have affected North Pacific and global conditions.); and (5) to constrain global models of subseafloor biomass and microbial respiration by quantifying subseafloor cell abundance and pore water chemistry in an extremely high productivity region of the ocean. We also aim to determine how subseafloor community composition is influenced by high productivity in the overlying water column.

Seven sites whose terrigenous and biogenic components capture the spatial and temporal evolution of the Bering Sea through the Pliocene and Pleistocene[*] were successfully drilled with a total core length of 5741 m during Exp 323 (Expedition 323 Scientists, 2010; Takahashi et al., 2011; Fig. 1; Table 1). Additionally, Exp 323 collected a rich archive of information regarding the role of microbes on biogeochemical cycles in ultra-high-productivity environments, the postdepositional processes that impact geochemical, lithologic, and physical properties of the sediment, and past oceanic chemistry preserved in pore waters. This paper presents background on environmental setting and important scientific questions concerning the Bering Sea, followed by pertinent highlights of the scientific findings of Exp 323 mostly obtained onboard *JOIDES Resolution* during the cruise.

> ***Note**: In this paper we opt to continue using the last major published timescale, in which the base of the Pleistocene is defined by the Global Boundary Stratotype Section and Point (GSSP) of the Calabrian Stage at 1.806 (1.8) Ma (Gradstein et al., 2004).

Geological and Physical Setting

With an area of 2.29×10^6 km^2 and a volume of 3.75×10^6 km^3, the Bering Sea is the third largest marginal sea in the world, surpassed only by the Mediterranean and South China seas (Hood, 1983). Approximately half of the Bering Sea is a shallow (0–200 m), neritic environment, with the majority of the continental shelf spanning the eastern side of the basin off Alaska from Bristol Bay to the Bering Strait (Fig. 1). The northern continental shelf is seasonally ice-covered, but little ice forms over the deep southwest areas. In addition to the shelf regions, two significant topographic highs have better CaCO$_3$ preservation than the deep basins. First is the Shirshov Ridge, which extends south of the Koryak Range in eastern Siberia along 170°E and separates the southwestern part of the Bering Sea into two basins, Komandorski (to the west) and Aleutian (to the east). Second is the Bowers Ridge, which extends 300 km north from the Aleutian Island arc (Fig. 1). The Aleutian Basin is a vast plain 3800–3900 m deep with occasional gradually sloping depres-sions as deep as 4151 m (Hood, 1983).

Table 1. Summary of drilled results for IODP Exp 323 in the Bering Sea.

Area IODP Site Number	Water Depth (mbsl)	Depth DSF (m)	Age (Ma)	Average Sedimentation Rate (cm k.y.⁻¹)
Umnak Plateau				
U1339	1868	200	0–0.8	28
Bowers Ridge				
U1340	1295	605	0–5.0	12
U1341	2140	600	0–4.3	12
U1342	819	128	0–1.2*	4.5
Bering Slope				
U1343	1953	745	0–2.4	35
U1344	3173	745	0–1.9	45
U1345	1008	150	0–0.5	29

*Bulk of sediment samples were <1.2 Ma (top 41 mbsf) in age except for the Middle Miocene diatoms located in the samples from directly above the basaltic basement rocks.

Three major rivers flow into the Bering Sea; the Kuskokwim and Yukon rivers drain central Alaska, and the Anadyr River drains eastern Siberia (Fig. 1). The Yukon is the longest of the three rivers and supplies the largest discharge into the Bering Sea. Its discharge peaks in August because of meltwater and is about equal to that of the Mississippi. It has a mean annual flow of 5×10^3 $m^3 s^{-1}$ (Hood, 1983).

Today, a substantial amount of water is transported in and out of the Bering Sea across the Aleutian Island arc and the Bering Strait through passes (Fig. 1). Water mass exchange with the Pacific through the Aleutian Islands, such as through the Kamchatka Strait, is significant, linking Bering Sea conditions to the Pacific climate. The Alaskan Stream, an extension of the Alaskan Current, flows westward along the Aleutian Islands and enters the Bering Sea partially through the Amchitka Strait and to significant extent through the Near Strait west of Attu Island in the eastern Aleutian Islands (Fig. 1). A part of the Subarctic Current also joins the Alaskan Stream, resulting in a combined volume transport of 11 Sv (0.011 $km^3 s^{-1}$) (Ohtani, 1965).

Bottom and intermediate water in the Bering Sea originates from the North Pacific. After flowing into the Bering Sea it is slightly modified by the mixing of relatively fresh, warm water with very small amounts of bottom water formed within the Bering Sea today (Warner and Roden, 1995). Nutrient concentrations of North Pacific origin are high compared to all other regions in the global oceans; this explains the high productivity in the surface layers and consequent very low oxygen concentrations in intermediate and deep water masses of the Bering Sea today (Fig. 2). The oxygen and nutrient composition of the Bering Sea waters is further modified by denitrification (Lehmann et al., 2005)

and respiration of organic matter in the water column (Nedashkovskiy and Sapozhnikov, 1999). Respiration and the development of an OMZ are particularly intense at water depths of ~1000 m (Fig. 2).

Much of the Pacific water entering the Bering Sea is matched by outflow through the Aleutian Islands. The most significant outflow is through the Kamchatka Strait, which has a maximum depth of 4420 m (Stabeno et al., 1999) (Fig. 1). If some component of NPIW or deep water formed in the Bering Sea in past times, particularly when sea level was lower, it would have flowed out through the Kamchatka Strait or a secondary outlet near the Commander-Near Strait at 2000 m (Fig. 1).

The unidirectional northward transport of water masses (0.8 Sv) from the Bering Sea through the Bering Strait to the Arctic Ocean contributes to the salinity and biogeochemical contrast between the Pacific and the Atlantic. The Bering Strait region is one of the most biologically productive regions in the world (Sambrotto et al., 1984). Much of this biologically produced organic matter and the associated nutrients flow into the Arctic Ocean because of the northward current direction. This may profoundly influence the present dominance of carbonate production in the Atlantic versus opal production in the Pacific, as described by models of basin-to-basin fractionation (Berger, 1970) and "carbonate ocean vs. silica ocean" (Honjo, 1990). Flow through the Bering Strait, which is ~50 m deep today (Fig. 1), was certainly different at times of lower sea level or enhanced perennial sea-ice cover. The closing of this gateway and the accompanying changes in ocean and river flow through time could have caused changes in global patterns of circulation or in nutrient and salinity distributions.

Ages and Sedimentation Rates

Among the three drill sites explored in the Bowers Ridge region, both of the deepest holes drilled—Hole U1340A (605 m uncompressed core depth below seafloor (CSF-A), hereafter meters below seafloor (mbsf) and Hole U1341B (600 m mbsf)—represent the time spans from the Holocene to the Pliocene, back to ~5 Ma and 4.3 Ma, respectively (Table 1; Fig. 3). Note that the 4.3 Ma bottom age of Hole U1341B has been revised by an onshore study from the shipboard data (~5 Ma; Expedition 323 Scientists, 2010; Takahashi et al., 2011). The expedition's initial goal of penetrating to ~5 Ma was adequately accomplished at Site U1340, despite the failure of the extended core barrel (XCB) cutting shoe in Hole U1340A, which target depth of penetration was 700 m. In the gateway region sites (at the Bering slope), two deep holes were drilled: Holes U1343E (744 m mbsf) and U1344A (745 m mbsf). Hole U1343E reached ~2.4 Ma, where as Hole U1344A reached ~1.9 Ma (Figs. 3 and 4). Also note that the 2.4 Ma bottom age of Hole U1344A has been revised by an onshore study from the shipboard data (2.1 Ma). At other drill sites, the bottom ages of

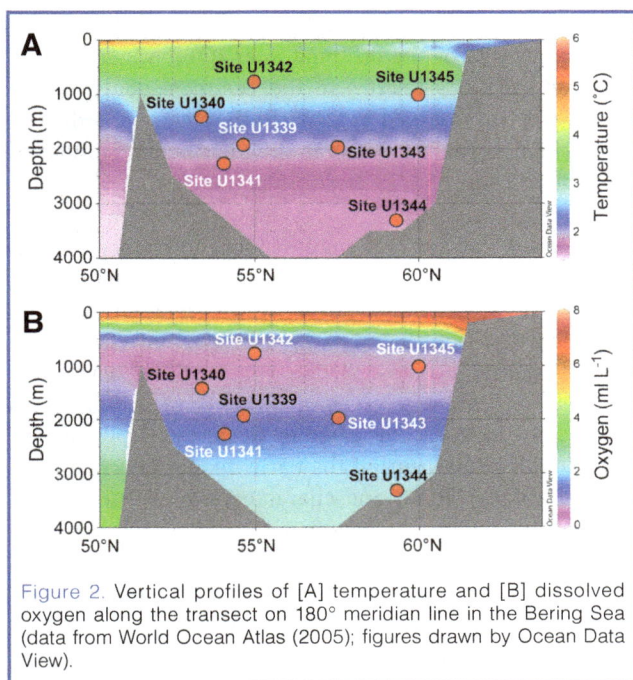

Figure 2. Vertical profiles of [A] temperature and [B] dissolved oxygen along the transect on 180° meridian line in the Bering Sea (data from World Ocean Atlas (2005); figures drawn by Ocean Data View).

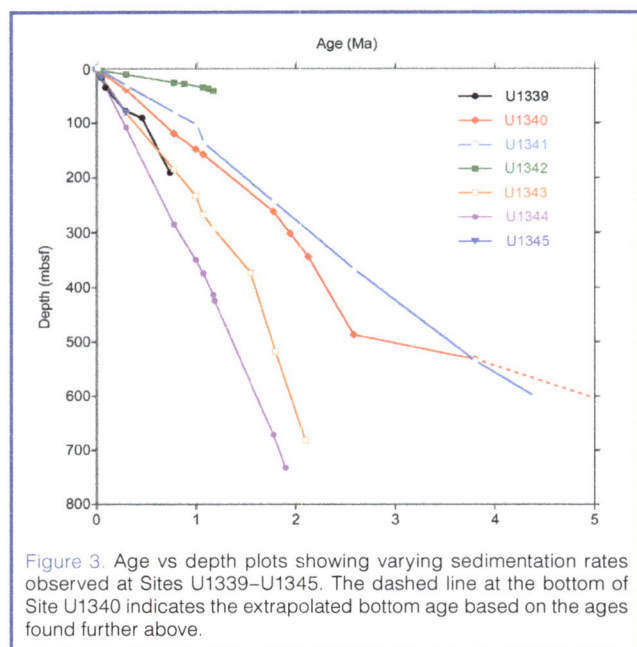

Figure 3. Age vs depth plots showing varying sedimentation rates observed at Sites U1339–U1345. The dashed line at the bottom of Site U1340 indicates the extrapolated bottom age based on the ages found further above.

the sedimentary sequences based on biomagnetostratigraphy are as follows: Site U1339 at Umnak Plateau reached 0.74 Ma, Site U1342 at Bowers Ridge reached 1.2 Ma (with the exception of the middle Miocene sediments just above the basement), and Site U1345 at Bering slope reached 0.5 Ma (Table 1).

The sediments recovered from Bowers Ridge display high mean sedimentation rates (~12 cm k.y.$^{-1}$ at Sites U1340 and U1341) without apparent hiatuses, and are generally appropriate for high-resolution Pliocene–Pleistocene paleoceanographic studies with adequate calcareous benthic foraminiferal preservation in the Pleistocene, but lower preservation in the Pliocene. On the other hand, sediments at these sites are generally barren of planktonic foraminifera and calcareous nannofossils except for the section between ~2.5 Ma and ~3 Ma. The abundance of all siliceous microfossils is generally high, enabling good biostratigraphy and paleoceanographic reconstruction. Furthermore, the upper part of Site U1340 (~20–150 m uncompressed core composite depth, CCSF-A) had obvious soft-sediment deformation due to mass movement possibly caused by local seismic activity. Although such deformation hinders the continuous reconstruction of late Pleistocene high-resolution paleoceanography at this site, information from other drill sites can readily fill the gap.

In the region of the Arctic gateway sites proximal to the Bering slope, the observed sedimentation rates were overwhelmingly high: Hole U1343E had sedimentation rates of 21–58 cm k.y.$^{-1}$ and Hole U1344A had rates of 29–50 cm k.y.$^{-1}$. Sedimentation rates were so high, in fact, that drilling reached ages of only 2.4 Ma and 1.9 Ma, respectively, despite penetration to 745 m mbsf at each site. Such high sedimentation rates stem from the deposition of silt and clay transported by the Yukon and other rivers as well as the terrigenous

sediments once deposited on the shelf. In spite of the high percentage of terrigenous material, pertinent biotic proxies including benthic foraminifera and siliceous microfossils are adequately preserved, enabling future paleoceanographic studies. Therefore, the overall coverage of excellent cores to ~5 Ma in the Bowers Ridge region and ~2 Ma in the gateway region allows detailed, continuous high-resolution paleoceanographic studies relevant to global climate change.

Depositional Environments and Lithology

The seven sites drilled during Exp 323 provide a continuous high-resolution record of the evolution of marine sedimentation in the marginal Bering Sea (Fig. 4). Overall, the sediments recovered in the Bering Sea are a mixture of three components: biogenic, siliciclastic, and volcaniclastic. Other accessory lithologies identified include authigenic carbonates (dolomite Fe-rich carbonates and Mg calcite), barite and sulfides. The most prominent sedimentary features observed were decimeter- to meter-scale bedded alternations of sediment color and texture, reflecting alternations in lithology between more siliciclastic and more biosiliceous deposits (Fig. 4). The sediments were generally highly bioturbated. However, fine laminations preserving alternations between millimeter-scale laminae of biogenic and terrigenous material were also present in several of the drilled sites (Fig. 5).

The distributions of the sedimentary components and sedimentary structures, and their variability both within and between the Exp 323 sites account for changes of the biogenic, glaciomarine, terrigenous, and volcanogenic sediment sources and the environmental conditions present during sediment deposition. The scales of these lithologic variations indicate that sedimentation in the Bering Sea has recorded long-term trends that include the critical period of reorganization of Earth's climate from the warm early Pliocene, and the transition into the ice ages. The physiographic settings of the different sites, their water depths, and their locations relative to the sediment source areas account for the marked regional differences in sediment composition, especially between the Pleistocene sections of the Bowers Ridge and the Bering slope sites.

The results of Exp 323 suggest that the history of sedimentation in the Bering Sea is broadly characterized by three main sedimentary phases that occurred between ~5 Ma and ~2.7 Ma, ~2.7 Ma and ~1.74 Ma, and ~1.74 Ma to recent (Fig. 4). The oldest portion of the sedimentary record (~5 Ma to ~2.7 Ma) was retrieved only at Bowers Ridge Sites U1340 and U1341. As illustrated by the age vs depth curves, sedimentation rates during the early middle Pliocene were relatively high (Fig. 3) and characterized by diatom ooze with minor amounts of diatom silt, sponge spicules, and vitric ash. Although the Pliocene sediment is commonly bioturbated, distinct intervals characterized by extensive lamination also occur. The oldest laminated intervals (<3.8 Ma) were observed at Site U1341, although the origin of the laminations, and whether they represent primary or secondary processes is unknown. Notably, especially in the deeper parts of the record, compaction or diagenetic phase transformations might have created secondary sedimentological features, overprinting the primary ones. Isolated ice rafted debris (IRD) pebbles were observed in sediments older than 3.8 Ma only at Site U1340. Limited dropstone occurrence prior to 2.7 Ma was also reported at two sites drilled in the northern Pacific during Leg 145 (ODP Sites 881 and 883) and in the Yakataga Formation in Alaska (Lagoe et al., 1993), which suggests the development of Alpine glaciers prior to the onset of Northern Hemisphere glaciation (NHG) (Krissek, 1995).

The middle section of Sites U1340 and U1341 (~2.7–1.74 Ma) is characterized by beds of diatom ooze with minor amounts of calcareous nannofossils and foraminiferal ooze alternating with diatom silt beds. The latter are composed of subequal proportions of siliciclastic (silt-sized quartz, feldspar, and rock fragments and/or clay) and biogenic components and minor volcaniclastic components. Dropstone occurrence is common—indicating a peak in siliciclastic deposition that has also been observed at Leg 145 sites—and coincides with the beginning of NHG. However, the dramatic drop in paleoproductivity recorded at Site 882 (Haug et al., 1999) is not present at the Bowers Ridge sites where, conversely, the biogenic component is high throughout the late Pliocene and Pleistocene.

All sites drilled during Exp 323 preserve a record of sedimentation ranging from the early Pleistocene through the Holocene (1.74 Ma to recent). Lithologies and sedimentation rates vary between the different sites, as indicated by a basin-wide comparison of the evolution of sedimentation in the Bering Sea during this period (Figs. 3, 4). The lowest sedimentation rates (only 4.5 cm k.y.$^{-1}$) were observed at Site U1342, where laminated foraminifera-rich diatom ooze beds alternate with silty clay beds at timescales ranging in the Milankovitch band (Fig. 5). The same temporal interval corresponds to a much thicker section at Sites U1340 and U1341, where the bedding alternations are less distinct and the abundance of IRD is higher. Although lamination is common at Sites U1342 and U1340, laminae are virtually absent at the deeper Site U1341. At the Bering Sea slope site, sedimentation rates are about three times higher than at the Bowers Ridge sites. At Sites U1339, U1343, and U1344, siliciclastic-rich beds and mixed siliciclastic-biogenic beds alternate cyclically. The sections are pervasively bioturbated, and laminated intervals are rare. Overall, sedimentation on the Bering slope is characterized by higher influence of (1) siliciclastic material delivered by ice sheets and (2) terrigenous sedimentation derived from the continental shelf and slope, which are indented by some of the largest submarine canyons in the world. However, because of their proximity to the continents, it is not clear whether the sediments characterized by high siliciclastic content are recording periods of ice sheet expansion (stadials) or increased runoff (interstadials). IRD is a common feature at all sites during this time period, and it increases significantly in sediments younger than 1 Ma, as is also observed in coeval sediments from the North Pacific based on the results of Leg 145 (Krissek, 1995).

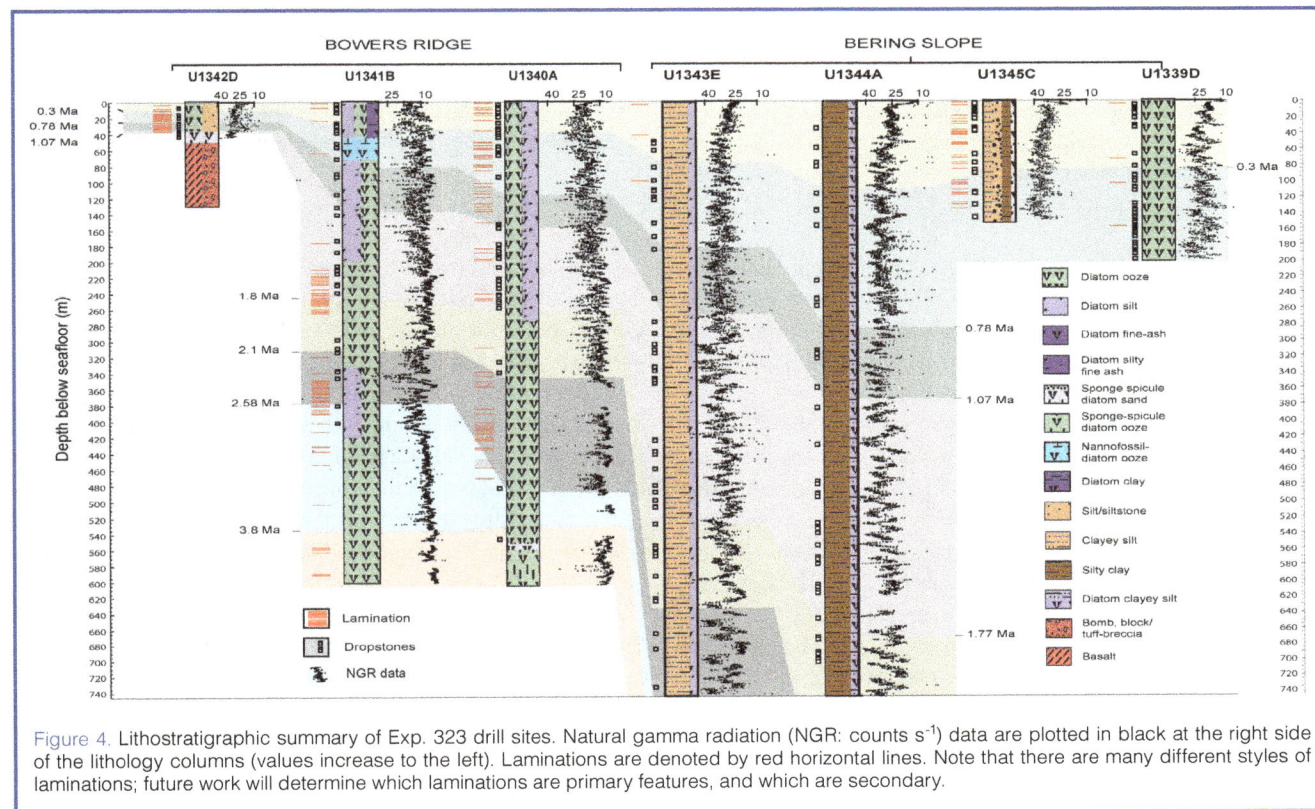

Figure 4. Lithostratigraphic summary of Exp. 323 drill sites. Natural gamma radiation (NGR: counts s^{-1}) data are plotted in black at the right side of the lithology columns (values increase to the left). Laminations are denoted by red horizontal lines. Note that there are many different styles of laminations; future work will determine which laminations are primary features, and which are secondary.

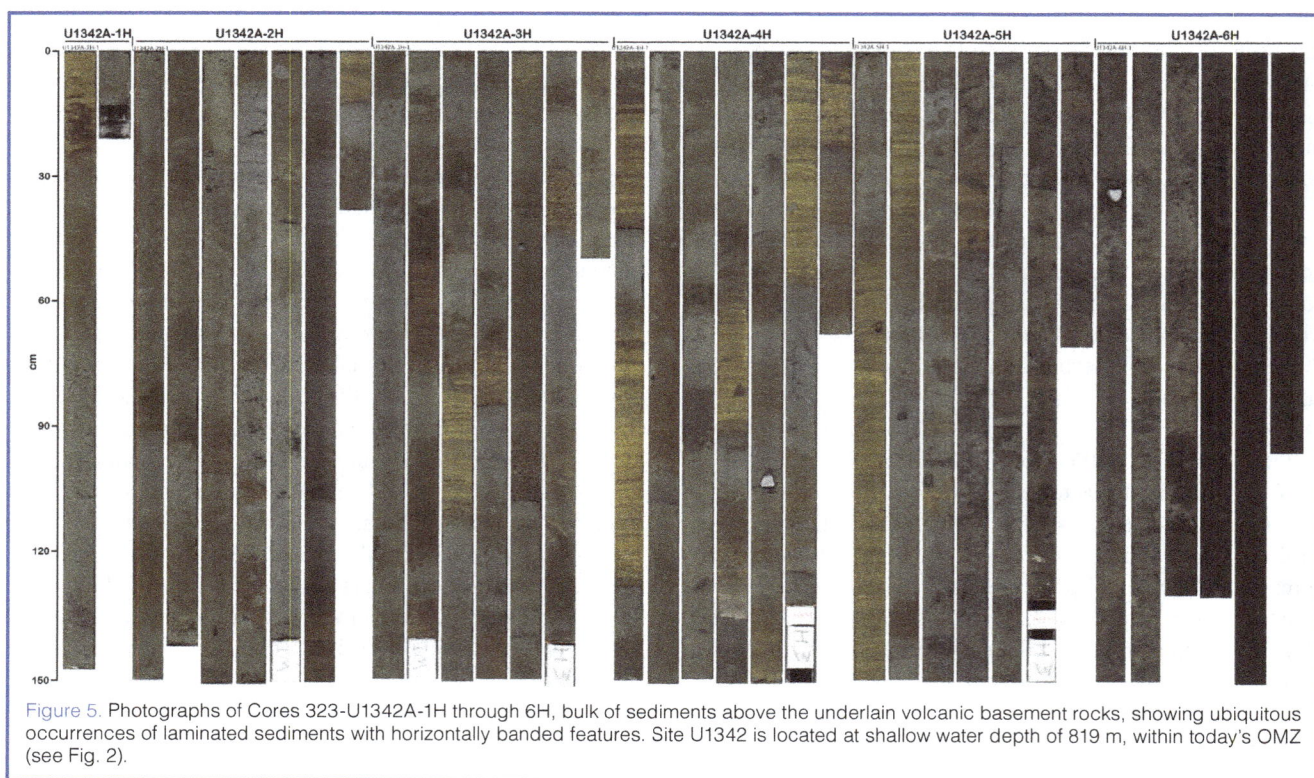

Figure 5. Photographs of Cores 323-U1342A-1H through 6H, bulk of sediments above the underlain volcanic basement rocks, showing ubiquitous occurrences of laminated sediments with horizontally banded features. Site U1342 is located at shallow water depth of 819 m, within today's OMZ (see Fig. 2).

History of Sea-Ice Development

One of the most striking findings of Exp 323 is the general sea-ice distribution history of the Bering Sea for the past 5 Ma. As described earlier, the first sign of sea ice is the presence of pebbles, which are thought to be transported as IRD starting at >3.8 Ma at Site U1340 (Fig. 4), indicating the formation of sea ice or iceberg transportation to the Bowers Ridge region. The bulk of the evolution of sea-ice distribution has been interpreted from shipboard analyses of sea-ice diatoms and sea-ice-related dinoflagellate taxa and to a lesser extent by other diatom taxa and intermediate water-dwelling radiolarians.

The details of sea-ice evolution are derived from changes in the relative abundance of sea-ice diatom taxa, which are represented mainly by *Thalassiosira antarctica* spores (Abelmann, 1992a) and sea-ice dinoflagellates. The first signs of sea ice diatoms and dinoflagellates are subtle increases in their abundances at Bowers Ridge Site U1340, starting at ~3.4 Ma for dinoflagellates and ~2.7 Ma for diatoms, coincident with NHG (Maslin et al., 1996). Later than ~2 Ma the sea-ice assemblage signals become progressively stronger into the present, up to values of ~10%–20% of the total respective assemblages. In contrast to the Bowers Ridge sites, sea-ice cover at the Bering slope sites is markedly severe, indicated by significantly higher sea-ice assemblage percentages. Sea-ice diatom values range from ~10% to 50% during the latest Pliocene and increase from ~30% to 70% during the Pleistocene. Notably, sea-ice diatom and dinoflagellate assemblages clearly show a significant increase in both abundance and amplitudes of variability around the

mid-Pleistocene Transition (MPT). Analogous to sea-ice-associated diatom and dinoflagellate taxa, a clear increasing trend in abundance of intermediate water-dwelling radiolarian taxa at the MPT is also observed at Sites U1343 and U1344. This is consistent with the interpretation that the surface water became gradually more affected by the formation of sea ice as climate progressively cooled; in the presence of sea ice, surface-dwelling radiolarians disappeared, and as a consequence, the relative percentages of intermediate water dwellers such as *Cycladophora davisiana* were higher (Abelmann, 1992b; Okazaki et al., 2003).

At Sites U1343 and U1344, which are located closer to the southern boundary of today's seasonal sea-ice maximum extent than the Bowers Ridge sites, a dramatic change in the dominance of dinoflagellate cyst assemblages from autotrophic to heterotrophic taxa is evident at ~1.2–1.5 Ma. This suggests that sea-ice formation occurred well before the time when the abundance of sea-ice taxa significantly increased at ~1 Ma. Along with significant increases of both sea-ice dinoflagellates and diatoms (e.g., *T. antarctica* spores) at ~1 Ma at both sites, there is a rather significant drop in the abundance of the typical pelagic diatom *Neodenticula seminae*. All of these biotic events are within the time interval of the MPT, which spans from ~1.2 Ma to 0.8 Ma and marks the transition from 41 k.y. obliquity ice volume cycles to longer ice age cycles that vary at ~100-k.y. frequencies.

As noted above, the Bowers Ridge and Bering slope regions show distinct differences in the extent of sea-ice cover throughout the last ~2.4 Ma. The extent of sea-ice cover of the latter was substantially greater than that of the former because of the proximal locations of the three Bering slope sites, which are most prone to perennial sea-ice cover in the Bering Sea. The spatial differences in sea-ice cover today are mainly attributed to the surface water circulation pattern; this spatial difference appears to have persisted for at least 2.4 Ma, implying that the surface water circulation patterns were comparable as well.

Changes in Biological Productivity and Subarctic Pacific Water Mass Entry

Based on the spatial distributions of long-term temporal changes of three diatom taxa (*Coscinodiscus marginatus*, *Neodenticula*, and *Actinocyclus curvatulus*), it is clear that the influence of subarctic Pacific waters, which are relatively warm and less eutrophic than Bering Sea waters, has typically been strongest at the Bowers Ridge sites, followed by the Umnak site. The weakest influence of this warm water mass has occurred at the Bering slope sites. The same pattern was found by Katsuki and Takahashi's (2005) study of past water mass circulation patterns, which they inferred from sea-ice distributions over the last glacial period. The wide-ranging records from Exp 323 indicate that as climate cooled through the Pleistocene, pelagic water influence at all the sites progressively weakened. Furthermore, the sites closest to straits through which pelagic water flows into the Bering Sea have consistently higher abundances of subarctic diatom species than those downstream of the counterclockwise circulation pattern of the surface water masses.

From the bottom of the holes upward at the Bowers Ridge sites, a marked drop in *C. marginatus* was seen at ~2.8 Ma at Site U1341 and at ~2.6 Ma at Site U1340. This can be interpreted as resulting from a sharp reduction in supply of nutrients due to the development of upper layer stratification. It is apparent that the diatom taxon *C. marginatus* requires a relatively high nutrient supply and tolerates low light intensity. This is substantiated by the fact that (1) today this diatom taxon dwells in the lower euphotic zone off Spain (Nogueira et al., 2000; Nogueira and Figueiras, 2005), and (2) it occurs during early winter (November–January) in the subarctic Pacific and the Bering Sea based on time-series sediment trapping (Takahashi, 1986; Takahashi et al., 1989; Onodera and Takahashi, 2009). This timing of 2.8–2.6 Ma coincides approximately with the so-called end of "opal dump" observed in the subarctic Pacific at ~2.7 Ma, which is coincidental with the onset of NHG (Maslin et al., 1996). Although the reduction in *C. marginatus* around the time of NHG persisted, an overwhelmingly continuous presence of diatom ooze and interbedded diatom ooze and silt sediments accumulated throughout the Pliocene–Pleistocene in the Bering Sea. This clearly suggests that a high amount of opal sedimentation continued after the onset of NHG well into the Pleistocene.

The 5-Ma long-term trend of *Neodenticula* (*N. kamtschatica*, *N. koizumii*, *N. seminae*, and *Neodenticula* sp.) in the Bowers Ridge region shows the following patterns. Generally higher percentages of *Neodenticula* in total diatoms are observed from the base of the holes towards younger ages until ~2.8–2.7 Ma. After that, there is a decline in *Neodenticula* with sizable fluctuations, indicating that surface water stratification progressively developed as the climate cooled from the Pliocene into the Pleistocene. As surface waters became increasingly stratified, especially after ~0.9 Ma with Milankovitch-scale 100-k.y. climatic cyclic regimes, *N. seminae* declined with the emerging sea-ice diatoms.

Changes in Bottom and Intermediate Water Conditions

In order to elucidate the history, temporal variability, and intensity of NPIW and deepwater formation in the Bering Sea and its links to surface water processes, the insights provided by the investigation on benthic foraminifera and midwater radiolarians are prerequisites. The Bering Sea sites ranged from 818 m to 3174 m in depth, and they allow for characterization of past vertical water mass distribution and for reconstruction of the history of the OMZ distribution in the region (Fig. 2). Shipboard analyses of sediment samples during Exp 323 show continuous recovery of Pliocene to Holocene deep-sea benthic foraminifera and midwater radiolarians at all sites, although calcareous benthic foraminifera appear to be rare in the Pliocene. The benthic foraminifera composition displays large assemblage changes, likely related to variability in local bottom water oxygen concentration in the bottom waters associated with surface water productivity and/or deepwater ventilation on Milankovitch and shorter timescales. For example, *Bulimina* aff. *exilis*, a common species in Bering Sea samples, is generally regarded as a low oxygen/deep infaunal species and has been found in samples associated with high productivity and low sea ice (Bubenshchikova et al., 2008; Kaiho, 1994).

Previous piston core studies showed a large increase in the intensity of the OMZ during the last deglacial at Umnak Plateau (Okazaki et al., 2005), suggesting a relationship between productivity and terrestrial nutrient supply from melting ice and increased river input. However, there is no information regarding the longer timescale relationship through the Pleistocene. Analysis of fauna from the newly drilled Bering Sea sites will be particularly important in extending this record through the entire Pliocene (at Bowers Ridge) and Pleistocene (at Bowers Ridge and the Bering slope). It will allow us to decipher the onset and evolution of the OMZ and provide further insight into NPIW production in this marginal sea. Furthermore, Site U1344 at ~3200 m (presently located below the OMZ) has the potential to provide records of past deepwater changes.

A striking finding of the expedition was the relatively low oxygen content of intermediate water mass conditions at most sites during the last 5 Ma, as indicated by the presence of episodic laminated sediment intervals throughout the entire record. The benthic foraminifera *Martinottiella communis* occurred persistently in the Pliocene. The co-occurrence of other low oxygen species (e.g., *Bulimina* aff. *mexicana*), together with the modern distribution of *M. communis* in OMZs, indicates low oxygen conditions persisted throughout the last 5 Ma. However *M. communis* is not recorded in the Bering Sea after ~2 Ma, suggesting changes to deepwater properties after this time. Abundant calcareous benthic species occur after ~2 Ma (e.g., *Bulimina, Globobulimina, Islandiella, Nonionella,* and *Valvulineria*) that are typically indicative of very low oxygen conditions (Bubenshchikova et al., 2008).

High sediment accumulation rate at Sites U1339 and U1345, located within the current OMZ (Figs. 2, 4), reveal high-amplitude variability in the relative abundance of the deep infaunal assemblage for the past 0.8 Ma. This appears to be associated with interglacial-deglacial cyclicity, represented by higher abundance of deep infaunal species (reflecting the lowest bottom water oxygen conditions) during interglacials. This particularly true during the strong interglacial-like Marine Isotope Stages (MIS) 1, 5, and 11. Higher bottom water oxygen concentrations appear to correlate with some glacial periods. Sites U1340, U1343, and U1344 contain well-preserved foraminifera over the last 2 Ma with increasing absolute abundances of benthic and planktonic taxa across the MPT (~0.8–1.1 Ma) in association with an increase in abundance of the polar planktonic foraminifera *Neogloboquadrina pachyderma*. This cooling trend was also observed as an increase in the abundance of sea-ice dinoflagellates and diatoms and coincided with increasing intermediate water-dwelling radiolarians (e.g., *C. davisiana*). Cooling of the surface waters would have enhanced ventilation of the intermediate waters during glacials and would have increased density stratification during interglacials, contributing to a drop in oxygen content in the intermediate

and bottom waters at these times. Such a decrease in oxygen content is supported by a possible increase in deep infaunal benthic foraminifera taxa at Sites U1343 and U1344 over the MPT, but higher resolution sampling from existing core material is needed to resolve this.

Microbiology and Geochemistry in High Surface Productivity Environments

The microbiological objectives of Exp 323 were to constrain global models of subseafloor biomass and microbial respiration by quantifying subseafloor cell abundance and pore water chemistry in an extremely high productivity region of the ocean. We also sought to determine how subseafloor community composition is influenced by high productivity in the overlying water column. To meet these objectives, high-resolution sampling for microbiological analyses and pore water chemistry took place at five sites throughout the Bering Sea. Each site was selected based upon its distance from land and its levels of marine productivity determined by annual chlorophyll-a concentrations in the water column.

The geochemical data obtained during the expedition show that the present-day microbial activity along the slope sites (Sites U1339, U1343, U1344, and U1345) is substantially higher and more diverse in terms of respiration pathways than at Bowers Ridge (Sites U1340, U1341, and U1342). At the slope sites, the concentrations of microbial respiration products such as dissolved inorganic carbon (DIC), ammonium, and phosphate are approximately an order of magnitude higher than at Bowers Ridge (Fig. 6). A shallow sulfate-methane transition zone (SMTZ) (~6–11 m mbsf) is also present, indicating that both methanogenesis and sulfate reduction based on methane oxidation occur in these sediments. Pore water data suggest the presence of microbially

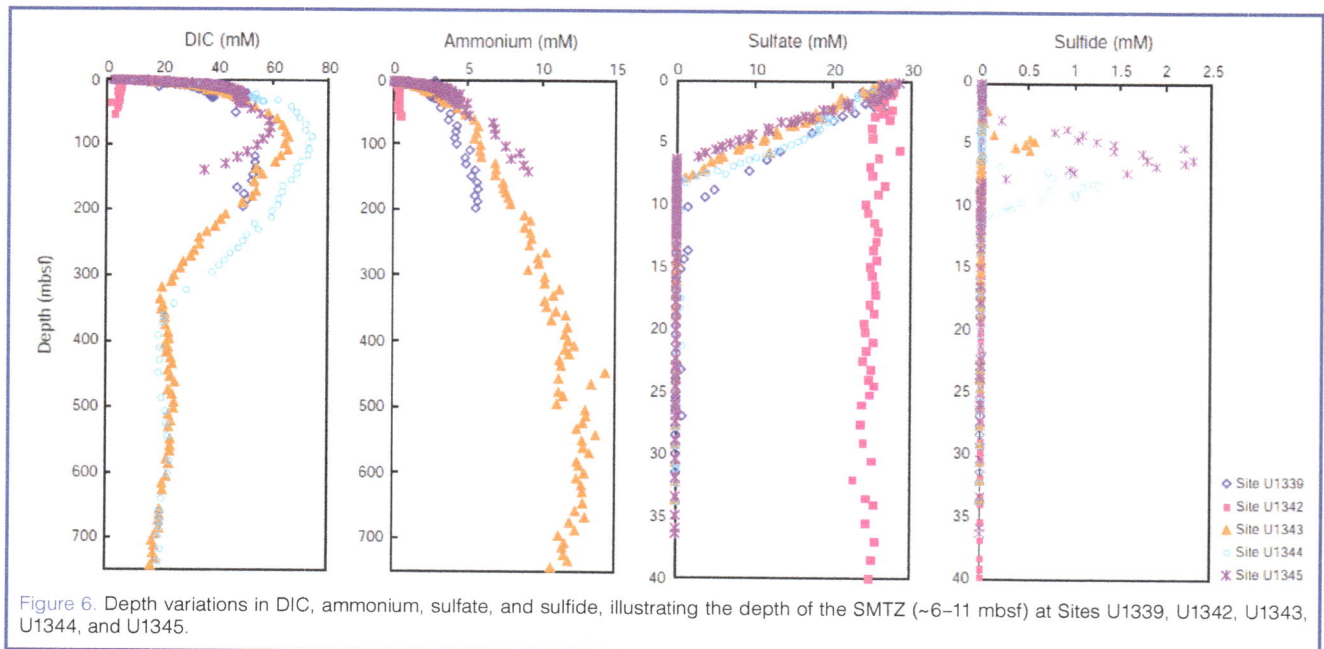

Figure 6. Depth variations in DIC, ammonium, sulfate, and sulfide, illustrating the depth of the SMTZ (~6–11 mbsf) at Sites U1339, U1342, U1343, U1344, and U1345.

mediated Fe and Mn reduction. The geochemical profiles also indicate significant microbial activity as deep as 700 mbsf. In contrast, at Bowers Ridge sulfate penetrates to the basement and is almost unaltered with depth, suggesting only very low rates of microbially mediated sulfate reduction. Methane is mostly below detection limit. The differences in microbial activity at these sites may be caused by differences in water column productivity and sedimentation patterns.

We expect that the differ-ences in the geochemical parameters between the slope and ridge sites will be reflected in microbial abundance and diversity. A larger and more diverse microbial community at the slope sites is likely. Specifically, we expect elevated cell density and an assemblade of bacteria and archaea at the SMTZ. At the slope sites, geochemical profiles suggest that methanogens, iron reducers, manganese reduc-ers, and sulfate reducers exist throughout the sediment column. At Bowers Ridge, geochemical profiles indicate that, at present, diagenetic processes are dominated by nitrate, manganese and iron reducers, while sulfate reducers and methanogens are of minor importance.

Acknowledgements

We thank the captain, operation superintendent, crew, and technicians who assisted us in drilling and sample analyses during Exp 323. Numerous people at IODP-TAMU as the USIO provided their dedicated effort supporting us, including preparation of the expedition and publication of the proceedings. We also thank the crucial members of ODP and IODP Proposal 477, which was initially written and submitted in 1995; without their cooperation this project could have never been materialized. This includes their effort during the site survey cruise on R/V Hakuho-Maru which took place in 1999 as well as the data processing performed afterwards. The curatorial team provided us with their able support during the two time sampling parties at the Kochi Core Center of JAMSTEC. We also thank the following reviewers for their constructive comments on the manuscript: Hans Christian Larsen, Ulrich Harms, Jamus Collier, Mika Saido, Renata Szarek, and Glen Hill.

The IODP Expedition 323 Scientists

K. Takahashi (Co-Chief Scientist), A.C. Ravelo (Co-Chief Scientist), C. Alvarez Zarikian (Staff Scientist), I. Aiello, H. Asahi, G. Bartoli, B. Caissie, M. Chen, E. Colmenero-Hidalgo, M. Cook, K. Dadd, G. Guèrin, Y. Huh, K. Husum, A. Ijiri, M. Ikehara, S. Kender, T. Liu, S. Lund, C. März, A. Mix, M. Ojha, M. Okada, Y. Okazaki, J. Onodera, C. Pierre, T. Radi, N. Risgaard-Petersen, T. Sakamoto, D. Scholl, H. Schrum, Z.N. Stroynowski, E.A. Walsh, and L. Wehrmann.

References

Abelmann, A., 1992a. Diatom assemblages in Arctic sea ice—indicator for ice drift pathways. *Deep-Sea Res., Part A Oceanogr. Res. Papers*, 39(2–1):S525–S538, doi:10.1016/S0198-0149(06)80019-1.

Abelmann, A., 1992b. Radiolarian flux in Antarctic waters (Drake Passage, Powell Basin, Bransfield Strait). *Polar Biol.*, 12(3–4):357–372, doi:10.1007/BF00243107.

Berger, W.H., 1970. Biogenous deep-sea sediments: fractionation by deep-sea circulation. *Geol. Soc. Am. Bull.*, 81(5):1385–1402, doi:10.1130/0016-7606(1970)81[1385:BDS-FBD]2.0.CO;2.

Bubenshchikova, N., Nürnberg, D., Lembke-Jene, L., and Pavlova, G., 2008. Living benthic foraminifera of the Okhotsk Sea: faunal composition, standing stocks and microhabitats. *Mar. Micropaleontol.*, 69(3–4):314–333, doi:10.1016/j.marmicro.2008.09.002.

Cook, M.S., Keigwin, L.D., and Sancetta, C.A., 2005. The deglacial history of surface and intermediate water of the Bering Sea. *Deep-Sea Res., Part II Topical Studies Oceanogr.*, 52(16–18):2163–2173, doi:10.1016/j.dsr2.2005.07.004.

Expedition 323 Scientists, 2010. Bering Sea paleoceanography: Pliocene–Pleistocene paleoceanography and climate history of the Bering Sea. *IODP Prel. Rept.*, 323, doi:10.2204/iodp.pr.323.2010.

Gorbarenko, S.A., 1996. Stable isotope and lithological evidence of late glacial and Holocene oceanography of the northwestern Pacific and its marginal seas. *Quat. Res.*, 46(3):230–250, doi:10.1006/qres.1996.0063.

Gradstein, F.M., Ogg, J.G., and Smith, A.G., 2004. *A Geologic Time Scale 2004*: Cambridge (Cambridge University Press).

Haug, G.H., Sigman, D.M., Tiedemann, R., Pedersen, T.F., and Sarnthein, M., 1999. Onset of permanent stratification in the subarctic Pacific Ocean. *Nature (London, U.K.)*, 401(6755):779–782, doi:10.1038/44550.

Hendy, I.L., and Kennett, J.P., 2003. Tropical forcing of North Pacific intermediate water distribution during late Quaternary rapid climate change? *Quat. Sci. Rev.*, 22(5–7):673–689, doi:10.1016/S0277-3791(02)00186-5.

Honjo, S., 1990. Particle fluxes and modern sedimentation in the polar oceans. *In* Smith, W.O., Jr. (Ed.), *Polar Oceanography (Pt. B): Chemistry, Biology, and Geology*: New York (Academic), 687–739.

Hood, D.W., 1983. The Bering Sea. *In* Ketchum, B.H. (Ed.), *Estuaries and Enclosed Seas*: The Netherlands (Elsevier), 337–373.

Kaiho, K., 1994. Benthic foraminiferal dissolved-oxygen index and dissolved-oxygen levels in the modern ocean. *Geology*, 22(8):719–722, doi:10.1130/ 0091-7613(1994)022<0719:BFDOIA>2.3.CO;2.

Katsuki, K., and Takahashi, K., 2005. Diatoms as paleoenvironmental proxies for seasonal productivity, sea-ice and surface circulation in the Bering Sea during the late Quaternary. *Deep-Sea Res., Part II Topical Studies Oceanogr.*, 52(16–18):2110–2130, doi:10.1016/j.dsr2.2005.07.001.

Krissek, L.A., 1995. Late Cenozoic ice-rafting records from Leg 145 sites in the North Pacific: late Miocene onset, late Pliocene intensification, and Pliocene–Pleistocene events. *In* Rea, D.K., Basov, I.A., Scholl, D.W., and Allan, J.F. (Eds.), *Proc. ODP, Sci. Results*, 145: College Station, TX (Ocean Drilling Program), 179–194. doi:10.2973/odp.proc.sr.145.118.1995.

Lagoe, M.B., Eyles, C.H., Eyles, N., and Hale, C., 1993. Timing of late Cenozoic tidewater glaciation in the far North Pacific. *Geol. Soc. Am. Bull.*, 105(12):1542–1560, doi:10.1130/ 0016-7606 (1993)105<1542:TOLCTG>2.3.CO;2.

Lehmann, M.F., Sigman, D.M., McCorkle, D.C., Brunelle, B.G., Hoffmann, S., Kienast, M., Cane, G., and Clement, J., 2005. Origin of the deep Bering Sea nitrate deficit: constraints from the nitrogen and oxygen isotopic composition of water column nitrate and benthic nitrate fluxes. *Global Biogeochem. Cycles*, 19(4):GB4005, doi:10.1029/2005GB 002508.

Maslin, M.A., Haug, G.H., Sarnthein, M., and Tiedemann, R., 1996. The progressive intensification of Northern Hemisphere glaciation as seen from the North Pacific. *Geol. Rundsch.*, 85(3):452–465, doi:10.1007/BF02369002.

Matsumoto, K., Oba, T., Lynch-Stieglitz, J., and Yamamoto, H., 2002. Interior hydrography and circulation of the glacial Pacific Ocean. *Quat. Sci. Rev.*, 21(14–15):1693–1704, doi:10.1016/ S0277-3791(01)00142-1.

Nedashkovskiy, A.P., and Sapozhnikov, V.V., 1999. Variability in the components of the carbonate system and dynamics of inorganic carbon in the western Bering Sea in summer. *In* Loughlin, T.R., and Ohtani, K. (Eds.), *Dynamics of the Bering Sea. A Summary of Physical, Chemical, and Biological Characteristics, and a Synopsis of Research on the Bering Sea*: Fairbanks (University of Alaska Sea Grant), 311–322.

Nogueira, E., and Figueiras, F.G., 2005. The microplankton succession in the Ria de Vigo revisited: species assemblages and the role of weather-induced, hydrodynamic variability. *J. Mar. Syst.*, 54(1–4):139–155, doi:10.1016/j.jmarsys.2004. 07.009.

Nogueira, E., Ibanez, F., and Figueiras, F.G., 2000. Effect of meteorological and hydrographic disturbances on the microplankton community structure in the Ría de Vigo (NW Spain). *Mar. Ecol. Prog. Ser.*, 203:23–45, doi:10.3354/meps203023.

Ohtani, K., 1965. On the Alaskan Stream in summer. *Bull. Fac. Fish., Hokkaido Univ.*, 15:260–273. (In Japanese).

Okazaki, Y., Takahashi, K., Asahi, H., Katsuki, K., Hori, J., Yasuda, H., Sagawa, Y., and Tokuyama, H., 2005. Productivity changes in the Bering Sea during the late Quaternary. *Deep-Sea Res., Part II Topical Studies Oceanogr.*, 52(16–18):2150–2162, doi:10.1016/j.dsr2.2005.07.003.

Okazaki, Y., Takahashi, K., Yoshitani, H., Nakatsuka, T., Ikehara, M., and Wakatsuchi, M., 2003. Radiolarians under the seasonally sea-ice covered conditions in the Okhotsk Sea: flux and their implications for paleoceanography. *Mar. Micropaleontol.*, 49(3):195–230, doi:10.1016/S0377-8398(03) 00037-9.

Onodera, J., and Takahashi, K. 2009. Long-term diatom fluxes in response to oceanographic conditions at Stations AB and SA in the central subarctic Pacific and the Bering Sea, 1990-1998. *Deep-Sea Research I*, 56(2):189–211. doi:10.1016/j. dsr.2008.08.006.

Sambrotto, R.N., Goering, J.J., and McRoy, C.P., 1984. Large yearly production of phytoplankton in the western Bering Strait. *Science*, 225(4667):1147–1150, doi:10.1126/science.225. 4667.1147.

Scholl, D.W., and Creager, J.S., 1973. Geologic synthesis of Leg 19 (DSDP) results; far North Pacific, and Aleutian Ridge, and Bering Sea. *In* Creager, J.S., Scholl, D.W., et al., *Init. Repts.*

DSDP, 19: Washington, DC (U.S. Govt. Printing Office), 897–913, doi:10.2973/dsdp.proc.19.137.1973.

Solomon, S., Qin, D., Manning, M., Marquis, M., Averyt, K., Tignor, M.M.B., Miller, H.L., Jr., and Chen, Z., 2007. *Climate Change 2007: The Physical Science Basis*: Cambridge (Cambridge University Press).

Stabeno, P.J., Schumacher, J.D., and Ohtani, K., 1999. The physical oceanography of the Bering Sea. *In* Loughlin, T.R., and Ohtani, K. (Eds.), *Dynamics of the Bering Sea: A Summary of Physical, Chemical, and Biological Characteristics, and a Synopsis of Research on the Bering Sea*: Fairbanks (University of Alaska Sea Grant), 1–28.

Takahashi, K., 1986. Seasonal fluxes of pelagic diatoms in the subarctic Pacific, 1982–1983. *Deep-Sea Res., Part A Oceanogr. Res. Papers*, 33:1225–1251, doi:10.1016/0198-0149(86)90022-1.

Takahashi, K., 2005. The Bering Sea and paleoceanography. *Deep-Sea Res., Part II Topical Studies Oceanogr.*, 52(16–18):2080–2091, doi:10.1016/j.dsr2.2005.08.003.

Takahashi, K., Honjo, S., and Tabata, S., 1989. Siliceous phytoplankton flux: interannual variability and response to hydrographic changes in the northeastern Pacific. *In* Peterson, D. (Ed.), *Aspects of Climate Variability in the Pacific and Western Americas*. Geophys. Monogr., 151–160.

Takahashi, K., Jordan, R.W., and Boltovskoy, D., 2005. *Deep-Sea Res., Part II Topical Studies Oceanogr.*, 52(16–18):2079–2364, doi:10.1016/j.dsr2.2005.08.002.

Takahashi, K., Ravelo, A.C., Alvarez Zarikian, C.A., and the Expedition 323 Scientists. 2011. *Proc. IODP*, 323: Tokyo Integrated Ocean Drilling Program Management International, Inc.), doi:10.2204/iodp.proc.323.2011.

Tanaka, S., and Takahashi, K., 2005. Late Quaternary paleoceanographic changes in the Bering Sea and the western subarctic Pacific based on radiolarian assemblages. *Deep-Sea Res., Part II Topical Studies Oceanogr.*, 52(16–18):2131–2149, doi:10.1016/j.dsr2.2005.07.002.

Warner, M.J., and Roden, G.I., 1995. Chlorofluorocarbon evidence for recent ventilation of the deep Bering Sea. *Nature (London, U.K.)*, 373(6513):409–412, doi:10.1038/373409a0.

Authors

Kozo Takahashi, Department of Earth & Planetary Sciences, Graduate School of Sciences, Kyushu University, Hakozaki 6-10-1, Higashi-ku, Fukuoka 812-8581, Japan, e-mail: kozo@geo.kyushu-u.ac.jp.

A. Christina Ravelo, Ocean Sciences Department, University of California, 1156 High Street, Santa Cruz, CA 95064, U.S.A.

Carlos Alvarez Zarikian, Integrated Ocean Drilling Program & Department of Oceanography, Texas A&M University, 1000 Discovery Drive, College Station, TX 77845-9547, U.S.A.

and the IODP Expedition 323 Scientists

Related Web Link

http://www.iodp.tamu.edu/scienceops/expeditions/bering_sea.html

15

Scientific Drilling Into the San Andreas Fault Zone —An Overview of SAFOD's First Five Years

by Mark Zoback, Stephen Hickman, William Ellsworth, and the SAFOD Science Team

Abstract

The San Andreas Fault Observatory at Depth (SAFOD) was drilled to study the physical and chemical processes controlling faulting and earthquake generation along an active, plate-bounding fault at depth. SAFOD is located near Parkfield, California and penetrates a section of the fault that is moving due to a combination of repeating microearthquakes and fault creep. Geophysical logs define the San Andreas Fault Zone to be relatively broad (~200 m), containing several discrete zones only 2–3 m wide that exhibit very low P- and S-wave velocities and low resistivity. Two of these zones have progressively deformed the cemented casing at measured depths of 3192 m and 3302 m. Cores from both deforming zones contain a pervasively sheared, cohesionless, foliated fault gouge that coincides with casing deformation and explains the observed extremely low seismic velocities and resistivity. These cores are being now extensively tested in laboratories around the world, and their composition, deformation mechanisms, physical properties, and rheological behavior are studied. Downhole measurements show that within 200 m (maximum) of the active fault trace, the direction of maximum horizontal stress remains at a high angle to the San Andreas Fault, consistent with other measurements. The results from the SAFOD Main Hole, together with the stress state determined in the Pilot Hole, are consistent with a strong crust/weak fault model of the San Andreas. Seismic instrumentation has been deployed to study physics of faulting—earthquake nucleation, propagation, and arrest—in order to test how laboratory-derived concepts scale up to earthquakes occurring in nature.

Introduction and Goals

SAFOD (the San Andreas Fault Observatory at Depth) is a scientific drilling project intended to directly study the physical and chemical processes occurring within the San Andreas Fault Zone at seismogenic depth. The principal goals of SAFOD are as follows: (i) study the structure and composition of the San Andreas Fault at depth, (ii) determine its deformation mechanisms and constitutive properties, (iii) measure directly the state of stress and pore pressure in and near the fault zone, (iv) determine the origin of fault-zone pore fluids, and (v) examine the nature and significance of time-dependent chemical and physical fault zone processes (Zoback et al., 2007).

Detailed planning of a research experiment focused on drilling, sampling, and downhole measurements directly within the San Andreas Fault Zone began with an international workshop held in Asilomar, California in December 1992. This workshop highlighted the importance of deploying a permanent geophysical observatory within the fault zone at seismogenic depth for near-field monitoring of earthquake nucleation. Hence, from the outset, the SAFOD project has been designed to achieve two parallel suites of objectives. The first is to carry out a series of experiments in and near the San Andreas Fault that address long-standing questions about the physical and chemical processes that control deformation and earthquake generation within active fault zones. The second is to make near-field observations of earthquake nucleation, propagation, and arrest to test how laboratory-derived concepts about the physics of faulting

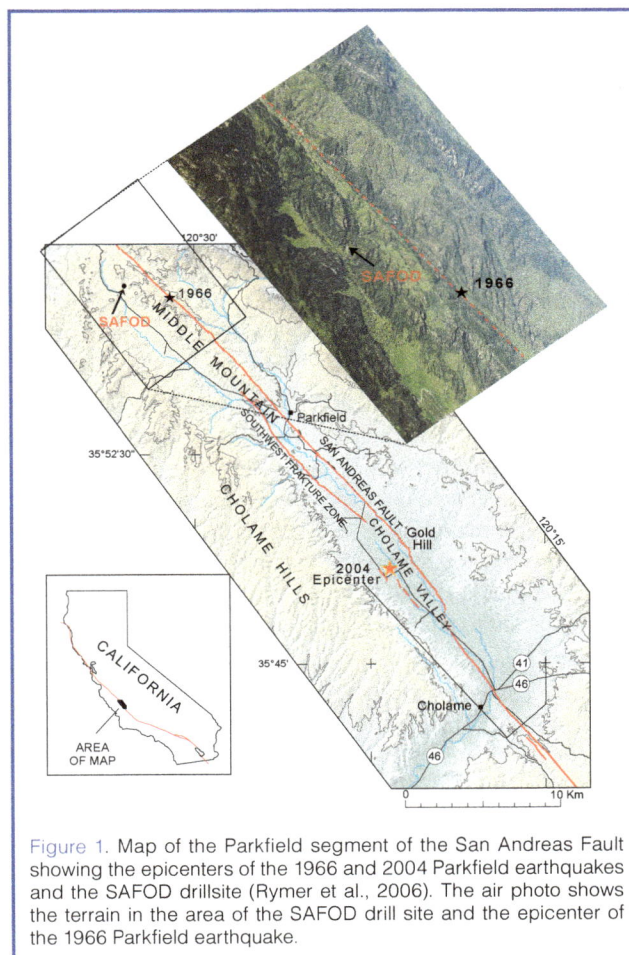

Figure 1. Map of the Parkfield segment of the San Andreas Fault showing the epicenters of the 1966 and 2004 Parkfield earthquakes and the SAFOD drillsite (Rymer et al., 2006). The air photo shows the terrain in the area of the SAFOD drill site and the epicenter of the 1966 Parkfield earthquake.

scale up to earthquakes occurring in nature. In the years following the Asilomar workshop, dozens of planning meetings were held to synthesize the research questions of highest scientific priority that were deemed to be operationally achievable. Numerous other meetings were also held related to site selection and to detailed operational plans for drilling, sampling, downhole measurements, and long-term monitoring.

When planning of the EarthScope initiative got underway at the National Science Foundation (NSF) in the late 1990s, the project was named SAFOD and became one of the three components of EarthScope along with the Plate Boundary Observatory (PBO) and USArray. In 2002, a 2.2-km-deep Pilot Hole was funded by the International Continental Scientific Drilling Program (ICDP) and was drilled at the SAFOD site. The main SAFOD project started when NSF funded the EarthScope proposal in 2003, with substantial cost sharing and operational support for SAFOD provided by the U.S. Geological Survey (USGS), ICDP, and other agencies.

The SAFOD operational plan was designed to address a number of first-order scientific questions related to fault mechanics in a hostile environment where the mechanically and chemically altered rocks in the fault zone are subject to high mean stress, potentially high pore pressure, and elevated temperature. Some of these questions are listed below.

- *What are the mineralogy, deformation mechanisms, and constitutive properties of fault gouge?* Why do some faults creep? What are the strength and frictional properties of recovered fault rocks at *in situ* conditions of stress, fluid pressure, temperature, strain rate, and pore fluid chemistry? What determines the depth of the shallow seismic-to-aseismic transition? What do mineralogical, geochemical, and microstructural analyses reveal about the nature and extent of water-rock interaction?

- *What is the fluid pressure and permeability within and adjacent to fault zones?* Are there super-hydrostatic fluid pressures within some fault zones, and through what mechanisms are these pressures generated and/or maintained? How does fluid pressure vary during deformation and episodic fault slip (creep and earthquakes)? Do fluid pressure seals exist within or adjacent to fault zones, and at what scales?

- *What are the composition and origin of fault-zone fluids and gases?* Are these fluids of meteoric, metamorphic, or mantle origin (or combinations of the three)? Is fluid chemistry relatively homogeneous, indicating pervasive fluid flow and mixing, or heterogeneous, indicating channelized flow and/or fluid compartmentalization?

- *How do stress orientations and magnitudes vary across fault zones?* Are principal stress directions and magni-

tudes different within the deforming core of weak fault zones compared to the adjacent (stronger) country rock, as predicted by some theoretical models? How does fault strength measured in the near field compare with depth-averaged strengths inferred from heat flow and regional stress directions? What is the nature and origin of stress heterogeneity near active faults?

- *How do earthquakes nucleate?* Does seismic slip begin suddenly, or do earthquakes begin slowly with accelerating fault slip? Do the size and duration of this precursory slip episode, if it occurs, scale with the magnitude of the eventual earthquake? Are there other precursors to an impending earthquake, such as changes in pore pressure, fluid flow, crustal strain, or electromagnetic field?

- *How do earthquake ruptures propagate?* Do they propagate as a uniformly expanding crack, as a slip pulse, or as a sequence of slipping high-strength asperities? What is the effective (dynamic) stress during seismic faulting? How important are processes such as shear heating, transient increases in fluid pressure, and fault-normal opening modes in lowering the dynamic frictional resistance to rupture propagation?

- *How do earthquake source parameters scale with magnitude and depth?* What is the minimum size earthquake that occurs on faults? How is long-term energy release rate partitioned between creep dissipation, seismic radiation, dynamic frictional resistance, and grain size reduction (determined by integrating fault zone monitoring with laboratory observations on core)?

- *What are the physical properties of fault-zone materials and country rock (seismic velocities, electrical resistivity, density, porosity)?* How do physical properties from core samples and downhole measurements compare with properties inferred from surface geophysical observations? What are the dilational, thermoelastic, and fluid-transport properties of fault and country rocks, and how might they interact to promote either slip stabilization or transient over-pressurization during faulting?

- *What processes control the localization of slip and strain?* Are fault surfaces defined by background microearthquakes and creep the same? Would active slip surfaces be recognizable through core analysis and downhole measurements in the absence of seismicity and/or creep?

In addition, a substantial body of evidence indicates that slip along major plate-bounding faults like the San Andreas occurs at much lower levels of shear stress than expected, based upon laboratory friction measurements on standard rock types and assuming hydrostatic pore fluid pressures (i.e., it is a weak fault). Yet, the cause of this weakness has remained elusive (Hickman, 1991). In the context of the San Andreas, two principal lines of evidence indicate that the fault has low frictional strength: the absence of frictionally-generated heat, and the orientation of the maximum principal stress in the crust adjacent to the fault. A large number of heat flow measurements show no evidence of frictionally generated heat adjacent to the San Andreas Fault (Lachenbruch and Sass, 1980, 1992; Williams et al., 2004), which implies that shear motion along the fault is resisted by shear stresses approximately a factor of five less than fric-tional strength of the adjacent crust. This observation is sometimes referred to as the San Andreas stress/heat flow paradox. Saffer et al. (2003) showed that it is highly unlikely that topographically driven fluid flow has an appreciable effect on these heat flow measurements, indicating that the lack of frictionally-generated heat in the vicinity of the San Andreas Fault is indeed indicative of low aver-age shear stress levels acting on the fault at depth. In addition to the heat flow data, the orientation of principal stresses in the vicinity of the fault also indicates that right-lateral strike slip motion on the fault occurs in response to low levels of shear stress (Zoback et al., 1987; Mount and Suppe, 1987; Oppenheimer et al., 1988).

Why Parkfield? SAFOD is located in central California (Fig. 1) near the town of Parkfield, at the transition between the locked (i.e., seismogenic) portion of the fault to the southeast and the segment of the fault to the northwest where slip dominantly occurs by aseismic creep. The fault is seismically active around SAFOD with numerous sites of repeating microearthquakes, M3 and smaller, occurring on the fault at depths of 2–12 km (Waldhauser et al., 2004). The Parkfield segment of the fault hosts the well-studied seven M6 earthquakes that have ruptured since 1857 (Bakun and McEvilly, 1984). Slip distributions for the last two Parkfield earthquakes—on 28 June 1966 and 28 September 2004—determined using geodetic measurements, indicate that the ruptures terminated a few kilometers southeast of SAFOD (Murray and Langbein, 2006; Harris and Arrowsmith, 2006 and papers therein).

Beginning at the Asilomar meeting, site selection committees winnowed down eighteen potential sites to four, and eventually the northwest end of Parkfield segment was selected. The geology seemed ideal since Salinian granite on the west side of the fault was expected to be juxtaposed against Franciscan melange on the east side, so a major geologic discontinuity was expected when crossing the fault at depth. Also, the San Andreas Fault is quite active in the area, exhibiting a combination of aseismic creep and frequent microearthquakes that would help define the exact location of the active fault trace at depth. In addition, more is known about this section of the San Andreas than any other, due to the intense interest in capturing a M~6 earthquake within a dense network of instrumentation.

After selecting the Parkfield segment of the San Andreas for the SAFOD experiment, the next question of particular importance was where exactly to site the borehole. The site chosen (Fig. 1) was selected near Middle Mountain because repeating microearthquakes could be reached at the shallowest depth possible close to the fault to limit the horizontal reach of the borehole. As shown in the photo inset of Fig. 1, the selected site is a broad, relatively flat area where a 5-acre drill pad could be constructed 1.8 km southwest of the surface trace of the fault. Once this area was identified, a number of detailed geophysical and geologic site studies were carried out to allow results from SAFOD to be placed in the appropriate geological and geophysical context and to assure that the drill site selected would not encounter any large-scale faults or struc-

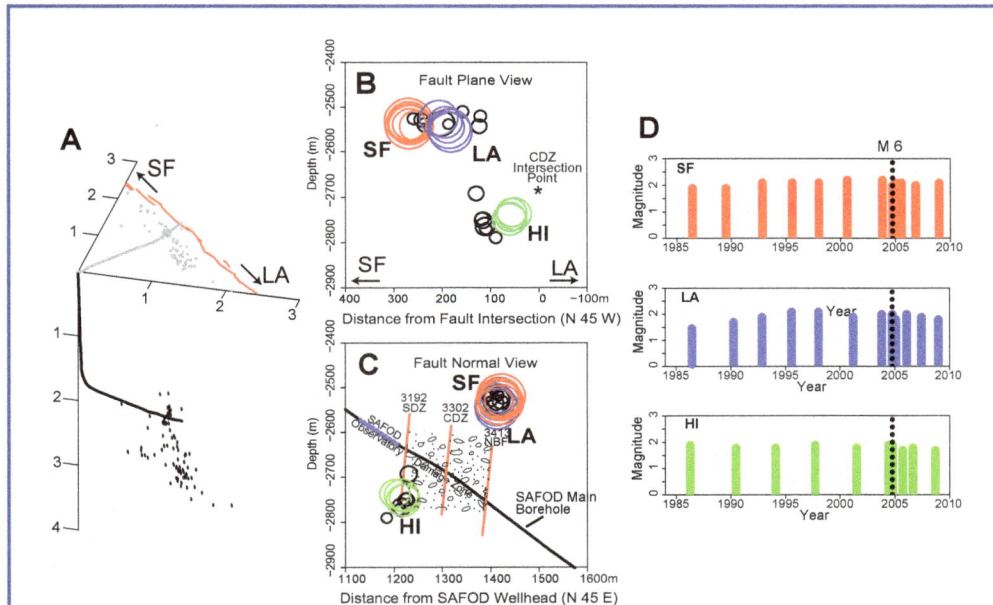

Figure 2. Microearthquakes selected for targeting with SAFOD. [A] 3-D perspective view of the seismicity with respect to the path of the SAFOD borehole, with north pointing up, east to the right, and depth down (all axes in km). [B] View of the plane of the San Andreas Fault at about 2.7-km depth looking to the northeast. The red, blue, and green circles represent seismogenic patches of the San Andreas Fault that produce nearly identical, regularly repeating microearthquakes termed the San Francisco (SF), Los Angeles (LA), and Hawaii (HI) clusters, respectively. The point at which the SAFOD borehole passes through the central deforming zone (CDZ) is shown by the asterisk. [C] Cross-sectional view of these earthquakes looking to the northwest, parallel to the San Andreas Fault, including the trajectory of the SAFOD borehole and the principal faults associated with the damage zone shown in Figs. 3 and 4. Note that the HI events occur about 100 m below the fault intersection at 3192 m (measured depth), indicating that the HI microearthquakes occur on the southwest deforming zone (SDZ). The SF and LA sequences occur on the northwest bounding fault (NBF), as discussed in the text.

tural complexities in the near surface. These studies included an extensive microearthquake survey, high-resolution seismic reflection/refraction profiling, magnetotelluric profiling, ground and closely-spaced aeromagnetic surveys, gravity surveys, and geologic mapping.

The repeating microearthquakes provide targets on the fault plane at depth to guide the drilling trajectory (Fig. 2A) into the microearthquake zone at less than 3 km depth. Another reason for choosing this site is that there are three sets of repeating M~2 earthquakes in the target area. Surrounding these patches, fault slip occurs through aseismic creep. In a view normal to the plane of the San Andreas Fault Zone at 2.65 km depth (Fig. 2B), we see the source zones associated with these three patches (scaled for a ~10-MPa stress drop). The seismograms from each of these source zones are essentially identical (Nadeau et al., 2004), and cross-correlation demonstrates that within ±10 m uncertainty these events are located in exactly the same place on the faults (F. Waldhauser, pers. comm.).

As shown in Fig. 2B, we refer to the shallower source zone in the direction of San Francisco as the SF events, and the adjacent source zone in the direction of Los Angeles as LA events. Note that the SF and LA patches are adjacent to each other; it is common for LA events to occur immediately after SF events as triggered events. As seen in Fig. 2B, the third cluster of events (in green) occurs on a fault plane to the southwest of that upon which the SF and LA events occur. As this cluster of events is to the southwest of the other two clusters, these are referred to as the Hawaii (HI) events.

The time sequences of the three clusters of repeating earthquakes are shown in Fig. 2C. Note that prior to the M6 Parkfield earthquake of September 2004, each of the three clusters produced an event every ~2.5–3.0 years. Following the Parkfield earthquake, the frequency of the events increased dramatically, apparently due to accelerated creep on this part of the fault resulting from stress transfer from the M~6 main-shock. Following this flurry of events the frequency of the repeaters slowed down and is presently in the process of returning to the background rate exhibited prior to the main shock. Similar behavior has been seen elsewhere along the San Andreas Fault system in California (Schaff et al., 1998).

Note in Fig. 2B that the HI events occur about 100 m below the fault intersection at 3192 m (measured depth), indicating that the HI microearthquakes occur on the southwestern-

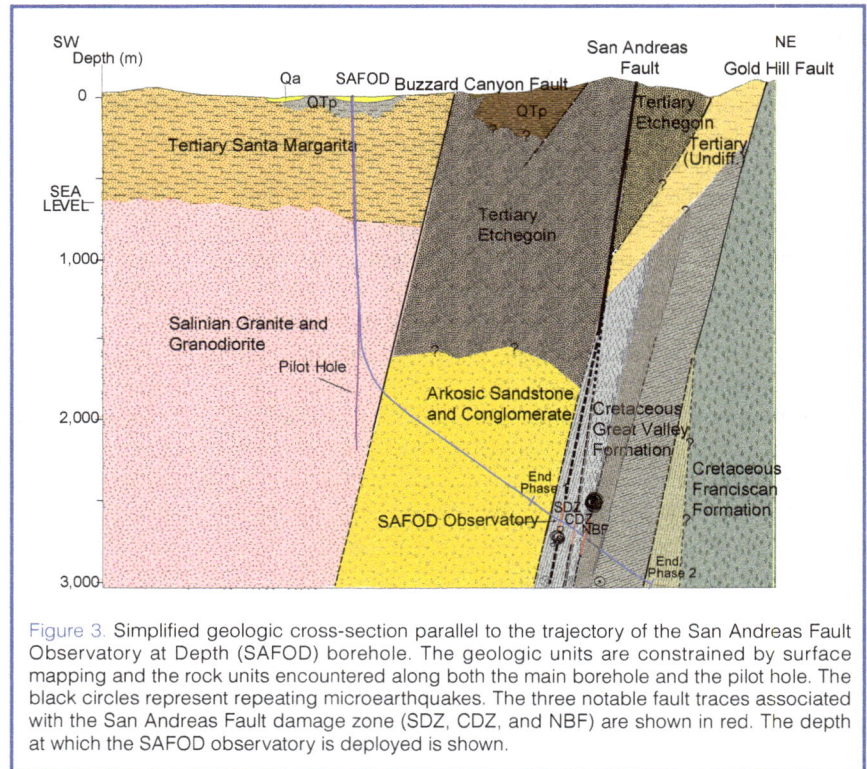

Figure 3. Simplified geologic cross-section parallel to the trajectory of the San Andreas Fault Observatory at Depth (SAFOD) borehole. The geologic units are constrained by surface mapping and the rock units encountered along both the main borehole and the pilot hole. The black circles represent repeating microearthquakes. The three notable fault traces associated with the San Andreas Fault damage zone (SDZ, CDZ, and NBF) are shown in red. The depth at which the SAFOD observatory is deployed is shown.

most of the two actively deforming fault traces identified in the SAFOD crossing. The microearthquake locations shown in Fig. 2 were determined utilizing subsurface recordings of these earthquakes from various geophone deployments in the SAFOD borehole along with surface recordings from the dense Parkfield Area Seismic Observatory (PASO; Thurber et al., 2004). This said, although the accuracy of location of HI is good (being determined by a seismometer deployed in SAFOD directly above the events), the location of SF and LA with respect to HI is relatively uncertain.

SAFOD Pilot Hole

In preparation for SAFOD, a 2.2-km-deep, near-vertical Pilot Hole was drilled and instrumented at the SAFOD site in the summer of 2002. The Pilot Hole was rotary drilled with a 22.2-cm bit, and cased with 17.8-cm outside diameter (OD) steel casing. The Pilot Hole is currently open to a depth of 1.1 km (explained below) and available for instrument testing, cross borehole experiments, and other scientific studies. Hickman et al. (2004) present an overview of the Pilot Hole experiment.

There were a number of important technical, operational, and scientific findings in the Pilot Hole. These include geologic confirmation of the depth at which the Salinian granites and granodiorites would be encountered (Fig. 3), and calibration of geophysical models with direct measurements of seismic velocities (Boness and Zoback, 2004; Thurber et al., 2004), resistivity (Unsworth and Bedrosian, 2004), density, and magnetic susceptibility (McPhee et al., 2004). In addition, stress measurements in the Pilot Hole were found to be consistent with the strong crust/weak fault model discussed above (Boness and Zoback, 2004; Hickman and Zoback, 2004). In other words, stress differences in the crust 1.8 km from the San Andreas were high and consistent with Byerlee's law, whereas the direction of maximum horizontal stress in the lower part of the hole was nearly orthogonal to the San Andreas Fault. Furthermore, heat flow measured to 2.2 km depth (Williams et al., 2004) was found to be consistent with shallower data in the region, confirming that the shallow measurements are not affected by heat transport and thus indicate no frictional heat being generated by slip on the San Andreas Fault. Hence, the Pilot Hole confirmed that the SAFOD site was indeed an appropriate site for examining possible explanations for the San Andreas stress/heat flow paradox.

After drilling and downhole measurements were completed, the Pilot Hole was used for deployment of a vertical seismic array to record naturally occurring microearthquakes and to image some of the large-scale structures at depth in the vicinity of the San Andreas (Chavarria et al., 2003; Oye and Ellsworth, 2007). This array was also used to record surface explosions as an important part of the effort to constrain seismic velocities in the vicinity of the borehole for achieving the best possible locations of the target earthquakes (Roecker et al., 2004). Use of the Pilot Hole for experiments such as cross-hole monitoring of time-varying shear velocity (Niu et al., 2008) will continue to produce interesting results for years to come. From an engineering perspective, by establishing the depth to basement and the conditions affecting drilling in the upper sedimentary section, the Pilot Hole helped establish key aspects of the engineering design of the upper part of the SAFOD Main Hole.

SAFOD Main Borehole

A great deal of engineering and operational planning went into SAFOD, since drilling, coring, and scientific measurements in the hostile environment of an active, plate-bounding fault zone had never been attempted before. A number of scientific workshops were held on drilling and downhole measurements, fault zone monitoring, and core handling. In addition, a formal advisory structure was established to take advantage of the knowledge and experience of scientists from universities, the USGS and U.S. Department of Energy (DOE) national labs, and the petroleum industry. A Scientific Advisory Board provided high-level scientific guidance for the project. Technical panels on drilling, coring, and safety, downhole measurements, core handling, and downhole monitoring provided invaluable advice on literally hundreds of issues affecting how the project was eventually carried out.

One of the most important aspects of the SAFOD operational plan that came out of this planning process was to carry out the project in three distinct phases. Phase 1,

carried out during the summer of 2004, involved rotary drilling vertically to a depth of ~1.5 km, then steering the well at an angle ~60° from vertical toward the repeating microearthquakes described above (Fig. 3). Note that these earthquakes occur to the southwest of the surface trace of the San Andreas, which indicates that at this location the fault dips steeply to the southwest. By design, Phase 1 ended just outside the San Andreas Fault Zone so that relatively large-diameter (24.4 cm) steel casing could be deployed and cemented in place prior to drilling through the active fault zone where substantial drilling problems might be encountered. Results from a number of scientific studies carried out during Phase 1 were needed to establish key engineering parameters (such as the optimal density of the drilling mud) for drilling through the San Andreas Fault during Phase 2 (Paul and Zoback, 2008).

Phase 2 was carried out during the summer of 2005. A relatively large-diameter (21.6 cm) hole was rotary drilled across the San Andreas Fault Zone (Fig. 3). While many of the key scientific objectives of SAFOD require recovery of core samples from the fault zone, we decided to rotary drill through the fault zone for several reasons. First, rotary drilling is far more robust than core drilling. If the borehole turned out to be unstable due to the rock being highly broken up and chemically altered by faulting (which turned out to be the case), and/or high pore pressure was encountered in the fault zone at depth (which was not the case), it would be much easier to deal with such problems and ensure that we would make it all the way across the fault zone with rotary drilling rather than core drilling. Second, rotary drilling produces a larger diameter hole than core drilling. This was needed to carry out a wide range of sophisticated geophysical measurements (especially well logs) in the fault zone with equipment developed for the petroleum industry. When drilling problems are encountered during coring, it is common for the drill rod to get stuck in the hole. When this happens, the sizes of drill bit and coring rods are reduced so that coring can continue through the bottom of the stuck coring rod. Consequently, the diameter of core holes start relatively small and potentially reduces rapidly. As illustrated below, these geophysical measurements proved to be critical for defining the nature of the overall fault zone as well as the active shear zones within it. The final reason for maintaining a relatively large-diameter hole was related to deployment of the observatory instrumentation in the fault zone after drilling. It was important to complete the well with sufficiently large-diameter casing (17.8 cm) to allow a suite of seismometers and accelerometers to be deployed in the borehole.

Phase 3 was carried out during the summer of 2007; it involved drilling multi-lateral holes which start by milling a hole in the side of the steel casing in the Main Hole. By using multilateral drilling to create secondary holes at optimal locations (a technology that is now commonplace in the petroleum industry), we could direct coring efforts within the most important intervals identified during Phase 2. By design, the samples and physical property measurements of the fault zone obtained during Phase 2 were not the only information available to us to guide Phase 3 coring operations. Due to accelerated fault creep following the 2004 earthquake, the casing deployed across the fault zone following Phase 2 was deformed at specific places which directly indicated the active strands of the San Andreas at depth.

Phase 1 and 2 Operational Overview. As mentioned above, Phases 1 and 2 were rotary drilled. In order to obtain as much scientific information as possible during drilling, a comprehensive real-time sampling of drill cuttings, drilling fluid, and formation gases in the drilling mud was carried out. Following each phase, a suite of geophysical measurements was obtained, and a limited amount of coring was done at each depth where casing was set.

As can be seen in Fig. 3, the Main Hole starts vertically and at approximately 1.5 km depth; directional drilling techniques were employed to slowly deviate the borehole (eventually at an angle ~60° from vertical) in order to intersect the San Andreas Fault in the vicinity of the repeating target earthquakes. A wide range of information is available online including that related to real-time operations (Table 1). One source of information that provides a convenient overview of Phases 1 and 2 are the Commercial Mud Logs, which deliver also lithologic descriptions of the drill cuttings. Numerous faults were observed in all of the rock units drilled through (Boness and Zoback, 2006). Bradbury et al. (2007) described the mineralogy of drill cuttings in terms of fault zone composition and geologic models.

The first geologic surprise that occurred during Phase 1 was that soon after deviating the borehole toward the San Andreas Fault, we drilled through a major fault zone at a vertical depth of 1.8 km (interpreted to be the Buzzard Canyon Fault, see Fig. 3) as we passed out of the Salinian granitic basement rocks and into previously unknown arkosic sandstones and conglomerates, with some interbedded shales (Boness and Zoback, 2006; Solum et al., 2006). In general, these are strongly cemented rocks that are likely derived from weathering of Salinian granites and granodiorites. Draper Springer et al. (2009) described this section in some detail and pointed to at least a dozen significant faults within

it. While they argued for this being a depositional unit formed proximal to the Salinian granite, they suggested that it may have been translated along strike by as much as ~300 km. One reason this unit had not been identified by geophysical surveys through the site area is that these rocks are so strongly cemented that their seismic velocities and resistivity do not vary significantly from the fractured Salinian granites and granodiorites (Boness and Zoback, 2006).

At a measured depth along the borehole of 1460 m (while still drilling in the granite/granodiorite), a planned pause in drilling took place to run steel casing into the hole before further drilling. Prior to casing the hole, a suite of geophysical logs was run. After running the casing into the hole and cementing it in place, 7.9 meters of fractured and faulted hornblende-biotite granodiorite core were obtained. In addition, fluid samples were taken at this depth, and a small-scale hydraulic fracturing experiment was done to constrain the magnitude of the least principal stress.

After drilling resumed, Phase 1 continued to a total vertical depth of 2507 m. As shown in Fig. 3, Phase 1 drilling ended in the arkosic sandstone/conglomerate section. At the end of Phase 1 drilling a second suite of geophysical logs was run. Boness and Zoback (2006) presented a summary of the Phase 1 lithologies and geophysical logs. After cementing steel casing into the wellbore, an 11.6-m core—composed of fractured and faulted arkosic sandstone and conglomerate—was obtained, and fluid sampling was then performed.

One mishap that occurred during Phase 1 was a collision between the Main Hole and the Pilot Hole at 1.1 km depth. Because of the respective layouts of the drilling equipment used for the Pilot and Main Holes, the wellheads of the two boreholes were located only 6.75 m apart. In an attempt to avoid collision of the two holes at depth, repeated gyroscopic surveys of both holes and directional drilling were used. This is commonplace in the oil industry where dozens of wells are often drilled from the same platform or drill site. After the incident, we learned that the collision was caused by poor calibration of one set of the gyroscopic survey instruments. The lasting impact of the hole collision is loss of access to the lower part of the Pilot Hole, as the casing is severely damaged at 1.1 km depth. The Pilot Hole seismic

Table 1. Accessing SAFOD Data Online.

Description	URL	
EarthScope Data Portal – Information about and access to all SAFOD EarthScope data and samples	http://www.earthscope.org	
IRIS DMC – SAFOD seismological data archive including assembled data sets	http://www.iris.edu/hq	
Northern California Earthquake Data Center – Earthquake catalogs and seismograms for all local networks including SAFOD, High-Resolution Seismic Network (HRSN) and NCSN	http://www.ncedc.org/safod/	
ICDP Web site – Direct access to all data obtained as drilling, logging and coring operations were underway. Bibliography of SAFOD papers.	http://safod.icdp-online.org	
Online Core Viewer – Photographs of all cores and samples taken for scientific study	http://www.earthscope.org/data/safod_core_viewer	
Phase 3 Core Atlas – High-resolution images of Phase 3 cores as well as preliminary lithologic and microstructural descriptions	http://www.icdp-online.org/upload/projects/safod/phase3/Core_Photo_Atlas_v4.pdf	
General information about the Parkfield Experiment	http://earthquake.usgs.gov/research/parkfield/index.php	

array was also lost as a consequence of the accident; the lowermost twenty-five levels were severed during the intersection, and the remaining seven levels were decommissioned in the spring of 2005 when an unsuccessful attempt was made to regain access to the Pilot Hole below the intersection.

During the nine-month hiatus (September 2004 to June 2005) between the end of Phase 1 and the beginning of Phase 2, a number of seismometers were deployed in the SAFOD Main Hole as part of an instrument testing program for eventual deployment of the SAFOD observatory. A number of shots were set off while the seismometers were in the borehole to better constrain the velocity model and reduce uncertainty in the location of the target earthquakes. In addition, an eighty-level, 240-component seismic array was made available by Paulsson Geophysical Services, Inc. (PGSI) and recorded by Geometrics at no cost to the project. This array was deployed in the borehole for a period of five weeks in order to test its suitability for recording microearthquakes and to record additional shots for structural imaging (Chavarria and Goerrtz, 2007). In addition to recording microearthquakes and shots during this period, a tectonic (i.e., non-volcanic) tremor was recorded on this array. The tremor occurred in the lower crust directly below the surface trace of the San Andreas Fault for at least 70 km to the northwest and 80 km to the southeast of SAFOD (Shelly and Hardebeck, 2010). The likely source of the tremor recorded by the PGSI array was in the vicinity of the energetic tremor source near Cholame (Nadeau and Dolenc, 2005) near the base of the crust (~25 km; Shelly and Hardebeck, 2010).

As shown in Fig. 3, Phase 2 drilling passed from the arkosic sandstones and conglomerates into mudstones and shales at a depth of 2600 m, and at a position ~500 m southwest of the surface trace of the San Andreas Fault. Microfossil evidence from core obtained at the bottom of the Phase 2 hole indicates that these formations are part of the Cretaceous Great Valley sequence, which was deposited on the North American plate in a forearc environment at a time when subduction was occurring along the western margin of California (K. McDugall, pers. comm., 2005). In the long-term geologic sense, the contact between the Salinian-derived arkosic sandstones and conglomerates and the Great Valley formation is the boundary between the Pacific and North American plates. As shown by progressive deformation of the casing discussed below (Fig. 4), the south-westernmost of the active traces of the San Andreas Fault Zone at depth is located several tens of meters to the northeast of this geologic boundary.

No evidence was found that we had encountered the Franciscan Formation in the borehole, even though it is exposed at the surface about 600 m east of the San Andreas Fault (Fig. 3), and was predicted by several of the geophysical surveys conducted in advance of drilling. However, there is evidence of serpentinite directly within the fault zone associated with either the Coast Range ophiolite or Franciscan formation. Hence, there is likely serpentinite in contact with the San Andreas along strike and/or at greater depth. A reasonable conceptual model is that slivers of Great Valley and the Franciscan are intermixed at depth along the fault, just as they are found in surface exposures at several locations in central California.

Rotary drilling through the San Andreas Fault during Phase 2 was accomplished with no small amount of difficulty—some caused by the fault zone, some caused by unrelated operational problems (for example, the top drive, an extremely important component of the drill rig, broke and was inoperable for two weeks). We also noted a considerable degree of time-dependent wellbore failure (Paul and Zoback, 2008), especially after passing through the active traces of the San Andreas Fault Zone. An appreciable amount of time was required to clean the hole through wash and ream operations. In fact, the combined result of time-dependent wellbore instabilities and a mistake by the drilling crew resulted in the drillstring being stuck in the hole for four days at a vertical depth of 2800 m. Despite these problems, drilling across the entire fault zone was successfully achieved. Comprehensive cuttings and gases were sampled over the entire Phase 2 interval (Table 2), and a number of geophysical measurements were made in real-time as drilling across the fault zone was underway (Run 4, Table 3). After the hole was drilled, a comprehensive suite of geophysical logs was obtained, and fifty-two 19-mm-diameter side-wall cores were obtained in the open hole (Run 4, Table 3). After the hole was cased and cemented, 3.9 meters of core (mudstones of the Great Valley formation, mentioned above) were obtained from the very bottom of the hole.

Phase 1 and 2 Real-time Sampling. Drill cuttings and formation gases were collected in real time as drilling was taking place. Drill cuttings were collected every 3 m and preserved in both washed and unwashed states, and larger volumes of cuttings were collected at less frequent intervals, as were samples of the drilling mud. Table 2 summarizes the cuttings samples, side-wall cores, and the three cores obtained after casing was cemented into place at various depths. Photographs, detailed descriptions, and other information about the extensive collection of cuttings are available online (Table 1). A summary of the lithologies encountered during Phases 1 and 2 is provided by Solum et al. (2006) and Bradbury et al. (2007), principally based on X-ray diffraction (XRD) analyses and optical analyses of mineralogy and texture of the cuttings, augmented by the spot and sidewall cores.

The near-continuous collection of cuttings revealed a number of lithologic changes along the trajectory of the hole that correlated very well with geophysical logs and other information. In addition, analysis of these cuttings revealed trace amounts of serpentine and a high level of clay minerals in the localized intervals that proved to be the active San

Table 2. Summary of Physical Samples Obtained from SAFOD.

Types of samples	Phase 1	Phase 2	Phase 3
Washed cuttings, small sample bags	3 sets, every 3 m	3 sets, every 3 m	intermittent depths
Washed cuttings, large (15 cm x 25 cm) sample bags	every 30 m	every 30 m	
Washed cuttings, large (25 cm x 43 cm) sampole bags	every 91 m	every 91 m	
Unwashed cuttings	every 3m	every 3 m	
Drilling mud	every 30 m	every 30 m	
Core	8.5 m at 1.5 km MD, 10 cm diameter	3.7 of 6.6 cm diameter core at 4 km MD	Core 1.1 run, 11.08 m / 3141.1–3153.6 m MD / 10 cm diameter
	11 m at 3.0 km MD, 10 cm diameter		Core 2 runs 1–3, 12.03 m, / 3186.7–3200.4 m MD / 10 cm diameter / Core 3 runs 4–5, 16.15 m, / 3294.9–3313.5 m MD / 10 cm diameter
Sidewall cores		52 small (2 cm dia. x 2.5 cm) side wall cores between 3.1 and 4.0 km MD	
Miscellaneous rock samples	3 samples	40 samples	

Andreas Fault Zone (Solum et al., 2006). Moore and Rymer (2007) demonstrated that some of the serpentinite in the fault zone has been altered to talc, an unusual mineral in that it has exceptionally low frictional strength and is thermodynamically stable over the range of depths and pressures characteristic of the upper crust in this region. They speculated that if talc is widespread in the fault zone, it could explain both the strength of the fault and its creeping behavior.

Gases coming into the well as the borehole was being drilled yielded a great deal of useful data. This technology, in which gas is separated from the drilling mud as it comes to the surface, was also used in the Pilot Hole where gas anomalies correlated with shear zones in the granite/granodiorite (Erzinger et al., 2004). During Phases 1 and 2, implementation of this technology showed a number of important correlations with major faults and geologic boundaries. One finding of particular interest reported by Wiersberg and Erzinger (2007) is that there is a marked difference in the concentration of $^3He/^4He$ across the San Andreas Fault. On the southwest side of the fault this ratio is ~0.4, whereas on the northeast side of the fault it is ~0.9. This data and differences in the relative concentrations of hydrogen, carbon dioxide, and methane on the two sides of the fault indicate that the San Andreas Fault has very low permeability and hydrologically separates the Pacific and North American plates (Wiersberg and Erzinger, 2008).

Downhole Measurements. A wide range of downhole measurements was carried out as part of SAFOD Phases 1 and 2 (Table 3). As the structure and properties of the San Andreas Fault Zone are of most importance, we show in Fig. 4A a summary of the geophysical logs from Phase 2 along with some of the main lithologic units encountered.

An approximately 200-m-wide damage zone of anomalously low P- and S-wave velocities and low resistivity (Fig. 4A) is interpreted to be the result of both physical damage and chemical alteration of the rocks due to faulting as well as the unusual, fault-related minerals (discussed above) that were noted during drilling. There are also a number of localized zones where the physical properties are even more anomalous. Repeated measurements of the shape of the steel casing deployed in the borehole revealed that the steel casing was being deformed by fault movement in at least two places. Figure 4C shows the casing radius (as measured using a 40-finger caliper) as a function of position around the hole. While the amount of deformation associated with the 3302-m shear zone is more pronounced than the

Table 3. SAFOD Geophysical Logging Data.

Run	Depth Range (Measured Depth)	Logging Technique	Parameters Measured
Run 1	602.5–1443.5 m	Open Hole, Wireline	Density, porosity, gamma, caliper, resistivity, cross-dipole sonic velocity, FMI
Run 2a	1368–2030 m	Open Hole, Wireline	Density, porosity, gamma, caliper, resistivity, sonic velocity, FMI, UBI, ECS
Run 2b	1890–3043 m	Open Hole, Pipe Conveyed	Density, porosity, gamma, caliper, resistivity, sonic velocity, FMI
Run 3	1356–3033 m	Cased Hole, Wireline	Sonic velociy, elemental chemistry, cement bond
Run 4	3045–3712 m	Open Hole, Logging While Drilling	Density, porosity, gamma, caliper, resistivity, FMI
Run 5	3045–3965 m	Open Hole, Pipe Conveyed	Density, porosity, gamma, caliper, resistivity, sonic velocity, FMI
Runs 6–11*	2953–3815 m	Cased Hole, Wireline	Caliper, direction, temperature

* Runs 6–11 include caliper logs run 6 different times between September 2005 and June 2007

3192-m shear zone, both of these zones represent portions of the overall San Andreas Fault Zone in which active creep deformation is occurring. We refer to the actively deforming zones at 3192 m as the Southwest Deforming Zone (SDZ) and 3302 m as the Central Deforming Zone (CDZ). Note the remarkable similarity of the anomalously low compressional (Vp) and shear (Vs) wave velocities and resistivity within these two deformation zones (Fig. 4B). These two shear zones were primary targets for coring during Phase 3.

The HI earthquake cluster occurs on the SDZ about 100 m below the point where the borehole passed through this fault

(Fig. 1). Recent relocations of the SAFOD target earthquakes indicate that the SF/LA cluster correlates with the fault at 3413 m, as shown in Fig. 2D (Thurber et al., 2010). This fault defines the northeastern edge of the damage zone and has geophysical characteristics very similar to the SDZ and CDZ (Fig. 2A); hence, it has been designated as the Northeast Boundary Fault (NBF). However, unlike the SDZ and CDZ, no casing deformation was detected on the NBF in any of the caliper logs run in 2005 through 2007 (Runs 6–11, Table 3).

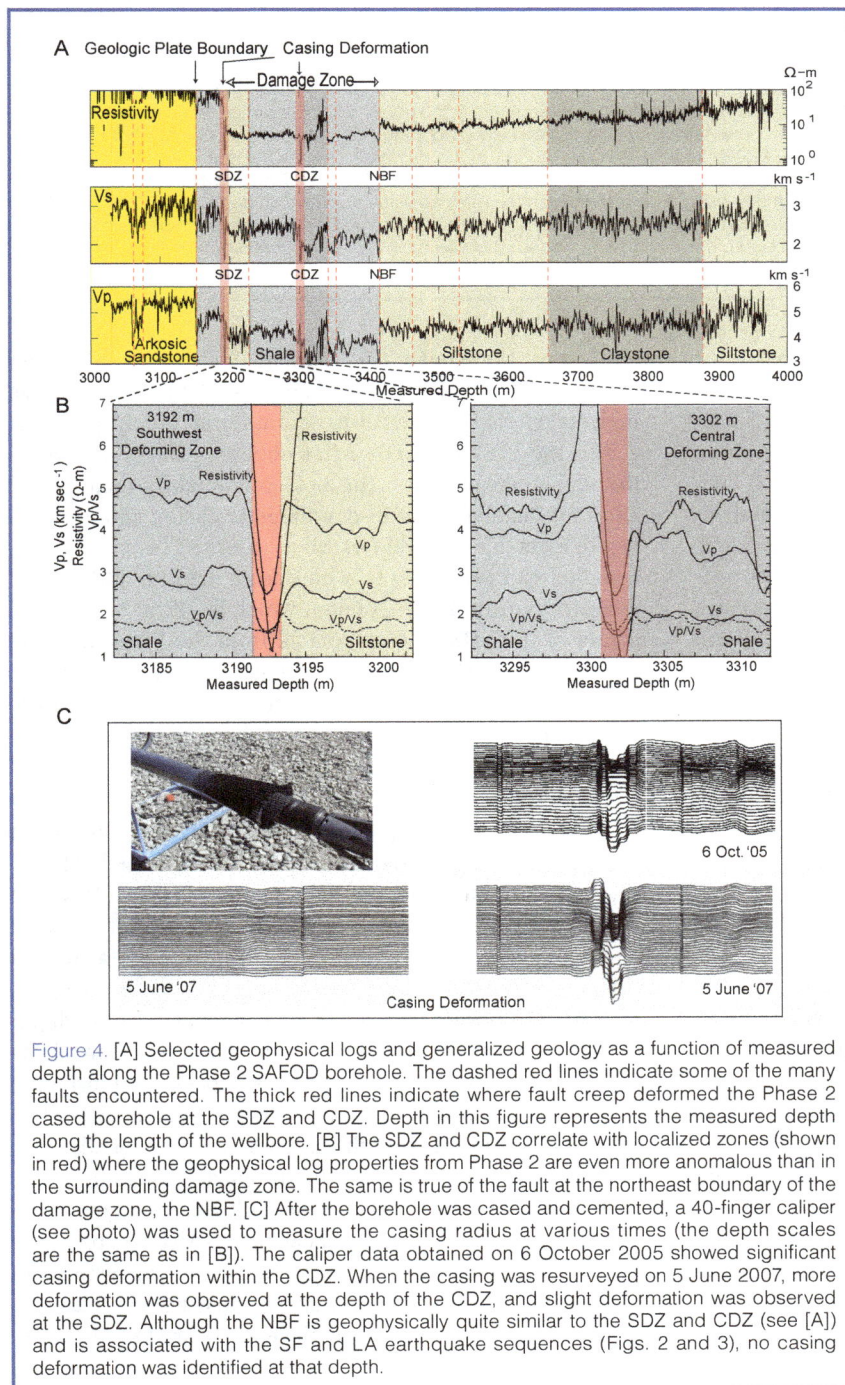

Figure 4. [A] Selected geophysical logs and generalized geology as a function of measured depth along the Phase 2 SAFOD borehole. The dashed red lines indicate some of the many faults encountered. The thick red lines indicate where fault creep deformed the Phase 2 cased borehole at the SDZ and CDZ. Depth in this figure represents the measured depth along the length of the wellbore. [B] The SDZ and CDZ correlate with localized zones (shown in red) where the geophysical log properties from Phase 2 are even more anomalous than in the surrounding damage zone. The same is true of the fault at the northeast boundary of the damage zone, the NBF. [C] After the borehole was cased and cemented, a 40-finger caliper (see photo) was used to measure the casing radius at various times (the depth scales are the same as in [B]). The caliper data obtained on 6 October 2005 showed significant casing deformation within the CDZ. When the casing was resurveyed on 5 June 2007, more deformation was observed at the depth of the CDZ, and slight deformation was observed at the SDZ. Although the NBF is geophysically quite similar to the SDZ and CDZ (see [A]) and is associated with the SF and LA earthquake sequences (Figs. 2 and 3), no casing deformation was identified at that depth.

A number of other important downhole measurements were made during Phases 1 and 2. Boness and Zoback (2006) reported that to within 200 m of the active trace of the fault, the direction of maximum horizontal stress remains at a high angle to the San Andreas Fault, consistent with measurements made in SAFOD at greater distances and with regional data that imply that fault slip occurs in response to low resolved shear stress. Zoback and Hickman (2007) reported that stress magnitudes are consistent with the prediction of high mean stress within the fault zone (Rice, 1992; Chery et al., 2004) and a classical Anderson/Coulomb reverse/strike-slip stress state outside it. Together with the stress state determined in the Pilot Hole (Hickman and Zoback, 2004), the results from the SAFOD Main Hole are consistent with a strong crust/weak fault model of the San Andreas. Almeida et al. (2005) carried out a paleostress analysis using slip directions on the faults encountered in the core obtained at the end of Phase 1 and also found a direction of maximum horizontal compression at a very high angle to the San Andreas Fault.

Further support for the low frictional strength of the San Andreas comes from temperature measurements in the SAFOD Main Hole. Heat flow data from the Pilot Hole were consistent with measurements made at relatively shallow depth and imply no frictionally generated heat by the San Andreas Fault (Williams et al., 2004). Heat flow measurements made in the Main Hole indicate no systematic change in temperature as a function of distance from fault. Hence, these data are also consistent with an absence of frictionally generated heat (Williams et al., 2005).

The possibility of extremely high pore pressure within the San Andreas Fault

Zone (near or above the weight of the overburden) has been one of the leading hypotheses to explain its low frictional strength (Rice, 1992). Two lines of evidence indicate an absence of severely elevated pore pressure (near-lithostatic, or greater) within the fault zone required to explain the low frictional strength of the San Andreas. Highly elevated fluid pressures were not observed during drilling in the fault zone. Such pressures would have resulted in influxes of formation fluid into the wellbore if the pore pressure was appreciably greater than the drilling mud pressure. While the density of the drilling mud

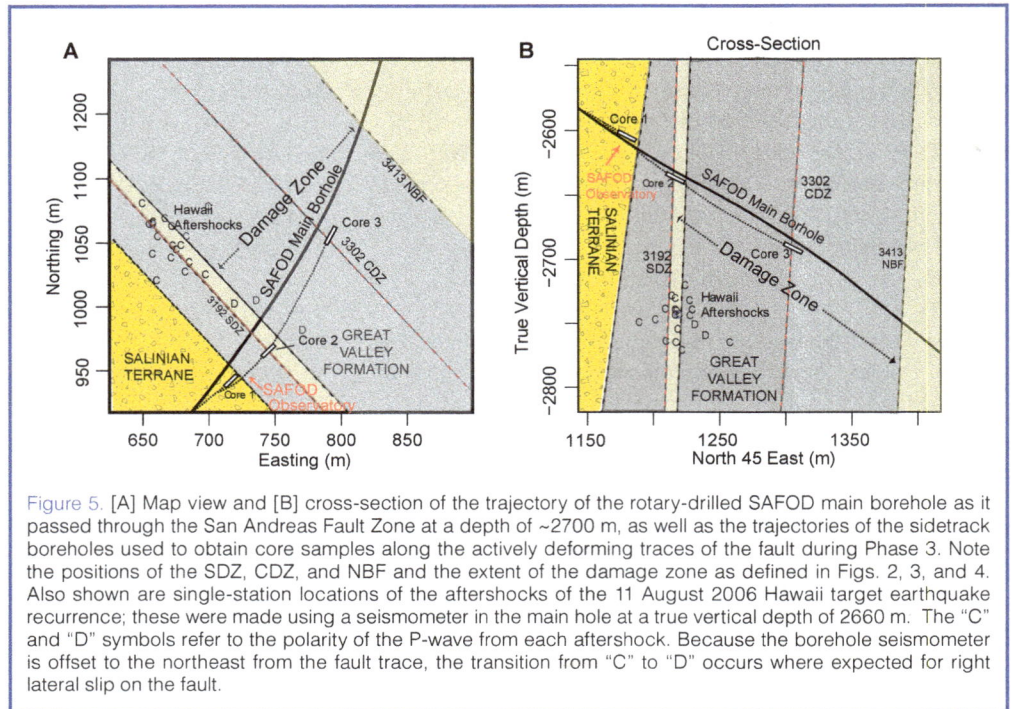

Figure 5. [A] Map view and [B] cross-section of the trajectory of the rotary-drilled SAFOD main borehole as it passed through the San Andreas Fault Zone at a depth of ~2700 m, as well as the trajectories of the sidetrack boreholes used to obtain core samples along the actively deforming traces of the fault during Phase 3. Note the positions of the SDZ, CDZ, and NBF and the extent of the damage zone as defined in Figs. 2, 3, and 4. Also shown are single-station locations of the aftershocks of the 11 August 2006 Hawaii target earthquake recurrence; these were made using a seismometer in the main hole at a true vertical depth of 2660 m. The "C" and "D" symbols refer to the polarity of the P-wave from each aftershock. Because the borehole seismometer is offset to the northeast from the fault trace, the transition from "C" to "D" occurs where expected for right lateral slip on the fault.

was about 40% greater than hydrostatic pore to stabilize the borehole, in the strike slip/reverse faulting stress state that characterizes the SAFOD area (Hickman and Zoback, 2004), pore pressures within the deforming fault zone would have to exceed the overburden stress in Rice's model (1992) for a weak fault in an otherwise strong crust. In addition, analysis of the rates of formation gas inflow during periods of no drilling (Wiersberg and Erzinger, submitted) shows no evidence of elevated pore pressure within the fault zone relative to the country rock, and the Vp/Vs ratio is relatively uniform (~1.7) across the ~200-m-wide damage zone and the localized shear zones within it (Fig. 4B). As Vp decreases severely at very elevated pore pressure (i.e., at very low effective stress), Vs would not be affected as much, and the Vp/Vs ratio would be expected to decrease (Mavko et al., 1998). Altogether, none of these observations indicate the presence of anomalously high pore pressure in the fault zone.

Phase 3 – Coring the San Andreas Fault Zone

During Phase 3 the SAFOD engineering and science teams successfully exhumed 39.9 meters of 10-cm-diameter continuous core, including cores from the two actively deforming traces of San Andreas Fault Zone (the SDZ and CDZ; Zoback et al., 2010). Figure 5 shows the sidetracks drilled laterally off the SAFOD main borehole in map and cross-sectional views. Note the position of the cores with respect to the various contacts and shear zones described above. As shown, Core 1 was obtained close to the contact between the arkosic sandstones and conglomerates of the Salinian Terrane and the shales, mudstones and siltstones associated with the Great Valley Formation. The first side-

track was abandoned and cemented off after retrieving Core 1 due to a drilling mishap. A second sidetrack was undertaken that enabled us to obtain Cores 2 and 3 (Table 2) across the SDZ and CDZ. After obtaining the cores across the active shear zones, the hole was slightly enlarged to allow for installation of 18-cm-diameter casing and eventual deployment of the SAFOD observatory. The casing was installed and cemented to a measured depth of 3214 m (as measured in the Phase 3 hole), which is ~17 m beyond the center of the SDZ as extrapolated from the Phase 2 to the Phase 3 holes. The casing could not be installed to greater depth in the Phase 3 hole due to progressive borehole instability and bridging.

When the cores reached the surface, they were carefully cleaned, labeled, and photographed, and they have been stored at 4°C to prevent desiccation and microbial activity. The core is currently stored at the IODP Gulf Coast Repository (GCR) at Texas A&M University. High-resolution photographs and descriptions of all Phase 3 cores (as well as supplemental information including thin-section analysis, results from preliminary XRD analysis and core-log depth integration) are presented in a comprehensive Core Atlas (Table 1). One page of the core atlas is presented in Fig. 6, which shows a section of the core that crosses the SDZ. The foliated gouge matrix is highly altered, both chemically (e.g., there is much less silica and different clay mineralogy than observed in the rocks outside the fault zone) and mechanically (e.g., there is pervasive shearing observed on planes of varied orientation within the core). Clasts of various types of rock are seen in the gouge matrix, most notably clasts of serpentinite including a large piece of sheared serpentinite with calcite veins.

To date, over 350 samples from the Phase 3 core have been distributed to investigators from around the world for laboratory analyses and testing; the latest results from these studies were discussed at two SAFOD special sessions of the 2010 annual meeting of the American Geophysical Union. These include studies of the mineralogy and chemical evolution of the fault zone, the physical properties of fault zone materials, the frictional strength of fault and country rock under a wide variety of loading conditions, and the evolution of deformation mechanisms and fluid-rock interaction within the fault zone over time. Procedures for requesting samples or gaining access to the SAFOD thin-section collection are available online (Table 1). The GCR staff is responsible for maintaining records of core, cuttings and fluid sample requests filled; names of people to whom these samples were provided; and the final disposition of samples (date samples returned and condition of samples). The GCR staff is also responsible for entering data and results from SAFOD sample investigations into the EarthScope Data Portal, which is currently under construction (Table 1).

SAFOD Observatory

In preparation for the establishment of a geophysical observatory deep within the fault, a series of nineteen temporary deployments of seismometers, accelerometers, and tiltmeters in the Main Hole and an additional eight deployments in the Pilot Hole were conducted between 2002 and 2008, leading up to the deployment of the SAFOD observatory in September 2008 (data available online, Table 1).

Seismic data collected during the temporary deployments are yielding important new findings on the structure of the San Andreas Fault and properties of the earthquakes that it produces. By combining surface and borehole observations of surface explosions and local earthquakes with double-difference tomography, Zhang et al. (2009) determined a detailed Vp, Vs, and Vp/Vs model for the SAFOD crustal volume. Their results refined earlier tomographic models for SAFOD to clearly image a deep low-velocity zone along the San Andreas Fault. This low-velocity zone supports the propagation of both P- and S-type fault zone guided waves. Observation of these waves on seismometers placed inside the fault zone places strong constraints on its geometry and continuity. Ellsworth and Malin (in press) determined that the low-velocity zone in which these waves propagate coin-cides with the zone of extensive rock damage seen in the downhole measurements (Fig. 4). The waveguide extends to the northwest and southeast of SAFOD for at least 8 km. Wu et al. (2010) used the dispersion properties of the S-type guided waves recorded in the Main Hole to show that the low-velocity wave-guide extends downward to near the base of the seismogenic zone at 10–12 km depth.

The short hypocentral distances and high-Q environment of the SAFOD boreholes make it possible to study source parameters to smaller magnitude than with data from instruments in shallow boreholes or on the surface. Only a small fraction (<1%) of the San Andreas Fault surface near SAFOD produces earthquakes, with the remainder of the fault moving through aseismic creep. The earthquakes that do occur are predominately located within clusters of repeating events. Static stress drops range from as low as 0.1 MPa to 100 MPa (Imanishi and Ellsworth, 2006). The upper limit is comparable to the laboratory-derived frictional strength of the country rock from outside of the damage zone (Lockner et al., in press). McGarr and Fletcher (2010) determined the yield stress for a repeat of the SF target earthquake of 64 MPa. These results suggest that the target events and other repeating earthquakes occur where the fault juxtaposes normal crustal rocks patches embedded within an otherwise weak, creeping fault. As a consequence, there is no contradiction between such high stress drop events and an intrin-sically weak, creeping San Andreas Fault in a strong crust, as indicated by the *in situ* stress and heat-flow measurements in the SAFOD Pilot Hole and Main Hole.

Figure 6. Photographs of the section of core 2 that crosses the SDZ (see Figs. 4 and 5) as they appear in the Photographic Atlas of the SAFOD Phase 3 (Table 1). The colored lettering indicates where samples were used for TS, XRD, and SEM presented in the Phase 3 Core Atlas. Note that the center and bottom photos are of the core sections split in half. Measured depths (in the sidetrack) are shown in feet (1 ft=30.48 cm).

The twenty-seven experimental deployments also guided the selection of sensors for the observatory and revealed mechanical and environmental issues that dictated the design of the observatory. The ambient temperature of up to 120°C at the planned depth of the observatory controlled the choice of downhole electronics and sensors. More seriously, the borehole fluid contains gases that penetrate past conventional O-rings and wireline insulation. Consequently, a design was selected that isolated all electrical and optical control lines and all sensors from contact with the wellbore fluid. The system was designed to be positively coupled to the casing and fully retrievable for maintenance when required.

The installation of the SAFOD observatory was completed on 28 September 2008. The observatory instruments were deployed approximately 100 m above the Hawaii target earthquake zone (Figs. 2, 5). As shown schematically in Fig. 7, the observatory instrumentation consisted of five pods containing different types of sensors. Pods 1 and 3 each contained a 3-component seismometer and a 3-component accelerometer, Pods 2 and 4 each contained a 2-axis tiltmeter, and Pod 5 contained a 3-component seismometer and accelerometer as well as a passive electromagnetic (EM) coil. The goal of the EM ex-periment was to determine if electromagnetic waves are radiated by the earthquake source. All of the instruments were housed in sealed steel pods that isolate them from contact with the wellbore fluids. The pods were attached to the outside of steel pipe (6-cm 'EUE' tubing) and coupled to the casing by decentralizing bow springs. The seismic and tilt systems were completely independent of each other, with separate power and data telemetry lines encapsulated in 6.4-mm-diameter stainless steel tubing with pressure-tight connections in and out of the pods.

The seismic system was based on the Oyo Geospace DS150 digital borehole seismometer with a set of 3-component, 15-Hz Omni-2400 geophones in each sonde. MEMS accelerometers replaced the geophones in additional DS150 units. The passive EM coil in Pod 5 was also digitized by a DS150. Fiber-optic telemetry was used to transmit the 4000-sample-per-second data from all seven DS150 units to the surface, where they were recorded on a USGS Earthworm computer system. The Earthworm system archived the data locally on LT3 tapes, downsampled selected channels to 250 samples per second and transmitted them to the Northern California Seismic Network (NCSN) where they were integrated into the real-time data system and archived at the Northern California Earthquake Data Center (NCEDC). Continuous full-sample-rate data are archived at the NCEDC and at the IRIS Data Management Center. The two borehole tiltmeters were manufactured by Pinnacle Technologies. Each tiltmeter produced two channels of tilt data—recorded at one sample per 3 seconds—which were transmitted to the NCEDC for processing and archiving.

An example of the data produced by the SAFOD observatory instruments is shown (Fig. 8) for an earthquake located

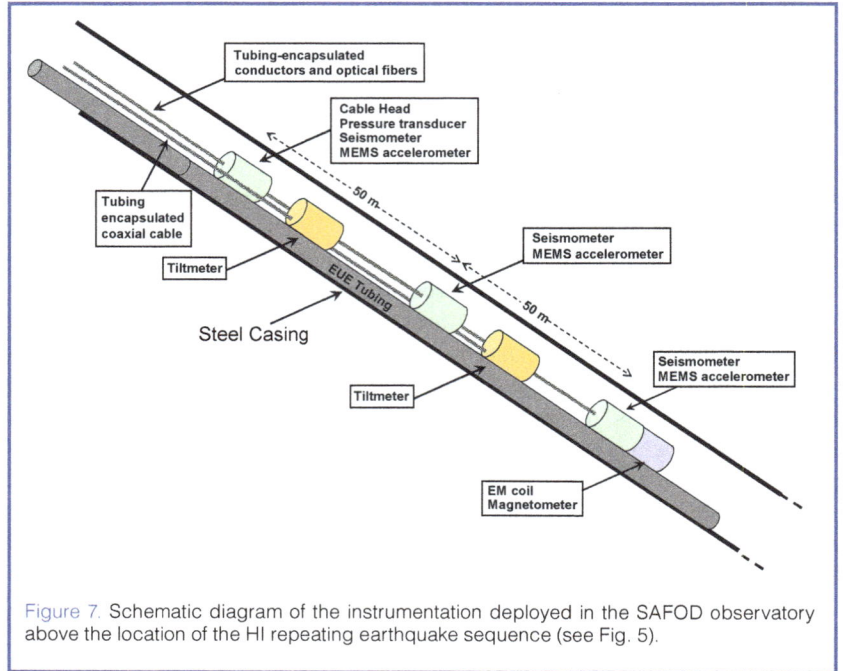

Figure 7. Schematic diagram of the instrumentation deployed in the SAFOD observatory above the location of the HI repeating earthquake sequence (see Fig. 5).

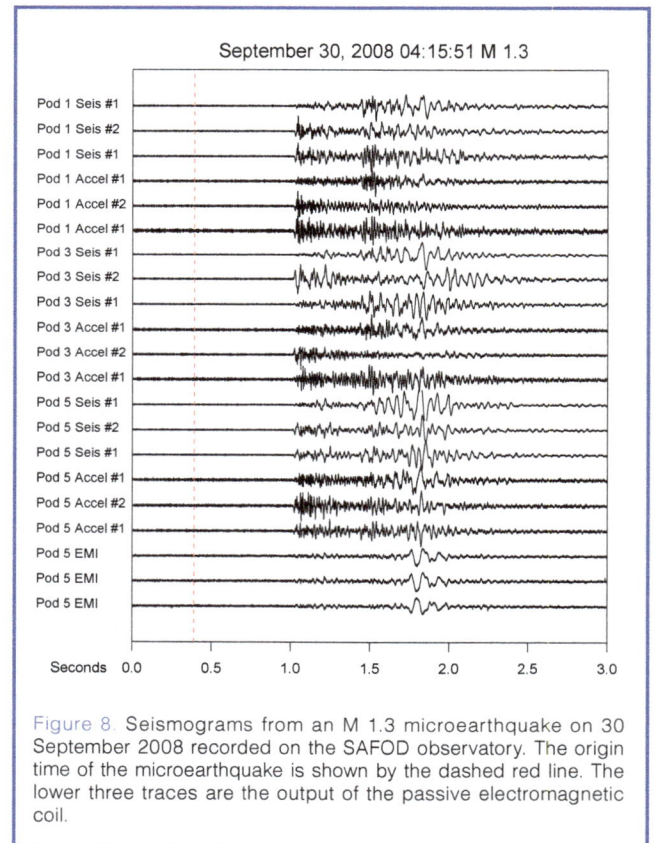

Figure 8. Seismograms from an M 1.3 microearthquake on 30 September 2008 recorded on the SAFOD observatory. The origin time of the microearthquake is shown by the dashed red line. The lower three traces are the output of the passive electromagnetic coil.

~2.5 km from the array. The EM trace appears three times because the EM signal was recorded at three different gains. Note that the EM signal appears at the same time as the seismic waves. Hence, the EM signal is the result of shaking of the coil within the Earth's magnetic field by the seismic waves as they pass the instrument.

Unfortunately, the SAFOD observatory instruments began to develop electronic problems soon after installation, and attempts to keep the instruments running were ultimately unsuccessful. An expert panel convened by NSF is currently in the process of examining the failed instrumentation. Leakage of water into pods was the probable cause of failure, although the actual failure point will not be known until the NSF panel report is completed. Fortunately, the SAFOD observatory was designed to permit ongoing access to the deepest part of the Main Hole through the inside of the EUE tubing (Fig. 7) to which the instrument pods were attached. A seismometer with three 15-Hz Omni-2400 geophones was deployed on wireline inside the EUE tubing in early December 2008, and this continues to operate as of March 2011. Data are digitized at the surface at 1000 samples per second and transmitted directly into the NCSN and are archived at the NCEDC (Table 1). While not a substitute for the observatory's full suite of digital seismometers and accelerometers, this interim instrument has allowed continuous observation of the target earthquakes to continue, and has produced important data including recordings of the SF and LA target earthquakes repeat in December 2008 (Fig. 9). The temporary geophone is planned to remain in operation until NSF develops a plan for installation of a new observatory.

In addition to the SAFOD observatory, an optical-fiber interferometric strainmeter was permanently installed at the conclusion of Phase 1 drilling in 2004 (Blum et al., 2010). Two optical-fiber loops were placed in the annulus formed by the 311-mm inside diameter (ID) initial casing and the 245-mm OD casing. The fiber sensors were attached to the outside of the inner casing as it was lowered into the well and then cemented in place. Each loop was anchored at the upper end at 9 m depth. One loop was anchored at the lower end at 864 m, and the other at 782 m, making strainmeters of 855 m and 773 m length, respectively. Although the longer loop failed in September 2007, vertical strain data continues to be produced from the shorter loop. Coseismic strain steps for local events have been reported by Blum et al. (2010) that are in general agreement with elastic dislocation theory.

Summary

We have already learned much about (i) the structure and physical properties of the fault zone at depth, (ii) the composition of fault zone rocks, (iii) the stress, temperature, and fluid pressure conditions under which earthquakes occur, and (iv) the absence of deep-seated fluids in fault zone processes. With the distribution of the Phase 3 core to researchers around the world now underway, we can expect new insights into the physical and chemical mechanisms controlling faulting and fault zone evolution within this major plate boundary fault. In addition, the observatory, even in its currently reduced state, is providing high-quality near-field seismograms that may lead to novel observations of rupture nucle-ation and other insights into the nature of the earthquake source and structure of the fault at seismogenic depth.

Acknowledgements

Scores of scientists, graduate students, engineers and technicians too numerous to name contributed immeasurably to the success of SAFOD. We would particularly like to thank Louis Capuano and Jim Hanson of ThermaSource, Inc., the prime drilling contractor, and the many individuals who served on the SAFOD Advisory Board and Technical Panels. We would especially like to thank Roy Hyndman of the Pacific Geoscience Center who served as Chair of the SAFOD Advisory Board. Funding for the project was provided by the NSF's EarthScope Program, with additional support from the USGS, the ICDP, Stanford University, and NASA. Any use of trade, product, or firm names is for descriptive purposes only and does not imply endorsement by the U.S. government.

References

Almeida, R., Chester, J., Chester, F., Kirschner, D., Waller, T., and Moore, D., 2005. Mesoscale structure and lithology of the SAFOD Phase I and II core samples. *Eos Trans. AGU*, 86 (52), Fall Meeting Suppl., Abstract T21A-0451.

Bakun, W., and McEvilly, T., 1984. Recurrence models and Parkfield, California, earthquakes. *J. Geophys. Res.*, 89(B5): 3051–3058.

Blum, J., Igel, H., and Zumberge, M., 2010. Observation of Rayleigh-wave phase velocity and coseismic deformation using an optical fiber, interferometric vertical strainmeter at the SAFOD Borehole, California. *Bull. Seismol. Soc. Am.*, 100(5A):1879–1891, doi:10.1785/0120090333.

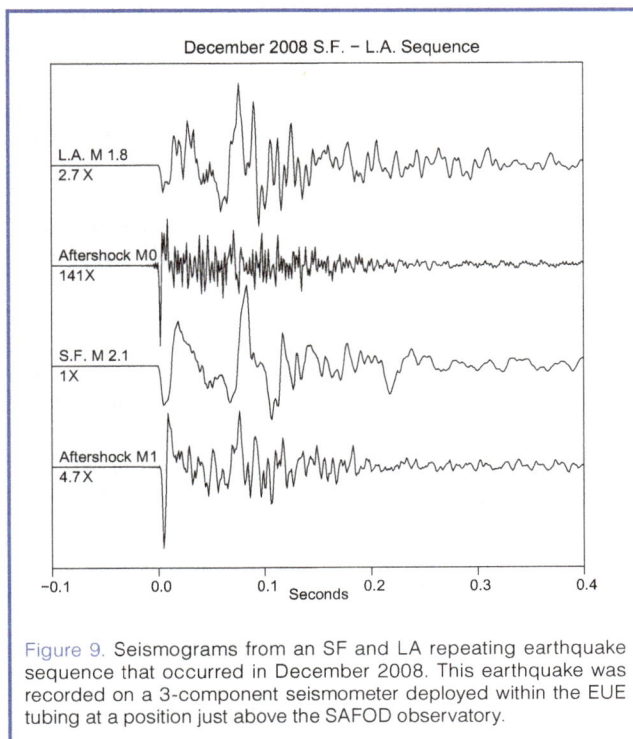

Figure 9. Seismograms from an SF and LA repeating earthquake sequence that occurred in December 2008. This earthquake was recorded on a 3-component seismometer deployed within the EUE tubing at a position just above the SAFOD observatory.

Boness, N., and Zoback, M.D., 2004. Stress-induced seismic velocity anisotropy and physical properties in the SAFOD Pilot Hole in Parkfield, CA. *Geophys. Res. Lett.*, 31:L15S17, doi:10.1029/2003GL019020.

Boness, N., and Zoback, M.D., 2006. A multi-scale study of the mechanisms controlling shear velocity anisotropy in the San Andreas Fault Observatory at Depth. *Geophysics*, 7(5):F131–F146, doi:10.1190/1.2231107.

Bradbury, K.K., Barton, D.C., Solum, J.G., Draper, S.D., and Evans, J.P., 2007. Mineralogic and textural analyses of drill cuttings from the San Andreas Fault Observatory at Depth (SAFOD) boreholes: initial interpretations of fault zone composition and constraints on geologic models. *Geosphere*, 3(5):299–318, doi:10.1130/GES00076.1.

Chavarria, J., and Goerrtz, A., 2007. The use of VSP techniques for fault zone characterization: an example from the San Andreas Fault. *The Leading Edge*, 26(6):770–776, doi:10.1190/1.2748495.

Chavarria, J., Malin, P., Catchings, R., and Shalev, E., 2003. A look inside the San Andreas fault at Parkfield through vertical seismic profiling. *Science*, 302(5651):1746, doi:10.1126/science.1090711.

Chery, J., Zoback, M.D., and Hickman, S., 2004. A mechanical model of the San Andreas fault and SAFOD pilot hole stress measurements. *Geophys. Res. Lett.*, 31(15):L15S13, doi:10.1029/2004GL019521.

Draper Springer, S.D., Evans, J.P., Garver, J.I., Kirschner, D., and Janecke, S.U., 2009. Arkosic rocks from the San Andreas Fault Observatory at Depth (SAFOD) Borehole, Central California: implications for the structure and tectonics of the San Andreas fault zone. *Lithosphere*, 1:206–226, doi:10.1130/L13.1.

Ellsworth, W.L., and Malin, P., in press. Deep rock damage in the San Andreas Fault revealed by P- and S-type fault zone guided waves. *In* Fagereng, A., Toy, V.G., and Rowland, J. (Eds), *Geology of the Earthquake Source: A Volume in Honor of Rick Sibson:* London (Geological Society of London).

Erzinger, J., Wiersberg, T., and Dahms, E., 2004. Real-time mud gas logging during drilling of the SAFOD pilot hole in Parkfield, CA. *Geophys. Res. Lett.*, 31(15): L15S18, doi:10.1029/2003GL019395.

Harris, R.A., and Arrowsmith, J.R., 2006. Introduction to the special issue on the 2004 Parkfield earthquake and the Parkfield earthquake prediction experiment. *Bull. Seismol. Soc. Am.*, 96(4B):S1–S10, doi:10.1785/0120050831.

Hickman, S., 1991. Stress in the lithosphere and the strength of active faults. *In* Shea, M.A. (Ed.), *U.S. National Report International Union Geodesy and Geophysics, 1987–1990: Contributions in Tectonophysics:* Washington, DC (American Geophysical Union), 759–775.

Hickman, S., and Zoback, M.D., 2004. Stress measurements in the SAFOD pilot hole: implications for the frictional strength of the San Andreas fault. *Geophys. Res. Lett.*, 31:L15S12.

Hickman, S., Zoback, M.D., and Ellsworth, W., 2004. Introduction to special issue: preparing for the San Andreas fault observatory at depth. *Geophys. Res. Lett.*, 31:L12S01, doi:10.1029/2004GL020688.

Imanishi, K., and Ellsworth, W.L., 2006. Source scaling relationships of microearthquakes at Parkfield, CA, determined using the SAFOD Pilot Hole seismic array. *In* Abercrombie, R.,

McGarr, A., Di Toro, G., and Kanamori, H. (Eds.), *Earthquakes: Radiated Energy and the Physics of Faulting, Geophysical Monograph Series 170:* Washington, DC (American Geophysical Union), 81–90.

Lachenbruch, A.H., and Sass, J.H., 1980. Heat Flow and Energetics of the San Andreas Fault Zone. *J. Geophys. Res.*, 85(11):6185–6223.

Lachenbruch, A.H., and J.H. Sass, 1992. Heat flow From Cajon Pass, fault strength and tectonic implications. *J. Geophys. Res.*, 97(B4):4995–5015, doi:10.1029/91JB01506.

Lockner, D.A., Morrow, C., Moore, D.E., and Hickman, S., in press. Low strength of deep San Andreas Fault gouge from SAFOD core. *Nature*.

Mavko, G., Mukerjii, T., and Dvorkin, J., 1998. *Rock Physics Handbook:* Cambridge, U.K. (Cambridge University Press).

McGarr, A., and Fletcher, J.B., 2010. Laboratory-based maximum slip rates in earthquake rupture zones and radiated energy. *Bull. Seismol. Soc. Am.*, 100(6):3250–3260, doi:10.1785/0120100043.

McPhee, D.K., Jachens, R.C., and Wentworth, C.M., 2004. Crustal structure across the San Andreas Fault at the SAFOD site from potential field and geologic studies. *Geophys. Res. Lett.*, 31(12):L12S03, doi:10.1029/2003GL019363.

Moore, D.E., and Rymer, M.J., 2007. Talc-bearing serpentinite and the creeping section of the San Andreas fault. *Nature*, 448(16):795–797, doi:10.1038/nature06064.

Mount, V.S., and Suppe, J., 1987. State of stress near the San Andreas fault: implications for wrench tectonics. *Geology*, 15:1143–1146, doi:10.1130/0091-7613(1987)15<1143:SOSNTS>2.0.CO;2.

Murray, J., and Langbein, J., 2006. Slip on the San Andreas fault at Parkfield, California, over two earthquake cycles, and the implications for seismic hazard. *Bull. Seismol. Soc. Am.*, 96:S283–S303, doi:10.1785/0120050820.

Nadeau, R., and Dolenc, D., 2005. Nonvolcanic tremors deep beneath the San Andreas fault. *Science*, 307(5708):389, doi:10.1126/science.1107142.

Nadeau, R.M., McEvilly, T.V., Michelini, A., Uhrhammer, R.A., and Dolenc, D., 2004. Detailed kinematics, structure and recurrence of micro-seismicity in the SAFOD target region. *Geophys. Res. Lett.*, 31:L12S08.

Niu, F., Silver, P.G., Daley, T.M., Cheng, X., and Majer, E., 2008. Preseismic velocity changes observed from active source monitoring the Parkfield SAFOD drill site. *Nature*, 454:204–208, doi:10.1038/nature07111.

Oppenheimer, D.H., Reasenberg, P.A., and Simpson, R.W., 1988. Fault-plane solutions for the 1984 Morgan Hill California earthquake sequence: evidence for the state of stress on the Calaveras fault. *J. Geophys. Res.*, 93:9007–9026, doi:10.1029/JB093iB08p09007.

Oye, V., and Ellsworth, W.L., 2007. Small-scale structures derived from microearthquake locations using SAFOD and HRSN data., *Eos, Trans. AGU*, 88(52), Fall Meet. Suppl., Abstract T53C-03.

Paul, P., and Zoback, M.D., 2008. Wellbore-stability study for the SAFOD borehole through the San Andreas Fault, SPE-102781-PA, *SPE Drilling and Completion*, 23(4):394–408, doi: 10.2118/102781-PA.

Rice, J.R., 1992. Fault stress states, pore pressure distributions, and the weakness of the San Andreas fault. *In* Evans, B., and

Wong, T.F. (Eds.), *Fault Mechanics and Transport Properties of Rocks:* San Diego, CA (Academic Press), 475–503, doi:10.1016/S0074-6142(08)62835-1.

Roecker, S., Thurber, C., and McPhee, D., 2004. Joint inversion of gravity and arrival time data from Parkfield: new constraints on structure and hypocenter locations near the SAFOD drill site. *Geophys. Res. Lett.*, 31:1–4, doi:10.1029/2003GL019396.

Rymer, M.J., Tinsley, J.C., Treiman, J.A., Arrowsmith, J.R., Clahan, K.B., Rosinski, A.M., Bryant, W.A., Snyder, H.A., Fuis, G.S., Toke, N.A., and Bawden, G.W., 2006. Surface fault slip associated with the 2004 Parkfield, California, earthquake. *Bull. Seismol. Soc. Am.*, 96(B4):S11–S27, doi:10.1785/0120050830.

Saffer, D.M., Bekins, B.A., and Hickman, S., 2003. Topographically driven groundwater flow and the San Andreas heat flow paradox revisited. *J. Geophys. Res.*, 108(B5):2274.

Schaff, D., Beroza, G., and Shaw B., 1998. Postseismic response of repeating aftershocks. *Geophys. Res. Lett.*, 25(24):4549–4552, doi:10.1029/1998GL900192.

Shelly, D.R., and Hardebeck, J.L., 2010. Precise tremor source locations and amplitude variations along the lower-crustal central San Andreas Fault. *Geophys. Res. Lett.*, 37:L14301, doi:10.1029/2010GL043672.

Solum, J.G., Hickman, S.H., Lockner, D.A., Moore, D.E., van der Pluijm, B.A., Schleicher, A.M., and Evans, J.P., 2006. Mineralogic characterization of protolith and fault rocks from the SAFOD main hole. *Geophys. Res. Lett.*, 33:L21314, doi:10.1029.2006GL027285.

Thurber, C., Roecker, S., Zhang, H., Baher, S., and Ellsworth, W.L., 2004. Fine-scale structure of the San Andreas fault zone and location of the SAFOD target earthquakes. *Geophys. Res. Lett.*, 31:L12S02, doi:10.1029/2003GL019398.

Thurber, C., Roecker, S., Zhang, H., Bennington, N., and Peterson, D., 2010. Crustal structure and seismicity around SAFOD: a ten-year perspective. *Eos, Trans. AGU*, 91, Fall Meeting Suppl., Abstract T52B-01.

Unsworth, M., and Bedrosian, P.A., 2004. Electrical resistivity structure at the SAFOD site from magnetotelluric exploration. *Geophys. Res. Lett.*, 31(12):L12S05, doi:10.1029/2003GL019405.

Waldhauser, F., Ellsworth, W.L., Schaff, D.P., and Cole A., 2004. Streaks, multiplets, and holes: High-resolution spatio-temporal behavior of Parkfield seismicity. *Geophys. Res. Lett.*, 31:L18608, doi:10.1029/2004GL020649.

Wiersberg, T., and Erzinger, J., 2007. A helium isotope cross-section study through the San Andreas Fault at seismogenic depths. *Geochem. Geophys. Geosyst.*, 8(1):Q01002, doi:10.1029/2006GC001388.

Wiersberg, T., and Erzinger, J., 2008. On the origin and spatial distribution of gas at seismogenic depths of the San Andreas Fault from drill mud gas analysis. *Appl. Geochem.*, 23:1675–1690, doi:10.1016/j.apgeochem.2008.01.012.

Wiersberg, T., and Erzinger, J., 2011. Chemical and isotope compositions of drilling mud gas from the San Andreas Fault Observatory at Depth (SAFOD) boreholes: Implications on gas migration and the permeability structure of the San Andreas Fault, *Chem. Geol.*, doi:10.1016/j.chemgeo.2011.02.016

Williams, C.F., Grubb, F.V., and Galanis, S.P., 2004. Heat flow in the SAFOD pilot hole and implications for the strength of the San Andreas fault. *Geophys. Res. Lett.*, 31:L15S14.

Williams, C.F., D'Alessio, M.A., Grubb, F.V., and Galanis, S.P., 2005. Heat flow studies in the SAFOD main hole. *Eos, Trans. AGU*, 86(52), Fall Meeting Suppl., Abstract T23E-07.

Wu, J., Hole, J.A., and Snoke, J.A., 2010. Fault zone structure at depth from differential dispersion of seismic guided waves: evidence for a deep waveguide on the San Andreas Fault. *Geophys. J. Int.*, 182:343–354.

Zhang, H., Thurber, C., and Bedrosian, P.A., 2009. Joint inversion for Vp, Vs, and Vp/Vs at SAFOD, Parkfield, California. *Geochem. Geophys. Geosyst.*, 10(11):Q11002, doi:10.1029/2009GC002709.

Zoback, M.D., and Hickman, S.H., 2007. Preliminary results from SAFOD Phase 3: implications for the state of stress and shear localization in and near the San Andreas Fault at depth in central California. *Eos, Trans. AGU*, 88(52), Fall Meeting Suppl., Abstract T13G-03.

Zoback, M.D., Hickman, S., and Ellsworth, W., 2007. The role of fault zone drilling. *In* Kanamori, H., and Schubert, G. (Eds.), *Earthquake Seismology—Treatise on Geophysics Vol. 4*: Amsterdam (Elsevier), 649–674.

Zoback, M.D., Hickman, S., and Ellsworth, W.L., 2010. Scientific drilling into the San Andreas Fault. *Eos, Trans. AGU*, 91(22):197–204, doi:10.1029/2010EO220001.

Zoback, M.D., Zoback, M.L., Mount, V.S., Suppe, J., Eaton, J.P., Healy, J.H., Oppenheimer, D., Reasenberg, P., Jones, L., Raleigh, C.B., Wong, I.G., Scotti O., and Wentworth, C., 1987. New evidence for the state of stress on the San Andreas fault system. *Science*, 238:1105–1111, doi:10.1126/science.238.4830.1105.

Authors

Mark Zoback, Department of Geophysics, Stanford University, Stanford, CA 94305-2215, U.S.A., e-mail: zoback@stanford.edu.
Stephen Hickman and William Ellsworth, U.S. Geological Survey, 345 Middlefield Road MS 977, Menlo Park, CA 94025-3591, U.S.A.
and the SAFOD Science Team

Figure Credits

Fig. 1: air photo courtesy of M. Rymer

16

The Lake El'gygytgyn Scientific Drilling Project – Conquering Arctic Challenges through Continental Drilling

by Martin Melles, Julie Brigham-Grette, Pavel Minyuk, Christian Koeberl, Andrei Andreev, Timothy Cook, Grigory Fedorov, Catalina Gebhardt, Eeva Haltia-Hovi, Maaret Kukkonen, Norbert Nowaczyk, Georg Schwamborn, Volker Wennrich, and the El'gygytgyn Scientific Party

Abstract

Between October 2008 and May 2009, the International Continental Scientific Drilling Program (ICDP) co-sponsored a campaign at Lake El'gygytgyn, located in a 3.6-Ma-old meteorite impact crater in northeastern Siberia. Drilling targets included three holes in the center of the 170-m-deep lake, utilizing the lake ice cover as a drilling platform, plus one hole close to the shore in the western lake catchment. At the lake's center. the entire 315-m-thick lake sediment succession was penetrated. The sediments lack any hiatuses (i.e., no evidence of basin glaciation or desiccation), and their composition reflects the regional climatic and environmental history with great sensitivity. Hence, the record provides the first comprehensive and widely time-continuous insights into the evolution of the terrestrial Arctic since mid-Pliocene times. This is particularly true for the lowermost 40 meters and uppermost 150 meters of the sequence, which were drilled with almost 100% recovery and likely reflect the initial lake stage during the Pliocene and the last ~2.9 Ma, respectively. Nearly 200 meters of under-lying rock were also recovered; these cores consist of an almost complete section of the various types of impact breccias including broken and fractured volcanic basement rocks and associated melt clasts. The investigation of this core sequence promises new information concerning the El'gygytgyn impact event, including the composition and nature of the meteorite, the energy released, and the shock behavior of the volcanic basement rocks. Complementary information on the regional environmental history, including the permafrost history and lake-level fluctuations, is being developed from a 142-m-long drill core recovered from the permafrost deposits in the lake catchment. This core consists of gravelly and sandy alluvial fan deposits in ice-rich permafrost, presumably comprising a discontinuous record of both Quaternary and Pliocene deposits.

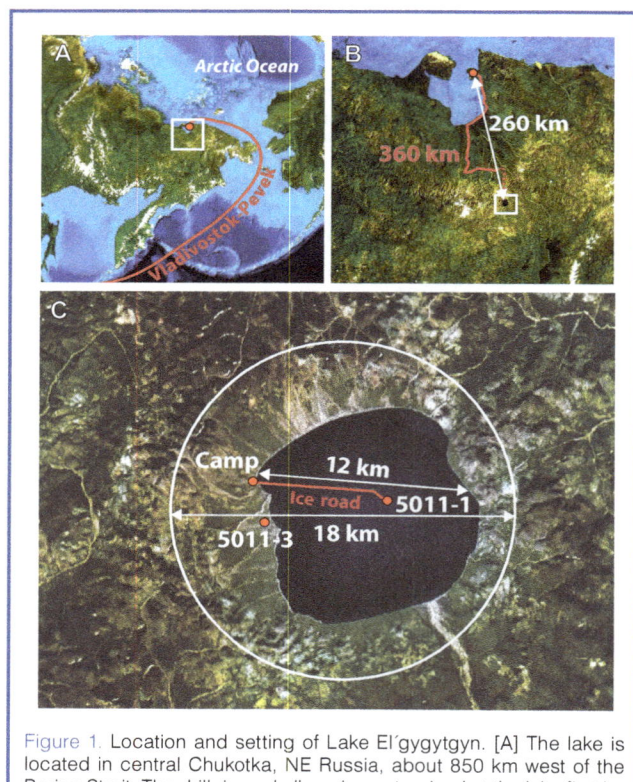

Figure 1. Location and setting of Lake El'gygytgyn. [A] The lake is located in central Chukotka, NE Russia, about 850 km west of the Bering Strait. The drill rig and all equipment arrived at the lake first by barge from Vladivostok along the route in red. [B] All logistics were based out of the town of Pevek, a gold mining center located on the coast of the East Siberian Sea. Helicopters were used to transport scientists, food and delicate equipment out to the drill site, whereas the 17 shipping containers with the drilling system were transported by truck. [C] Satellite image of Lake El'gygytgyn Crater showing dimensions, regional relief, the locations of ICDP Sites 5011-1 and 5011-3, and the location of crater rim (white circle).

Introduction

Lake El'gygytgyn is located 100 km to the north of the Arctic Circle in remote Chukotka, northeastern Russia (67°30' N, 172°05' E; Fig. 1). The lake lies within a meteorite impact crater measuring 18 km in diameter (Gurov et al., 1978, 2007) that was created 3.6 million years ago in volcanic target rocks (Layer, 2000). Today, the lake is 170-m-deep and has a roughly circular shape with a diameter of 12 km. Higher sediment supply from the western and northern reaches of the crater over time has caused the displacement of the lake toward the southeastern part of the basin. The sediments in the surrounding lake catchment are derived from slope processes and fluvial activity. Regionally these sediments are thought to contain permafrost to a depth of 500 m (Yershov, 1998). A seismic survey on the lake floor detected more than 300 meters of lacustrine sediments above an impact breccia and brecciated volcanic bedrock (Gebhardt et al., 2006), confirming the assumption that the basin had escaped continental-scale glaciations since the time of the impact (Glushkova, 2001).

Because of its unusual origin and high-latitude setting in western Beringia, scientific drilling at Lake El'gygytgyn offered unique opportunities across three geoscientific disciplines. These include (i) paleoclimate research, allowing the time-continuous reconstruction of the climatic and environmental history of the terrestrial Arctic back into the mid-Pliocene for the first time; (ii) permafrost research, promising a better understanding of the history and present behavior of the Arctic's frozen surficial materials; and (iii) impact science, providing new insights into planetary cratering processes and the response of volcanic target rocks. This report summarizes aspects of the pre-site surveys which provided the impetus for drilling, highlights the challenging, sometimes gut-wrenching drilling logistics, and outlines some results and first interpretations from the limited on-site and ongoing off-site analyses of the lake sediments, impact rocks, and permafrost deposits.

Pre-site Surveys

A first international expedition was carried out on Lake El'gygytgyn early in spring 1998. Using the lake ice as a platform, six participants from Germany, Russia, and the U.S.A. conducted initial shallow coring in the deepest part of the lake. Succeeding expeditions in summer 2000 and spring and summer 2003 included eleven and sixteen participants, respectively (Melles et al., 2005). These projects provided a more comprehensive understanding of the modern setting and processes operating in the crater, the Late Quaternary climatic and environmental history of the region, and the structure of the impact crater and the thickness and architecture of its lacustrine sediment infill (Brigham-Grette et al., 2007, and references therein).

The climate at Lake El'gygytgyn is cold, dry, and windy. In 2002, the mean annual air temperature was -10.3°C, with extremes ranging from -40°C in winter to +26°C in summer (Nolan and Brigham-Grette, 2007). The annual precipitation amounted to ~200 mm water equivalent. Dominant wind directions were either from the north or from the south. The mean hourly wind speed was 5.6 m s⁻¹, with strong winds above 13.4 m s⁻¹ occurring every month but more frequently in winter. The modern vegetation in the catchment of Lake El'gygytgyn is herb-dominated tundra with rare local patches of low shrubs, particularly willow (*Salix*) and dwarf

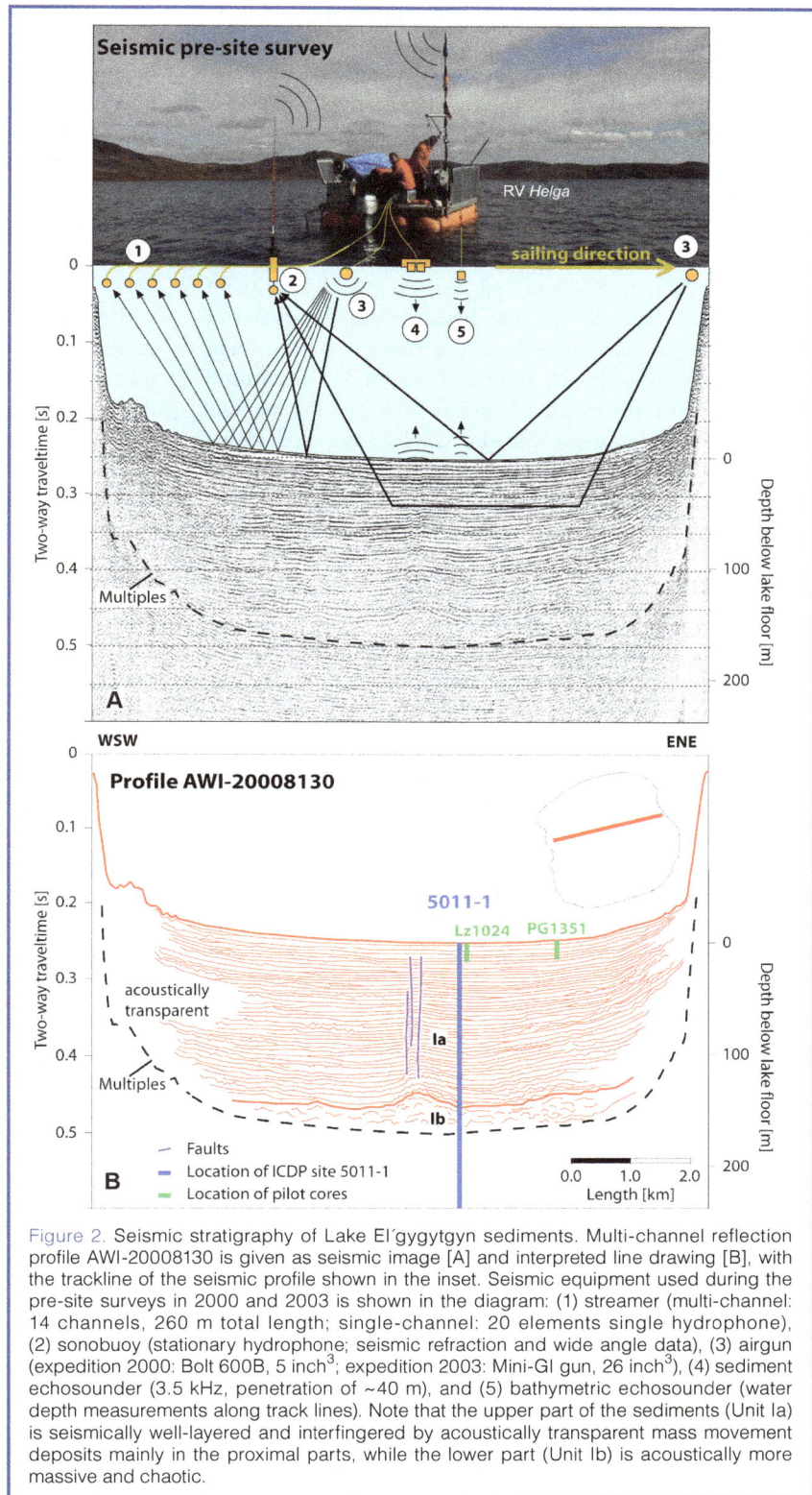

Figure 2. Seismic stratigraphy of Lake El'gygytgyn sediments. Multi-channel reflection profile AWI-20008130 is given as seismic image [A] and interpreted line drawing [B], with the trackline of the seismic profile shown in the inset. Seismic equipment used during the pre-site surveys in 2000 and 2003 is shown in the diagram: (1) streamer (multi-channel: 14 channels, 260 m total length; single-channel: 20 elements single hydrophone), (2) sonobuoy (stationary hydrophone; seismic refraction and wide angle data), (3) airgun (expedition 2000: Bolt 600B, 5 inch³; expedition 2003: Mini-GI gun, 26 inch³), (4) sediment echosounder (3.5 kHz, penetration of ~40 m), and (5) bathymetric echosounder (water depth measurements along track lines). Note that the upper part of the sediments (Unit Ia) is seismically well-layered and interfingered by acoustically transparent mass movement deposits mainly in the proximal parts, while the lower part (Unit Ib) is acoustically more massive and chaotic.

birch (*Betula nana*) (Lozhkin et al., 2007a). Ice formation on Lake El'gygytgyn usually starts in October (Nolan et al., 2003). The blanketing snow cover melts in May/June, whereas the lake ice, which reaches a maximum thickness of 1.5–2.0 m, starts disintegration with the formation of moats at the shore in June/July and culminates in open water by mid-July/August. Biogenic primary production in this ultra-oligotrophic lake is concentrated in the short ice-free period in summer, but considerable phytoplankton growth also takes place beneath the ice cover (Cremer and Wagner, 2003; Cremer et al., 2005). Lake El'gygytgyn today is a cold-monomictic system with slightly acidic pH. The water column down to 170 m is stratified in winter but completely mixed in summer, though never exceeding 4°C (Nolan and Brigham-Grette, 2007). About fifty streams enter the lake at 492 m above sea level (a.s.l.) from the catchment that extends to the crater rim up to 935 m a.s.l.; however, fluvial sediment supply to the lake is very low, because the watershed of 293 km² is less than three times the lake's surface area of 110 km². In addition, much of the sediment today is captured and deposited at the mouth of the inflows in shallow lagoons that are dammed by gravel bars formed by wave and lake ice action. The restricted fluvial input together with the low primary production produces remarkably clear surface waters, giving a Secchi transparency depth of 19 m in summer.

The first 13- and 16-m-long sediment cores from central Lake El'gygytgyn yielded basal ages of ~250 kyr and 340 kyr before present (BP), respectively, confirming that very low and relatively constant sedimentation rates are characteristic of both interglacial and glacial times (Forman et al., 2007; Juschus et al., 2007; Nowaczyk et al., 2007). The highly variable characteristics of the sediment underscore the sensitivity of this lacustrine system to regional climatic and environmental change (Asikainen et al., 2007; Brigham-Grette et al., 2007; Cherepanova et al., 2007; Lozhkin et al., 2007a, 2007b; Melles et al., 2007; Minyuk et al., 2007). Shallow cores were also taken of sub-recent mass movement deposits first identified in seismic profiles as originating from the steep (up to 30°) lake slopes (Niessen et al., 2007). This case study demonstrated that debris and density flows can be associated with significant erosion on the lake slopes, but these processes usually do not reach the lake center, where suspension clouds produced by these events in most cases accumulate as non-erosive turbidites (Juschus et al., 2009). Complementary information concerning Late Quaternary lake-level fluctuations, cryogenic weathering, and landscape development was obtained by ground-penetrating radar surveys and investigations of sediment stratigraphic sections exposed in the catchment of Lake El'gygytgyn (Schwamborn et al., 2006, 2008a, 2008b; Glushkova and Smirnov, 2007; Glushkova et al., 2009).

During summer seismic surveys conducted in 2000 and 2003, a 3.5-kHz echosounder with high spatial resolution (up to 40 m penetration) was combined with single-channel and multi-channel airgun seismic systems to provide the clearest information possible of the deeper lacustrine sediments and the structure of the impact crater underneath (Gebhardt et al., 2006; Niessen et al., 2007). Both systems were run simultaneously for efficiency from a small open platform resting on four inflatable pontoons (Fig. 2). Sonobuoy refraction data from the lake center formed the basis of a five-layer velocity-depth model. The results show that the El'gygytgyn Crater

has an uplifted central ring structure with its top in about 330 m depth below lake floor (mblf); the structure was built by impact breccia and buried by alluvial deposits in the northwestern part of the basin. Above this structure, two lake sediment units were identified based on seismic characteristics. According to the air gun reflection data, the upper unit down to 170 mblf appeared to be well stratified, while the lower unit appeared to be more massive (Fig. 2). Draping of the uplift structure is visible and inferred in the lower part of the upper unit. Both units were shown to be intercalated with thick mass movement deposits largely confined to marginal areas. Because these units also lack seismic discontinuities suggestive of glacial overriding or lake desiccation, a nearly time-continuous sediment record following the impact event was expected from parts of the central lake.

Drilling Operation

Drilling in remote northeastern Russia was a massive logistical undertaking. In summer 2008, the majority of the technical equipment and field supplies were transported in fifteen shipping containers from Salt Lake City, U.S.A. to Pevek, Russia by way of Vladivostok and the Bering Strait (Fig. 1). Additional freight from Germany (two containers) joined the cargo in Vladivostok via the Trans-Siberian Railway. In Pevek, the combined cargo was loaded onto trucks driven with bulldozer assistance more than 350 km

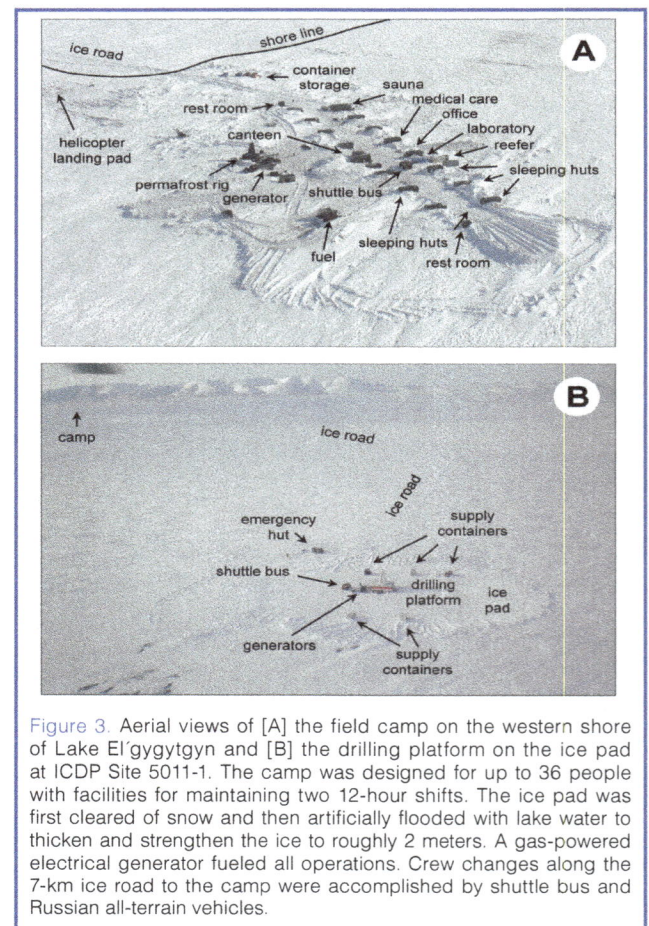

Figure 3. Aerial views of [A] the field camp on the western shore of Lake El'gygytgyn and [B] the drilling platform on the ice pad at ICDP Site 5011-1. The camp was designed for up to 36 people with facilities for maintaining two 12-hour shifts. The ice pad was first cleared of snow and then artificially flooded with lake water to thicken and strengthen the ice to roughly 2 meters. A gas-powered electrical generator fueled all operations. Crew changes along the 7-km ice road to the camp were accomplished by shuttle bus and Russian all-terrain vehicles.

Table 1. Penetration, drilling and core recovery at ICDP Sites 5011-1 and 5011-3 in the El'gygytgyn Crater (all data given in field depth).

Site	Hole	Type of Material	Penetrated (mblf)	Drilled (m)	Recovered (m)	Recovery (%)
5011-1	1A	lake sediment	146.6	143.7	132.0	92
	1B	lake sediment	111.9	108.4	106.6	98
	1C	total	517.3	431.5	273.8	63
		lake sediment		225.3	116.1	52
		impact rocks		207.5	157.4	76
5011-3		permafrost deposits	141.5	141.5	129.9	91

over winter roads and cross-country to Lake El'gygytgyn. There, the operation was supported by a temporary winter camp that was designed for up to thirty-six people and set up on the western lake shore (Fig. 3A). In the camp, a laboratory container for whole-core measurements of magnetic susceptibility stood next to a reefer in which the sediment cores were kept from freezing.

The project completed one borehole into permafrost deposits in the western lake catchment (ICDP Site 5011-3) and three holes at 170 m water depth in the center of the lake (Site 5011-1; Figs. 1, 4; Table 1). Permafrost drilling at Site 5011-3 was conducted from 23 November until 12 December 2008. Using a mining rig (SIF-650M) employed by a local drilling company (Chaun Mining Corp., Pevek), the crew reached a depth of 141.5 m with 91% recovery. After drilling, the borehole was permanently instrumented with a thermistor chain for future ground temperature monitoring as part of the "Global Terrestrial Network for Permafrost" (GTN-P) of the International Permafrost Association (IPA), thus contributing to our understanding of future permafrost behavior in light of contemporary rapid change.

In January/February 2009 an ice road between the camp and Site 5011-1 on Lake El'gygytgyn was established (Figs. 1, 3B). Subsequently, an ice pad of 100 m diameter at the drill site was artificially thickened to 2.3 m to allow for lake drilling operations from the 100-ton drilling platform. The Russian GLAD 800 was developed for extreme cold and operated by the U.S. consortium DOSECC. It consisted of a modified Christensen CS-14 diamond coring rig positioned on a mobile platform that was weather-protected by insulated walls and a tent on top of the 20-m-high derrick. The system was permanently imported into Russia, where it is now available for scientific drilling projects at no cost until 2014.

Drilling at Site 5011-1 was conducted from 16 February to 26 April 2009. The drill plan included the use of casing anchored into the sediment to allow drilling to start at a field depth of 2.9 mblf. Holes 1A and 1B had to be abandoned after twist-offs at 147 mblf and 112 mblf, respectively. In Hole 1A the Hydraulic Piston Corer (HPC) system was used down to 110 mblf, followed by the Extended Nose Corer (EXC) below. The recovery achieved with these tools was 92%. Similarly, drilling with HPC down to 100 mblf and with EXC below provided 98% recovery in Hole 1B. Hole 1C was first drilled by HPC between 42 mblf and 51 mblf, in order to recover gaps still existing in the core composite from Holes 1A and 1B, and was then continued from 100 mblf. Due to the loss of tools during the twist-offs, further drilling had to be performed with the Alien Bit Corer (ALN). The employment of this tool may at least partly explain a much lower recovery of the lake sediments in Hole 1C (total 52%), although this could also be due to the higher concentration of gravel and sand in these deeper lake sediments. The recovery jumped up to almost 100% again at a depth of 265 m, when the tool was changed to a Hard rock Bit Corer (HBC), which has a smaller

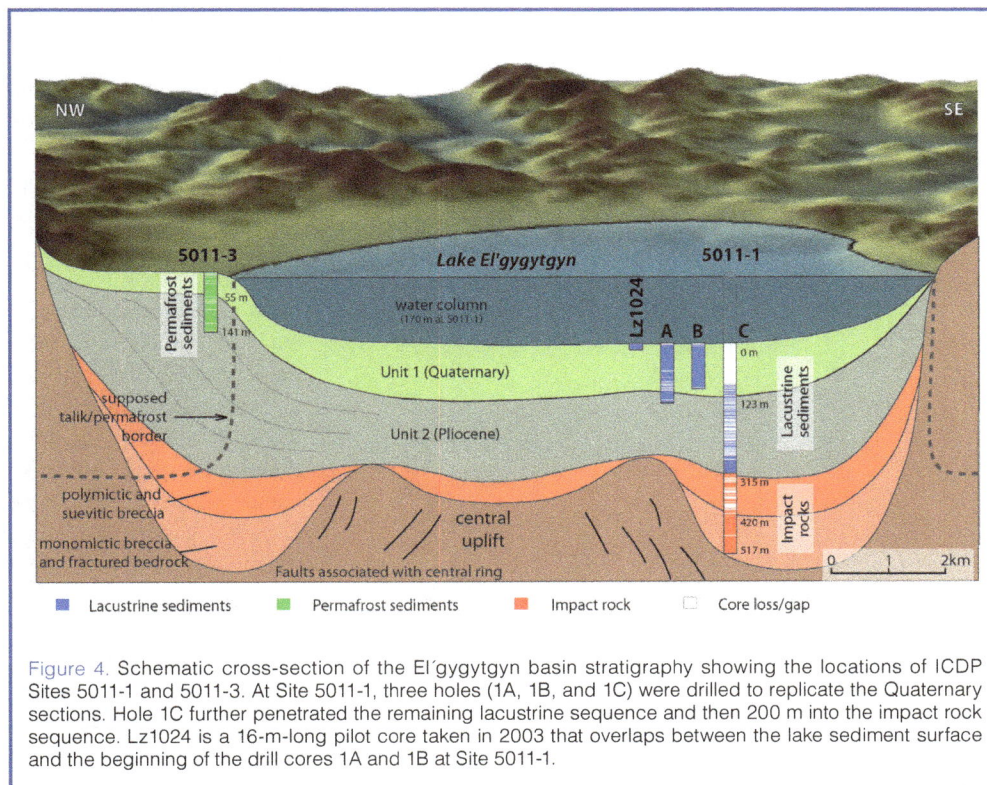

Figure 4. Schematic cross-section of the El'gygytgyn basin stratigraphy showing the locations of ICDP Sites 5011-1 and 5011-3. At Site 5011-1, three holes (1A, 1B, and 1C) were drilled to replicate the Quaternary sections. Hole 1C further penetrated the remaining lacustrine sequence and then 200 m into the impact rock sequence. Lz1024 is a 16-m-long pilot core taken in 2003 that overlaps between the lake sediment surface and the beginning of the drill cores 1A and 1B at Site 5011-1.

diameter than the tools employed before. The boundary between lake sediments and impact rocks was encountered at 315 mblf. Further drilling into the impact breccia and brecciated bedrock down to 517 mblf by HBC took place with 76% recovery.

On-site processing of the cores recovered at Site 5011-1 involved magnetic susceptibility measurements with a multi-sensor core logger (MSCL, Geotek Ltd.) down to a depth of 380 mblf. Initial core descriptions were conducted based on macroscopic and microscopic investigations of the material contained in core catchers and cuttings (lake sediments) and on the cleaned core segments not cored with liners (impact rocks). Additionally, downhole logging was carried out in the upper 394 m of Hole 1C by the ICDP Operational Support Group (OSG), employing a variety of slim hole wireline logging sondes. Despite disturbance of the electrical and magnetic measurements in the upper part of the hole, due to the presence of metal after the twist-offs at Holes 1A and 1B and to some technical problems, these data provide important information on the *in situ* conditions in the hole (e.g., temperature, natural gamma ray, U, K, and Th contents) and permit depth correction of the individual core segments.

Lake Sediments

Based on the whole-core magnetic susceptibility measurements on the drill cores from ICDP Site 5011-1, the field team was able to confirm that the core composite from Holes 1A to 1C provided nearly complete coverage of the uppermost 150 m of the sediment record in central Lake El'gygytgyn (Fig. 5), and that the gap between the top of the drill cores and the sediment surface was properly recovered by the upper part of a 16-m-long sediment core (Lz1024) taken during the 2003 site survey (Fig. 6). The construction of a final composite core record was completed during core processing and subsampling, which began in September 2009 at the University of Cologne, Germany, with the involvement of scientists from Russia and the U.S. The cores were first split lengthwise, and both core halves were macroscopically described and documented by high-resolution line scan images (MSCL CIS Logger, Geotek Ltd.). One core half was then used for measurements of color spectra and magnetic susceptibility in 1-mm increments (SCL2.3 Logger, GFZ Potsdam). This same core was then scanned using X-ray fluorescence (XRF) analyses of light and heavy elements and X-radiography in steps of 2.0 mm and 0.2 mm, respectively (ITRAX Core scanner, Cox Analytical Systems). Measurements of P-wave velocity and gamma-ray density (MSCL Logger, Geotek Ltd.) were then conducted in steps of 2 mm at the Alfred Wegener Institute in Bremerhaven, Germany, before the cores were continuously subsampled back in Cologne with u-channels for paleomagnetic and rock magnetic measurements. Subsequently, 2-cm-thick slices were continuously sampled from the core composite, excluding deposits from mass movement events, and split into eight aliquots of different sizes for additional biological and geochemical analyses. These aliquots, along with some irregular samples from replicate cores (e.g., for luminescence dating or tephra analyses), were subsequently sent to the science team members responsible for their analyses. In addition, thin sections were

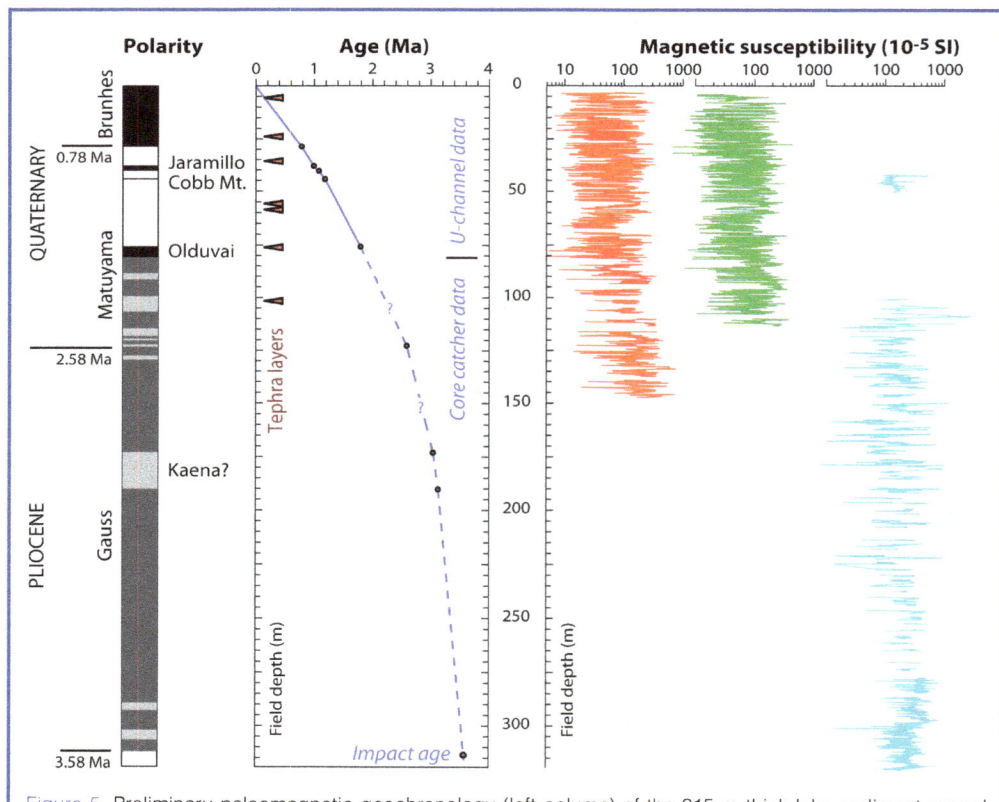

Figure 5. Preliminary paleomagnetic geochronology (left column) of the 315-m-thick lake sediment record from ICDP Site 5011-1 in the central part of Lake El'gygytgyn (for location see Figs. 1 and 4). Black and dark gray refer to normal polarity, white and light gray to reversed polarity, with gray shades representing uncertain interpretations. The age model presented still needs to be confirmed by ongoing paleomagnetic measurements on u-channels and dating of volcanic ash (tephra) layers. The data available so far indicate significantly decreasing sedimentation rates from the Pliocene into the Quaternary. The magnetic susceptibility (MS) data measured in the field (right columns) illustrate the high variability throughout the sediment succession, and that a core composite from Holes 1A, 1B, and 1C provides an almost complete record down to about 150 mblf, representing the uppermost Pliocene and the entire Quaternary.

prepared from representative sections of the cores to facilitate microanalysis of the various lithologies identified during visual core descriptions. In 2011, the remaining untouched core halves will be shipped to the U.S. National Lacustrine Core Repository (LacCore) at the University of Minnesota for long-term archiving.

The current chronological information from the lake sediments relies predominantly on paleomagnetic measurements, which were continuously carried out on the u-channel samples at the GFZ, Potsdam, Germany. In the uppermost 78 m of the core composite, as yet based on uncorrected field depths, magnetozones with normal/reversed polarity can clearly be related to established polarity chrons and subchrons (Ogg and Smith, 2004), including the boundary between the Brunhes and Matuyama chrons (0.781 Ma; ~28.5 mblf), the Jaramillo subchron (1.072–0.988 Ma; ~40.5–38.0 mblf), the termination of the Cobb Mountain subchron (1.173 Ma; ~44.2 mblf; its onset is masked by sediment disturbances at this level), and the termination of the Olduvai

subchron (1.778 Ma; 75.5 mblf; Fig. 5). Below 78 mblf, paleomagnetic information is currently (December, 2010) restricted to initial measurements on semi-oriented discrete samples, which were taken roughly every ~3 m from the core catcher samples in Holes 1A, 1B, and 1C. While these results need to be confirmed by the ongoing u-channel measurements, they suggest that the boundary between the Gauss and Matuyama chrons (2.581 Ma) is located at ~123 mblf, where initial palynological data is consistent with an age close to the Pliocene/Pleistocene boundary. The sediment section with reversed polarity found in the upper part of the normal polarity Gauss chron is tentatively interpreted as the Kaena subchron (3.116–3.032 Ma; ~190–173 mblf). Based on these results, and the assumption that Lake El'gygytgyn was formed shortly after the impact event at 3.58 ± 0.04 Ma (Layer, 2000), long-term sediment accumulation rates are found to be at a maximum in the early part of the record and to decrease in more recent deposits. Specifically, sedimentation rates decrease from ~270 mm ka^{-1} below the Kaena subchron, to ~110 mm ka^{-1} between the termination of the Kaena

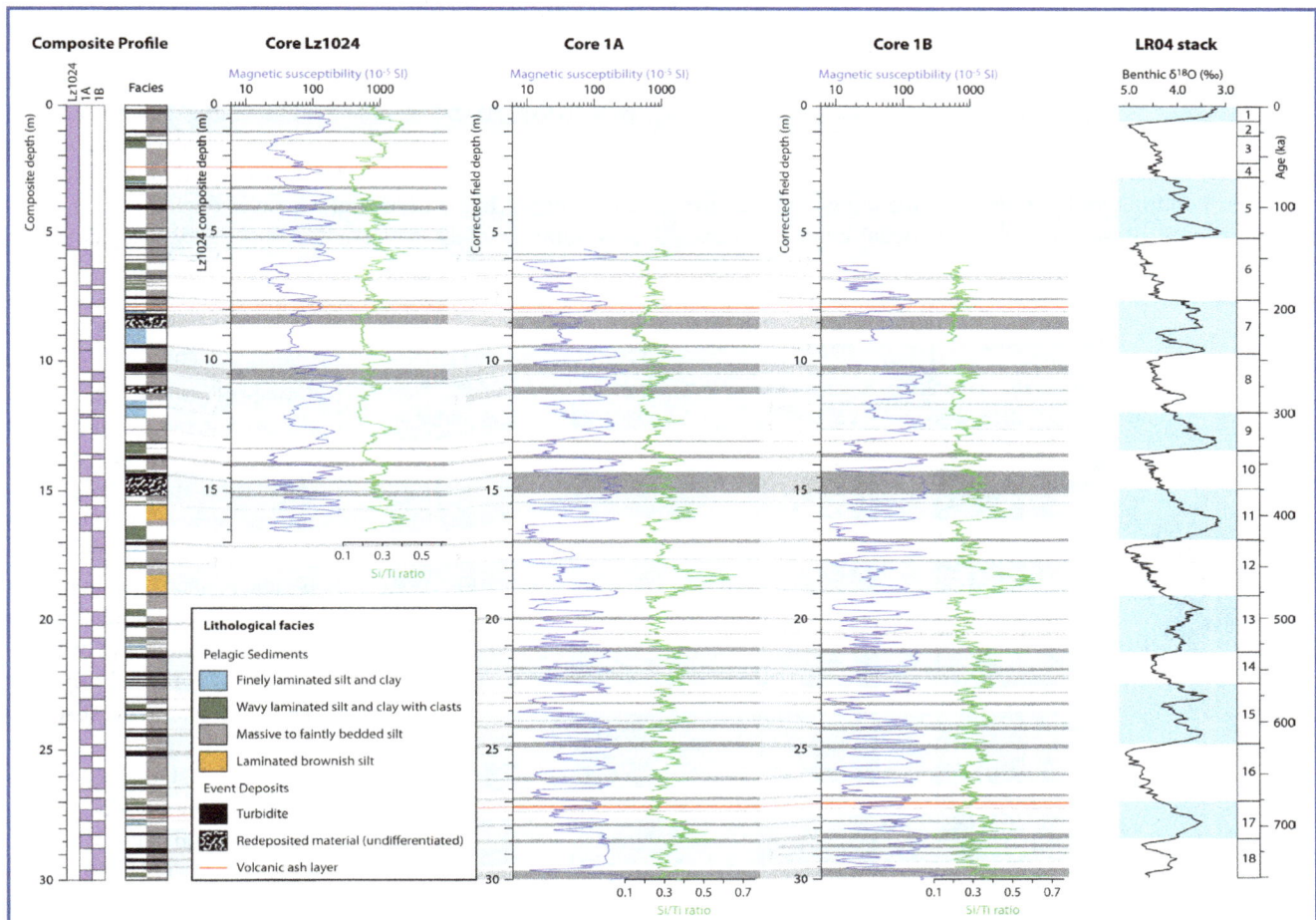

Figure 6. Initial results from the upper 30 meters of the lacustrine record from ICDP Site 5011-1 demonstrate the reasonable fidelity of the high-resolution data throughout the last ~750 ka. Note that the field depths of cores 1A and 1B were corrected by 2.7 m downward, based on the correlation with pilot core Lz1024. Left column with purple boxes shows the sections used to create the composite record. Sediment lithologic facies are shown according to the key in the legend. Columns for cores Lz1024 and ICDP 5011-1A and 1B show on-site data of whole-core magnetic susceptibility next to Si/Ti ratios from XRF scanning, used as proxies for magnetite dissolution during anoxic lake stages and bioproduction vs. clastic sedimentation, respectively. Interglacials, highlighted in light blue bars, are temporally correlated with the marine isotopic record of Lisiecki and Raymo (2005), supported by the Brunhes/Matuyama boundary (780 ka) at 31.2 corrected mblf (referring to 28.5 mblf uncorrected field depth).

subchron and the Gauss/Matuyama boundary, to ~50 mm ka⁻¹ during the Quaternary. These preliminary chronological data are encouraging, because they confirm the feasibility of one of the major objectives of the El'gygytgyn Drilling Project, which is to investigate climatic developments during the Pliocene/Pleistocene transition and within the course of the Quaternary glacial/interglacial cycles. This can now be accomplished based on the core composite that was recovered from the uppermost 150 m of the lake sediment record.

Additional age control is expected to come from luminescence age estimates in the upper ~30 m and from the numerical ages of seven tephra layers identified so far in the sediment record (Fig. 5); both are still in progress. The paleomagnetic, luminescence, and tephra ages will provide chronological tie points for a more detailed age model that will be derived from the correlation of high-resolution proxy measurements with regional insolation variations. This approach was successfully employed by Nowaczyk et al. (2007) on a 13-m-long core from a different location in central Lake El'gygytgyn. Preliminary results from the uppermost 30 m of the core composite at Site 5011-1 (Fig. 6) demonstrate that this approach should also function in the deeper sediments of Lake El'gygytgyn.

The lacustrine sediment succession at Site 5011-1 is remarkably heterogeneous. Changes in lithology occur every few centimeters to decimeters throughout the entire record (Fig. 6). Based primarily on visual characteristics, qualitative grain-size information, and sedimentary struc-

tures observed in the split core halves and radiographs, the sediments are currently subdivided into four distinct facies reflecting continuous, pelagic deposition. Facies 1 is characterized by dark gray to black, finely laminated (<5 mm) silt and clay (Fig. 7A). Laminations are defined by alternating grain-size variations and are clearly observed as density variations in radiographs. Facies 2 is comprised of wavy laminations (<5 mm thick) of alternating silt and clay layers with an abundance of elongate sediment clasts made up of silt-sized particles (Fig. 7B). These clasts are typically several millimeters thick and have long axes up to ~1 cm lying parallel to the bedding plane. The most abundant unit is classified as Facies 3 and is characterized by massive to faintly banded silt of various colors (Fig. 7C). Banding typically occurs as color variations 2–5 cm thick observed in split core halves, with only slight density variations observed in radiographs. Facies 4 consists of red or brownish silt-sized sediment with distinct fine laminations (<5 mm; Fig. 7D). In contrast to other laminated sections of the core, these laminae are not associated with any obvious grain-size variations and less pronounced density variations in the radiographs, possibly reflecting a biological or chemical origin.

Facies 1 through 3 are also present in a 13-m-long pilot core (PG1351) from which additional geochemical data are available and were related to different climate modes by Melles et al. (2007). According to that study, Lake El'gygytgyn had a perennial ice cover during glacial times and some stadials of the Late Quaternary. On one hand, this restricted light penetration and thus biogenic production in the surface waters, as reflected by low biogenic silica deposition (low Si/Ti ratios, Fig. 6). On the other hand, the perennial ice cover also hampered mixing of the water column, leading to anoxic H_2S-bearing bottom waters and no bioturbation. Anoxic bottom waters dur-ing glacial/stadial times also lead to magnetite dissolution, reflected in low values of magnetic susceptibility (Nowaczyk et al., 2007). The glacial/stadial sediments in part consist of finely laminated silt and clay of Facies 1 (Fig. 7A). This facies is thought to reflect cold and relatively moist climates, when blanketing snow on the ice cover led to a reduction in biogenic primary production. In contrast, wavy laminated clast-containing sediments of Facies 2 (Fig. 7B) are pre-

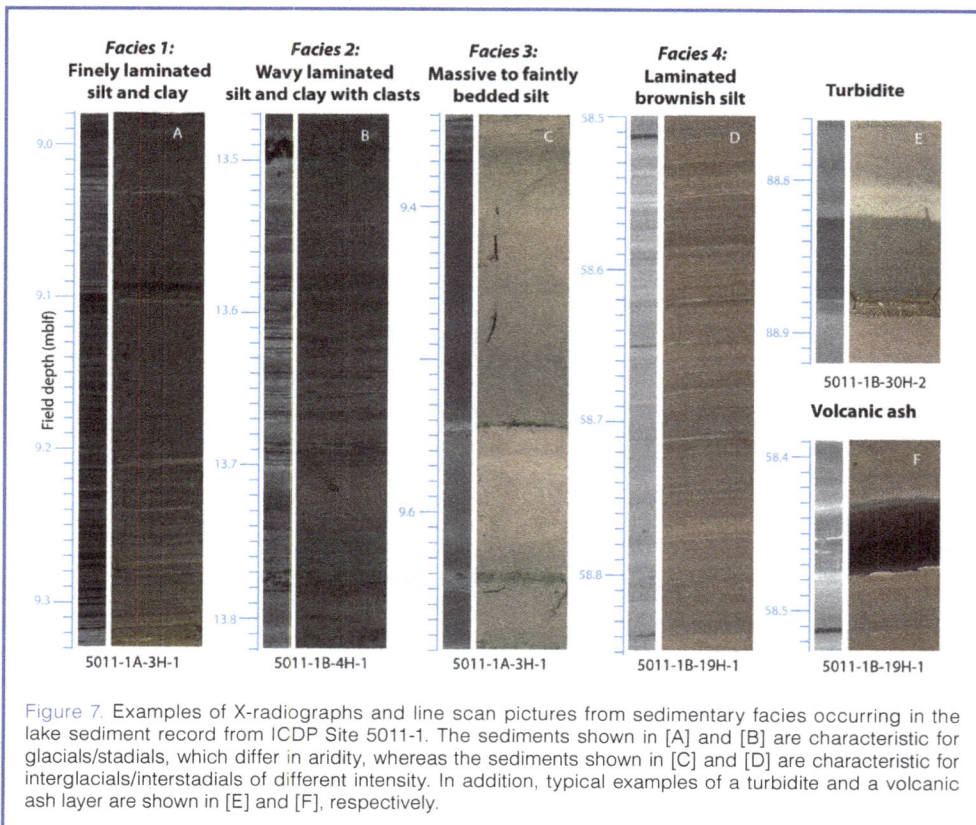

Figure 7. Examples of X-radiographs and line scan pictures from sedimentary facies occurring in the lake sediment record from ICDP Site 5011-1. The sediments shown in [A] and [B] are characteristic for glacials/stadials, which differ in aridity, whereas the sediments shown in [C] and [D] are characteristic for interglacials/interstadials of different intensity. In addition, typical examples of a turbidite and a volcanic ash layer are shown in [E] and [F], respectively.

sumed to reflect drier glacial intervals, when the absence of snow cover on the perennial ice allowed for a higher primary production. The clasts associated with Facies 2 may reflect enhanced deposition of aeolian material on the ice surface (forming cryoconites) followed by the agglomeration of these particles during their transport through the ice along vertical conduits. However, the precise origin and composition of these clasts are still being investigated.

Interglacial/interstadial sedimentation in Lake El'gygytgyn is to a degree reflected by the massive to faintly banded olive-gray to brownish sediments of Facies 3 (Fig. 7C). This facies reflects a semi-permanent ice cover that allows for a higher primary production, as evident in high Si/Ti ratios (Fig. 6). A complete mixing of the water column at the end of ice breakup leads to oxygenation of the bottom waters, as indicated by the preservation of high magnetic susceptibilities. The predominantly massive appearance of the deposits is likely due to some homogenization by bioturbation. In contrast, fine laminations in Facies 4 (Fig. 7D) are indicative of a lack or strong limitation of bioturbation. These sediments also include dark organic-rich layers and are associated with distinct maxima in Si/Ti ratios (Fig. 6). These characteristics are best attributed to deposition during an extraordinarily warm climate, with a prolonged ice-free period and enhanced nutrient supply from the catchment leading to exceptionally high rates of primary production and the exclusion of bioturbation due to depletion in oxygen content caused by enhanced decomposition of organic matter. Facies 4 last occurred at Lake El'gygytgyn during Marine Isotope Stages (MIS) 11 and 9.3 (Fig. 6). The details of this interpretation will require further study.

Pelagic sedimentation in Lake El'gygytgyn is irregularly interrupted by short-term sedimentary events (Fig. 6). These include gravitational mass movements and volcanic ash fallouts (Fig. 7E), which have formed up to 7.4-cm-thick tephra layers in the sediment record of Site 5011-1. Mass movement events are predominantly observed as turbidites, characterized by sharp basal contacts followed by a fining upwards sequence of sand to clay (Fig. 7F). Juschus et al. (2009) described the origin of the turbidites as resulting from sediment settling from suspension clouds produced by debris and density flows that originate on the lake slopes and occasionally penetrate into the center of the lake. Altogether, fifty-three graded beds have been identified in the upper 30 m of the Site 5011-1 record, and most are a few centimeters thick. Only three intervals with debrites, densites (Gani, 2004), or other re-deposited material related to mass movement events occur in this part of the core. Significant erosion by these mass movement events of the pelagic sediment record is not evident according to the age model for this part of the record, which is based on the correlation of sediment facies and proxy measurements with the global MIS stack (Fig. 6), constrained by the Brunhes/Matuyama boundary just below (Fig. 5).

The first information concerning the Pliocene history recorded in Lake El'gygytgyn relies on multi-proxy analyses of small samples taken from the core catchers (every ~3 m) and core cuttings (every meter). Based on these very preliminary data, we observe that the concentration of biogenic silica (BSi), total organic carbon (TOC), and total nitrogen (TN) is significantly lower in the Pliocene than in the Quaternary. Presumably, this is due to a much higher clastic input associated with significantly higher sedimentation rates during the Pliocene (Fig. 5). The Pliocene pollen assemblages are so far dominated mostly by tree pollen. Repeated changes in the plant assemblages through time reflect variations in forests of pine (*Pinus*), larch (*Larix*), spruce (*Picea*), fir (*Abies*), alder (*Alnus*), and hemlock (*Tsuga*). The tree pollen significantly decreases during the presumed Kaena subchron (3.116–3.032 Ma), concomitant with an increase in the relative abundance of wormwood (*Artemisia*) pollen—spores of rock spike-moss (*Selaginella rupestris*)—and coprophilous fungi. This pollen composition suggests treeless glacial environments - over some intervals - which can be described as tundra-steppe. The transition from the Pliocene to the Pleistocene still needs to be studied in detail, but it is broadly marked in the widely spaced samples studied so far by a distinct change from predominantly coniferous assemblages to pollen spectra dominated by dwarf birch, shrub alder, and herbs at ~123 mblf.

Impact Rocks

The El'gygytgyn Crater represents the only currently known impact structure on Earth formed in siliceous volcanic rocks including tuffs. The impact melt rocks and target rocks provide an excellent opportunity to study shock metamorphism of volcanic rocks. The shock-induced changes observed in porphyritic volcanic rocks from El'gygytgyn can be applied to a general classification of shock metamorphism of siliceous volcanic rocks. That El'gygytgyn is an impact crater was confirmed in the late 1970s by Gurov and co-workers (Gurov et al., 1978, 2007; Gurov and Koeberl, 2004), who found shocked minerals and impact glasses in samples at the crater. However, the impact rocks on the surface have been almost totally removed by erosion, and so the El'gygytgyn Drilling Project provides the unique opportunity to study the crater-fill impactites *in situ* and determine their relations and succession. The investigations are expected to provide information on the shock behavior of the volcanic target rocks, the nature and composition of the asteroid that formed the crater, and the amount of energy that was involved in the impact event. This will also allow us to constrain the effects this impact event had on the regional and circumarctic environment.

The impact portion of the drill core from ICDP Site 5011-1, spanning the interval ~315–517 mblf, was handled differently from the lake sediment portion to the extent that no special storage and temperature requirements were necessary for sample export from Pevek to our laboratories. The cores were initially shipped to the Natural History Museum in Berlin, where detailed basic core characterization was done (including complete photographic documentation) during the time period November 2009 to May 2010. In May 2010 an international sampling party was held in Berlin, and samples were allocated to about half a dozen research groups around the world.

The upper part of the impact core directly underneath the lake sediments consists of a unit of so-called suevitic breccia, with a thickness of ~100 m. A suevite is a glass-bearing polymict impact breccia, which contains fragments of a variety of rocks that represent different layers in the target rocks, cemented in a fine-grained matrix (Fig. 8). The glasses, on the other hand, were formed by the melting of the target rocks at very high temperatures. Such breccias are uniquely characteristic of impact craters on Earth and not found in any other geological setting. The green color of the suevite is due to alteration and the abundance of sheet minerals

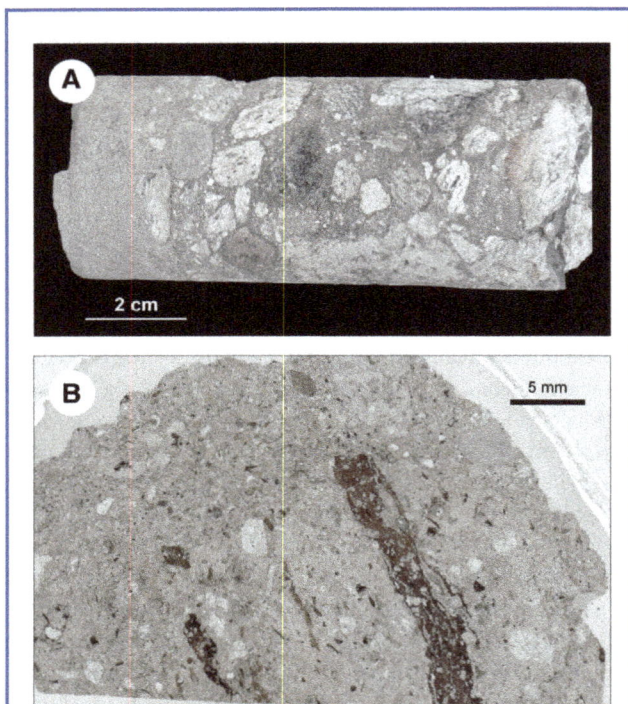

Figure 8. Typical suevitic impact breccia from the El'gygytgyn impact crater. [A] Core segment retrieved from a field depth of about 320 mblf at ICDP Site 5011-1. The rock fragments that occur in this breccia are a mixture of lithologies from the target that represent a pre-impact stratigraphic range of many hundred meters. [B] Thin-section of impact core sample 5011-107Q-4 from 343.8 mblf field depth, showing the so-called "red" suevite, which contains abundant large melt clasts, elongated and parallel to each other, comprising a general fluidal texture. The suevite is an impact breccia containing clastic components and melt clasts, and has a general light brownish color due to alteration into clay minerals. The largest melt clasts show internal fluidal texture. The groundmass of the melt clasts is glassy and brown. Clasts included are composed of fractured quartz, altered feldspar, and a minor amount of partially altered twinned amphibole. Lithic clasts include mainly fragmented quartz, fragmented and altered feldspars, finely twinned calcite, and fragments of volcanic rocks, which might be part of the target. The matrix of the suevite is microcrystalline, mainly composed of clay minerals that show a "fluidal" texture around the clasts.

in the matrix; it contains abundant black melt clasts. The suevite shows a strong anisotropic fabric with fluidal texture. Fractures crosscutting the suevite are common, as are green clay and/or white-reddish carbonate veins. The suevites continue through a highly fractured transition zone with breccia intercalated to suevite and pass into a unit of shocked and locally brecciated volcanic target rocks (which may also contain suevitic breccias), which was uplifted during the impact event. During the formation of such central peaks, which are typical for impact craters of this size (also called "complex craters"), deeper layers of target rock rebound towards the surface and then solidify; thus, a mountain several kilometers in diameter is uplifted over 1000 m vertically in less than a minute. This is truly a spectacular geological process.

Permafrost Deposits

The single deep core obtained at ICDP Site 5011-3, from the western catchment of Lake El'gygytgyn, was entirely frozen when recovered. This confirmed modeling results, which suggested that the unfrozen talik alongside the lake descends with more or less a vertical boundary until the permafrost base is reached at a depth of a few hundred meters (Fig. 4). On site, the permafrost cores were initially described and photographically documented. They were kept frozen in the field and during transport to the ice laboratory at the Alfred Wegener Institute in Bremerhaven, Germany. There, the cores were cleaned, the documentation was completed, and subsamples were taken from the sediment and ice for ongoing laboratory analyses.

The permafrost core contains ground ice throughout and largely consists of sandy gravels with volcanic clasts embedded in a sandy matrix. In the uppermost 75 m processed so far, pollen has only been found in the uppermost 10 m and within a few thin intervals below (Fig. 9). Comparison of the pollen assemblages in these cores with those in nearby permafrost and lake sediment cores retrieved during the site surveys (Lozhkin et al., 2007a, 2007b; Glushkova et al., 2009; Shilo et al., 2008; Matrosova, 2009) remains somewhat speculative, because the reworking of slope materials from the alluvial fans still needs to be assessed. However, the pollen assemblages suggest that the upper 9 m at Site 5011-3 represent a discontinuous record back to the Allerød period, with the Holocene being restricted to the upper 1.8 m and the Younger Dryas represented by the interval 1.8–2.5 m, and that the sediments at ~20 m depth were formed during MIS 5.5 or 7. While the pollen assemblages at ~36 m and ~51 m also indicate Pleistocene ages, those at about 62–65 m depth strongly indicate a Pliocene age for lower portions of the core, based on high pollen counts of pine (*Pinus* subgenus *Haploxylon*) as well as some of larch (*Larix*), fir (*Abies*), spruce (*Picea*), and hemlock (*Tsuga*; Fradkina, 1983). Hence, the Pliocene/Pleistocene boundary in core 5011-3 probably has to be placed somewhere between 51 m and 62 m depth, but additional study is required to test this conjecture.

Organic matter occurs in significant amounts (>1%) only in the Holocene sediments. The inferred climate oscillations for the transition from the Allerød via the Younger Dryas into the Holocene are also suggested in the water isotope record of the ground ice (Fig. 9). There, $\delta^{18}O$ minima and maxima support the inferred vegetation history indicated by the pollen record. Below the Allerød, the $\delta^{18}O$ values of the ground ice show less variation and tend toward more negative values, but not as negative as one would expect for full glacial values, like those seen in other regions of relict permafrost. The values observed here are currently interpreted to be due to a change in ice sources. While the ground ice in the Allerød and younger sediments likely originates from meteoric precipitation, the ice below could have been formed by freezing of lake sediment pore waters following a basinward migration of the talik boundary with lake lowering at some point in the past. Hence, it seems likely that the ground ice in these sediments is much younger than the enclosing sediments. A marginal lake environment prior to the Allerød is also indicated by the occasional occurrence of distinctly rounded pebbles, suggesting shore-line processes, and well-sorted sandy layers, possibly deposited on the upper lake slope. If this preliminary interpretation is confirmed by ongoing core analyses and modeling of the freezing front migrations, then the permafrost core 5011-3 may contribute to the reconstruction of the lake-level changes in the El'gygytgyn Crater since Pliocene times.

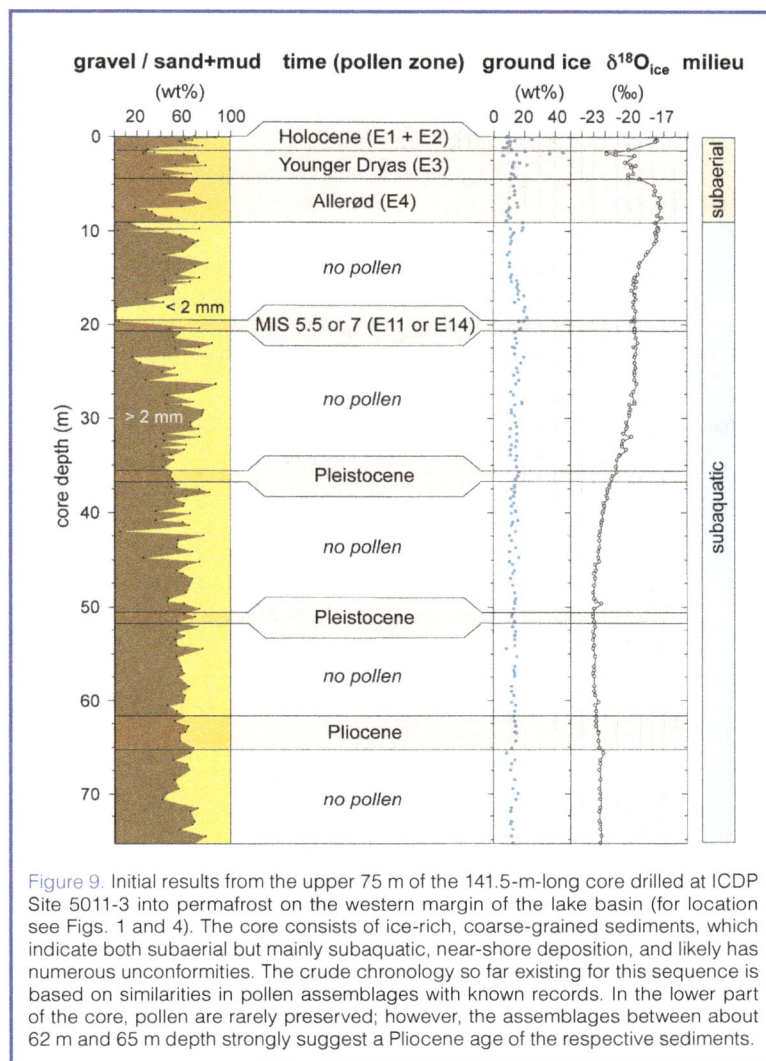

Figure 9. Initial results from the upper 75 m of the 141.5-m-long core drilled at ICDP Site 5011-3 into permafrost on the western margin of the lake basin (for location see Figs. 1 and 4). The core consists of ice-rich, coarse-grained sediments, which indicate both subaerial but mainly subaquatic, near-shore deposition, and likely has numerous unconformities. The crude chronology so far existing for this sequence is based on similarities in pollen assemblages with known records. In the lower part of the core, pollen are rarely preserved; however, the assemblages between about 62 m and 65 m depth strongly suggest a Pliocene age of the respective sediments.

Acknowledgements

Marianna Voevodskaya (RAS/CRDF) provided a smooth transition for the field parties through Moscow. In Pevek, Dmitry Koselov from Chukotrosgidromet arranged all of the necessary paperwork required by local authorities. He also managed all of the in-town shuttle services and contacts with the airport. On-site, Nikolai Vasilenko and his colleagues from the Chaun Mining Corporation provided a friendly, cooperative atmosphere for daily operations in the base camp and during drilling crew shifts. Contracts for the camp were organized by CH2MHill. We are particularly grateful to the personnel in the Pevek office of the Kinross Gold Corporation for their assistance. Ice pad preparation and safety monitoring were carried out by EBA Engineering Consultants Ltd., Canada. The Russian GLAD 800 drilling system was developed and operated by DOSECC Inc., the downhole logging was performed by the ICDP-OSG, and LacCore at the University of Minnesota handled core curation.

Funding for this research was provided by the International Continental Scientific Drilling Program (ICDP), the U.S. National Science Foundation (NSF), the German Federal Ministry of Education and Research (BMBF), Alfred Wegener Institute (AWI) and GeoForschungsZentrum Potsdam (GFZ), the Russian Academy of Sciences Far East Branch (RAS FEB), the Russian Foundation for Basic Research (RFBR), and the Austrian Federal Ministry of Science and Research (BMWF).

The El'gygytgyn Scientific Party

P. Anderson, A. Andreev, I. Bindeman, D. Bolshiyanov, V. Borkhodoev, K. Brady, J. Brigham-Grette (Principal Investigator), L. Brown, S. Burns, B. Chabligin, M. Cherepanova, T. Cook, R. Deconto, G. Fedorov, S. Forman, A. Francke, D. Froese, C. Gebhardt, O. Glushkova, V. Goette, J. Griess, A. Hilgers, A. Holland, E. Haltia-Hovi, H.-W. Hubberten, O. Juschus, J. Karls, C. Koeberl (Principal Investigator), S. Koenig, C. Kopsch, M. Kukkonen, P. Layer, A. Lozhkin, K. Mangelsdorf, T. Martin, T. Matrosova, H. Meyer, C. Meyer-Jacob, M. Melles (Principal Investigator), P. Minyuk (Principal Investigator), K. Murdock, F. Niessen, M. Nolan, A. Noren, N. Nowaczyk, N. Ostanin, S. Petsch, L.

Pittarello, V. Ponomareva, M. Portnyagin, F. Preusser, V. Pushkar, U. Raschke, J. Reed, P. Rosén, G. Schwamborn, V.S. Sakhno, T. Sapelko, N. Savva, L. Schirrmeister, V. J. Snyder, V.N. Smirnov, C. Van den Bogaard, H. Vogel, B. Wagner, D. Wagner, V. Wennrich, K. Wilkie.

References

Asikainen, C.A., Francus, P., and Brigham-Grette, J., 2007. Sedimentology, clay mineralogy and grain-size as indicators of 65 ka of climate change from El′gygytgyn Crater Lake, northeastern Siberia. *J. Paleolimnol.*, 37:105–122, doi:10.1007/s10933-006-9026-5.

Brigham-Grette, J., Melles, M., Minyuk, P., and Scientific Party, 2007. Overview and significance of a 250 ka paleoclimate record from El′gygytgyn Crater Lake, NE Russia. *J. Paleolimnol.*, 37:1–16, doi:10.1007/s10933-006-9017-6.

Cherepanova, M.V., Snyder, J.A., and Brigham-Grette, J., 2007. Diatom stratigraphy of the last 250 ka at Lake El′gygytgyn, northeast Siberia. *J. Paleolimnol.*, 37:155–162, doi:10.1007/s10933-006-9019-4.

Cremer, H., and Wagner, B., 2003. The diatom flora in the ultra-oligotrophic Lake El′gygytgyn, Chukotka. *Polar Biol.*, 26:105–114, doi:10.1007/s00300-002-0445-0.

Cremer, H., Wagner, B., Juschus, O., and Melles, M., 2005. A microscopical study of diatom phytoplankton in deep crater Lake El′gygytgyn, northeast Siberia. *Algol. Studies*, 116:147–168, doi:0342-1120/04/0157-147.

Forman, S.L., Pierson, J., Gomez, J., Brigham-Grette, J., Nowaczyk, N.R., and Melles, M., 2007. Luminescence geochronology for sediments from Lake El′gygytgyn, northwest Siberia, Russia: constraining the timing of paleoenvironmental events for the past 200 ka. *J. Paleolimnol*, 37:77–88, doi:10.1007/s10933-006-9024-7.

Fradkina, A.F., 1983. *Neogene Palynofloras of Northeast Asia*. Moscow (Nauka), [in Russian].

Gani, M.R., 2004. From turbid to lucid: a straightforward approach to sediment gravity flows and their deposits. *The Sedimentary Record*, 2:4–8.

Gebhardt, A.C., Niessen, F., and Kopsch, C., 2006. Central ring structure identified in one of the world's best-preserved impact craters. *Geology*, 34:145–148, doi:10.1130/G22278.1.

Glushkova, O.Y., 2001. Geomorphological correlation of Late Pleistocene glacial complexes of Western and Eastern Beringia. *Quat. Sci. Rev.*, 20:405–417, doi:10.1016/S0277-3791(00)00108-6.

Glushkova, O.Y., and Smirnov, V.N., 2007. Pliocene and Holocene geomorphic evolution and paleogeography of the El′gygytgyn Lake region, NE Russia. *J. Paleolimnol.*, 37:37–47, doi:10.1007/s10933-006-9021-x.

Glushkova, O.Y., Smirnov, V.N., Matrosova, T.V., Vazhenina, L.N., and Braun, T.A., 2009. Climatic-stratigraphic characteristics and radiocarbon dates from the terrace complex in the El′gygytgyn Lake basin. *Vestnik FEB RAS*, 2:31–43 [in Russian].

Gurov, E.P., and Koeberl, C., 2004. Shocked rocks and impact glasses from the El′gygytgyn impact structure (Russia). *Meteorit.*

Planet. Sci., 39:1495–1508, doi:10.1111/j.1945-5100.2004.tb00124.x.

Gurov, E.P., Koeberl, C., and Yamnichenko, A., 2007. El′gygytgyn impact crater, Russia: structure, tectonics, and morphology. *Meteorit. Planet. Sci.*, 42:307–319, doi:10.1111/j.1945-5100.2007.tb00235.x.

Gurov, E.P., Valter, A.A., Gurova, E.P., and Serebrennikov, A.I., 1978. Meteorite impact crater El′gygytgyn in Chukotka. *Dokl. Akad. Nauk SSSR+*, 240:1407–1410 [in Russian].

Juschus, O., Melles, M., Gebhardt, A.C., and Niessen, F., 2009. Late Quaternary mass movement events in Lake El′gygytgyn, north-eastern Siberia. *Sedimentol.*, 56:2155–2174. doi:10.1111/j.1365-3091.2009.01074.x.

Juschus, O., Preusser, F., Melles, M., and Radtke, U., 2007. Applying SAR-IRSL methodology for dating fine-grained sediments from Lake El′gygytgyn, north-eastern Siberia. *Quat. Geochron.*, 2:187–194, doi:10.1016/j.quageo.2006.05.006.

Layer, P., 2000. Argon-40/argon-39 age of the El′gygytgyn impact event, Chukotka, Russia. *Meteorit. Planet. Sci.*, 35:591–599, doi:10.1111/j.1945-5100.2000.tb01439.x.

Lisiecki, L.E., and Raymo, M.E., 2005. A Pliocene-Pleistocene stack of 57 globally distributed benthic $\delta^{18}O$ records. *Paleoceanography*, 20:PA1003, doi:10.1029/2004PA001071.

Lozhkin, A.V., Anderson, P.M., Matrosova, T.V., and Minyuk, P., 2007a. The pollen record from El′gygytgyn Lake: implications for vegetation and climate histories of northern Chukotka since the late middle Pleistocene. *J. Paleolimnol.*, 37:135–153, doi:10.1007/s10933-006-9018-5.

Lozhkin, A.V., Anderson, P.M., Matrosova, T.V., Minyuk, P.S., Brigham-Grette, J., and Melles, M., 2007b. Continuous record of environmental changes in Chukotka during the last 350 thousand years. *Russ. J. Pac. Geol.*, 1:550–555, doi:10.1134/S1819714007060048.

Matrosova, T.V., 2009. Vegetation and climate change in northern Chukotka during the last 350 ka (based on lacustrine pollen records from Lake El′gygytgyn). Vestnik FEBRAS, 2:23–30 [in Russian].

Melles, M., Brigham-Grette, J., Glushkova, O.Y., Minyuk, P., Nowaczyk, N.R., and Hubberten, H.-W., 2007. Sedimentary geochemistry of a pilot core from El′gygytgyn Lake—a sensitive record of climate variability in the East Siberian Arctic during the past three climate cycles. *J. Paleolimnol.*, 37:89–104, doi:10.1007/s10933-006-9025-6.

Melles, M., Minyuk, P., Brigham-Grette, J., and Juschus, O., 2005. The Expedition El′gygytgyn Lake 2003 (Siberian Arctic). Ber. Polarforsch. Meeresforsch. 505:139 pp.

Minyuk, P., Brigham-Grette, J., Melles, M., Borkhodoev, V.Y., and Glushkova, O.Y., 2007. Inorganic geochemistry of El′gygytgyn Lake sediments, northeastern Russia, as an indicator of paleoclimatic change for the last 250 kyr. *J. Paleolimnol.*, 37:123–133. doi:10.1007/s10933-006-9027-4.

Niessen, F., Gebhardt, A.C., Kopsch, C., and Wagner, B., 2007. Seismic investigation of the El′gygytgyn impact crater lake (Central Chukotka, NE Siberia): preliminary results. *J. Paleolimnol.*, 37:17–35, doi:10.1007/s10933-006-9022-9.

Nolan, M., and Brigham-Grette, J., 2007. Basic hydrology, limnology, and meteorology of modern Lake El′gygytgyn, Siberia. *J. Paleolimnol.*, 37:17–35, doi:10.1007/s10933-006-9020-y.

Nolan, M., Liston, G., Prokein, P., Brigham-Grette, J., Sharpton, V., and Huntzinger, R., 2003. Analysis of lake ice dynamics and morphology on Lake El'gygytgyn, Siberia, using SAR and Landsat. *J. Geophys. Res.*, 108(D2):8062, doi:10.1029/2001JD000934.

Nowaczyk, N.R., Melles, M. and Minyuk, P., 2007. A revised age model for core PG1351 from Lake El'gygytgyn, Chukotka, based on magnetic susceptibility variations correlated to northern hemisphere insolation variations. *J. Paleolimnol.*, 37:65–76, doi:10.1007/s10933-006-9023-8.

Ogg, J.G., and Smith, A.G., 2004. The geomagnetic polarity scale. *In* Gradstein, F.M., Ogg, J.G., and Smith A.G. (Eds.), *A Geologic Time Scale 2004.* Cambridge (Cambridge University Press), 63–86.

Schwamborn, G., Fedorov, G., Schirrmeister, L., Meyer, H., and Hubberten, H.-W., 2008a. Periglacial sediment variations controlled by Late Quaternary climate and lake level change at Elgygytgyn Crater, Arctic Siberia. *Boreas*, 37:55–65. doi:10.1111/j.1502-3885.2007.00011.x.

Schwamborn, G., Förster, A., Diekmann, B., Schirrmeister, L., and Fedorov, G., 2008b. Mid to Late Quaternary cryogenic weathering conditions in Chukotka, northeastern Russia: inference from mineralogical and microtextural properties of the Elgygytgyn Crater Lake sediment record.In Kane, D.L., and Hinkel, D.M. (Eds.), *Ninth International Conference on Permafrost*, Fairbanks (Institute of Northern Engineering, University of Alaska), 1601–1606.

Schwamborn, G., Meyer, H., Fedorov, G., Schirrmeister, L., and Hubberten, H.-W., 2006. Ground ice and slope sediments archiving late Quaternary paleoenvironment and paleoclimate signals at the margins of El'gygytgyn Impact Crater, NE Siberia. *Quaternary Res.*, 66:259–272, doi:10.1016/j.yqres.2006.06.007.

Shilo, N.A., Lozhkin, A.V., Anderson, P.M., Vazhenina, L.N., Stetsenko, T.V., Glushkova, O.Y., and Matrosova, T.V., 2008. First data on the expansion of Larix gmelinii (Rupr.) into arctic regions of Beringia during the early Holocene. *Dokl. Akad. Nauk+*, 423:680–682.

Yershov, E.D., 1998. *General Geocryology. (Studies in Polar Research).* Cambridge (Cambridge University Press), 580 pp, doi:10.1017/CBO9780511564505.

Authors

Martin Melles, Institute of Geology and Mineralogy, University of Cologne, Zuelpicher Str. 49a, D-50674 Cologne, Germany, e-mail: mmelles@uni-koeln.de.

Julie Brigham-Grette, Department of Geosciences, University of Massachusetts, 611 North Pleasant Street, Amherst, MA 01003, U.S.A.

Pavel Minyuk, North-East Interdisciplinary Scientific Research Institute, FEB RAS, 16 Portovaya St., 685000, Magadan, Russia.

Christian Koeberl, Department of Lithospheric Research, University of Vienna, Althanstrasse 14, A-1090 Vienna, Austria (and: Natural History Museum, A-1010 Vienna, Austria).

Andrei Andreev, Institute of Geology and Mineralogy, University of Cologne, Zuelpicher Str. 49a, D-50674 Cologne, Germany.

Timothy Cook, Department of Geosciences, University of Massachusetts, 611 North Pleasant Street, Amherst, MA 01003, U.S.A.

Grigory Fedorov, Arctic and Antarctic Research Institute, Bering Street, 199397 St. Petersburg, Russia.

Catalina Gebhardt, Alfred Wegener Institute for Polar and Marine Research, Am Alten Hafen 26, D-27568 Bremerhaven, Germany.

Eeva Haltia-Hovi, GFZ German Research Centre for Geosciences, Potsdam, Telegrafenberg C321, D-14473 Potsdam, Germany.

Maaret Kukkonen, Institute of Geology and Mineralogy, University of Cologne, Zuelpicher Str. 49a, D-50674 Cologne, Germany.

Norbert Nowaczyk, GFZ German Research Centre for Geosciences, Potsdam, Telegrafenberg C321, D-14473 Potsdam, Germany.

Georg Schwamborn, Alfred Wegener Institute for Polar and Marine Research, Telegrafenberg A43, D-14473 Potsdam, Germany.

Volker Wennrich, Institute of Geology and Mineralogy, University of Cologne, Zuelpicher Str. 49a, D-50674 Cologne, Germany.

and the El'gygytgyn Scientific Party

Related Web Links

http://elgygytgyn.icdp-online.org
http://www.elgygytgyn.uni-koeln.de
http://www.geo.umass.edu/lake_e/index.html
http://www.dfg-science-tv.de/en/projects/polar-archive
http://www.polartrec.com/geologic-climate-research-in-siberia

Figure Credits

Fig. 1: satellite images from NASA WorldWind
Fig. 3A: photo by Tim Martin, Greensboro Day School
Fig. 3B: photo by Jens Karls, University of Cologne

17

Twenty Years of Drilling the Deepest Hole in Ice

by Nikolay I. Vasiliev, Pavel G. Talalay, and Vostok Deep Ice Core Drilling Parties

Introduction

Ice sheets and glaciers contain stratified ancient ice that fell as snow years to millions of years ago. The dust particles, soluble chemicals, and gases trapped in the ice can be used to study how Earth's climate system operated in the past. However, this requires deep ice coring. The retrieved glacial ice can be utilized for an accurate measure of past greenhouse gases with climate clearly documented in the same core. Therefore, ice core data have become crucial to our understanding of past climate change and to making assessments about future climate.

The Soviet Antarctic research station Vostok was founded at the center of the East Antarctic Ice Sheet (78°28'S, 106°48'E, 3488 m.a.s.l.) in 1957 (Fig. 1). This place turned out to be the coldest on Earth; the lowest reliably measured temperature of -89.2°C was recorded on 21 July 1983. In addition, by good fortune Vostok was set above the southern end of the largest subglacial lake in Antarctica, discovered in 1996 by Russian and British scientists (Kapitsa et al., 1996) while drilling deep boreholes.

Deep ice core drilling at Vostok station began in 1970. In the 1970s a set of open uncased holes were drilled by a thermal drill system suspended on cable. The deepest dry hole in ice reached 952.4 m (Hole #1, May 1972). It was concluded that for drilling at greater depths it is necessary to prevent hole closure by filling of the borehole with a fluid. Thus, from 1980 on new thermal and electromechanical drill systems working in fluid were used. Two boreholes reached depths of more than 2000 m. Hole #3G-2 was deepened to 2201.7 m depth in 1985 (Kudryashov, 1989) and Hole #4G-2 to 2546.4 m depth in 1989 (Kudryashov et al., 1994). Drilling of both holes was aborted because of stuck tools.

Drilling Operations 1990–2010

Drilling a new deep Hole #5G started in February 1990, using a TELGA-14M thermal drill for dry coring to a depth of 120 m (Table 1). Thereafter, the thermal drill TBZS-152M for fluid-filled holes was used down to 2502.7 m, at which point it became stuck during tripping out due to hole closure caused by insufficient fluid pressurization. As recovery attempts failed, the cable was pulled out of the top of the drill (Tchistiakov et al., 1994).

About 35 meters of artificial core was dropped on top of the stuck drill, creating a base for a new offset hole. The TBZS-132 thermal drill was used to sidetrack and drill Hole #5G-1 (Fig. 2). The main difference between the thermal drills TBZS-132 and TBZS-152M was the outer diameter of the drill head and the tubing used for the core barrel and water tank (Kudryashov et al., 1998). In 1993 Hole #5G-1 reached 2755.3 m depth, a new record for thermal drilling in ice.

During the summer season 1993–1994, the borehole diameter was enlarged from 180 mm to 220 mm in the upper portion using an electromechanical reaming technique. The hole was cased with fiberglass tubing with a thermal shoe at the bottom, and was sealed at a depth of 120 m to prevent fluids from entering the hole through an upper permeable zone.

In November 1994, drilling operations in Hole #5G-1 were resumed with the KEMS-135 electromechanical drill (Fig. 3), reaching 3350 m depth by January 1996. An average penetration of 2.8 m was achieved per run, but at the depth greater than 2930 m, progress decreased dramatically because of frequent sticking and jamming (Kudryashov et al., 2002).

With season 1996–97 the drilling operations were reduced to the short Antarctic summer because a vehicle traverse failed to reach Vostok in the previous season. When

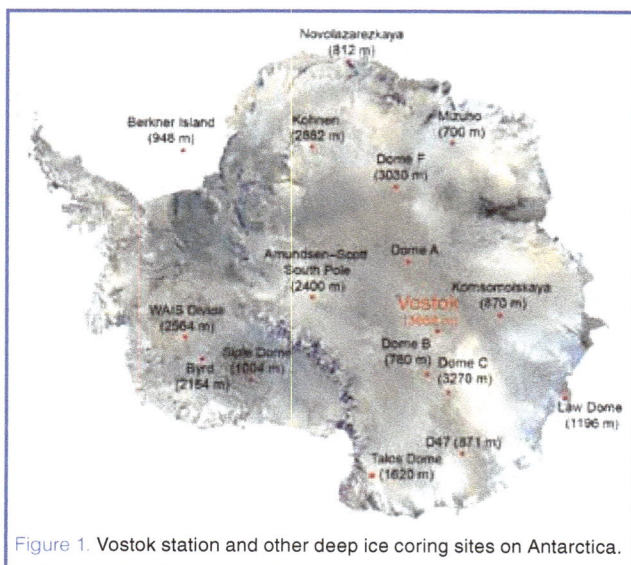

Figure 1. Vostok station and other deep ice coring sites on Antarctica.

drilling of Hole #5G-1 was continued until January 1998 (reaching 3623 m depth), several design changes allowed an increase in the efficiency of ice destruction and drilling chips removal from the bottom of the hole.

After an eight-year hiatus, this hole was reopened in the summer 2005–06 (Vasiliev et al., 2007). A new geometry of cutters for penetration of "warm ice" improved the drilling process. As a result, Hole #5G-1 was deepened to 3658 m (January 2007), with an average core length of 0.7–0.8 m per run (Fig. 4). At this depth the drill became stuck at the bottom of the hole. A drill team that remained at Vostok station over winter filled the lower hole with eighty liters of an anti-freeze agent using a special fluid barrel with an electro-magnetic valve. The drill was captured with an overshot gripper and was lifted to surface in the first attempt before the water-glycol solution was removed from the hole. In May 2007 drilling continued, and different lengths of core barrels were applied to ensure proper functioning of an anti-torque system in cavity intervals. A total of fifty-five runs were required between 3658 m and 3668 m.

Unfortunately, during the hole enlargement in October 2007, the core barrel suddenly dropped to the bottom of the hole. All attempts to recover it failed, and operations did not resume before December 2008. A new deviated hole was drilled using thermal directional drilling to bypass the stuck tools (Vasiliev et al., 2007). The new deviated Hole #5G-2 was drilled without using any special whipstock because the drill bit is usually pushed into a vertical position.

Deviated drilling started at the depth of 3580 m using the electromechanical drill KEMS-135 with a special drill head and cutters. The hole was successfully deviated and drilled to 3600 m where the core with a normal circular cross-section was pulled out. Sidetracking of Hole #5G-2 for testing purpose showed the high efficiency of this technology.

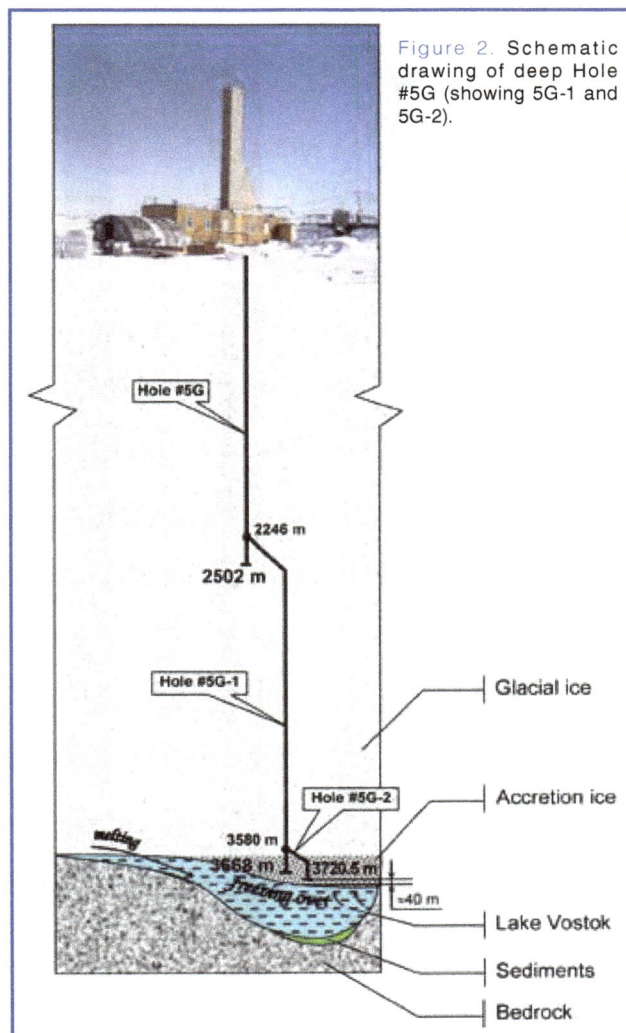

Figure 2. Schematic drawing of deep Hole #5G (showing 5G-1 and 5G-2).

In summer 2009–2010 drilling at Vostok continued, and Hole #5G-2 was deepened from 3600 m to 3650 m, nearly achieving the depth reached previously. According to our estimate the distance between the 5G-1 and 5G-2 holes at the

Table 1. Deep drilling of the Hole #5G (5G-1, 5G-2) at Vostok station.

Expedition # (Year)	Leader of the drilling team	Hole #	Interval of drilling (m)	Type of drill	Mean rate of penetration (m h^{-1})	Mean length of run (m)
35 (1990)	A.A. Zemtsov	5G	0–120	TELGA-14M	1.8	1.9
			120–1279.8	TBZS-152M	2.1	2.9
36 (1991)	A.V. Krasilev		1279.8–2502.7		2.3	3.0
37 (1992)	B.S. Moiseev	5G-1	2232–2249.5	TBZS-132	2.0	1.0
			2249.5–2270.7		2.0	2.0
38 (1993)	V.K. Chistyakov		2270.7–2755.3		1.8	2.5
40 (1993)	N.I. Vasiliev		2755.3–3109	KEMS-135	8.0	2.5
41* (1995)	N.I. Vasiliev		3109–3350		8.0	2.2
42* (1995/96)	N.I. Vasiliev		3350–3523		8.0	2.1
43* (1997/98)	N.I. Vasiliev		3523–3623		8.0	1.8
51* (2005/06)	N.I. Vasiliev		3623–3650		5.0	0.8
52 (2007)	N.I. Vasiliev		3650–3668		5.0	0.7
54* (2009/09)	N.I. Vasiliev	5G-2	3580–3600		-	-
55* (2009/10)	N.I. Vasiliev		3600–3650		5.0	0.7
56* (2010/11)	N.I. Vasiliev		3650–3720.5		5.0	0.9

*The drilling was conducted during the austral summer only.

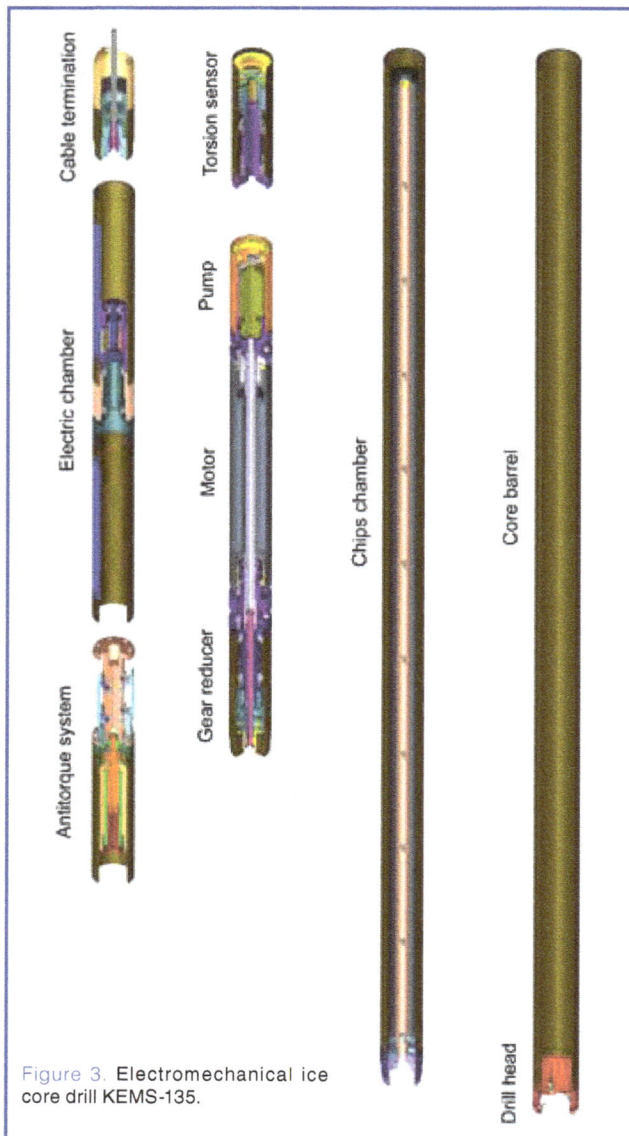

Figure 3. Electromechanical ice core drill KEMS-135.

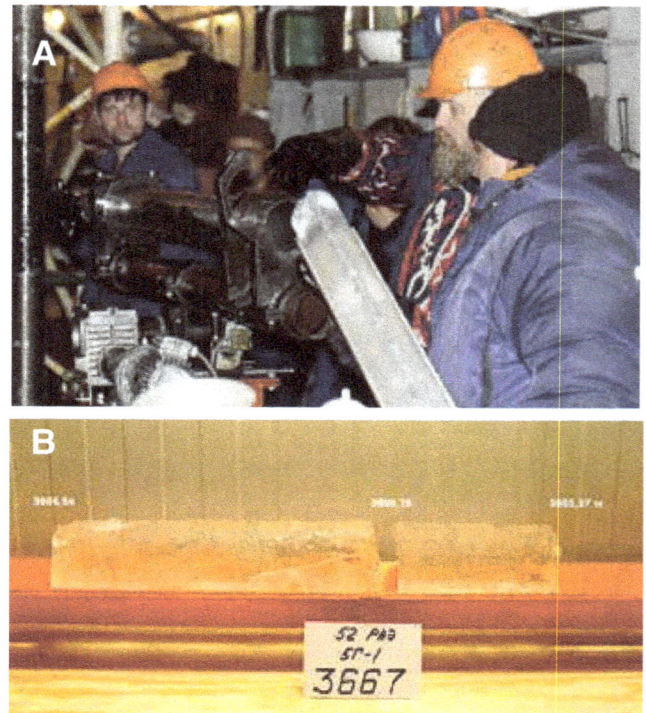

Figure 4. [A] Core recovering from the core barrel (January 2007). [B] Ice core from the deepest layers.

bottom of Hole #5G-2 is nearly 1.5 m. In the summer season of 2010-11 the Hole #5G-2 reached the depth of 3720.5 m. The penetration into Lake Vostok will start in the coming years.

Summary of the Scientific Results

The upper 3310 meters of the Vostok ice core has provided a detailed paleoclimate record for the past four glacial-interglacial cycles occurring about every 100,000 years (Petit et al., 1999). The ice core record extends over the last 420,000 years (Fig. 5). The succession of changes through each climate cycle and termination was similar for the parameters shown; atmospheric and climate properties oscillated within stable limits. In contrast, interglacial periods differ in temporal evolution and duration.

Between 3310 m and 3539 m, the glacial core is disturbed by bedrock deformation (Souchez et al., 2003). Information on microparticles, crystal sizes, and chemical element distri-

butions in that part of the core shed new light on this deformation process—the ice deformation occurred when the ice was still grounded upstream from Vostok station in a region with subfreezing temperatures.

The deepest portion of the ice core (from 3539 m to 3668 m) has a chemistry, isotopic composition, and crystallography distinctly different from the overlying glacial ice. Geochemical and physical data indicate that it originated from the accretion of subglacial lake water to the underside of the ice sheet (Jouzel et al., 1999). Together with data on ionic chemistry, these ice core data favor an origin of the lake ice by frazil ice generation in a supercooled water plume existing in the lake, followed by accretion and consolidation from subsequent freezing of the host water.

Microbiological studies of the Vostok glacial and accreted ice have indicated that low, but detectable, concentrations of prokaryotic cells (Fig. 6) and DNA are present (Bulat et al., 2004). Many of the bacterial cells are associated with non-living organic and inorganic particulate matter. Some of the viable bacteria were deposited more than 400,000 years ago (Bobin et al., 1994).

According to petro-fabric investigation, development of shear zones in the Antarctic ice sheet is linked with global increase of the dust concentration in the atmosphere during past glacial maxima (Lipenkov et al., 2007). The ice strata forming in these periods are characterized by high impurities of microparticles, small ice-grain sizes, single-maximum c-axis orientation, and low ice viscosity.

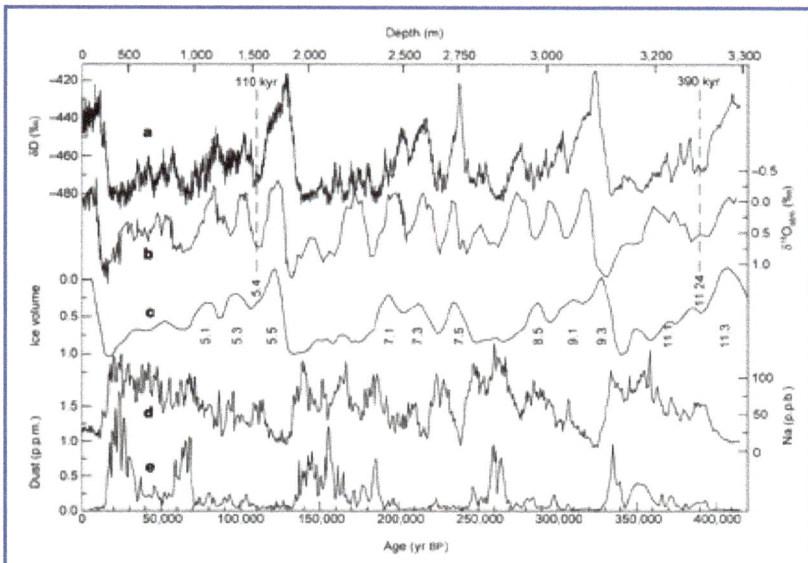

Figure 5. 420,000-year profiles of (1) deuterium, (2) $\delta^{18}O_{atm}$, (3) seawater $\delta^{18}O$ (ice volume proxy), (4) sodium, and (5) dust (Petit et al., 1999).

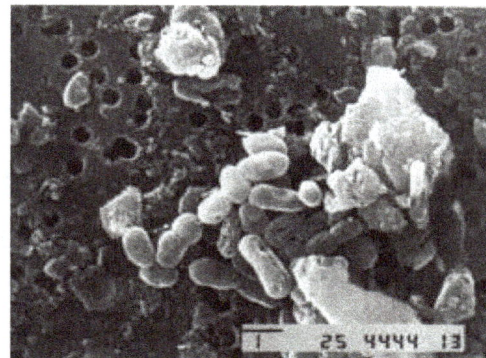

Figure 6. Microbes from Vostok ice core, depth of 2395 m, magnification x14,000 (Bobin et al., 1994).

consisting of two parts: a 2-m-long pilot micro-drill generating a 50-mm-diameter hole, followed by the main drill equipped with a 132-mm-diameter thermal drill bit. With this system an average penetration rate of 3–4 m hr^{-1} is expected. The drill will be cleaned by the produced melt water which creates a second clean layer separating the bottom of the hole from the drilling fluid. Once the tip of the thin pilot drill punctures the lake surface, the packer will be automatically turned on, and the drill heating and ad-vance will stop. After determination of the pressure differ-ence between hole and lake and maintaining it in a range of 3–4 bars, the thermal drill TBPO-132 will be pulled up. This will allow lake water to enter into the hole and to fill up its lower 30–40 meters.

The third stage could be conducted after checking if the water has frozen in the hole. Then, the frozen lake water will be sampled with electromechanical drill KEMS-135 to a level of about 15–20 m above the ice-water interface.

In recent years, advances have been made in understanding and predicting the physical and chemical environment of Lake Vostok based on modeling efforts that set boundary conditions for various attributes. Future studies of the subglacial water properties and searching for ancient life are now important parts of the project at Lake Vostok.

Planning Penetration into Lake Vostok

With dimensions of 280 km x 50 km and water depth reaching 1200 m (Fig. 7) beneath an almost 4-km-thick ice sheet, Lake Vostok is the largest among more than 145 subglacial lakes identified by radar surveys in Antarctica (Siegert et al., 2005). Independent data sources indicate the ice-water interface at Vostok is at 3760±15 m depth. The remaining ice between the bottom of Hole #5G-2 and the lake is about 40 m thick. We plan to access the lake and sample its waters in three stages (Verkulich et al., 2002).

On the first stage an ecologically inert liquid (e.g., polydimethylsiloxane) will be injected to the hole bottom using a special tanker. It is anticipated that, being heavier than the drilling fluid and lighter than water, this hydrophobic liquid will create a 100-m-thick "buffer-layer" at the lower part of the hole. The hydrostatic pressure at the bottom of Hole #5G-2 should be slightly lower than the overburden ice pressure.

In a second stage, Hole #5G-2 will be deepened down to the ice-water interface. The access to the lake will be completed with the coreless thermal drill system TBPO-132

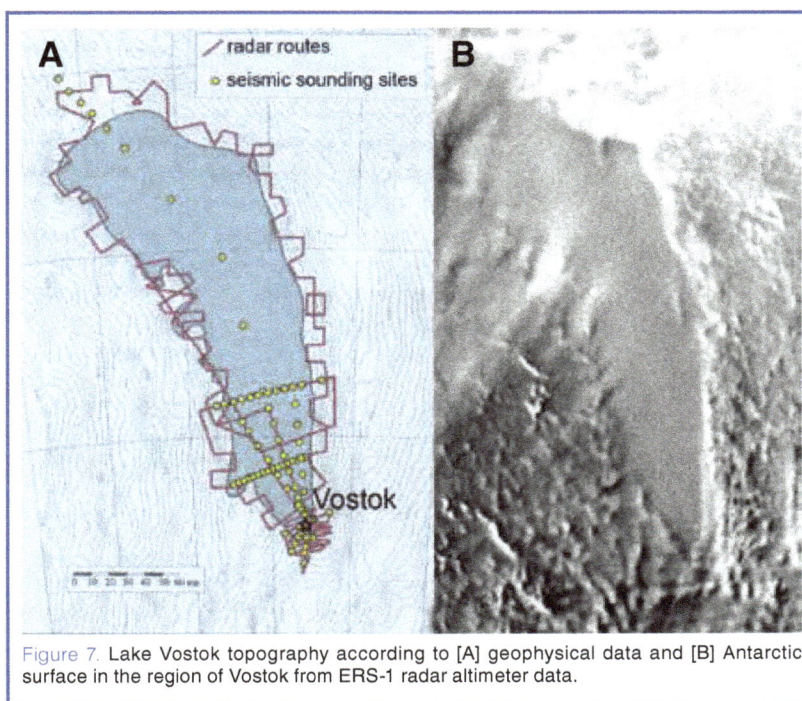

Figure 7. Lake Vostok topography according to [A] geophysical data and [B] Antarctic surface in the region of Vostok from ERS-1 radar altimeter data.

The Lake Vostok penetration approach discussed above has already been tested, though unintentionally, in the course of implementing two international glacial drilling projects in Greenland (North Greenland Ice core Project, NGRIP, 2003) and Antarctica (EPICA drilling at Kohnen station, 2006). In both cases the drilling was performed with routine electromechanical coring of ice without special precautions in the holes filled with kerosene-based drilling fluid similar to that currently used at Vostok. When the base of the ice sheet was reached, the sub-ice water flooded the hole up to several tens of meters above the bed. The water frozen in the NGRIP borehole was drilled in 2004 and later analyzed.

References

Bobin, N.E., Kudryashov, B.B., Pashkevitch, V.M., Abyzov, S.S., and Mitskevich, I.N., 1994. Equipment and methods of microbiological sampling from deep levels of ice in central Antarctica. *Mem. Natl. Inst. Polar Res. Spec. Issue*, 49:184–191.

Bulat, S.A., Alekhina, I.A., Blot, M., Petit, J.R., de Angelis, M., Wagenbach, D., Lipenkov, V.Ya., Vasilyeva, L.P., Wloch, D.M., Raynaud, D., and Lukin, V.V., 2004. DNA signature of thermophilic bacteria from the aged accretion ice of Lake Vostok, Antarctica: implications for searching life in extreme icy environments. *Int. J. Astrobiol.*, 3(1):1–12.

Jouzel, J., Petit, J.R., Souchez, R., Barkov, N.I., Lipenkov, V.Ya., Raynaud, D., Stievenard, M., Vassiliev, N.I., Verbeke, V., and Vimeux, F., 1999. More than 200 meters of lake ice above subglacial Lake Vostok, Antarctica. *Science*, 286(5447): 2138–2141, doi:10.1126/science.286.5447.2138.

Kapitsa, A.P., Ridley, J.K., de Robin, Q.G., Siegert, M.J., and Zotikov, I.A., 1996. A large deep freshwater lake beneath the ice of central East Antarctica. *Nature*, 381:684–686, doi:10.1038/381684a0.

Kudryashov, B.B., 1989. Soviet experience of deep drilling in Antarctica. *In* Bandopadhyay, S., and Skudrzyk, F.J. (Eds.), *Mining in the Arctic: Proc. 1st Int. Symp. Fairbanks*, London (Taylor & Francis), 113–122.

Kudryashov, B.B., Vasiliev, N.I., and Talalay, P.G., 1994. KEMS-112 electromechanical ice core drill. *Mem. Natl. Inst. Polar Res.*, 49:138–152.

Kudryashov, B.B., Krasilev, A.V., Talalay, P.G., Tchistyakov, V.K., Vassiliev, N.I., Zubkov, V.M., and Lukin, V.V., 1998. Drilling equipment and technology for deep ice coring in Antarctica. *In* Hall, J. (Ed.) *Proc. 7th Symp. Antarctic Logistics and Operations,* Cambridge, U.K. (British Antarctic Survey), 205–210.

Kudryashov, B.B., Vasiliev, N.I., Vostretsov, R.N., Dmitriev A.N., Zubkov, V.M., Krasilev, A.V., Talalay, P.G., Barkov, N.I., Lipenkov, V.Ya., and Petit, J.R., 2002. Deep ice coring at Vostok station (East Antarctica) by an electromechanical drill. *Mem. Natl. Inst. Polar Res. Spec. Issue*, 56:91–102.

Lipenkov, V.Ya., Polyakova, E.V., Duval, P., and Preobrazhenskaya, A.V., 2007. Osobennosti stroenia Antarktticheskogo lednikovogo pokrova v raione stantsii Vostok po rezul'tatam petrostruktnikh issledovanyi ledyanogo kerna (Structure of Antarctic Ice Sheet in the region of Vostok station according to petro-fabric investigation of ice core). *Problemi Arktiki i Antarktiki (Arctic and Antarctic Problems)*, 76:68–77. (in Russian).

Petit, J.-R., Jouzel, J., Raynaud, D., Barkov, N.I., Barnola, J.-M., Basile, I., Bender, M., Chappellaz, J., Davis, M., Delaygue, G., Delmotte, M., Kotlyakov, V.M., Legrand, M., Lipenkov, V.Y., Lorius, C., Pepin, L., Ritz, C., Saltzman, E., and Stievenard, M., 1999. Climate and atmospheric history of the past 420,000 years from the Vostok ice core, Antarctica. *Nature*, 399:429–436, doi:10.1038/20859.

Siegert, M.J., Carter, S., Tabacco, I.E., Popov, S., and Blankenship, D.D., 2005. A revised inventory of Antarctic subglacial lakes. *Antarct. Sci.*, 17(3):453–460.

Souchez, R., Jean-Baptiste, P., Petit, J.R., Lipenkov, V.Ya., and Jouzel, J., 2003. What is the deepest part of the Vostok ice core telling us? *Earth-Sci. Rev.*, 60:131–146, doi:10.1016/S0012-8252(02)00090-9.

Tchistiakov, V.K., Kracilev, A., Lipenkov, V.Ya., Balestrieri, J.Ph., Rado, C., and Petit, J.R., 1994. Behavior of a bore hole drilled in ice at Vostok station. *Mem. Natl. Inst. Polar Res. Spec. Issue*, 49:247–255.

Vasiliev, N.I., Talalay, P.G., Bobin, N.E., Chistyakov, V.K., Zubkov, V.M., Krasilev, A.V., Dmitriev, A.N., Yankilevich, S.V., and Lipenkov, V.Ya., 2007. Deep drilling at Vostok station, Antarctica: history and last events. *Annal. Glaciol.*, 47:10–23.

Verkulich, S.R., Kudryashov, B.B., Barkov, N.I., Vasiliev, N.I., Vostretsov, R.N., Dmitriev, A.N., Zubkov, V.M., Krasilev, A.V., Talalay, P.G., Lipenkov, V.Ya., Savatyugin, L.M., and Kuz'mina, I.N., 2002. Proposal for penetration and exploration of sub-glacial Lake Vostok, Antarctica. *Mem. Natl. Inst. Polar Res. Spec. Issue*, 56:245–252.

Authors

Nikolay I. Vasiliev, Drilling Department, St. Petersburg State Mining Institute, 2, 21 Line, St. Petersburg 199106, Russia, e-mail: vasilev_n@mail.ru.

Pavel G. Talalay, Polar Research Center, Jilin University, No. 6 Ximinzhu Street, Changchun City, Jilin Province 130026, China, e-mail: ptalalay@yahoo.com.

and Vostok Deep Ice Core Drilling Parties (Vladimir M. Zubkov, Valery K. Chistyakov, Andrey N. Dmitriev, Vladimir Ya. Lipenkov, and others)

Related Web Links

NGRIP: http://www.gfy.ku.dk/~www-glac/ngrip/hovedside_eng.htm

EPICA: http://www.awi.de/en/research/research_divisions/geosciences/glaciology/projects/epica/

Figure Credits

All photos were provided by members of Vostok drilling team.

18

Deep Drilling at the Dead Sea

by Mordechai Stein, Zvi Ben-Avraham, Steve Goldstein, Amotz Agnon, Daniel Ariztegui, Achim Brauer, Gerald Haug, Emi Ito, and Yoshinori Yasuda

At the lowest point on Earth, the Dead Sea, a unique scientific project, the Dead Sea Deep Drilling Project (DSDDP), is being conducted to establish a late Quaternary paleoenvironmental, tectonic, and seismological archive. Scientific groups from Germany, Israel, Japan, Jordan, Norway, Palestine, Switzerland, and the U.S.A. gathered for the first time on 21 November 2010 to perform scientific drilling at the floor of the deep basin of the Dead Sea (Fig. 1). With current lake level of 423 m below sea level and water depth of 300 m, coring started at 723 m below mean sea level. During the first three weeks of drilling ~460 meters of sediment cores were recovered. As expected from shallow piston cores and on-land deposits from the lake level highstands, the cores are composed of alternating intervals of marly units and salts (Fig. 2). The sedimentary intervals represent several glacial and interglacial cycles spanning an estimated interval of ~200,000 years. Two coarse-grained sections imply almost complete dry-out phases of the Paleo-Dead Sea, meaning that twice the lake surface was several hundred meters below present day sea level.

Drilling is being conducted with the Large Lake Drilling Facility (see front cover of this issue) of DOSECC (Drilling Observation and Sampling of the Earth's Continental Crust, Inc.). The upper 30 meters were cored by using a hydraulic piston coring system that is capable of penetrating several salt layers with high core recovery, while the deeper section was retrieved with the extended-nose bit coring tools (Fig. 3).

Why Drill the Dead Sea?

The Dead Sea Basin (DSB) is located between the Mediterranean and desert climate zones. In the late Neogene the basin was invaded by Mediterranean marine water that formed the Sedom lagoon. The evaporated ingressing seawater led to the deposition of thick sequences of salt and formation of the calcium chloride brine that dominated the subsequent geochemical-limnological evolution of water bodies in the basin. After disconnection of the Sedom lagoon from open sea, the basin was filled with several lakes that captured in their sedimentary filling the hydrological regime of large drainage area of the DSB, reflecting the Levant paleoclimate. The lakes expanded during ice ages and contracted during interglacials. During the last glacial (~70–14 ka ago) Lake Lisan rose up to 250 m above the Holocene Dead Sea and extended from south of the modern Dead Sea northward to the Sea of Galilee. This configuration illustrates the dramatic changes in the regional hydrology and lake configuration and reflects global climate conditions. Moreover, the formation of the DSB is associated with the tectonic activity along the Dead Sea Transform Fault (Ben-Avraham, 1997); thus, its sediments preserve the history of earthquakes (Migowski et al., 2004). The DSB is also the locus of humankind's migration out of Africa, and the home of people from Paleolithic to modern times (Goren-Inbar et al., 2000). Studies of the sedimentary sections exposed on the Dead Sea margins have been applied to issues with global and region-al implications associated with paleoclimate, tectonics, paleoseismology, paleomagnet-ism, and human history (Stein, 2001; Enzel et al., 2006; Waldmann et al., 2010).

The lacustrine sections exposed in the marginal terraces of the modern Dead Sea contain only the sedimentary archives deposited during lake highstands (e.g., the Lisan Formation). The main operational purpose of the DSDDP is to recover long, continuous, high-resolution cores that will provide a com-

Figure 1. The Dead Sea basin and the location of the ICDP-DSDDP drilling site.

Figure 2. Drill cores in liners recovered at depth around 300 m below lake floor. From top to bottom the cores show change from gravel (top core) to marls interlaced with fine-grained salt layers (middle cores). This suggests that at that time the shore was very close to the drilling site, and the wavy salt patterns were interpreted as a result of salt flow.

plementary record to the sections recovered from the marginal terraces, particularly for time interval of low lake stands when the lake retreated from the marginal terraces. The calcium chloride brine that was produced during the ingression of the Sedom lagoon is poor in bicarbonate and sulfate, and therefore, when freshwater enters the lake, primary aragonite is deposited. The aragonite provides an excellent and unique means to achieve calendar chronology down to a few hundred thousand years. This illustrates a major advantage of the sedimentary archive of the Dead Sea and allows comparison to global records such as ice cores and deep sea cores. Data from core samples will establish a pattern of abrupt hydrological events in the drainage area, and the brine-freshwater relations during the different stages will be explored to evaluate effects of long-term climatic trends versus short-term fluctuations.

Figure 3. Extended-nose coring tool with outer and inner core bit. Most of the cores were retrieved with this tool.

The project is performed under the wings and support of the International Continental Scientific Drilling Program (ICDP). The drilling operation is performed by DOSECC.

References

Ben-Avraham, Z., 1997. Geophysical framework of the Dead Sea: structure and tectonic. *In* Niemi, T.M., Ben-Avraham, Z., and Gat, J.R. (Eds.), *The Dead Sea: The Lake and Its Settings*: New York (Oxford University Press), Oxford Monographs on Geology and Geophysics, 36:22–35.

Enzel. Y., Agnon A. and Stein M. 2006. *New Frontiers in Dead Sea Paleoenvironmental Research, GSA Spec. paper 401:* Boulder, CO (The Geological Society of America).

Goren-Inbar, N., Feibel, C.S., Verosub, K.L., Melamed, Y., Kislev, M., Tchernov, E., and Saragusti, I., 2000. Pleistocene milestones on the out-of-Africa corridor at Gesher Benot Ya'aqov, Israel. *Science*, 89:944–947.

Migowski, C., Agnon, A., Bookman, R., Negendank, J.F.W., and Stein, M., 2004. Recurrence pattern of Holocene earthquakes along the Dead Sea transform revealed by varve-counting and radiocarbon dating of lacustrine sediments. *Earth Planet. Sci. Lett.*, 222:301–314.

Stein, M., 2001. The history of Neogene-Quaternary water bodies in the Dead Sea Basin. *J. of Paleolimnology* 26: 271-282.

Waldmann, N., Torfstein, A., and Stein, M., 2010. Northward migration of monsoon activity across the Saharo-Arabian desert belt during the last interglacial: evidence from the Levant. *Geology*, 38:567–570.

Authors

Mordechai Stein, Geological Survey of Israel, 30 Malkhe Israel St. Jerusalem, 95501, Israel, e-mail: motistein@gsi.gov.il.

Zvi Ben-Avraham, Department of Geological Sciences, Louis Ahrens Building, Library Road, University of Cape Town, Rondebosch, 7700, Republic of South Africa.

Steve Goldstein, Lamont-Doherty Earth Observatory, 213 Comer, 61 Route 9W - P.O. Box, 1000 Palisades, NY 10964-8000, U.S.A.

Amotz Agnon, Institute of Earth Sciences, Hebrew University of Jerusalem, Edmond J. Safra campus, Givat Ram, 91904, Jerusalem.

Daniel Ariztegui, Department of Geology and Paleontology, University of Geneva, 13, Rue des Maraîchers, CH-1205 Genève, Switzerland.

Achim Brauer, Helmholtz Centre Potsdam, GFZ German Research Centre for Geosciences, Section 5.2, Climate Dynamics and Landscape Evolution, Telegrafenberg, C 323 D-14473 Potsdam, Germany.

Gerald Haug, ETH Zürich, Geologisches Institut NO G 51.1, Sonneggstrasse 5, 8092 Zürich, Switzerland.

Emi Ito, Geology and Geophysics, Room 108, PillsH 0211, 310 Pillsbury Drive SE, Minneapolis, MN 55455, U.S.A.

Yoshinori Yasuda, International Research Center for Japanese Studies, 3-2 Oeyama-cho, Goryo, Nishikyo-ku, Kyoto 610-1192, Japan.

Related Web Links

http://deadsea.icdp-online.org
http://www.dosecc.org

Figure Credits

Fig. 1: The National Aeronautics and Space Administration, (NASA)

Fig. 2: Michael Lazar, Department of Marine Geosciences, University of Haifa, Israel

Fig. 3: Uli Harms, ICDP, GFZ Potsdam, Germany

19

SAFOD Phase III Core Sampling and Data Management at the Gulf Coast Repository

by Bradley Weymer, John Firth, Phil Rumford, Frederick Chester, Judith Chester, and David Lockner

Introduction

The San Andreas Fault Observatory at Depth (SAFOD) project is yielding new insight into the San Andreas Fault (Zoback et al., 2010; Zoback et al., this issue). SAFOD drilling started in 2002 with a pilot hole, and proceeded with three phrases of drilling and coring during the summers of 2004, 2005, and 2007 (Fig. 1). One key component of the project is curation, sampling, and documentation of SAFOD core usage at the Integrated Ocean Drilling Program's (IODP) Gulf Coast Repository (GCR) at Texas A&M University. We present here the milestones accomplished over the past two years of sampling Phase III core at the GCR.

Research Themes

Several research themes rely heavily on SAFOD core samples. These are focused on understanding the structure, composition, and a variety of physical and mechanical properties of the San Andreas Fault. Structural studies are concerned with characterizing the geometry of the fault zone, the distribution of shear displacement, the process by which rocks are deformed, and the microscopic features that record past occurrences of earthquake slip and creeping deforma-tion. Core samples are ideal for structural study at macroscopic and microscopic scales, using non-destructive techniques such as mapping the surface of the core and X-ray computed tomography (CT scanning) to image the interior of the core, as well as destructive techniques such as cutting and polishing small sub-samples for optical and electron microscopy.

Compositional studies focus on the elemental, chemical, and mineral content of the fault zone to understand the processes and conditions which occur during faulting, such as rates of chemical reactions between minerals and pore fluids, the origin of the rock and fluids in the fault, and the pressure and temperature at depth. These studies involve a variety of subsample processing techniques and analytical instruments such as X-ray diffraction and fluorescence, electron microprobe, and mass spectrometers.

Laboratory experimentation with core samples is an important way to quantify a host of physical and mechanical properties that are important to faulting processes. Sub-samples are used to determine thermal conductivity, porosity and permeability, and seismic velocity within the fault. Mechanical testing explores the fracture, friction, and flow properties of the fault rocks in the actively deforming portions of the fault zone.

Inter-Laboratory Comparison Project

The main objective of the Inter-Laboratory Comparison Project was to run tests on standardized sample materials to provide a baseline for the comparison of results from experiments conducted on SAFOD Phase III core samples (Lockner et al., 2009). A total of eighteen laboratories from nine countries conducted the following tests: intact rock strength, frictional strength, permeability, electrical resistivity, ultrasonic wave speed, thermal conductivity, and related properties such as porosity and density. Standardized protocols were established, and test samples were distributed to participating laboratories in the fall of 2008. Friction tests were conducted on crushed and sieved samples of

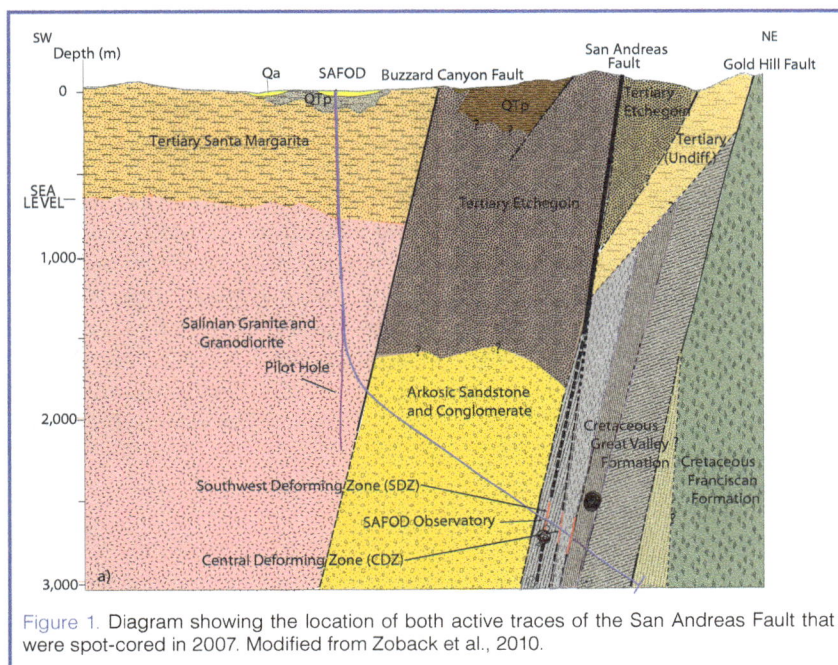

Figure 1. Diagram showing the location of both active traces of the San Andreas Fault that were spot-cored in 2007. Modified from Zoback et al., 2010.

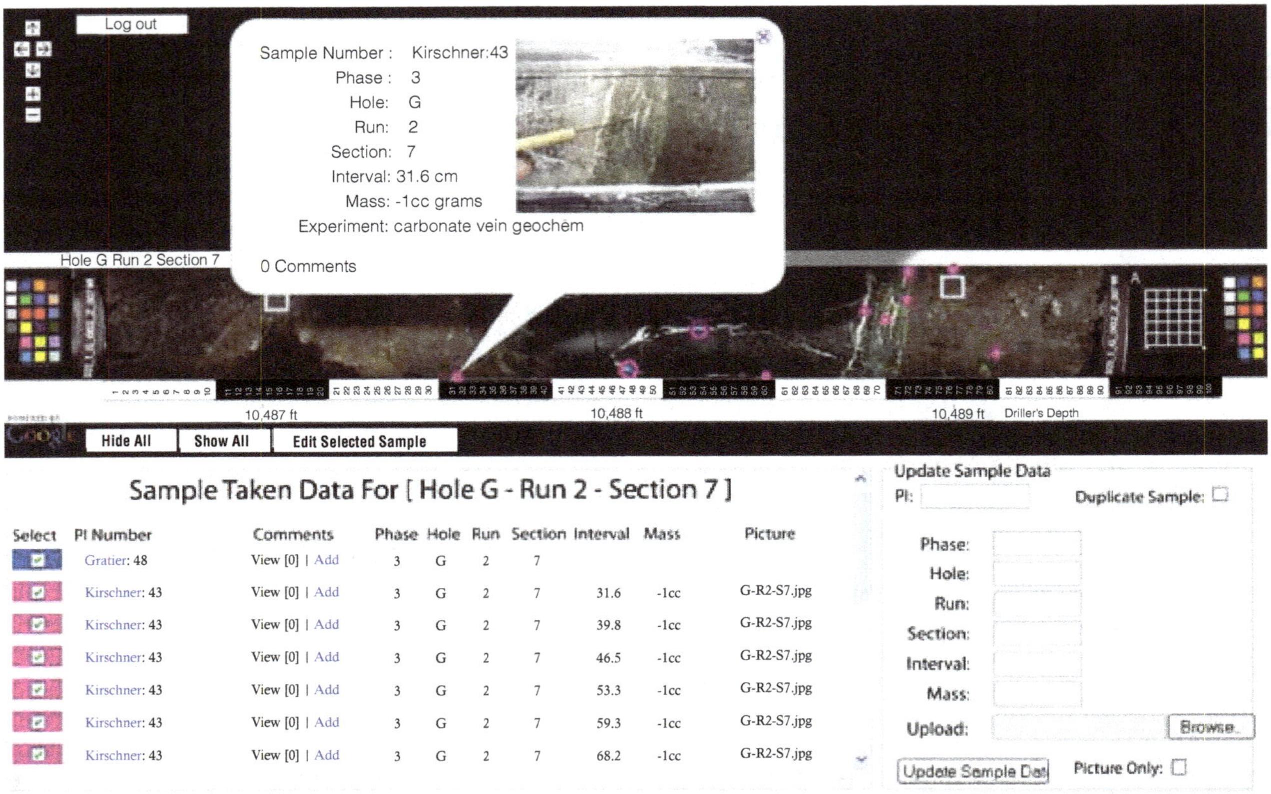

Figure 2. Example of SAFOD Core Viewer sample information taken from Round 1.

quartz, granite, talc, a quartz/smectite mixture, and SAFOD cuttings. Other physical property measurements were conducted on marble, granite, and three types of sandstones. Preliminary results of the inter-lab comparisons were discussed at the EarthScope meeting in Boise, Idaho in 2009 (Lockner et al., 2009), and further results were presented at the AGU 2010 Fall Meeting.

SAFOD Sampling at the Gulf Coast Repository

After the arrival of Phase III cores, the GCR assumed full curatorial responsibilities of all SAFOD samples (Phases 1 to 3). All SAFOD sampling follows the guidelines outlined by the SAFOD Sample Policy. This requires Principal Investigators (PIs) to submit all analytical data and metadata to the GCR for placement on the SAFOD web site in a timely manner. Currently, data are available to the public via the SAFOD Core Viewer.

The first cycle of sample requests (Round 1) were for 790 samples from twenty-eight PI groups comprising ninety-eight scientists from around the world. The approved sample distribution was 190 samples for twenty PI groups, comprising fifty-eight scientists. Round 1 sampling started on 28 June 2008 and was completed on 17 December 2008. After the completion of Round 1 sampling, the SAFOD Sample Committee (SSC) began accepting requests for Round 2.

Round 2 requests were for 344 samples from thirteen PI groups, comprising thirty-nine scientists, and all of them were approved. About 20% of the requesting PIs from Round 2 were from institutions outside the U.S. Round 2 sampling commenced on 21 July 2009 and remains a work in progress. To date, a total of 164 (as of December 2010) samples have been taken.

The SAFOD Core Viewer

A major component of the SAFOD project involved the creation and maintenance of an interactive, online core sample database and administration tool by staff at EarthScope/UNAVCO (University NAVstar COnsortium), with design input originally from Charley Weiland at Stanford University and now the IODP-GCR curatorial staff. Building upon the flexibility of Google Maps API, the SAFOD Core Viewer serves as a repository for all information related to the project from sample inventories and photo libraries to data storage. The Core Viewer has three views for each core section: (1) Samples Requested, (2) Samples Approved, and (3) Samples Taken. The Core Viewer provides the following functions (Fig. 2):

1. It allows all interested PI groups to submit their sample requests to be reviewed online by the SSC.

2. It provides the GCR curatorial staff the ability to view all approved sample requests and aids in the sampling process.

3. After samples are taken, any information pertaining to each sample, such as the type of experiment, expected results, pictures, and subsequent data, can be uploaded into the Core Viewer (Figs. 3A, 3B).

4. After review, data are also uploaded and made available on the public Core Viewer.

Acknowledgements

We thank the SAFOD Co-PI's Mark Zoback, Stephen Hickman, and William Ellsworth, as well as the support of Charley Weiland, for their guidance and support during the sampling and data management phase of the project. A special thanks to Brian Blackman, Michael Jackson, and Adrian Borsa at UNAVCO for the development and continued improvement of the Core Viewer. We also thank Kaye Shedlock and Greg Anderson at NSF, and the SAFOD Sample Committee.

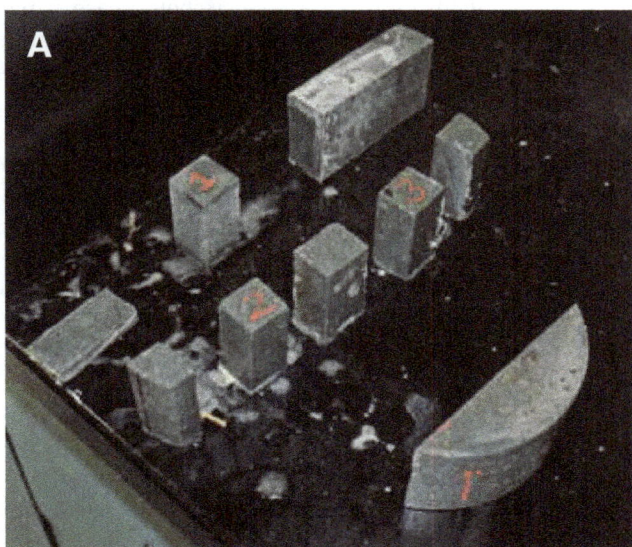

References

Lockner, D., Marone, C., and Saffer, D., 2009. SAFOD inter-laboratory comparisons - a progress report [poster presented at the EarthScope 2009 meeting in Boise, Idaho, 12–15 May 2009].

Zoback, M., Hickman, S., and Ellsworth, W., 2010. Scientific drilling into the San Andreas Fault zone. *Eos, Trans. AGU*, 91(22):197–204, doi:10.1029/2010EO220001.

Authors

Bradley Weymer, SAFOD Curatorial Specialist and Graduate Assistant Researcher, Integrated Ocean Drilling Program and SAFOD, Texas A&M University, 1000 Discovery Drive, College Station, TX 77845-9547, U.S.A, e-mail: weymer@iodp.tamu.edu.

John Firth, Curator, and **Phil Rumford**, GCR Superintendent, Integrated Ocean Drilling Program and SAFOD, Texas A&M University, 1000 Discovery Drive, College Station, TX 77845-9547, U.S.A.

Frederick M. Chester, Professor, and **Judith S. Chester**, Associate Professor, Department of Geology and Geophysics, TAMU, Center for Tectonophysics and Department of Geology & Geophysics, Texas A&M University, College Station, TX 77843-3115, U.S.A.

David Lockner, U.S. Geological Survey, Menlo Park, U.S. Geological Survey Earthquake Science Center, 345 Middlefield Road, MS/977Menlo Park, CA 94025, U.S.A.

Related Web Links

http://safod.icdp-online.org
http://www.earthscope.org/publications
http://www.earthscope.org/observatories/safod
http://www.earthscope.org/data/safod_core_viewer
http://www.earthscope.org/es_doc/safod/SAFOD_Core_Sample_Distribution.pdf
http://www.earthscope.org/data/safod_core_samples

Photo Credit

Fig. 3: Photos by Bradley Weymer, IODP and SAFOD, Texas A&M University

Figure 3. [A] Image of precise cuts for friction studies during Round 1 that documents the sampling process and was uploaded into the Core Viewer. [B] Example of Round 2 image within the SAFOD Core Viewer. Section E-R1-S5, showing a fault contact between dark grayish-black siltstone and grayish-red pebbly sandstone.

20

Executive Summary: "Mantle Frontier" Workshop

by Workshop Report Writing Group

Introduction

The workshop on "Reaching the Mantle Frontier: Moho and Beyond" was held at the Broad Branch Road Campus of the Carnegie Institution of Washington on 9–11 September 2010. The workshop attracted seventy-four scientists and engineers from academia and industry in North America, Asia, and Europe.

Reaching and sampling the mantle through penetration of the entire oceanic crust and the Mohorovičić discontinuity (Moho) has been a longstanding goal of the Earth science community. The Moho is a seismic transition, often sharp, from a region with compressional wave velocities (Vp) less than 7.5 km s^{-1} to velocities ~8 km s^{-1}. It is interpreted in many tectonic settings, and particularly in tectonic exposures of oceanic lower crust, as the transition from igneous crust to mantle rocks that are the residues of melt extraction. Revealing the *in situ* geological meaning of the Moho is the heart of the Mohole project. Documenting ocean-crust exchanges and the nature and extent of the subseafloor biosphere have also become integral components of the endeavor. The purpose of the "Mantle Frontier" workshop was to identify key scientific objectives associated with inno-

vative technology solutions along with associated timelines and costs for developments and implementation of this grand challenge.

Background: Ocean Drilling and the Mantle Target

Scientific ocean drilling started from the first excitement of Mohole Phase I that penetrated 180 m in 3300 m water depth off Guadalupe Island (west of Baja California, Mexico) in April 1961 (Bascom, 1961; Steinbeck, 1961; Cromie, 1964), although the Mohle project was abandoned soon after (Greenberg, 1966). Fifty years after Mohole Phase I, the deepest hole into the oceanic crust is located on the Nazca Plate in the eastern equatorial Pacific (ODP Hole 504B) to 2111 m below the seafloor (mbsf) within the sheeted dikes. The second deepest hole in the Pacific, 1256D (1507 mbsf), is on the Cocos Plate northwest of 504B; it penetrates the transition zone between the upper and the lower crust, in the upper gabbroic rocks below the sheeted dike complex. Other significantly deep holes over 1000 m deep beneath the seafloor include ODP Hole 735B (1508 mbsf) in the Indian Ocean (Atlantis Bank) and IODP Hole U1309D (1415 mbsf) at the Atlantis Massif in the Atlantic Ocean. These achievements of relatively deep crustal penetration were made with the available riserless drilling technology. The deep holes outside the Pacific Ocean were drilled in uplifted fault blocks where lower crustal rocks are exhumed at shallow depths, in heterogeneous slow-spread ocean lithosphere.

In 2007, a riser-equipped drilling ship was introduced to IODP (D/V *Chikyu*, owned and operated by JAMSTEC). Riser technology significantly improves the deep drilling capability as proven by oil industry experience. The science plan of IODP thus includes 21st Century Mohle as one of its initiatives (IODP, 2001). We are in an era where drilling technology is rapidly advancing to realize deep drilling (>6 km below seafloor) in deep waters (industry drilling in >3000 m water depth in the Gulf of Mexico). Scientific and industry drilling

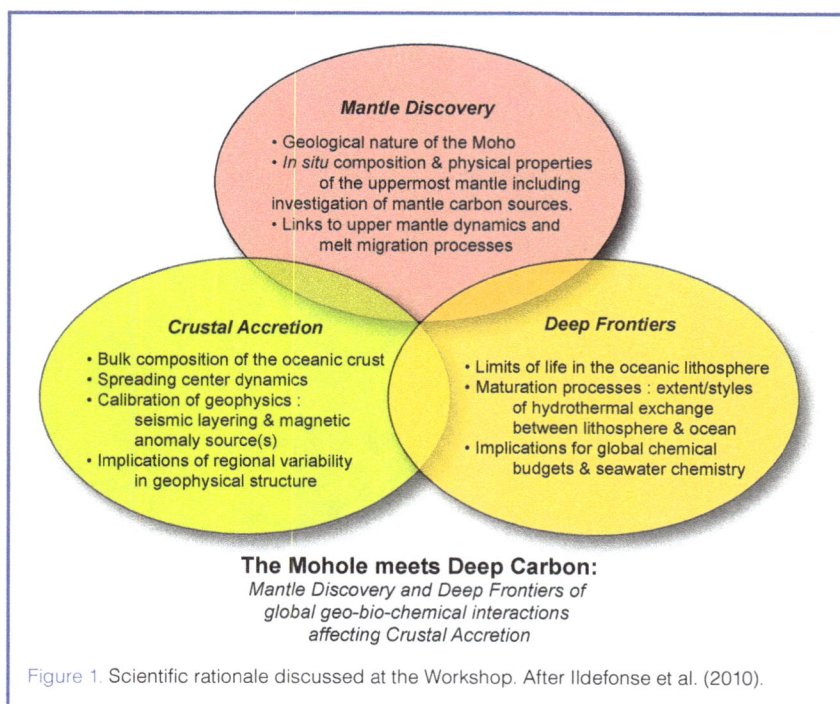

The Mohole meets Deep Carbon:
Mantle Discovery and Deep Frontiers of global geo-bio-chemical interactions affecting Crustal Accretion

Figure 1. Scientific rationale discussed at the Workshop. After Ildefonse et al. (2010).

have come a long way, and we can now seriously consider scientific drilling to the mantle.

Deep Carbon Observatory and Carbon Reservoirs

The Deep Carbon Observatory (DCO) is a multidisciplinary, international initiative dedicated to achieving a transformational understanding of Earth's deep carbon cycle. Key areas of study include the following:

- deep carbon mantle reservoirs and fluxes
- the nature and extent of the deep biosphere
- the physical and chemical behavior of carbon under extreme conditions
- the unexplored influences of the deep carbon cycle on energy, environment, and climate

The DCO's goal to advance understanding on these frontiers requires an integrated approach—incorporating field-based global sampling efforts, laboratory experiments, analytical methodology, and theoretical modeling, as well as establishing new research partnerships. Much of the DCO's work will be experimental, but much will also depend on deep Earth samples recovered using the framework of established programs like IODP and the International Continental Scientific Drilling Program (ICDP).

The present IODP and the current vision for the future International Ocean Discovery Program share numerous similar goals for understanding Earth processes and systems. Discoveries of microbial life deep in the crust beneath the oceans and continents indicate a rich subsurface biota that by some estimates may rival all surface life in total biomass. Much work also remains to understand how life adapts to deep environments, what novel biochemical pathways sustain life at high pressure and temperature, and what the extreme limits of life are. How does biological carbon link to the slower deep physical and chemical cycles? Is biologically processed carbon represented in deep Earth reservoirs? The nature and full extent of carbon reservoirs and fluxes in Earth's deep interior are not well known. The subduction of tectonic plates and volcanic outgassing are primary vehicles for carbon fluxes to and from deep in the Earth, but the processes and rates of these fluxes—as well as their variation throughout Earth's history—remain poorly understood. Likewise, there is evidence for abiogenic hydrocarbons in some deep crustal and mantle environments, but the nature and extent of deep organic synthesis is unknown. Last but not least, what are the impacts of deep carbon on energy and the surface carbon cycle?

The DCO recognizes a longstanding goal in the ocean drilling community to reach and sample *in situ* pristine mantle and—in the process—penetrate the entire ocean crust and the Moho. Samples obtained en route to and across the Moho will complement the DCO's other research efforts and may address some of the DCO questions above. Such samples and their subsequent study may also ground truth existing hypotheses and, perhaps the findings will inspire entirely new hypotheses and studies regarding the nature of Earth's upper mantle and lithosphere. Undoubtedly, the interest and participation of portions of the DCO community in such a monumental drilling project will expand the scope of the ocean drilling community with its own scientific goals related to carbon cycling deep in the Earth.

In Relation to Previous Workshops

The ocean drilling science community has met in numerous workshops over the course of Deep Sea Drilling Project (DSDP), Ocean Drilling Program (ODP), and IODP (Ildefonse et al., 2007; Teagle et al., 2009; Ildefonse et al., 2010). An international workshop on "The Mohle: a Crustal Journey and Mantle Quest" was held in Kanazawa, Japan in June 2010; it reaffirmed the scientific rationale, considered technological realities and opportunities, and identified potential drilling sites for site surveys planning (Ildefonse et al., 2010).

The "Mantle Frontier" workshop was planned to make a natural step forward in technological discussion, but the emphasis of the scientific discussion was to expand the scope irrespective of specific sites, to emphasize the mantle portion of the targeted section, and to ask the general and fundamental questions of interest to the broader scientific community, such as the DCO.

Scientific Presentations

The DCO overview was given by Robert Hazen from a program-wide perspective, by Constance Bertka from a program management perspective, and by Erik Hauri from the carbon reservoirs and fluxes viewpoint (in his presentation "REFLEX: Deep Carbon Reservoirs, Fluxes and Experiments"). We are accustomed to thinking about the carbon cycle near the Earth's surface, but we know so little about Earth's deep carbon that we lack estimates of carbon quantity or chemical structure, and the effects of carbon on mantle (or core) behavior. The nature and extent of the deep microbial biosphere also need to be investigated.

REFLEX's interests in deep carbon include 1) the pathways and fluxes of carbon exchange between the surface and deep Earth; 2) the nature and variability of carbon compounds in the deep Earth; 3) the interactions between carbon concentration and the dynamics of the Earth's interior; and 4) the ultimate origins of mantle carbon. From these perspectives, REFLEX can use the IODP database to inventory carbonate and organic carbon content in deep-sea sediment cores; analyze a complete ocean crustal section for full understanding of $CaCO_3$ addition to the mantle at subduction zones; and determine carbon flux from pore fluid release in subduction zones.

An illuminating keynote address was given by Donald Beattie, who oversaw the Apollo lunar rock sampling project (Beattie, 2001). A proper project management system to manage a project of this scope from the beginning to end is the key and challenge to success.

Benoit Ildefonse gave a summary of Mohle history and outlined the scientific rationale for the Mohle in three categories (based on the outcome of the previous recent workshops, and as summarized in the Kanazawa workshop report): mantle discovery, crustal accretion, and deep frontiers (Ildefonse et al., 2010). An anticipated timetable for the new Mohle project will enable complete preparations by 2017 and reach the mantle by 2022. Three candidate sites are being considered for reaching the mantle: Cocos Plate site (including Site 1256), off southern Baja California, and north of Hawaii. Site surveys are being planned to gather data to make the final selection.

Shuichi Kodaira presented recent high-resolution seismic profiles of oceanic Moho and mantle from active source seismic studies in the western Pacific that can help extrapolate drilling observations to mantle dynamics from ridges to trenches. Seismic images of the Moho can vary from sharp to diffuse boundaries, which may correspond to the geologic variety found at the crust/mantle transition in ophiolites. Strong seismic azimuthal anisotropy can be expected to start immediately beneath the Moho, such as measured in the NW Pacific (Oikawa et al., 2010). Lower crustal dipping reflectors matching fast Vp directions may be manifestations of basal shear near the Moho.

Donna Blackman showed how grain-scale deformation due to mantle asthenospheric flow, with melt and recrystallization overprints, may be linked to seismological observations. So far, such inferences have been made without in situ knowledge of crystallographic fabric. Mantle samples will document structures and ground truth petrophysical properties. Borehole experiments will provide high-resolution information to be extrapolated to kilometers beyond the hole.

Yoshiyuki Tatsumi showed how drilling could contribute to the understanding of mantle dynamics and geochemical cycles. He emphasized the important roles of water and carbon in creation-destruction cycles in the ocean lithosphere, including arc and continental crust genesis. Deep

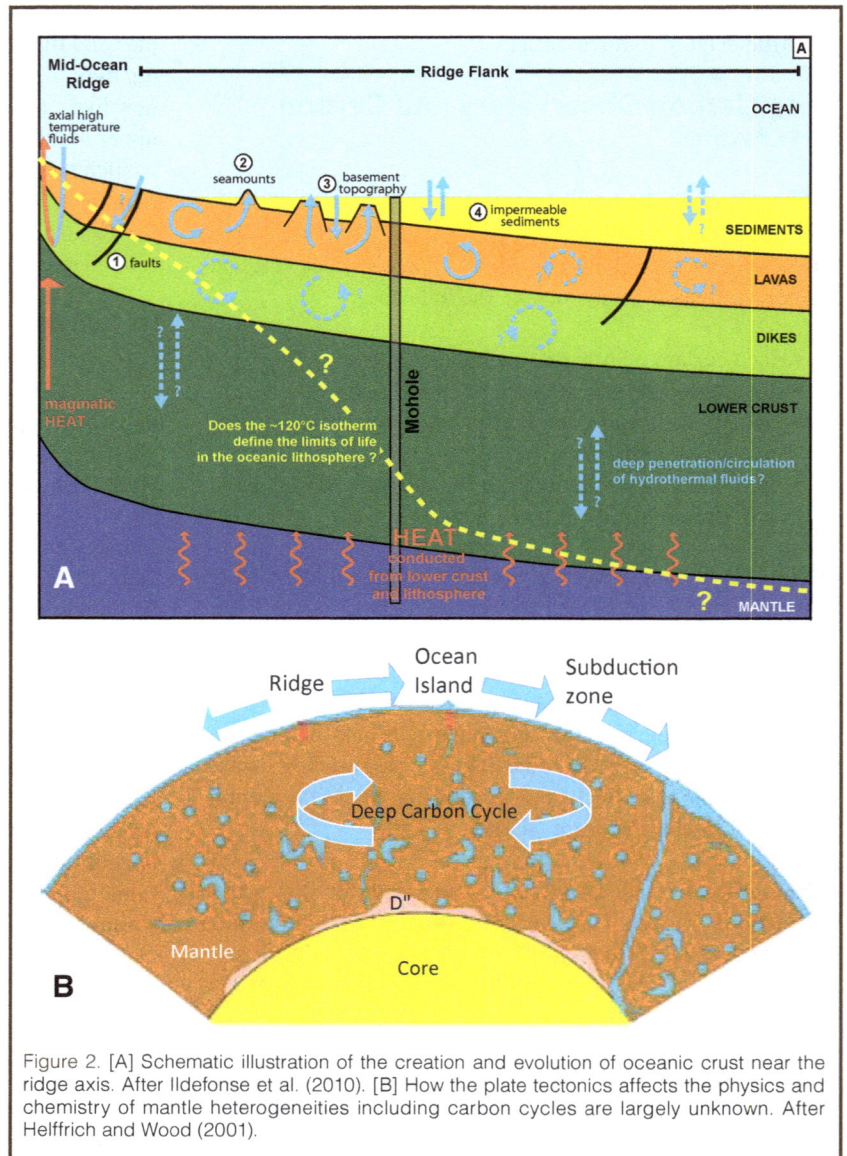

Figure 2. [A] Schematic illustration of the creation and evolution of oceanic crust near the ridge axis. After Ildefonse et al. (2010). [B] How the plate tectonics affects the physics and chemistry of mantle heterogeneities including carbon cycles are largely unknown. After Helffrich and Wood (2001).

drilling at key sites along the ocean lithosphere pathway will contribute to a better constrained global geochemical model including the explanation of mantle geochemical heterogeneities.

Matt Schrenk presented how drilling can be used to discover the extent of microbial life in the deep biosphere. The subseafloor biosphere may host one-third to one-half of all prokaryotic cells on Earth, and contain biomass equivalent to that of all plant life at the Earth's surface. Furthermore, the deep biosphere is dependent upon energy in the form of chemical disequilibria and not directly coupled to photosynthesis; it is sometimes referred to as the dark energy biosphere (DEB). However, the absolute extent, the nature, and controls upon the subseafloor biosphere are not completely known. Fluid circulation (hydrology) is considered a key to nourishing the DEB; drilling and associated hydrological experiments can provide direct observations of cell density, together with quantitative measurements of permeability and time-integrated fluid/rock ratios. Drilling through the ocean crust means penetrating from the life to the non-life regime, and it provides an opportunity to explore the connectivity and flow between deep and surface chemical reservoirs. Developing technologies to overcome contamination-by-drilling as well as distinguishing signal from "noise" introduced by drilling fluids is crucial to interpreting the results of this portion of the project.

Peter Kelemen made a presentation on deep energy, environment, and climate. Carbon is present in the mantle as a result of hydrothermal interaction followed by subduction, and perhaps as a primordial component. Shallow interaction yields fluids containing hydrogen and methane as well as more complex hydrocarbons similar to those stable at greater depth, and these could be a future fuel source. ODP data yield an average of 0.6 wt% CO_2 in altered peridotites, extrapolated to ~0.3 wt% CO_2 over a 7-km depth where mantle peridotite is exposed at the seafloor. A mass equivalent to all dissolved CO_2 in the oceans is added to altered mantle peridotite every ten million years. Optimizing this near-surface weathering holds great potential for carbon storage. Each ton of mantle peridotite can permanently store up to 600 kg of CO_2 in the form of inert, non-toxic solid carbonates. Kinetic data show that a rate of one billion tons CO_2 per km^3 of rock per year can be achieved under optimal conditions (Kelemen and Matter, 2008; Kelemen et al., 2011). There are tens of thousands of cubic kilometers of peridotite near the surface on land, and millions near the seafloor along slow spreading ridges. A Mohle would provide crucial data on the depth of natural CO_2 uptake.

Engineering Presentations

Greg Myers gave a comprehensive presentation on current technological capabilities and limitations. The uniqueness of mantle drilling is the water depth/hole depth combination and the rocks (not sediments) to be drilled. The oil and gas industry already drills deeper holes, yet in water depths less than 3300 m. An integrated approach utilizing all available IODP platforms will reduce the overall cost. Engineered mud must be circulated continuously as part of a comprehensive plan to drill and core effectively. Improved borehole pressure control for deep drilling can be achieved by utilizing dual gradient drilling, which applies mud pressure from the seafloor rather than the platform or vessel. Discussions of continuous coring vs. spot coring and downhole equipment (drilling/coring/logging) are necessary. Myers emphasized the definition of success must be clear and understood by all.

Randy Normann supplied a presentation introducing electronics, batteries, and tools that withstand very high temperatures continuously (>250°C). Michael Freeman lectured on drilling fluid and making deep holes. John Cohen introduced a riserless mud recovery (RMR) method as an applica-

tion of the dual gradient drilling concept. At present, there is technology qualified for 1500 m water depth.

Michael Ojovan introduced a totally different approach to investigating Earth's interior with the use of self-sinking capsules. The capsule melts the rocks and creates acoustic signals to be detected at the surface, thus yielding information about the nature of the rocks through which the capsule and the signals pass (Ojovan et al., 2005). In their design the probe reaches the Moho in about five months (100 km depth in 35 years).

Larry Karl introduced Remotely Operated Vehicles (ROV) for deep-water applications (depth rated to 10,000 ft) used in offshore oil and gas fields. Also presented were unique and robust techniques for resupply at sea. John Kotrla made a presentation on blow-out preventers (BOP) and seafloor isolation devices. The standing water depth record well is at 3051 m in the Gulf of Mexico (Transocean "Discoverer Deep Seas"). In order to go deeper, utilizing a surface BOP and an environmental safe guard (ESG) on the seafloor was introduced.

John Thorogood presented the management aspect of mantle drilling. There are multiple technology and operational options available to achieve project goals, and yet new technologies may arrive to alter the direction of the project. Subsurface conditions may differ profoundly from the prognosis. Effective operations will involve multiple contingencies, defined rules, and protocols for changing the rules. These indicate the project is not a "normal" project, but will require skillful management from project scoping to execution.

Outcome of the Workshop

The participants agreed on the following:

1. IODP and DCO recognize the potential for synergy towards a comprehensive understanding of carbon-water cycle in the deep Earth system, including consequences of microbial activities.

2. The workshop participants endorse the following outline of the Mohle project scientific rationale (Figs. 1 and 2).

3. The participants agree that the scale of mantle drilling—which is not just drilling but requires long-term commitment before and after the drilling—needs to be recognized by the wider IODP entities from the decision making level.

4. The workshop participants propose to establish a Mohle scoping group. The group will review and refine the science goals, identify technology, and review plans to meet the science goals. Also recognized was the need to establish a management structure, estimate the total cost of the project, and seek funding

along with outreach and communication activities within a broad IODP umbrella.

Acknowledgements

The Integrated Ocean Drilling Program (IODP) and the Deep Carbon Observatory (DCO) initiative jointly supported the workshop with primary financial support from the Alfred P. Sloan Foundation.

References

Bascom, W., 1961. A Hole in the Bottom of the Sea, *The Story of the Mohle Project*: New York (Doubleday & Co.).

Beattie, D.A., 2001. *Taking Science to the Moon: Lunar Experiments and the Apollo Program*: Baltimore MD (The Johns Hopkins University Press).

Cromie, W.J., 1964. *Why the Mohle: Adventures in Inner Space*: Boston (Little, Brown and Co.).

Greenberg, D.S., 1966. News and Comment. *Science*, 152:895–896, doi:10.1126/science.152.3724.895.

Helffrich, G.R., and Wood, B.J., 2001. The Earth's mantle. *Nature*, 412:501–507.

Ildefonse, B., Christie, D.M., and Mission Moho Workshop Steering Committee, 2007. Mission Moho workshop: drilling through the oceanic crust to mantle. *Sci. Drill.*, 4:11–18, doi:10.2204/iodp.sd.4.02.2007.

Ildefonse, B., Abe, N., Blackman, D., Canales, J.P., Isozaki, Y., Kodaira, S., Myers, G., Nakamura, K., Nedimovic, M., Skinner, A.C., Seama, N., Takazawa, E., Teagle, D.A.H., Tominaga, M., Umino, S., Wilson, D.S., and Yamao, M., 2010. The MoHole: a crustal journey and mantle quest, workshop in Kanazawa, Japan, 3–5 June 2010. *Sci. Drill.*, 10:56–63.

IODP, 2001. *Earth, Oceans and Life, Initial Science Plan*, 2003-2013, IODP Planning Subcommittee, pp. 110.

Kelemen, P.B., and Matter, J., 2008. *In situ* carbonation of peridotite for CO2 storage. *Proc. Natl. Acad. Sci.*, 105:17295–17300, doi:10.1073/pnas.0805794105.

Kelemen, P.B., Matter, J., Streit, E.E., Rudge, J.F., Curry, W.B., and Blusztajn, J., 2011. Rates and mechanisms of mineral carbonation in peridotite: natural processes and recipes for enhanced, *in situ* CO2 capture and storage. *Ann. Rev. Earth Planet. Sci.*, in press.

Oikawa, M., Kaneda, K., and Nishizawa, A., 2010. Seismic structure of the 154–160 Ma oceanic crust and uppermost mantle in the Northwest Pacific Basin. *Earth Planets Space*, 62:e13–e16, doi:10.5047/eps.2010.02.011.

Ojovan, M.I., Gibb, F.G.F., Poluetkov, P.P., and Emets, E.P., 2005. Probing the interior layers of the Earth with self-sinking capsules. *Atomic Energy*, 99:556–562, doi:10.1007/s10512-005-0246-y.

Steinbeck, J., 1961. High drama of bold thrust through ocean floor. *Life*, 50(15):111–122.

Teagle, D., Ildefonse, B., Blackman, D., Edwards, K., Bach, W., Abe, N., Coggon, R., and Dick, H., 2009. Melting, magma, fluids and life: challenges for the next generation of scientific ocean drilling into the oceanic lithosphere. [Workshop Report, National Oceanographic Center, Southampton, 27–29 July 2009], http://www.interridge.org/files/interridge/MMFL_wkshp_rpt_2009_final.pdf.

Workshop Report Writing Group (in alphabetical order)

Constance Bertka, Deep Carbon Observatory, Senior Consultant, Geophysical Laboratory, 5251 Broad Branch Road NW, Washington, DC 20015, U.S.A., e-mail: cbertka@ciw.edu.

Donna K. Blackman, IGPP, Scripps Institution of Oceanography, University of California San Diego, 9500 Gilman Drive, La Jolla, CA 92093-0225, U.S.A., e-mail: dblackman@ucsd.edu.

Benoit Ildefonse, CNRS, Géosciences Montpellier, Université Montpellier 2, CC 60, 34095 Montpellier Cédex 05, France, e-mail: benoit.ildefonse@univ-montp2.fr.

Peter B. Kelemen, 211 Comer, 61 Route 9W – P.O. Box 1000, Palisades, NY 10964-8000, U.S.A., e-mail: peterk@ldeo.columbia.ed.

Andrea Johnson Mangum, Deep Carbon Observatory, Program Associate, Geophysical Laboratory, 5251 Broad Branch Road NW, Washington, DC 20015, U.S.A., e-mail: amangum@ciw.edu.

Greg Myers, Senior Technical Expert: Engineering and Technology, Department: Ocean Drilling, Consortium for Ocean Leadership, 1201 New York Avenue NW, 4th Floor, Washington, DC 20005, U.S.A., e-mail: gmyers@oceanleadership.org.

Jason Phipps-Morgan, Professor, Earth and Atmospheric Sciences, Cornell University, Snee Hall, Room 2122, Ithaca, NY 14853, U.S.A., e-mail: jp369@cornell.edu.

Matthew Schrenk, Assistant Professor, Howell Science S301B, Department of Biology, East Carolina University, Greenville, NC 27858, U.S.A., e-mail: schrenkm@ecu.edu.

Kiyoshi Suyehiro (corresponding author), President & CEO, IODP-MI, Tokyo University of Marine Science and Technology, Office of Liaison and Cooperative Research 3rd Floor, 2-1-6, Etchujma, Koto-ku, Tokyo 135-8533, Japan, e-mail: ksuyehiro@iodp.org.

Yoshiyuki Tatsumi, Institute for Research on Earth Evolution (IFREE), Japan Agency for Marine-Earth Science and Technology, 2-15 Natsushima-cho, Yokosuka-city, Kanagawa 237-0061, Japan, e-mail: tatsumi@jamstec.go.jp.

Jessica Warren, Assistant Professor, School of Earth Sciences, Department of Geological and Environmental Sciences, Stanford University, 450 Serra Mall, Stanford, CA 94305, U.S.A., e-mail: warrenj@stanford.edu.

Related Web Links

Integrated Ocean Drilling Program: http://www.iodp.org
Deep Carbon Observatory Initiative: http://dco.gl.ciw.edu/
Mission Moho Proposal: http://www.missionmoho.org
Kanazawa MoHole Workshop Report: http://www.mohole.org

21

Postglacial Fault Drilling in Northern Europe: Workshop in Skokloster, Sweden

by Ilmo T. Kukkonen, Maria V.S. Ask, and Odleiv Olesen

Introduction

The majority of Earth's earthquakes are generated along plate margins, and the theory of plate tectonics provides the explanation for the occurrence of these earthquakes. However, a minority of earthquakes occurs within continental plates, and the theoretical understanding for these earthquakes is largely lacking (Stein and Mazzotti, 2007). The general assumption is that intraplate earthquakes tend to be relatively small in size. This report summarizes a workshop devoted to a special type of intraplate earthquake-generating faults—postglacial (PG) faults—that so far have been observed only in northern Europe.

Altogether, there are fourteen well-known PG fault structures in northern Sweden, Finland, and Norway with fault scarps up to 160 km in length and up to 30 m in height (Figs. 1, 2; Olesen et al., 1992; Lagerbäck and Sundh, 2008; Kukkonen et al., 2010). Assuming that these distinct faults were formed in single events, they would represent earthquakes with magnitudes of up to 7–8 (Bungum and Lindholm, 1997; Kuivamäki et al., 1998). This estimate is supported by numerous observations of massive landslides associated with these structures and dated to have occurred at the last stages of the glaciation. PG faults represent earthquakes with considerable contrast to the present seismic activity in continental northern Europe, where earthquakes are usually smaller than magnitude 4.

All known PG faults are located in old reactivated zones of weakness in crystalline rocks and are usually SE dipping, SW-NE oriented thrusts. The last major reactivation of these faults is believed to have occurred during the last stages of the Weichselian glaciation (~9,000–15,000 years B.P.). The earthquakes are believed to have been triggered by the combined effects of tectonic background stresses and rapidly changing stresses from glacial loading by the shrinking Weichselian ice sheet (Johnston, 1989; Wu et al., 1999; Lund, 2005; Lund et al., 2009).

From what is known today, large-scale types of PG faults appear to be restricted in occurrence to northern Fennoscandia. In other previously glaciated areas, such as Canada, postglacial faults are significantly smaller in size (Adams, 1989). Seismological data reveal that the PG faults are currently seismically active, and that small earthquakes are associated with these structures over a significant depth range (down to 37 km depth; Bungum and Lindholm, 1997; Arvidsson, 1996). They are obviously structures of crustal dimensions and relevance, but not thoroughly understood at the moment (Arvidsson, 1996).

Postglacial faulting has important implications for predicting the behavior of fault zones during future glaciations. Therefore, PG fault research is expected to contribute significantly in planning the disposal of spent nuclear fuel, CO_2 and toxic waste into bedrock that currently is prepared in the Nordic countries. Other fields of applied geoscience which may benefit from PG fault research are mineral exploration and estimation of mine stability, as some of the faults are located in areas which host gold, copper, and nickel mineralizations in northern Fennoscandia. Major hydropower and tailing dam complexes may also be influenced by PG faults and their current earthquake activity. An improved understanding of the prevailing *in situ* stress, erosion, uplift, and sedimentation also has implications for the understanding of offshore petroleum reservoirs on the Lofoten-Barents margin.

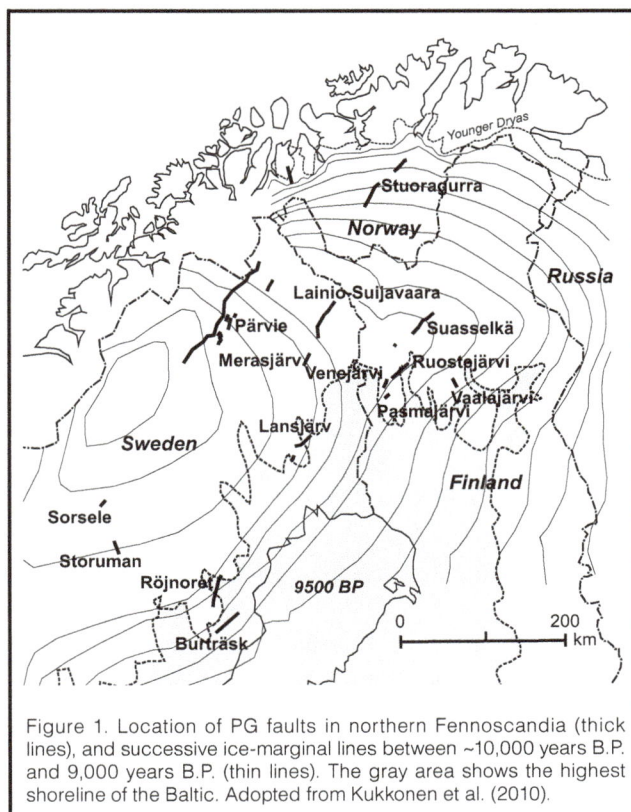

Figure 1. Location of PG faults in northern Fennoscandia (thick lines), and successive ice-marginal lines between ~10,000 years B.P. and 9,000 years B.P. (thin lines). The gray area shows the highest shoreline of the Baltic. Adopted from Kukkonen et al. (2010).

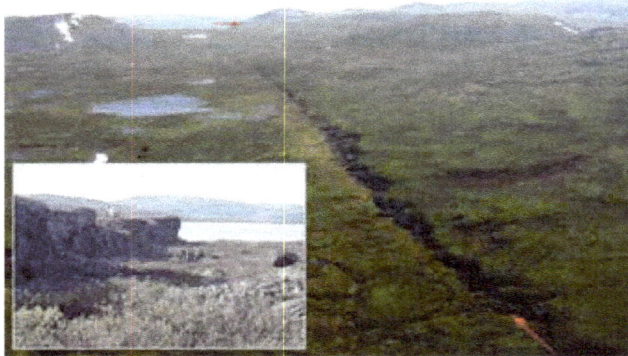

Figure 2. Helicopter view of the southwestern part of the Pärvie PG fault (see Fig. 1 for location). The red arrows show the trace of the fault scarp. The insert shows the fault scarp from the ground surface, about 85 km to the northeast of the location of the large photo, including a helicopter for scale.

Methods applied in PG fault research so far include bedrock and Quaternary field geology, trenching, seismicity, airborne and ground geophysics, and shallow drilling to about 500 m (Kukkonen et al., 2010). Revealing the mechanisms and processes related to PG faulting is highly relevant for understanding seismicity in these intraplate areas. Several disciplines and approaches can be used to improve our understanding of PG faults, for example, through earthquake seismology, stress field measurement and modeling, as well as geodetic surface monitoring of fault activity. Scientific drilling and coring is the only way to obtain direct core samples from PG faults at depth, and the resulting boreholes provide direct access to the fault structures for geophysical, hydrogeological, and biological sampling, monitoring and *in situ* experiments. We organized the ICDP-supported international workshop "Postglacial Fault Drilling in Northern Europe" in Skokloster, Sweden on 4–7 October 2010; thirty-nine participants represented basic research, applied geosciences, industry and authorities from eight countries. At the workshop, the status of PG fault research was discussed, and plans were made towards developing a realistic drilling plan.

Major Scientific Issues/Problems

The major scientific tasks of PG fault research were identified as follows:

1. What is the tectonic style, deep structure and depth extent of the PG faults?

2. Are PG faults still active?

3. What are the paleoseismic implications of postglacial faults?

4. Did PG faults reactivate more than once? Is it possible to provide quantitative ages of the tectonic systems hosting PG faults?

5. What are the present and paleostress fields and pore pressure of PG faults?

6. How has the faulting affected the rock properties, structure, and deformation in and near the fault surface?

7. What are the hydraulic properties of PG faults, and how did they control fresh glacial meltwater recharge?

8. What is the composition of groundwater (chemistry, salinity, pH, Eh, gas content) in PG faults?

9. Is there a deep biosphere in PG faults?

One of the relevant issues of PG faulting is whether their current appearances really are the result of single earthquake events. The risk and implications of PG faulting to intraplate seismicity in general, and waste disposal repositories in particular, is highly dependent on this. Previous investigations by Lagerbäck and Sundh (2008) suggest that massive landsliding and seismites in soft sediments occurred concurrently with the faulting. They based their arguments mainly on the relatively small erosion of the Weichselian glaciation, and stated that such dramatic faulting which generated the great PG faults in northern Sweden very probably did not occur in glaciations earlier than the Weichselian.

Workshop Discussions

The workshop presentations can be subdivided roughly into four sub-groups: (1) geology, tectonics, age determination studies; (2) seismic structures, seismicity and other geophysics; (3) stress field, land uplift and plate tectonic forces; and (4) hydrogeology, hydrochemistry, geothermics, and deep biosphere. The participants subsequently discussed the major scientific tasks within these four sub-groups.

The main aim for drilling is to penetrate a fault which presently is seismically active. It is also commonly agreed that it would be useful to compare an active fault with an inactive one. When defining drilling targets it will be important to locate the fault exactly at depth, but this may be difficult. Even in the shallow drilling of the Lansjärv fault in the 1980s, it was not easy to decide where the PG fault actually was because the rock was generally very fractured and broken (Bäckblom and Stanfors, 1989). One of the goals is to drill into the seismogenic zone of a PG fault. Although the macroseismic activity in Fennoscandia seems to be characterized by focal depths of 10–20 km (Ahjos and Uski, 1992; Bungum and Lindholm, 1997), the present seismic activity of PG faults seems to start from surface, at least in the case of the Pärvie fault. The need of seismic monitoring of several faults was considered relevant before the best candidate for drilling can be identified.

A major issue that may be addressed before the start of drilling is if surface studies can reveal whether the fault scarps were formed by one big earthquake or by several smaller ones. Closer inspection of the fault scarps themselves, as well as investigations of the sediment cover using traditional trenching coupled with ^{14}C-dating, would help address this question. Bungum and Lindholm (1997) and Kuivamäki et al. (1998) did comparative studies of the relationship between fault length and fault scarp height of PG faults in Scandinavia, and compared the data with recent large earthquakes. Their approach may be pursued to investigate what the scale effect is for the fault (i.e., if there is a relationship between the size of the earthquake and the size of the fault scarp). Furthermore, drill cores would reveal whether there have been paleodynamic weakening effects (thermal pressurization, frictional melting, etc.) related to major periods of faulting in the geological history.

Site survey data that needs to be collected include 2D and 3D reflection seismic surveys to identify the geometry of PG faults, and passive seismic network data to identify earthquake activity and tomography studies. In addition, ground penetrating radar, 2D resistivity measurements, gravity data, magnetotelluric soundings, and high-resolution topographic surveys with laser scanning (LIDAR) are needed. Drilling of shallow and relatively inexpensive pilot holes may allow characterization and identification of the fault at shallow depths as well as installation of instruments monitoring microseismicity. It is important to expand the seismic surveys and seismic networks to as many of the remaining faults as possible to allow the selection of the best candidates for drilling.

In addition, use of existing data may also improve an ICDP drilling proposal (e.g., synthesizing results on existing cores and their mechanical properties, and reinterpretation of the state of *in situ* and paleo-stresses). Finally, site investigation data can be utilized to calibrate and improve viscoelastic ice sheet numerical modeling within the site survey areas.

Challenges for Drilling

Different strategies of drilling geometry were outlined in the workshop. Assuming that drilling takes place on one fault site only, the alternatives would be (1) to drill only shallow (<1 km) boreholes located on a profile perpendicular to the fault plane, (2) drill one deep borehole (2–5 km) penetrating the fault at great depth, (3) drill a deep borehole with several shorter boreholes deviating from the main borehole at 1.5–2 km depth, or (4) combine 1–3 shallow boreholes and a deep (2–5 km) one. Option no.4 would allow learning while drilling (i.e., modification of drilling plans of the main borehole would be possible from the experience gained in the shallow ones). Such a drilling geometry would also allow cross-borehole experiments and various sampling and monitoring activities *in situ* and would provide good control of fault properties with depth.

We identified a range of criteria that is helpful to determine the best site for drilling. At the site of drilling, the selected PG fault should (1) be seismically active over a depth interval that can be reached by drilling and beyond; (2) reveal contrasting geology across the fault to allow unambiguous determination of the fault location; and (3) be a site with good logistics capacity. In addition, pre-drilling investigations should suggest the site has a very good scientific capacity (i.e., the majority of research hypotheses should have a good chance to be tested with drilling).

In order to address the scientific problems, a detailed drilling and testing program needs to be developed. The program should include the collection of oriented cores, borehole logging, fluid sampling, stress measurements, and long-term monitoring of strain/tilt, microseismicity, fluid pressure, and temperature. Preferred core tests include physical properties (petrophysics), rock mechanical determinations, deformation microstructures, mineralogy and geochemistry, and dating. Good quality downhole logging data will be required to allow as complete characterization of the fault as possible, including image logs, density, resistivity/induction, magnetic, full waveforms, and spectral gamma.

After drilling, the most important measurements are stress measurements, strain/tilt and microseismic monitoring, fluid pressure and temperature monitoring, borehole image logging, and geophysical logging. Hydrogeological and microbial studies require post-drilling time for long-term pumping of fluid and gas. Important laboratory investigations include geological logging, petrophysical measurements, rock mechanic testing, and core studies of deformation and fault related microstructures. They also include the capacity to link such data to geochemical studies of the core (e.g., fluid inclusions, if they exist) and geochronology. These data would help improve the models and quality of viscoelastic ice sheet modeling within the site survey area. The possibility for induced seismicity tests should be investigated.

Potential Drilling Targets

The workshop participants could already identify several potential drilling targets. At the moment the most promising ones would be structures which have long surface scarps, thus indicating crustal scale relevance. The targets should preferably be seismically active, and they should have structures which have been sufficiently imaged with various geophysical techniques. Seismicity has been monitored already in a number of faults with arrays designed for PG faults, but many major faults lack monitoring at the moment. An interesting option would be to compare two structures, one showing seismic activity and one devoid of any activity.

Identification of the scientifically most optimal drilling targets was not possible without more site-specific studies such as seismic arrays to be run for about one to two years. In addition, geodetic monitoring should be started to observe any creep. Previous geodetic leveling and GPS measurements in Finland (Kuivamäki et al., 1998, Poutanen and Ollikainen, 1995) did not show any measurable movement.

Conclusions and Road Map Forward

The workshop community considered drilling into PG faults a feasible scientific initiative which would lead to a research project with important societal implications. The present state of the art in PG fault studies is very promising for developing an ambitious new ICDP project "Postglacial Fault Drilling Project" (PFDP).

Many PG faults are seismically active, and they may represent structures which release the current plate tectonic stresses accumulating in the Fennoscandian continental plate. A concept for the project would be to define an active target fault where the preliminary results of seismic monitoring may suggest that the upper parts of the seismogenic zone could be reached with boreholes shallower than about 3 km. The fault would be investigated with both shallow boreholes (<1 km) and a deep borehole (max 2–5 km). Core drilling is essential for a representative sampling of the rocks at least in the expected depth levels of the fault. Furthermore, a combination of several boreholes would allow a variety of downhole experiments, logging, samplings, and monitoring after drilling.

Existing shallow cores (Kukkonen et al., 2010) should be re-examined with modern mineralogical and isotope methods. Pre-drilling science should also include re-analysis of stress field measurements (Bäckblom and Stanfors, 1989; Bjarnason et al., 1989). Pre-drilling science and gathering of site-specific data sets are estimated to take 2–3 years before a well-defined drilling proposal can be compiled. Meanwhile, information will be disseminated on the PFDP in international conferences, and working group meetings are planned to be organized in association with the EGU and AGU conferences. A session "Intraplate faulting and seismicity with special reference to the Fennoscandian postglacial fault province" is currently arranged at the EGU in Vienna, Austria, in April 2011.

References

Adams, J., 1989. Postglacial faulting in eastern Canada: nature, origin and seismic hazard implications. *Tectonophysics*, 163:323–331.

Ahjos, T., and Uski, M., 1992. Earthquakes in northern Europe 1375-1989. *Tectonophysics*, 203:1–23.

Arvidsson, R., 1996. Fennoscandian earthquakes: whole crustal rupturing related to postglacial rebound. *Science*, 274:744–746.

Bäckblom, G., and Stanfors, R., 1989. *Interdisciplinary Study of Post-Glacial Faulting in the Lansjärv area, northern Sweden, 1986-1988*. Swedish Nuclear Fuel and Waste Management Co., Stockholm, Technical Report 89-31.

Bjarnason, B., Zellman, O., and Wikberg, B., 1989. Drilling and borehole description. *In* Bäckblom, G., and Stanfors, R. (Eds.), *Interdisciplinary Study of Post-Glacial Faulting in the*

Lansjärv area, northern Sweden, 1986–1988, 7:1–7:14. Swedish Nuclear Fuel and Waste Management Co., Stockholm, Technical Report 89-31, 7:1–7:14.

Bungum, H., and Lindholm, C., 1997. Seismo- and neotectonics in Finnmark, Kola Peninsula and the southern Barents Sea. Part 2: seismological analysis and seismotectonics. *Tectonophysics*, 270:15–28.

Johnston, A., 1989. The effect of large ice sheets on earthquake genesis. *In* Gregersen, S., and Basham, P. (Eds.), *Earthquakes at North-Atlantic Passive Margins: Neotectonics and Postglacial Rebound:* Dordrecht (Kluwer Academic Publishers), 581–599.

Kuivamäki, A., Vuorela, P., and Paananen, M., 1998. Indications of postglacial and recent bedrock movements in Finland and Russian Karelia. Geological Survey of Finland, Nuclear Waste Disposal Research, Report YST-99, 92 p.

Kukkonen, I.T., Olesen, O., Ask, M.V.S., and the PFDP Working Group, 2010. Postglacial faults in Fennoscandia: targets for scientific drilling. *GFF*, 132:71–81.

Lagerbäck, R., and Sundh, M., 2008. Early Holocene faulting and paleoseismicity in northern Sweden. *SGU Research Paper C836*, 80 pp.

Lund, B., 2005. *Effects of Deglaciation on the Crustal Stress Field and Implications for Endglacial Faulting: A Parametric Study of Simple Earth and Ice Models*. Swedish Nuclear Fuel and Waste Management Co., Stockholm. Technical Report TR-05-04, 68 pp.

Lund, B., Schmidt, P., and Hieronymus, C., 2009. *Stress evolution and fault instability during the Weichselian Glacial Cycle*. Swedish Nuclear Fuel and Waste Management Co., Stockholm. Technical Report TR-09-15, 106 pp.

Olesen, O., Henkel, H., Lile, O.B., Mauring, E., and Rønning, J.S. 1992. Geophysical investigations of the Stuoragurra postglacial fault, Finnmark, northern Norway. *J. Appl. Geophys.*, 29:95–118.

Poutanen, M., and Ollikainen, M., 1995. *GPS Measurements at the Nuottavaara Postglacial Fault*. Finnish Geodetic Institute, Report 95, 6 pp.

Stein, S., and Mazzotti, S., 2007. Continental intraplate earthquakes: science, hazard, and policy issues. *GSA Special Paper 425*, Boulder, Colo., (The Geological Society of America, Inc.), 402 pp.

Wu, P., Johnston, P., and Lambeck, K., 1999. Postglacial rebound and fault instability in Fennoscandia. *Geophys. J. Int.*, 139:657–670.

Authors

Ilmo T. Kukkonen, Geological Survey of Finland, Espoo, P.O. Box 96, FI-02151 Espoo, Finland, e-mail: ilmo.kukkonen@gtk.fi.

Maria V.S. Ask, Luleå University of Technology, SE-971 87 Luleå, Sweden.

Odleiv Olesen, Geological Survey of Norway, NO-7491 Trondheim, Norway.

Figure Credits

Fig. 2: Björn Lund, Uppsala University, Sweden (large photo); and Roger Lagerbäck, Geological Survey of Sweden (insert photo).

22

The Scandinavian Caledonides—Scientific Drilling at Mid-Crustal Level in a Palaeozoic Major Collisional Orogen

by Henning Lorenz, David Gee, and Christopher Juhlin

Introduction

The Caledonides of western Scandinavia and eastern Greenland have long been recognized to have been part of a collisional orogen of Alpine-Himalayan dimensions, essentially the result of the closure of the Iapetus Ocean during the Ordovician, with development of island-arc systems, and subsequent underthrusting of continent Laurentia by Baltica in the Silurian and Early Devonian during Scandian collisional orogeny. Several hundreds of kilometers of thrust

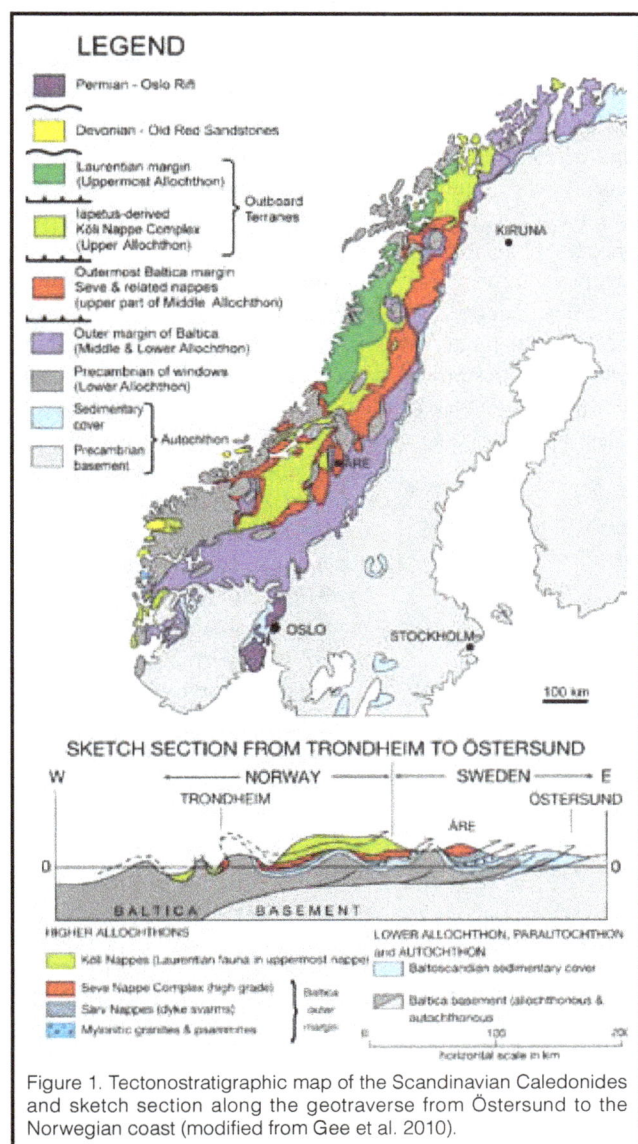

Figure 1. Tectonostratigraphic map of the Scandinavian Caledonides and sketch section along the geotraverse from Östersund to the Norwegian coast (modified from Gee et al. 2010).

emplacement of allochthons have been demonstrated, E-directed in the Scandes and W-directed in Greenland.

In Scandinavia, major allochthons (Fig. 1) were derived from Baltica's outer shelf, dyke-intruded continent-ocean transition zone (COT), Iapetus oceanic domains and (uppermost) from the Laurentian margin. On the western side of the North Atlantic, exposed along the eastern edge of the Greenland ice cap, there are major thrust sheets, all derived from the Laurentian continental margin and transported at least two hundred kilometers westwards onto the platform. In the Scandinavian and Greenland Caledonides, the major allochthons that were derived from the outer parts of the continent margins have been subject to high-grade metamorphism and apparently were emplaced hot onto the adjacent platforms.

The International Continental Scientific Drilling Program (ICDP) workshop in Sweden provided an opportunity to examine the evidence for Caledonian collisional orogeny in Scandinavia and to discuss its relevance for understanding other orogens, particularly Himalaya-Tibet, and also the subduction systems along the margin of the western Pacific. The Scandinavian Caledonides are one of the best places on the planet to study the emplacement not only of highly ductile allochthons generated in an outer continental margin subduction complex, but also of associated hot (granulite facies paragneisses with leucogranites) extruding nappes that were the potential heat source for the metamorphism of underlying and overlying long-transported allochthons.

COSC Project Rationale

The Collisional Orogeny in the Scandinavian Caledonides (COSC) project focuses on the transport and emplacement of subduction-related high-grade COT assemblages (the Seve Nappe Complex) onto the Baltoscandian platform and their influence on the underlying allochthons and basement. Research will be performed by an international working group with experience from studying fossil and active mountain belts. Orogenic processes and their development over time will be investigated by scientific drilling in the deeply eroded (mid-crustal levels) Scandinavian Caledonides, and the results will be compared from this unique locality with a modern analogue of similar size, the Himalaya-Tibet mountain belt, and the arc collisional systems between the Eurasian and Pacific plates.

The Workshop

The workshop was attended by about sixty participants— half from outside of the Nordic countries—on 21–25 June 2010 in Åre, Sweden, close to the planned drilling sites. The workshop was divided into two parts, separated by a full day's excursion on 23 June (Fig. 2). The Scandinavian Caledonides, a modern analogue (the Himalaya-Tibet mountain belt), and the highly successful ICDP project in the Sulu ultra-high pressure belt (the Chinese Continental Scientific Drilling Program) were presented on the first day. Presentations on the emplacement of hot allochthons were made on the second day, followed by geological and geophysical workshop sessions on orogen-scale processes. The afternoon session focused on the Scandian hinterland, before taking a rapid ascent to the "hot allochthon" on cold and snowy Åreskutan mountain. Evening lectures provided introductions to ICDP and the Swedish Deep Drilling Program, in particular the purchase of a mobile drilling rig capable of coring down to depths of at least 2.5 km. The mid-workshop excursion concentrated on the rock units (excluding the unknown basement) through which we plan to drill, from the amphibolite facies Seve Nappe Complex to underlying greenschist facies metasedimentary units of the Middle Allochthon and Cambro-Silurian Lower Allochthon. On the fourth day, lectures were held about drillhole-related geophysics and western Pacific subduction systems, particularly the Izu-Bonin-Mariana Arc; workshop sessions covered details of the science related to the drillholes and drill cores; and general discussion of COSC science was conducted. The day ended with presentations on the hydrological and geothermal aspects of the Scandian mountain belt and drilling program. The last morning of the workshop was spent winding up the COSC science plan and defining a road map for the coming six months, with preparation of a comprehensive drilling proposal to ICDP and applications to funding agencies.

COSC Drilling Program

Two drillholes, each ~2.5 km deep, are planned to core a composite profile from the "hot" Seve nappes downwards, through the underlying lower grade allochthons, into the Fennoscandian basement. They are located near the towns of Åre and Järpen (Fig. 3).

The geology of the Åre area is renowned for classical studies of vast overthrusting (Törnebohm, 1888), with high-grade metamorphic rocks (granulite facies) on the top of Åreskutan mountain emplaced over Cambro-Silurian sedimentary rocks in the valley below. Drilling will start in the lower part of the well-exposed section and continue through less exposed amphibolite and greenschist facies units in the underlying nappes. The second hole, near Järpen, will continue the section through the underlying autochthonous cover and deep into the Fennoscandian basement.

The high spatial resolution provided by continuous drillcore will allow a detailed study of the metamorphism and its changes through and across tectonic contacts, from the high-grade allochthons into the underlying less metamorphosed nappes and basement. Oriented drillcore will serve as a basis for understanding deformation and thrust emplacement, as well as heat transport and fluid migration during metamorphism, in time and space. COSC drilling will then penetrate the lower allochthonous units and the basal décollement, most likely in Cambrian alum shale, and enter Precambrian crystalline basement. Prominent basement seismic reflectors will be studied in detail. Investigation of the apparent deformation pattern in the autochthonous basement that is observed in the seismic data will be achieved by drillcore studies and in-hole measurements. *In situ* and drill core investigations are also necessary to study the amount of Caledonian and older deformation and metamorphism in the basement.

Two coreholes to ~2.5 km instead of one deep hole (~5 km) will make the COSC project economically feasible. The second drillhole will be located further towards the foreland of the Caledonides, starting in the tectonostratigraphy just above the base of the first hole (Fig. 3). These holes will be drilled with a diamond coring drill rig to maximize core recovery and minimize costs. Both boreholes will investi-

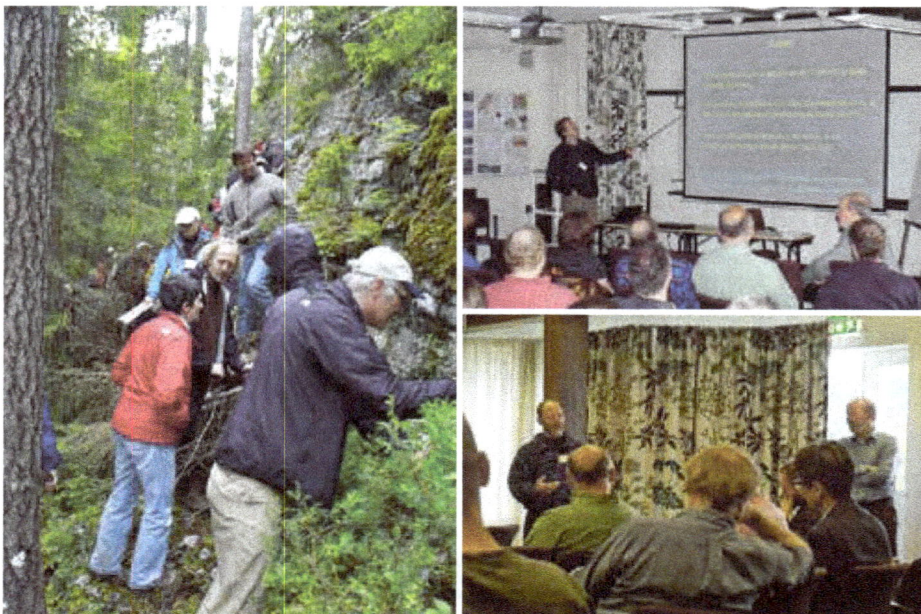

Figure 2. Scenes from the COSC science workshop presentations, discussions, and excursions. Participants on one of the key localities of the lower, poorly exposed tectonostratigraphic units.

gate regional heat flow, water circulation patterns, the deep biosphere, and the mineral potential of the area. They will also allow calibration of high-quality surface geophysical data at depths which are not normally accessible to drilling in this tectonic environment.

Working Groups

Working groups have been established for tectonics, geophysics, geothermics, hydrogeology, and the deep biosphere. Technical operations and research will be performed by the drilling management and technology working group, utilizing the Swedish scientific drilling infrastructure.

The main objectives of the tectonics working group concern the

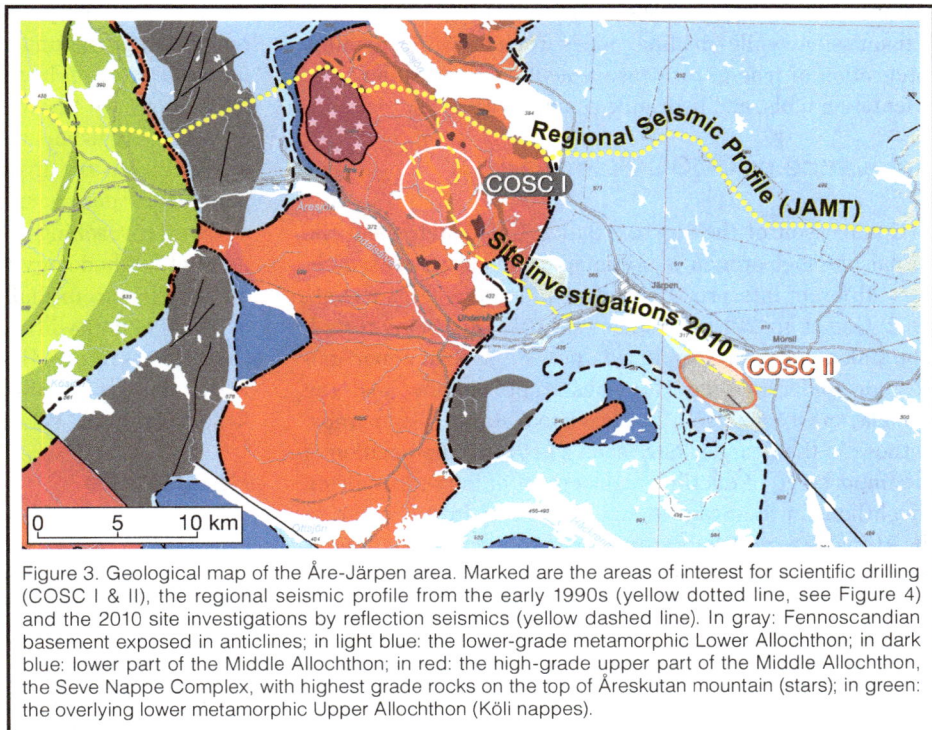

Figure 3. Geological map of the Åre-Järpen area. Marked are the areas of interest for scientific drilling (COSC I & II), the regional seismic profile from the early 1990s (yellow dotted line, see Figure 4) and the 2010 site investigations by reflection seismics (yellow dashed line). In gray: Fennoscandian basement exposed in anticlines; in light blue: the lower-grade metamorphic Lower Allochthon; in dark blue: lower part of the Middle Allochthon; in red: the high-grade upper part of the Middle Allochthon, the Seve Nappe Complex, with highest grade rocks on the top of Åreskutan mountain (stars); in green: the overlying lower metamorphic Upper Allochthon (Köli nappes).

mechanisms of emplacement of hot allochthons and the establishment a coherent model of mid-Palaeozoic (Scandian) mountain building in the North Atlantic Caledonides. Orogenic processes will be considered in ongoing and older orogens, with underthrusting of continents, doubling (even trebling) of continental thicknesses, elevation of a high plateau, partial melting of hinterland regions and ductile extrusion of allochthons many hundreds of kilometers onto adjacent platforms. In addition, new insights will be applied to the interpretation of modern analogues, in particular the Himalaya-Tibet mountain belt. The geophysics working group will map the large-scale geological structure around the boreholes and relate it to surface measurements. Cores and in-hole observations will allow determination of the origin of the observed seismic reflections during site investigations and regional seismic profiling across the mountain belt

(Fig. 4). The geothermics working group focuses on assessing the heat flow and temperature in the crystalline bedrock—in the Fennoscandian Shield basement and the overlying Caledonian allochthons—and on investigating shallow boreholes (300–400 m) that have been drilled for mineral exploration purposes several decades ago. The hydrogeology working group will model the large-scale groundwater circulation patterns in the mountain belt and their influence on the hydrosphere of the Fennoscandian Shield basement beneath other parts of Sweden. For this purpose it is necessary to build a regional geological model which is as yet an unaddressed problem in shield areas, with importance beyond the borders of Scandinavia. The microbiology working group will investigate microbial life in highly metamorphosed sedimentary and crystalline bedrock; this study will introduce challenges different from those dealt with in sediments. Anticipated differences include distribution of biota within the rocks formations and nutrient sources to the deep biosphere, where microbial life must utilize inorganic, geological sources such as energy-rich gases.

The availability of a diamond core drill rig, recently funded by the Swedish Research Council primarily for the Swedish Deep Drilling Program, will allow the drilling management and technology working group to develop and test new drilling technologies and tools. The emphasis will be on

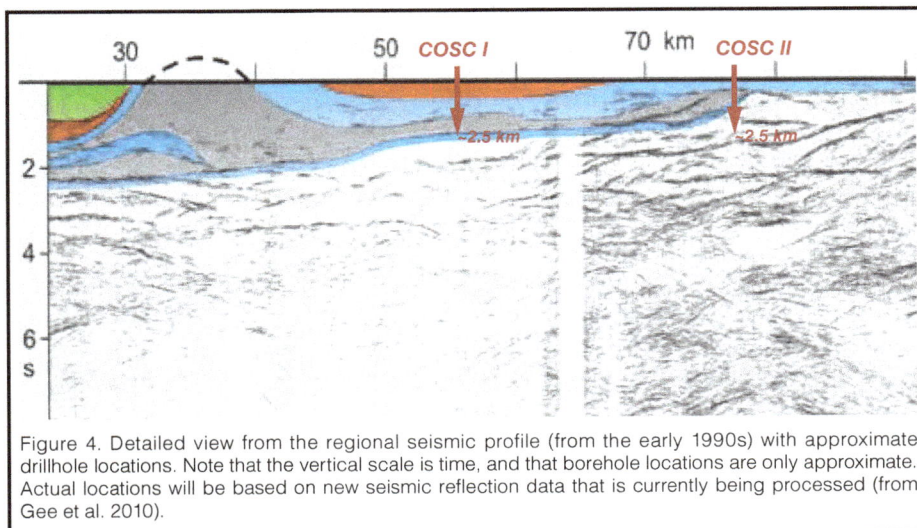

Figure 4. Detailed view from the regional seismic profile (from the early 1990s) with approximate drillhole locations. Note that the vertical scale is time, and that borehole locations are only approximate. Actual locations will be based on new seismic reflection data that is currently being processed (from Gee et al. 2010).

enhanced sampling methods in fractured formations, data transmission while drilling, measurement while drilling, integration of true-gyro measurements while drilling, core orientation tools, and hydraulic conductivity tools.

Relevance of COSC Science

Comparison of the North Atlantic Caledonides and the Himalaya-Tibet orogen is stimulating much new research. Very different interpretations of both orogens (Soper et al., 1992; Gee et al., 2010; Streule et al., 2010) are being tested. The Caledonides in Scandinavia provide special opportunities for understanding Himalayan-type orogeny and the Himalayan Orogen itself. The last two distances comparable to those in the Scandes have seen a growing appreciation of the importance of ductile emplacement of long-transported allochthons in the Scandes and elsewhere. In particular, in the Himalayas vast lateral transport of ductile allochthons over distances exceeding those in the Scandes has been demonstrated. The Izu-Bonin-Mariana arc system, target of the ODP leg 125 (Fryer et al., 1990), takes the comparison of collisional systems a step further—to a fore-arc system where subduction, collision, thrusting, and related igneous activity and exhumation have been studied in a currently existing smaller framework.

Geological processes along active continental margins, followed by collisional tectonics and mountain building, have a profound influence on human society. Massive mountain belts like the Himalayas influence climate and weather; natural disasters are common for settlements on its steep slopes and narrow valleys. Active collisional systems are known for inflicting earthquakes on inhabitants. COSC takes a comprehensive approach to mountain building processes and their development through geological time by integrating the drilling project in the fossil orogen of the Scandinavian Caledonides with research on the Himalayas and the Izu-Bonin-Mariana arc collisional systems. The project will contribute to the ICDP themes "Collision Zones and Convergent Margins", "Active Faulting and Earthquake Processes" and "Climate Dynamics and Global Environments".

After the COSC drilling project, the Jämtland transect across the Scandinavian Caledonides will be one of the best investigated profiles across a Palaeozoic mountain belt. Calibrated geophysical investigations will give insight into the structure of the shield basement and the overlying allochthons. Detailed geological studies will cover the section from the upper allochthons into the Precambrian basement (Fig. 1), including ore-bearing horizons. Intraterrestrial life, its activity, nature and origin are much less studied in crystalline bedrock than in sedimentary environments. This is also true for the hydrogeological conditions. An integrated geological-geophysical-hydrogeological model is envisaged based on new knowledge concerning the structure of the thrust sheets and the underlying basement. Results will be of importance for all kinds of underground infrastructure projects, in particular when very long-term resistance to the underground environment is central, like for waste storage. Heat flow studies will increase our knowledge about the thermal regime in the allochthons and the crystalline basement of the Fennoscandian Shield and, together with the hydrogeological results, will assess the potential for energy extraction in the Åre-Järpen region. For a more far-reaching approach, the evaluation and development of methodology to more reliably predict the geothermal gradient from shallow drill holes is important. Inversion of heat flow data will also provide valuable information about palaeotemperature. Hence, COSC will also contribute to the ICDP themes "Geobiosphere and Early Life", "Natural Resources", and "Volcanic Systems and Thermal Regimes".

References

Fryer, P., Pearce, J.A., Stokking, L.B., et al., 1990. *Proc. ODP, Init. Repts.*, 125: College Station, TX (Ocean Drilling Program).

Gee, D.G., Juhlin, C., Pascal, C., and Robinson, P., 2010. Collisional Orogeny in the Scandinavian Caledonides (COSC). *GFF*, 132:29–44.

Soper, N.J., Strachan, R.A., Holdsworth, R.E., Gayer, R.A., and Greiling, R.O., 1992. Sinistral transpression and the Silurian closure of Iapetus. *J. Geol. Soc.*, 149:871–880.

Streule, M.J., Strachan, R.A., Searle, M.P., and Law, R.D., 2010.. Comparing Tibet-Himalayan and Caledonian crustal architecture, evolution and mountain building processes. *Geol. Soc. London Spec. Pub.*, 335:207–232.

Törnebohm, A.E., 1888. Om fjällproblemet. *GFF*, 10(5):328–336. (in Swedish)

Authors

Henning Lorenz, David Gee, and Christopher Juhlin, Uppsala University, Department of Earth Sciences, Villavägen 16, 752 36 Uppsala, Sweden, e-mail: henning.lorenz@geo.uu.se.

Related Web Links

http://www.sddp.se/COSC
http://are-jarpen.icdp-online.org/

Figure Credits

Fig. 3: Geological map, copyright Geological Survey of Sweden (SGU).

23

U.S. Continental Scientific Drilling Community Looks to the Future

by Anthony W. Walton

Continental scientific drilling in the U.S.A. may be poised to take a significant step forward as a result of two recent workshops that laid out the possibilities for the future. The meetings, in June 2009 in Denver, Colorado and in June 2010 in Arlington, Virginia, brought together about 100 members of the community. The first meeting stressed the themes and topics of important science for which drilling is a necessary means of collecting samples and data. The second workshop developed recommendations for implementation of a strong U.S. program including its position as a necessary component of the International Continental Scientific Drilling Program (ICDP).

The June 2009 workshop reviewed the range of scientific interests that continental drilling alone enables and specified possible interactions between continental and ocean drilling. Four overarching themes emerged: (i) global environmental and ecological change (emphasizing Earth history), (ii) geodynamics (broadly defined), (iii) the geobiosphere, and (iv) natural resources and environmental concerns (Table 1). Within each theme are a number of topics. Each topic has enough intellectual coherence for a consensus to be developed that reviews the field, identifies subjects for future growth, and suggests the means to reach goals. Most of these topics are familiar ones that have been expounded previously. Progress constantly brings new topics to the drilling community; for example, it has recently emerged that lake sediments preserve records of rates, processes, and triggers of evolutionary events, so that a whole community of evolutionary biologists will have interests in drilling projects.

The two main problems identified were (i) the thematic breadth of scientific drilling allowing no single focus and (ii) the path to funding being hindered by obstacles and delays (Fig. 1). To strengthen the U.S. community an enlarged Science Planning Committee of DOSECC has been charged with overseeing overall and topical scientific planning, considering advances in equipment or facilities that are necessary for the drilling community, and communicating internally, to the broader scientific community as well as to key funding agencies and to the ICDP.

Both workshops concluded that scientific planning should be a bottom-up effort, with communities gathering to reflect, assess, propose, consider, and develop consensus. Three special considerations emerged. First, planning efforts should be inclusive and international, including participants who address the same questions through different means. Where appropriate, they should include ocean drillers. Second, these efforts should be broadly announced and their results communicated so that members of other communities who might profitably participate in projects are fully informed of the opportunities. For example, study of the deep biosphere can be a part of many investigations. Third, any plan should be a guide, not a limit. The seemingly infinite creativity of investigators should not be discounted simply because their proposal is not in line with a pre-existing document.

Currently, the DOSECC office acts to bind the U.S. community together and inform the broader Earth science profession through annual workshops, newsletters, and booths at large professional meetings. It also has a very successful but poorly known program of internships for students and schoolteachers. Workshop participants recommended that these efforts should be expanded and supplemented by the wealth of modern communication modes.

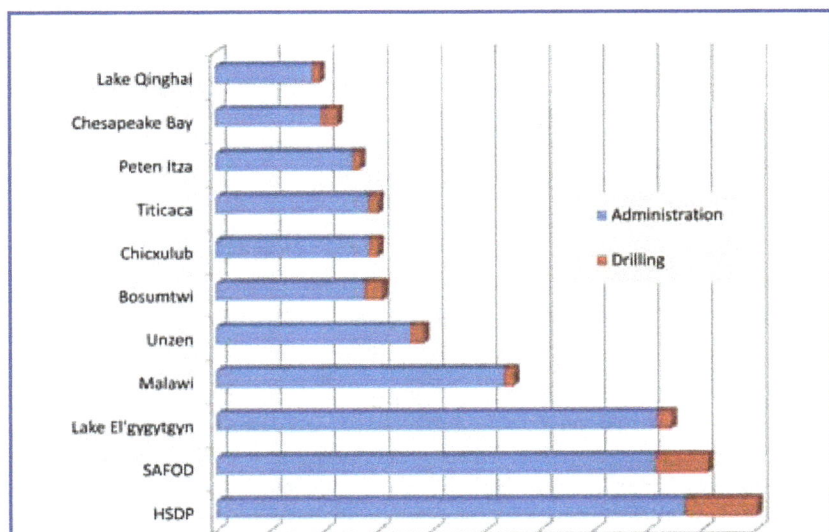

Figure 1. Comparison of drilling time to preparation time for representative projects. Administrative time includes the time from first workshop or first contact with DOSECC office until drilling actually begins. Projects undergo a year or more of planning and refinement before that occurs. The spectacle of 4–10 years of delay before operations begin effectively precludes young U.S. investigators from undertaking continental scientific drilling efforts (courtesy of Dennis Nielson).

Broadening the community is an important goal. An open planning process will do much to involve more investigators in drilling activities. The internship program should inform younger professionals of the potential rewards of drilling to gather necessary samples for their investigations and should enlist new members of the community. An important task will be to explore other ways to encourage investigators to undertake projects where drilling promises substantial rewards, despite the costs in money and time. Furthermore,

Table 1. Themes and topics in continental scientific drilling.

Themes		Topics	
Global environmental and ecological change	High-resolution time-series records	Plio-Pleistocene climate records Evolution in isolated lake systems Climate and evolution of hominins and associated faunas (History of the magnetosphere)	
	Deep-time records	Climate history Sea-level history Paleoceanography Atmospheric history and early life Cryospheric history from near-field sub-ice records (Stratigraphic architecture and crustal deformation) Evoluton and extinction Dynamics of the solar system (History of the magnetosphere) Antarctic deep-time records	
Geodynamics		Crustal evolution (Stratigraphic architecture and crustal deformation) Hotspots, mantle plumes, and large igneous provinces Processes and hazards at volcanoes Fault mechanics (History of the magnetosphere) Ice-sheet history and dynamics	
Geobiosphere		Microbiology, including ichnofossils Biogeochemistry	
Natural resource systems and related environmental concerns		Hydrothermal resources and core deposits Groundwater Hydrocarbons CO_2 sequestration	

the community will be looking at ways to mentor neophyte drilling scientists and to provide timely guidance to strengthen their proposals and projects.

For international projects, the ICDP remains a key source of funding. Currently the principal source of funds in the U.S.A. is the National Science Foundation (NSF). However drilling activities are supported by other federal agencies and private sources. The 2010 workshop recommended several steps to deal with funding issues.

1. The 2010 workshop encouraged the NSF to identify a central internal point of contact and to secure funding arrangements for the costs of continental drilling, much as it supports telescopes for astronomers and ships for oceanographers. Ideally the central point of contact would be a formal program at NSF with a director and budget. NSF should also coordinate with scientific drilling efforts in other agencies.

2. Workshop participants strongly favored maintaining an appropriately funded facility, the current DOSECC office or a similar agency, to serve the community and provide drilling services coupled with a formal program at NSF with a director and budget.

3. The workshop recommended that the allocation of funding for drilling operations be based upon a set amount each year or a set number of drilling days, with some flexibility to deal with significant opportunities in a timely fashion. Funds from other agencies would extend the level of activity. This arrangement would remove the severe obstacle of including drilling costs in proposals.

4. One of the most pressing perceived obstacles to developing drilling projects is the need for funds to do preliminary site and feasibility studies. Consequently, the workshops recommended development of a system of funding necessary preliminary studies.

To implement the recommendations of the workshops, the continental scientific drilling community must work together, justify its science, plan its future, and work with funding agencies to develop mutually satisfactory arrangements. An enhanced continental scientific drilling effort in the U. S. A. requires an active community, thoughtful planning, and a clear pattern of funding to synergistically interact with related organizations and overlapping communities, and it will strengthen the international drilling communities and the Earth science effort as a whole.

Acknowledgements

The workshops were funded by the National Science Foundation through grants EAR 1039441 and EAR 0923056 to the University of Kansas with additional contributions from DOSECC, Inc. Comments by Walter Snyder, Wilfred Elders, and Ulrich Harms improved the manuscript.

Author

Anthony W. Walton, Chair, Board of Directors, DOSECC Inc., Department of Geology, The University of Kansas, Lawrence, KS 66045, U.S.A., e-mail: twalton@ku.edu.

24

Ultra-Deep Drilling through 3.5-Billion-Year-Old Crust in South Africa

by Maarten de Wit

Introduction

The Makhonjwa Mountains of South Africa and Swaziland comprise some of the most sought-after geo-real estate in the world. It is priceless—that is, for geoscientists—because the rocks of this approximately 120 km by 60 km corner of southern Africa, also known as the Barberton Greenstone Belt, date back to 3.2–3.6 billion years (Ga), representative of Earth in early Archean times when it was still ~1 Ga years young. They are not the very oldest rocks on Earth (those occur in Greenland and Canada), but they are the oldest best-preserved ones; thus, this stretch of land is without equal for research into the early history of our Earth. It is home to some of the earliest fragments of island arc, oceanic crust, and vestigial tracts of continent covered with sedimentary and volcanic rocks. So well-preserved are these rocks that unless one radiometrically dates them, it is near impossible to distinguish them from many modern rocks. This exceptional preservation has ensured that the

Table 1. Workshop participants.

Name	Affiliation
L Ameglio	Exige, Geophysical Services, RSA
R Armstrong	Australian National University, Australia
N Arndt	University of Grenoble, France
N Banerjee	University of Western Ontario, Canada
A Biggin	University of Liverpool, UK
M de Wit	AEON, University of Cape Town, RSA
T Dhansay	student, AEON and Council of Geoscience, RSA
M Doucouré	AEON, University of Cape Town, RSA
J Ebbing	Geological Survey of Norway, Norway
J Erzinger	GFZ German Research Centre for Geosciences, Potsdam, Germany
C Fourie	TUT, Tshwane University of Techology, RSA
D Frei	Geol Survey of Denmark, Denmark
C Jaupart	IPGP, Institute de Physique du Globe, Paris, France
C Langereis	University of Utrecht, Holland
S MacLennan	student, AEON, University of Cape Town, RSA
M Mesli	Geological Survey of Norway, Norway
C Rice	Drilling Technology SA, RSA
O Ritter	GFZ German Research Centre for Geosciences, Potsdam, Germany
P Robinson	Dept Earth Sciences, Dalhousie University, Canada
G Stevens	Stellenbosch University, Dept Earth Science, RSA
A van Wyk	Drillers in Training CC, RSA
U Weckman	GFZ German Research Centre for Geosciences, Potsdam, Germany
A Wilson	University of Witwatersrand, RSA

Makhonjwa rocks yield the oldest directly dated and undisputed signs of life on Earth, and compared to our present biosphere they also provide detailed clues about the hostile nature of the paleoenvironments under which this life struggled to persist. One severe challenge entailed coping with more potent solar radiation to which life is particularly sensitive, when Earth's magnetic field was too weak to efficiently shield the surface from the relentless solar wind of lethal charged particles. Another is to explore for paleo-suture zones that can help establish when plate tectonics first emerged as the dominant solid earth recycling process to nurture the only sustainable habitable zone in our solar system. These then represent some of the targets of a new deep drilling project, on which an ICDP workshop was focused and held on 13–19 April 2010.

The workshop was attended by two students and twenty-one international scientists from four continents (Table 1), each with a different expertise and perspective with which to contemplate an 8–10 km drillhole through this unique terrain, as part of building an Early Earth Conservatory. The workshop was held at Travelport, the 'Cradle of Life' Conference-Conservation center, some 15 km from the town of Emanzana (formerly Badplaas), South Africa. The site is within walking distance from the world's oldest identified suture zone, the prime drilling target for this project (Fig. 1).

The project is both scientific and applied in scope. It is meant to characterize Earth's oldest subduction/suture zone and its paleoenvironments, to study the deep ancient and modern biosphere in pristine Archean crust, to establish a permanent 'on-site' early Earth laboratory-museum-educatorium in rural Africa, and to link these facilities to an African college of drilling technology.

Scientific Background

Tectonics

The existence and especially the onset of early Archean (>3.0 Ga) present-day style plate tectonics remains controversial, despite many studies having addressed this topic. Alternative models include plume dominated processes and crustal delamination during which vertical motions controlled Archean tectonics (Van Kranendonk, 2007; Hamilton,

Figure 1. Workshop participants in the field at a potential drill area.

2007). This controversy on the nature of Archean tectonics has been extensively debated over the last two decades without reaching consensus. Recent field-based research has provided some evidence for plate tectonics as early as 3.1 Ga and possibly as early as 3.8 Ga, but this is not generally accepted as conclusive (Schoene and Bowring, 2010; de Wit et al., 2011; Furnes et al., 2009). Geochemical analysis of Archean rocks shows that between 3.5 Ga and 3.8 Ga, Archean crust formation can, with apparent equal validity, be interpreted to have been generated during mantle plume magmatism or through subduction processes similar to that associated with plate tectonics (Bédard, 2006). Numerical modeling based on high mantle temperatures and geotherms, as is generally assumed for the Archean, is consistent with whole mantle plume tectonics (Davies, 2007). Similar modeling, particularly with a hydrous mantle, shows that plate tectonics is also capable of removing the required excess heat produced in the Archean at a rate of operation comparable to, and possibly even lower than, its current rate (Grove and Parman, 2004). In any case, recent thermochronology and petrology have questioned the existence of ubiquitous higher geothermal gradients everywhere during the Archean (Moyen et al., 2006; Diener et al., 2005).

A fundamental difference between plate tectonics and other scenarios is the occurrence of large horizontal lithosphere motion. Geological observations have revealed early Archean horizontal crustal motion. Extension and formation of sedimentary basins as early as 3.49 Ga and 3.45 Ga, as well as significant horizontal shortening episodes between 3.4 Ga and 3.2 Ga, suggest significant horizontal tectonic processes that possibly, but not definitively, reflect plate tectonic motions. The shortening episodes include associated high-pressure, low-temperature metamorphism in the Barberton Greenstone Belt at 3.2 Ga. Attempts at establishing extents and rates of horizontal motions of Archean terrains using paleomagnetism, have been suc-cessful only in terranes younger than 3.0 Ga (Strik et al., 2003; de Kock et al., 2009). Thus, a unified tectonic model for the early Archean Earth remains elusive. The interpretations and models remain controversial largely because of lack of geo-

physical data and robust structural/paleomagnetic analyses of tectonic events without precise thermochronology and pristine borehole samples.

Early life and ancient life-support systems

It has long been argued that understanding Archean tectonic processes provides fundamental keys to unraveling the origin and formation of Earth's earliest continents (cratons), its paleoenvironments, early ecosystems, and life.

Several decades have passed since the first description of recognizable early Archean microfossils (de Wit, 2010), yet morphology-focused imaging techniques of fossil-like objects and stable isotope (C, N, S) compositions of putative organisms have repeatedly failed to pose limits on the interpretation of the biogenic origin of the microstructures. Additionally, several abiologic metamorphic and hydrothermal reactions have been identified that can produce kerogen and graphite, and specific abiologic processes have been described that can generate complex structures that resemble microfossils (McLoughlin et al., 2007). In view of these uncertainties and controversies, it is clear that elucidating how and when life may have originated on Earth requires first to understand the conditions that prevailed early in Earth's history and the environments in which life may have appeared and later evolved. The recent discovery (Furnes et al., 2004) and *in situ* dating of ichnofossils in the rims of the world's oldest pillow lavas in Barberton (Fliegel et al., 2010) has dramatically shown that rocks previously ignored in studies of early life (e.g., basaltic igneous rocks) now offer a new paleoenvironment as habitats for early life. This holds great potential to track life back even further in time and must be considered a promising focus for such early life studies in places like the Barberton Greenstone Belt.

What the Makhonjwa Mountains can offer Archean science

The lower rock sequences of the Barberton Greenstone Belt and its surrounding granitoid terranes comprise the best well-preserved Paleo-Archean section of continental crust in the world (Fig. 2). The area contains rocks that have never been deeply buried, except within a limited zone in the southwest part of the belt where high-pressure, low-temperature metamorphism at 3.2 Ga has been recorded. This zone—part of the Inyoka fault system—has recently been suggested to represent a 3.2-Ga suture zone, separating two low-grade continental arc/back-arc/oceanic terranes of slightly different ages and geological history (Moyen et al., 2006). A similar second zone has been identified on the basis of thermochronology and structural mapping (Schoene et al., 2008, 2009) flanking the southeast margin of the belt, separating the central Barberton belt from a continental arc terrane, the Ancient Gneiss Complex. This implies that the two oldest sutures of the world are present in this area. Recent paleomagnetic studies on these older sequences of the Barberton terranes provide intriguing preliminary evidence that a stable and reversing geomagnetic field was up and running at ~3.5 Ga, and that horizontal motions were on the order of ~12 cm yr^{-1}—fast by today's standards, but well within the range of plate velocities observed in the Phanerozoic (Biggin et al., 2011).

Key scientific questions analyzed during the workshop

- Did plate tectonics operate 3.5 billion years ago?
- What is the geophysical character/image of the world's oldest suture zones?
- Do the proposed suture zones of the Barberton Greenstone Belt, which separate at least three different terranes, penetrate the entire crust, and how do they affect the old underlying lithospheric mantle?
- Are paleomagnetic reconstruction of plate motions fast or slow, and are we dealing with large or small plates?
- What was the intensity of earliest geomagnetic field in relation to inner core growth?
- What is the nature/age of the crust beneath the oldest preserved terranes?
- How did the earliest continental fragments of the Kaapvaal craton form and amalgamate to create Earth's first stable continent? What was its geothermal gradient?
- Are we dealing with a 'hot/dry' mantle or a 'wet/cool' Archean mantle? What were the geothermal gradients within different Archean terranes?
- How did early suture zone tectonics and related thermo-chemical fluid processes, including serpentinization, influence early life and ecosystems and gold metallogenesis?
- What is the depth distribution and biochemistry of extremophiles in the deep biosphere of the Archean compared to that of today in the same rock sequences?
- What was the optimum temperature window for preservation of microfossils in different Archean terranes?
- Can we define chemical fingerprints of interactions between fluids, rocks and microbes?
- Were Archean ocean/atmosphere temperature and composition hot or cold?
- Is the atmosphere redox state reduced or oxidized, or episodically both?
- Can microbial contamination be defined and quantified?

What the Makhonjwa Mountains can offer rural development in Africa

The Barberton Greenstone Belt is a geological hotspot that is presently being considered as a UNESCO world heritage site. The region has been a 'mecca' for countless generations of Earth and life scientists and has been a key location where significant new scientific ideas have emerged. This remains so to this very day, with new research programs and at least three shallow scientific drilling projects having been completed recently and/or planned for completion soon. The Barberton Greenstone Belt is a well-known region for teaching of field geology and studies in early Earth processes to undergraduates and research students from South Africa and other countries. Tourist routes are now also starting to include the region, but few local people benefit from its rich history. The area under investigation is rural and poor, without adequate schooling and health facilities in crowded townships. Education opportunities for young people are scarce and uninspiring. Field schools and excursions (national and international) are frequent, but few if any engage with local youth. The plans for a deep drill site will be dovetailed with outreach and education requirements of the local, rural communities. Several scientists at the workshop cooperate closely with local nature reserves (Songimvelo and Nkomazi), the Mpumalanga Parks Board, and local tourist agencies. In addition, in-depth discussions have been held with local farmers and entrepreneurs, traditional leaders, and regional and national government representatives about the vision of linking a deep drilling site to a local center for early Earth studies attractive to schoolteachers, school-learners, undergraduate students, and research scientists alike. These discussions have been welcomed by all these stakeholders.

Key socioeconomic & education questions addressed during the workshop

- How can we develop a long-lasting scientific interest in the early Earth that will also benefit the local rural communities, and in particular develop science and engineering skills related to geo-technology in rural Africa?
- How can we best dovetail scientific research with science and environmental outreach programs for general public awareness and youth education?
- Can we develop a local training center for drilling and related mining technology?
- Can we develop a rural center for early Earth studies, with open access for all researchers and learners to relevant materials and literature?

Workshop summary

Talks were presented on the regional geology and geophysics of the Barberton Greenstone Belt and surrounding regions, together with detailed overviews about the petrology and thermodynamics of the rocks found within and flanking the Inyoka Shear Zone (ISZ). These data form the backbone for models that represent the ISZ as a 3.2–3.3 Ga paleo-suture zone, within which evidence is preserved for a low Archean geothermal gradient of 10°C–20°C km^{-1} that was subsequently overprinted by higher temperatures at lower pressures, indicative of collision and exhumation tectonics. This is contrary to conventional theories that all Archean environments had high geotherms (Hamilton, 2007). The drilling through the ISZ is thus a prime target for the study of a range of early Earth processes in an environment similar to those in modern subduction and suture zones. Geologically, the ISZ coincides with a number of highly deformed serpentinized peridotites and tectono-sedimentary melange rocks similar to those found along Phanerozoic suture zones and active plate boundary faults such as the Alpine Fault in New Zealand. Midway through the workshop, participants visited a potential drill site near the surface exposure of the ISZ (Fig. 1) flanked on one side by serpentinites and on the other by a sequence of sandstones and conglomerates, not unlike that found at the San Andreas Fault Observatory at Depth across the San Andreas Fault.

Overviews of the geophysics and a preliminary 3-D model of the greenstone belt indicates that a 10-km-deep drillhole also has the potential to pierce the base of the belt and thus allow a detailed examination of the contact with the underlying rocks of the middle Archean crust. These contacts are invariably interpreted as deep tectonic boundaries that have been explicitly implicated (de Wit and Ashwal, 1997) as incubators for epi-mesothermal gold deposits, hallmarks of greenstone belts throughout the world. In view of the relatively poor surface outcrop, a complete section (assuming high core recovery) will allow systematic changes to be recorded through the com-

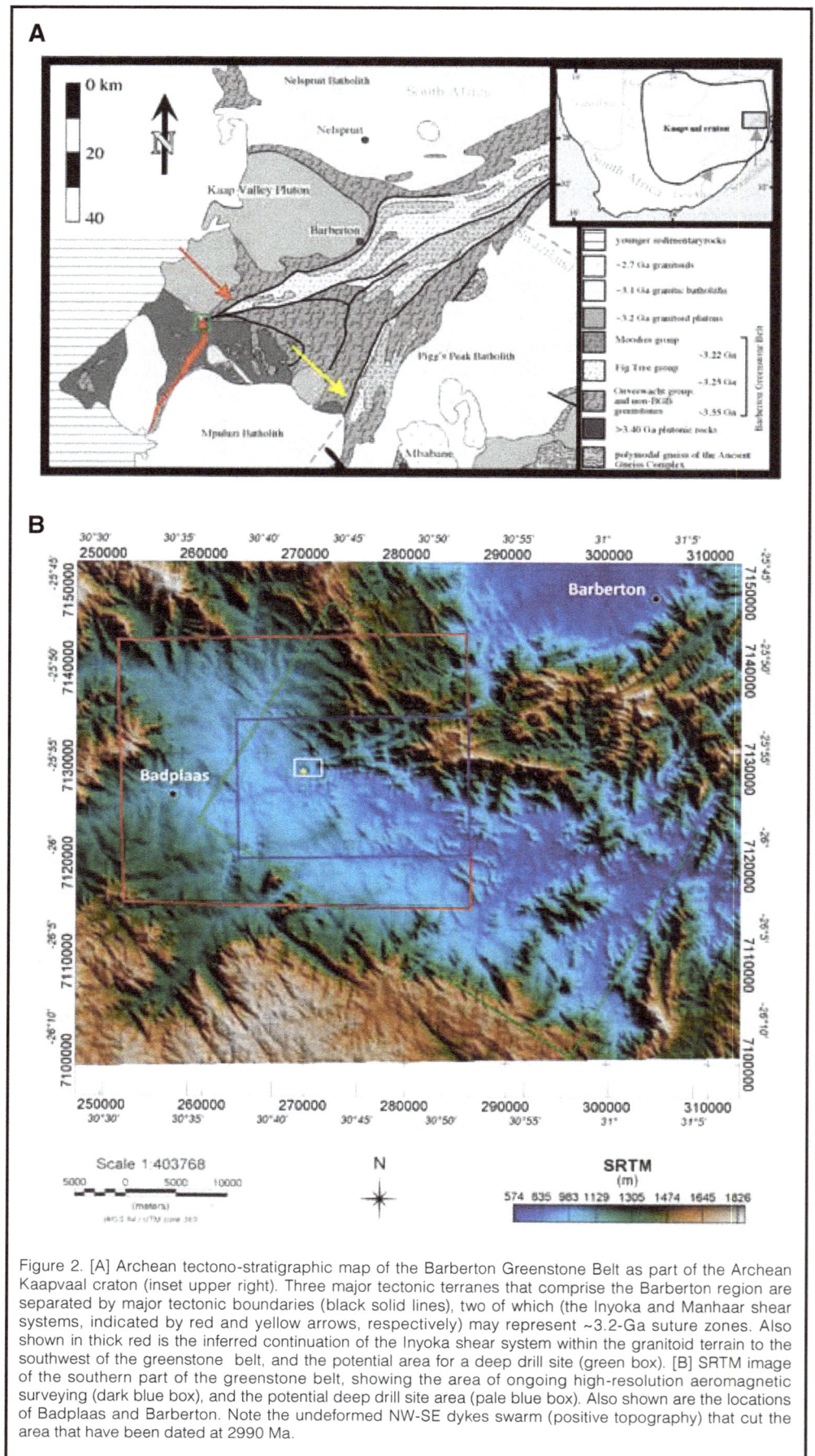

Figure 2. [A] Archean tectono-stratigraphic map of the Barberton Greenstone Belt as part of the Archean Kaapvaal craton (inset upper right). Three major tectonic terranes that comprise the Barberton region are separated by major tectonic boundaries (black solid lines), two of which (the Inyoka and Manhaar shear systems, indicated by red and yellow arrows, respectively) may represent ~3.2-Ga suture zones. Also shown in thick red is the inferred continuation of the Inyoka shear system within the granitoid terrain to the southwest of the greenstone belt, and the potential area for a deep drill site (green box). [B] SRTM image of the southern part of the greenstone belt, showing the area of ongoing high-resolution aeromagnetic surveying (dark blue box), and the potential deep drill site area (pale blue box). Also shown are the locations of Badplaas and Barberton. Note the undeformed NW-SE dykes swarm (positive topography) that cut the area that have been dated at 2990 Ma.

plex rock sequences with tectonic zones, and small features that are likely to be hidden in even the best outcrops will be much easier to interpret. Measurements on detailed chemical and physical parameters of the core are needed to ground truth geophysical profiles. Misapplication of seismic models developed for sedimentary sequences to metamorphic basement in the German Continental Deep Drilling Project, for example, resulted in erroneous interpretations (Emmermann and Lauterjung, 1997).

Workshop talks were presented also on how to collect fluids and gases, past and present, at all levels through a drilled sequence of this nature, and to measure changes in these over time and depth. The most abundant volatiles in common crustal rocks are water and carbon dioxide. However, little is known about the distribution and behavior of hydrocarbons, hydrogen, nitrogen, and noble gases in ancient continental crust. Generally these elements are minor components in crystalline rocks and, hence, do not significantly influence the physical or thermodynamic properties of a rock, but they have a large potential in tracing mass and heat transport processes. Moreover, noble gases (^{4}He, ^{40}Ar) and N in natural gases, crustal fluids, and fluid inclusions can be used as indicators of the fluid sources, and they are thus helpful in trying to solve questions of fluid generation, flow, and evolution in the deep crust.

These talks were complemented by biogeochemical views of how such a deep laboratory can further probe the present and past deep biosphere (microbiota) in rocks that may have harbored life as long ago as 3.4 Ga. The paleomagnetists also emphasized the need for careful magnetic measurements to constrain magnetic field strength variability, and the heat-flow modelers recommended *in situ* measurements of heat flow, conductivities, heat-producing elements, and high-resolution thermochronology to constrain variations in paleogeotherms.

The value and pitfalls of different types of geophysical surveys prior and during deep drilling projects, including the German KTB borehole, were presented and deliberated extensively during the workshop. In addition, an overview of the technical drilling capacity and training in South Africa was given by professional drilling consultants to the African mining industry. In 2009 the Mining Qualifications Authority estimated that there was a shortage of some 1200 drillers in South Africa, and the requirement for a steady stream of trained drillers into the broader African drilling industry will always be large. South African mining houses have for many years drilled some of the deepest cored boreholes in the world. In 2010 approximately fourteen boreholes were being drilled to depths in excess of 3500 m, but all of this drilling is still done using drilling systems that were developed many years ago. The need for an innovative approach to deep level core drilling is very great indeed.

The workshop also included an open 'town hall' meeting for the public, land owners, local school teachers and learners, non-governmental organization (NGO) representatives, and the media. Clearly, the workshop was a success judging not only by the interest in this project from a curiosity driven perspective, but also from the perspective of developing new drilling technology and the dire need for a sustainable education/training facility to ensure drilling expertise from Africa. It was perhaps surprising to learn that despite a severe shortage of drilling expertise and the great number of ongoing drilling projects in the exploration and extraction industries throughout onshore and off-shore Africa, there is nowhere in Africa for young people to pursue a career in drilling other than on-site learning on the job. The establishment of a training college focused on improving drilling (and possibly mining) skills would advance the goals of developing educational opportunities and drilling capabilities.

Recommendations

There was strong consensus at the workshop that we need to firmly establish whether more can be learned from two 5-km holes or several shallower holes, instead of one 10-km hole. Before further deliberations on this, and before honing in on a potential area, let alone a precise drill-site, there was unanimous agreement that a number of detailed surveys need to be completed. For example, more detailed surface mapping of the ISZ is required, in particular through higher resolution structural mapping and analyses. However, because of limited exposure a number of geophysical surveys are also prerequisites before the project can move into a drill-planning stage.

While preliminary 3-D gravity and magnetic models of the Makhonjwa Mountains were presented, their present utility is severely hampered by the lack of sufficiently high-resolution gravity, magnetic, and borehole data. Moreover, no crustal seismic reflection data are available. Although a teleseismic experiment has yielded a crustal thickness in this region of ~43 km from converted P-S wave receiver-function analyses, this experiment failed to provide any significant insights into the internal crustal structures (Nguuri et al., 2001). Current aeromagnetic data is too coarse to resolve the geology of the area. Additional geophysical methods (magnetotelluric magnetic, seismic) are therefore required, and only high-resolution data will improve the reliability of 3-D models required to understand surface structures with depth.

Developing plans for on-site, real-time mud-gas analysis during drilling—similar to those developed during drilling of the German KTB borehole, and in numerous scientific drilling projects since then—was proposed as essential at an early stage. Hydrocarbons, helium, radon, and (with limitations) carbon dioxide and hydrogen are the most suitable gases for the detection of fluid-bearing horizons, shear zones, open fractures, and sections of enhanced permeability. These will provide critical samples and analyses of ephemeral gas/fluid pockets penetrated during drilling that might otherwise escape unnoticed, and will provide essential guidance for decisions related to later fluid sampling and *in situ* hydrologic testing. Subsequent off-site isotope studies on mud gas samples help reveal the origin and evolution of deep-seated crustal fluids. Studies of crustal scale fluid transport over large distances and times indicate that fluid transport rates are significantly in excess of predictions based on simple theory (Erzinger and Stober, 2005). This implies that fluid flow in the deep crust is mechanically enhanced and/or episodic. The specific rare gas components will indicate the relative proportions of fluids arising from meteoric, magmatic, metamorphic, and mantle sources. Information about the evolution of fluids in space and time should result from investigations of the chemical and isotopic fingerprints of rocks and minerals, which were influ-

enced by fluid/rock interaction and fluid inclusions trapped as remnants of past fluids but also from the chemical-isotopic composition of fresh fluids present in open cavities and fractures. Therefore, such studies are fundamental to the success of a deep drilling project. Thus, while drilling campaigns provide unique opportunities to sample indigenous fluids/gases continuously from a section of the upper crust, site survery work needs to be completed prior to actual drilling.

As drill sites are selected, it is necessary to evaluate existing information on the local hydrology, hydrochemistry, and the occurrence of aquifers. A science team will plan to measure hydrologic properties at several levels by packing off favorable sections and to collect water samples (e.g., for tritium and noble gas isotope analyses [He, Ne, Ar, Kr, and Xe], stable isotope analyses [H, C, O, and S]), and for complete chemistry of dissolved constituents.

Ensuring successful drilling deep into the oldest suture zone will require the early cooperative efforts of many nations and experts, and good coordination is essential. Prior to drilling, a long lead time is required to establish a precise location where the suture will occur at depth and how its local dip might vary. Besides detailed geophysics, it will be important to obtain additional information through a number of shallow reconnaissance pilot holes at relatively low cost. Both partial core recovery and downhole geophysical logging will provide crucial information to improve 3-D modeling.

Ongoing Work

As part of laying further foundations for this project, ongoing work has focused on a detailed magnetotelluric (MT) survey (Weckmann et al., 2009) across the Inyoka paleo-suture zone and surrounding rocks to obtain high-resolution images of the shear zones. Over two consecutive years (2009–2010), two large MT experiments were carried out. To gain good 3-D coverage, 5-component MT data were recorded in a frequency range from 0.001 s to 1000 s at almost 200 sites (at an average spacing of ~2 km) arranged along a 110-km-long transect and five shorter transects covering an area of ~300 km^2 (Fig. 3). This setup provides good areal coverage of the ISZ and also a vertical resolution on lithospheric scale. The main difficulties for electromagnetic experiments in the Barberton area are the various man-made noise sources (e.g., electric fences, power lines, mining

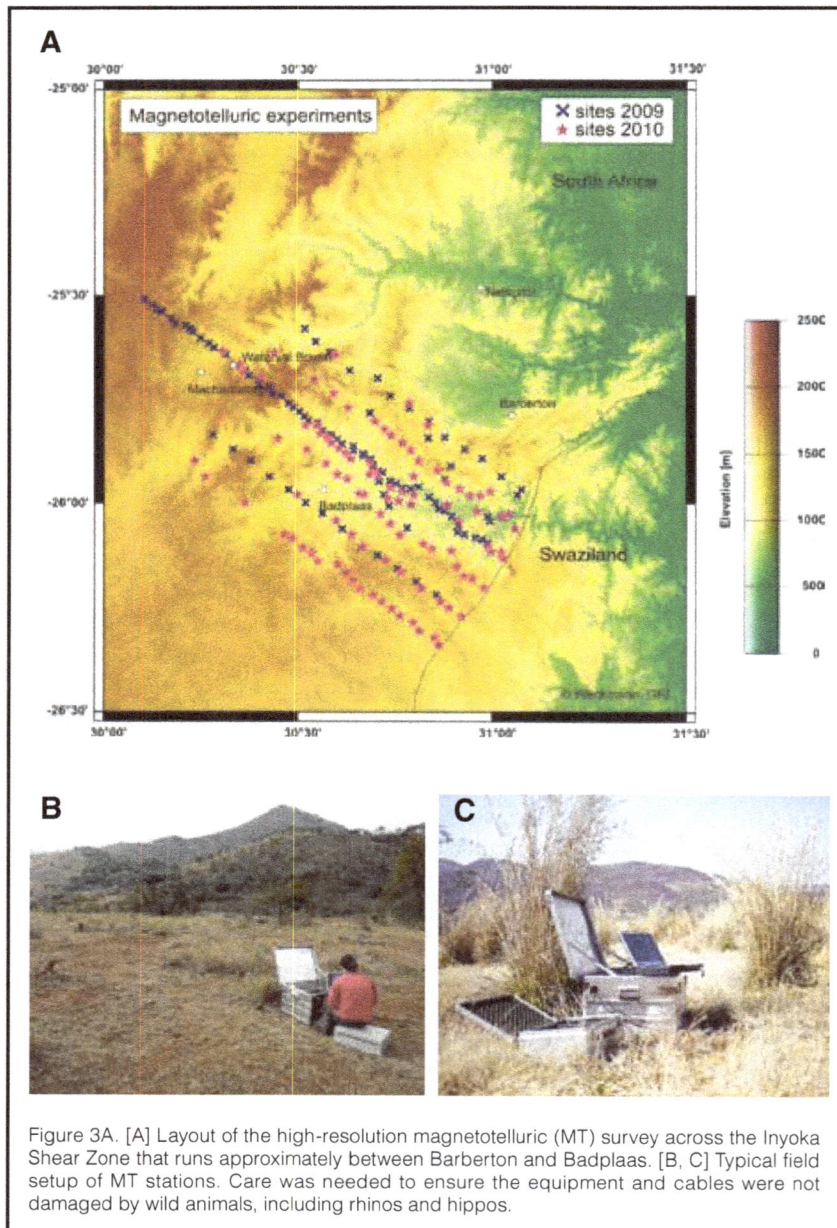

Figure 3A. [A] Layout of the high-resolution magnetotelluric (MT) survey across the Inyoka Shear Zone that runs approximately between Barberton and Badplaas. [B, C] Typical field setup of MT stations. Care was needed to ensure the equipment and cables were not damaged by wild animals, including rhinos and hippos.

activities and a major DC railway line). Hence, the natural electromagnetic field variations are overprinted by these strong electromagnetic signals. Nevertheless, the first 2-D inversion tests along the 110-km transect with a reduced data set already show strong correlation with subsurface geology, and zones of high electrical conductivities appear to correlate well with the surface location of known faults. The results of the MT work is being further integrated with ongoing laboratory conductivity measurements on representative rock samples collected across the suture zone during detailed structural mapping of a well exposed part of the ISZ.

A high-resolution aeromagnetic and radiometric survey is planned for March 2011, using the low flying Gyrocopter Kreik IIB from GyroLAG (Gyrocopter Light Airborne Geophysics), to complement the MT work. A special feature of this light airborne geophysics platform, which requires no formal landing strip, is its capability of performing safely at survey heights as low as 5 m above ground level at relatively slow speeds (75–100 km hr^{-1}), resulting in a significant improvement in quality of data (equivalent of 2–3 m ground equivalent sampling intervals).

Plans to establish an Africa college for drilling technology are in progress with the Tswana University of Technology (TUT) and relevant government agencies. Local property owners have identified several suitable sites where such a rural extension of TUT might be built. As part of a new drilling technology development initiative, an early start on developing a new type of high-speed coring turbine drill bit has begun at TUT. Although the design has not been finalized, the proto-drill includes a fluid-powered and cooled rotating drill head with a stationary drillstring, and a mechanism for core to be brought up via a core-mouse inside the drilling stem. Also part of this initiative is design of new drill bits (based on recent developments in synthetic diamond manufacturing techniques), face discharge designs, and hybrid bit designs.

Conclusion

We are confident that the proposed geophysics transects in the Barberton area will yield high quality depth profiles down to Moho and possibly deeper. This will allow imaging of the proposed suture zones, the bottom of the greenstone belt, and possibly other features not yet identified. The proposed suture zones are also principle zones of structurally controlled gold mineralization, allowing for significant spin-off for understanding links between these sutures and Archean metallogenesis. The suture zones are also the focus of significant serpentinization that must have been the source of large-scale fluid flow and hydrogen production, both important ingredients for the emergence of primitive life and thereafter to sustain it to the present day.

The Makhonjwa Mountain treasure chest continues to yield unique observations with which to model how our planet transformed from a near molten ball to a plate tectonic driven recycling plant. There is always a ripple of excitement at scientific meetings whenever the lid of the Makhonjwa Mountain chest is pierced further open, ever so slightly. It is hoped that a deep geoscientific drillhole with associated science and technology related infrastructure will provide new scientific opportunities and also add significant value to the local communities.

Acknowledgements

Funding for the workshop was provided through the ICDP, the SA National Research Foundation (NRF), the Africa Earth Observatory Network (AEON), the bilateral South Africa-German program Inkaba yeAfrica project, and Travelport through Mr. Fred Daniel. These support structures are gratefully acknowledged. The MT work is financed through the GFZ, the DFG, and Inkaba yeAfrica program, and is led by Ute Weckmann and Oliver Ritter and their MT Group at Potsdam. The high-resolution aeromagnetic survey was conducted by Laurent Ameglio of GyroLAG and EXIGE (EXpertise In GEophysics), South Africa.

Nazla Hassen and UCT students are thanked for logistic support, and the staff at Travelport for their efficient and friendly hospitality, and for arranging transport to the potential drill site. I would in particular like to thank all participating scientists. This is AEON contribution number 94.

References

Bédard, J.H., 2006. A catalytic delamination-driven model for coupled genesis of Archaean crust and sub-continental lithospheric mantle. *Geochimica et Cosmochimica Acta*, 70:1188–1214, doi:10.1016/j.gca.2005.11.008.

Biggin, A.J., de Wit, M.J., Langereis, C.G., Zegers, T.E., Voute, S., Dekkers, M.J., and Drost, K. 2011. Palaeomagnetism of Archaean rocks of the Onverwacht Group, Barberton Greenstone Belt (southern Africa): Evidence for a stable and potentially reversing geomagnetic field at ca. 3.5 Ga. *Earth Planet. Sci. Lett.*, (2011), doi:10.1016/j.epsl.2010.12.024.

de Kock, M.O., Evans, D.A.D., and Beukes, N.J., 2009. Validating the existence of Vaalbara in the Neoarchean. *Precambrian Res.*, 174(1–2):145–154.

de Wit, M.J and Ashwal, L.D., 1997. *Greenstone belts*: Oxford, U.K. (Oxford University Press).

de Wit, M.J., 2010. The deep-time treasure chest of the Makhonjwa Mountains. *S. Afr. J. Sci.*, 106(5/6), Art. #277, 2 pages, doi:10.4102/sajs.v106i5/6.277.

de Wit, M.J., Furnes, H., and Robins, B., 2011. Geology and tectonostratigraphy of the Onverwacht Suite, Barberton Greenstone Belt, South Africa. *Precambrian Res.*, in press, doi:10.1016/j.precamres.2010.12.007.

Davies, G. F. (2007), Controls on density stratification in the early mantle, *Geochem. Geophys. Geosyst.*, 8, Q04006, doi:10.1029/2006GC001414.

Diener, J.F.A., Stevens, G., Kisters, A.F.M., and Poujol, M., 2005. Metamorphism and exhumation of the basal parts of the Barberton greenstone belt, South Africa: constraining the rates of Mesoarchaean tectonism. *Precambrian Res.*, 143:87–112, doi:10.1016/j.precamres.2005.10.001.

Emmermann, R., and Lauterjung, J., 1997. The German Continental Deep Drilling Program KTB: overview and major results. *J. Geophys. Res.*, 102:18179–18201.

Erzinger, J., and Stober, I., 2005. Long-term fluid production in the KTB pilot hole, Germany. *Geofluids, Special Issue*, 5:1–7, doi:10.1111/j.1468-8123.2004.00107.x.

Fliegel, D., Kosler, J., McLoughlin, N., Simonetti, A., de Wit, M.J., Wirth, R., and Furnes, H., 2010. *In situ* dating of the Earth's oldest trace fossil at 3.34 Ga. *Earth Planet Sci. Lett.*, 299:290–298, doi:10.1016/j.epsl.2010.09.008.

Furnes, H., Banerjee, N.R., Muehlenbachs, K., Staudigel, H., and de Wit, M., 2004 Early life recorded in Archean pillow lavas. *Science*, 304:578–581, doi:10.1126/science.1095858.

Furnes, H., Rossing, M., Dillik, Y., and de Wit., M.J., 2009. Isua supracrustal belt (Greenland) — a vestige of a 3.8 Ga suprasubduction zone ophiolite, and the implications for Archean geology. *Lithos*, 113:115–132, doi:10.1016/j.lithos.2009.03.043.

Grove, T.L., and Parman, S.W., 2004. Thermal evolution of the Earth as recorded by komatiites. *Earth Planet. Sci. Lett.*, 219:173–187.

Hamilton, W.B., 2007. Earth's first two billion years —the era of internally mobile crust. *GSA Memoir*, 200:233–296.

McLoughlin, N., Brasier, M.D., Wacey, D., Green, O.R., and Perry, R.S., 2007. On biogenicity criteria for endolithic microborings on early Earth and beyond. *Astrobiology*, 7(1):10–26, doi:10.1089/ast.2006.0122.

Moyen, J-F., Stevens, G., and Kirsters, A., 2006. Record of mid-Archean subduction from metamorphism in the Barberton terrain, South Africa. *Nature*, 422:559–562, doi:10.1038/nature04972.

Nguuri, T.K., Gore, J., James, D.E., Webb, S.J. and the Kaapvaal Seismic Group, 2001. Crustal structure beneath southern Africa and its implications for the formation and evolution of the Kaapvaal and Zimbabwe cratons. *Geophys. Res. Lett.*, 28:2501–2504, doi:10.1029/2000GL012587.

Schoene, B., and Bowring, S.A., 2010. Rates and mechanisms of Mesoarchean magmatic arc construction, eastern Kaapvaal craton, Swaziland. *Geol. Soc. Am. Bull.*, 122(3/4):408–429, doi:10.1130/B26501.1.

Schoene, B., de Wit, M.J., and Bowring, S.A., 2008. Mesoarchean assembly and stabilization of the eastern Kaapvaal craton: a structural-thermochronological perspective. *Tectonics*, 27:TC5010, doi: 10.1029/2008TC002267.

Schoene, B., Dudas, F.O.L., Bowring, S.A., and de Wit, M.J., 2009. Sm-Nd isotopic mapping of lithospheric growth and stabilization in the eastern Kaapvaal craton. *Terra Nova*, 21:219–228.doi:10.1111/j.1365-3121.2009.00877.x.

Strik, G.H.M.A., Blake, T.S., Zegers, T.E., White, S.H., and Langereis, C.G., 2003. Palaeomagnetism of flood basalts in the Pilbara Craton, Western Australia: late Archaean continental drift and the oldest known reversal of the geomagnetic field. *J. Geophys. Res.*, 108:(B12), EPM 2–1–EPM 2–21.

Weckmann, U., Nube, A., Chen, X., Ritter, O., and de Wit, M. 2009. Overview and preliminary results of a magnetotelluric experiment across the southern Barberton greenstone belt. [11th SAGA Biennial Technical Meeting and Exhibition, Swaziland, 16–18 September], 583–586.

Van Kranendonk, M.J., 2007. Tectonics of early Earth. In Van Kranendonk, M.J., Smithies, R.H., and Bennet, V. (Eds.), *Earth's Oldest Rocks. Developments in Precambrian Geology, 15*: Amsterdam (Elsevier), 1105–1116.

Author

Maarten de Wit, Africa Earth Observatory Network, University of Cape Town, Rondebosch 7701, South Africa. e-mail: maarten.dewit@uct.ac.za.

Figure Credits

Fig. 1: Moctar Doucouré, AEON, University of Cape Town
Fig. 3b, 3c: Dr. Ute Weckmann, GFZ-Potsdam, Germany

IODP-Canada to Exhibit at GAC-MAC 2011

25–27 May 2011, Ottawa, Canada

IODP-Canada will have an exhibition booth at the upcoming joint annual meeting of the Geological Association of Canada, the Mineralogical Association of Canada, the Society of Economic Geologists and the Society for Geology Applied to Mineral Deposits (GAC-MAC-SEG-SGA) to be held at the University of Ottawa on 25–27 May 2011. Over a thousand Earth sciences specialists from Canada, Europe, and the U.S. will be present. Ottawa 2011's motto— *Navigating Past & Future Change*— highlights this meeting's commitment to exploring both the scientific and the societal aspects of Earth sciences.

For more details on IODP-Canada's activities please visit www.iodpcanada.ca or contact the IODP-Canada Coordinator, Diane Hanano at coordinator@mail.iodpcanada.ca.

ECORD Summer School Topic: Subseafloor Fluid Flow and Gas Hydrates

12–13 September 2011, Bremen, Germany

The 5th ECORD Summer School in Bremen, to be held on 12–23 September 2011 at the MARUM—Center for Marine Environmental Sciences—at the University of Bremen, Germany aims to bring PhD students and young postdocs in touch with IODP at an early stage of their careers, inform them about the actual research within this international scientific program, and to prepare them for future participation in IODP expeditions. Such training will be achieved by taking the summer school participants on a "virtual ship" utilizing the unique facilities linked to the IODP Bremen Core Repository where they get familiarized with a wide spectrum of state-of-the-art analytical technologies and core description methods including core logging/scanning according to the high standards on IODP expeditions. In addition, the topic "Subseafloor Fluid Flow and Gas Hydrates" will be covered by lectures and discussions with leading researchers in the field. A one-day field trip on a research vessel will round out the program.

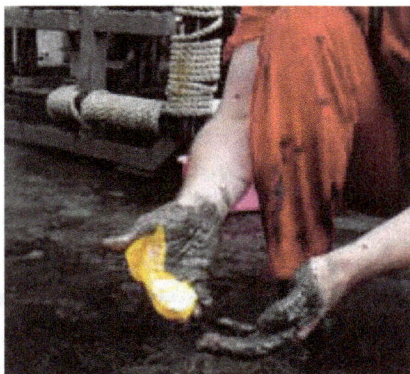

This comprehensive approach—combining scientific lectures with practicums on IODP-style "shipboard" measurements—is the blueprint for the Bremen ECORD summer school covering the three major topics of the IODP Initial Science Plan. The Summer School will be organised by Dierk Hebbeln, Director of the Bremen International Graduate School for Marine Sciences (GLOMAR), Gerhard Bohrmann, Head of Department of Marine Geology, Heiner Villinger, Head of Department of Marine Sensors, and Ursula Röhl, Head of the IODP Core Repository, all at the University of Bremen, Germany. For detailed information visit www.glomar.uni-bremen.de/ECORD_Summer_School.html.

ECORD Invites You to Host a Lecture

Since 2007, the European Consortium for Ocean Research Drilling has sponsored the ECORD Distinguished Lecturer Programme, an initiative for a lecture series to be given by leading scientists involved in the Integrated Ocean Drilling Program. The program is designed to bring the exciting scientific discoveries of the IODP to the geosciences community in ECORD and non-ECORD countries.

2010–2012 Lectures:

Kai-Uwe Hinrichs, MARUM, University of Bremen, Germany
Title: "Benthic archaea – the unseen majority with importance to the global carbon cycle revealed by IODP drilling."
Dominique Weis, Pacific Center for Isotopic and Geochemical Research, University of British Columbia, Canada
Title: "What do we know about mantle plumes and what more can we learn by IODP drilling?"
Helmut Weissert, ETH Zürich, Switzerland
Title: "Carbon cycle, oceans and climate in the Cretaceous: lessons from ocean drilling (DSDP to IODP) and from records on continents."

Applications to host a Distinguished Lecturer are accepted from any college, university or non-profit organization in all European countries and Canada. Applications from non-traditional IODP and ECORD audiences within the European Community are especially welcome. Apply via e-mail to essac.office@awi.de. Further information at http://www.essac.ecord.org/index.php?mod=education&page=dlp.

The ESF Magellan Workshop Series Program

The ESF Magellan Workshop Series Program, launched in 2006 with the aim of nurturing and coordinating innovative marine scientific drilling proposals for European scientists, is in its last phase of operation. The program will run until 31 July 2011. A decision was made in Burkheim, Germany in August 2010 to propose a successor program, The Magellan Plus Program. Currently, a committee lead by Lucas Lourens, (The Netherlands), Marit-

Solveig Seidenkrantz (Denmark), Ales Spicak (Czech Republic), and representatives from seven other European countries are developing a program proposal which, if funded, will support both marine and continental scientific drilling and coring.

To date the ESF Magellan Workshop Series Program has provided opportunities for senior level researchers and young scientists as well as students to contribute to ocean drilling research goals in Europe. More than eighteen workshops, fifteen short visit grants and one educational activity have been supported. Three workshops were held in 2010: Volcanic Basins: Scientific, economic and environmental aspects in Vienna (AU) by N. Arndt; RAMBO (Real-time Amphibic Monitoring & Borehole Observations) held in Bremen (GE) by A. Kopf; and The GOLD project, drilling in the Western Mediterranean Sea held in Banyuls (FR) by M. Rabineau. The most recent supported workshop, CCS (Carbon Capture & Storage) Oman 2011 convened by M. Godard, was held in the Sultanate of Oman in January 2011.

The final Magellan workshop, Arctic Ocean drilling and the site survey challenge, will be held in early November 2011 in Copenhagen (DK) and is being organized by N. Mikkelsen, Denmark.

Although no more workshops will be supported, there is currently a call open for short visit travel grants to support both young scientists and keynote speakers to attend meetings. This call will remain open until the end of the program. All scientists and students who are interested are encouraged to submit a proposal for a short visit grant. Priority will be given to proponents from ESF Magellan member countries and/or workshops to be held in member countries. ESF Magellan member countries are: Austria, Belgium, Denmark, Finland, France, Germany, Ireland, The Netherlands, Norway, Portugal, Sweden, and Switzerland.

For more information and to apply to the ESF Magellan Workshop Program, please see www.esf.org/

magellan, or contact the ESF program administrator at edegott@esf.org, or the Chair of the Program Jochen Erbacher Jochen.Erbacher@bgr.de.

The First IODP Meeting at AOGS in India

India is an associate member of IODP and has been regularly participating in various IODP expeditions around the world aimed at addressing geo-scientific issues. In order to showcase various IODP-related activities in India, a parallel poster session was organized in conjunction with the IODP-MI at Asia Oceania Geosciences Conference in Hyderabad, India, 5–9 July 2010. The objective of this stall was to promote overall awareness about the benefits of deep-sea scientific drilling and its role in researching various scientific questions.

A dedicated session was also organized to discuss thrust areas of research in the Indian Ocean that could be potentially addressed through ocean drilling. The meeting was chaired by the Secretary, Ministry of Earth Sciences, Government of India and attended by delegates from numerous institutes/organizations across the country, such as Physical Research Laboratory, National Institute of Oceanography, National Geophysical Research Institute, and National Centre for Antarctic and Ocean Research. The meeting was open to all the participants of the AOGS meeting to receive feedback for developing a comprehensive drilling proposal for the northern Indian Ocean. The meeting was highly significant in terms of collecting valuable suggestions related to the IODP and scientific ocean drilling interests in India.

Report of the 4th ECORD Summer School 2010

The ECORD Summer School 2010 on "Dynamics of Past Climate Changes" was held at the MARUM (Center for Marine

Environmental Sciences) Bremen University, Germany, on 13–24 September 2010. It was organized by Prof. Dierk Hebbeln, Director of the Bremen International Graduate School for Marine Sciences "Global Change in the Marine Realm" (GLOMAR), by Prof. Dr. Michael Schulz, Head of the Geosystem Modelling Group at the University of Bremen, and by Dr. Ursula Röhl, IODP Curator at the Bremen Core Repository (BCR). Twenty-eight PhD students and postdoctoral fellows from several European countries and Canada participated in the two-week course which combined lectures, interactive discussions, practical exercises on a "virtual ship" (i.e. in the lab and in the facilities of the IODP core repository), and a field trip to the Late Quaternary Landscapes in the vicinity of B9.25

Successful Port Call of *JR* at Victoria, B.C., Canada

Taking advantage of the *JOIDES Resolution* port call in the Victoria harbor between the Juan de Fuca Hydrogeology Expedition 327 and the Cascadia CORK Expedition 328 on 5–9 September 2010, lectures and guided tours on the ship were organized for the public by Ocean Leadership in collaboration with IODP-Canada and Ocean Networks Canada.

Kiyoshi Suyehiro, President of IODP-MI, Catherine Mével, Chair of the ECORD Managing Agency and Anne de Vernal, Chair of IODP-Canada, participated in the event, which was covered by the local press.

During the port call, about 150 people had the opportunity to get acquainted with IODP by visiting the

JOIDES Resolution, and more than seventy people attended the public lectures.

The lecture by Earl Davis focused on deep-ocean boreholes for long-term observation of crustal temperature and pressure along active seismogenic margins. The other lecture by Michael Riedel addressed the question of gas hydrates in marine sediments as a potential energy resource and cause of geohazards.

Workshop about MELAGUS at Burgos, Spain

icdp ▌

Intramontane basins have the potential for providing unique, continuous sedimentary records of paleoclimate and paleoenvironmental changes. The Guadix-Baza Basin in southern Spain —the largest, southernmost paleolake in Europe—is a particularly important example of such valuable sedimentary archives. Its rich and extensive depositional sequence are key to understanding the Neogene Mediterranean-Atlantic seaway and provides an unprecedented paleoclimatic, paleogeographic, and fossiliferous record of the region throughout the Neogene and Pleistocene. A program of drilling, which is about the Mediterranean-Atlantic seaway and Lacustrine strata Guadix-Baza Basin, Spain (MELAGUS), is seen as the key for obtaining continuous sedimentary records from this intramontane basin, since its deposits are otherwise inaccessible or only partially exposed along degraded outcrops. On 21 October 2010, a group of twenty-five scientists met at the Centro Nacional de Investigación sobre la Evolución Humana (CENIEH), Burgos, Spain, to discuss an initial blueprint for drilling and obtaining sediment cores from the Guadix-Baza Basin. The attendees included researchers from Spain, the United Kingdom and Italy: they specialize in a wide variety of disciplines, including geophysics, paleontology, sedimentology, geochemistry, geochronology, paleopedology, mineralogy, and palynology. At present, there are no precedents of lacustrine drilling programs in any of the major mid-latitude paleolakes of Europe. The proposed drilling project would furnish an unparalleled southernmost reference framework for understanding past environmental changes during the Pliocene and Pleistocene, and their implications for human evolution. Contacts: Josep M. Parés (Josep.pares@cenieh.es) and César Viseras (viseras@ugr.es).

Sub-Seafloor Microbes and Wandering Hotspots Meet in Auckland, New Zealand

Australian and New Zealand IODP Consortium

Perhaps it is more accurate to say that two deep-sea drilling expeditions with *JOIDES Resolution* (*JR*) "crossed over" in Auckland last mid December.

The scientific ocean drilling ship *JR* undertook two expeditions in the southwest Pacific, northeast of New Zealand: one to learn more about the limits to life deep beneath the seafloor (Expedition 329), and the other to test if and how much the Louisville hotspot has moved over the past eighty million years (Expedition 330).

In the intervening time, as *JR* was moored in Auckland, several activities had been organized by the New Zealand IODP Office (GNS Science), the Auckland Museum Institute, and the University of Auckland, with the support of the Integrated Ocean Drilling Program and the Consortium for Ocean Leadership.

These included eight ship tours, a lunch reception, talks by expedition co-chief scientists Steven d'Hondt (329: University of Rhode Island) and Anthony Koppers (330: Oregon State University), and evening public lectures. Thus, many Aucklanders learned about IODP and the importance and relevance of its programs.

Port call activities and the two cross-over expeditions generated interest in the local media, including TV3NZ, RadioNZ, the New Zealand Press Agency, Australia ABC Science and the Australian Science Media Centre.

The *JR* visited Auckland again in mid February, and visitors to the Auckland Museum followed the Louisville expedition through an interactive exhibit.

DFDP, Alpine Fault, New Zealand

icdp ▌

The Deep Fault Drilling Project (DFDP) completed its first two boreholes through the Alpine Fault in early February. DFDP-1A penetrated fault gouge at 91 m and reached a total depth of 101 m in gravel. DFDP-1B penetrated fault gouge at 128 m, reached a total depth of 152 m, and collected the first continuous record of cataclasites on both sides of the fault. Initial results include the discovery of a large fluid pressure difference across the fault. Fluid pressures are hydrostatic above the fault, with a water level at 7 m below ground surface. In contrast, the water level in the sampling tube from beneath the fault is 40 m below ground surface. This pressure difference decreased borehole stability in the highly fractured fault rocks,

but the fault has now been resealed in both boreholes with a bentonite-cement grout. All aspects of the project were successful. High-quality cores and a comprehensive suite of wireline logs were collected from both boreholes. DFDP-1A has a seismometer installed at a depth of 83 m, just below steelcasing. DFDP-1B has a seismometer, 4 piezometers, and 24 temperature sensors installed within it, and a 25-mm fluid sampling tube to a depth of 133 m. An additional seismometer and piezometer will be installed later. Additional information can be found at wiki.gns.cri.nz/DFDP. DFDP-1 drilling was managed by Dr. Rupert Sutherland, GNS Science, Lower Hutt, NZ (r.sutherland@gns.cri.nz). It was funded by Germany (DFG, University of Bremen), New Zealand (Marsden Fund; GNS Science; and Victoria, Otago, Auckland, and Canterbury Universities), and the United Kingdom (NERC, University of Liverpool). Reference: Townend, J., Sutherland, R., and Toy, V., 2009. Deep Fault Drilling Project—Alpine Fault, New Zealand. *Sci. Drill.*, 8:75–82, doi: 10.2204/iodp.sd.8.12. 2009.

Towards an Integrated Biochronology for the Cenozoic

One of the magnificent legacies of ocean drilling is the recovery of abundant marine microfossils. These microfossils provide an excellent evolutionary record that can be readily utilized in biostratigraphy. From the earliest days of the Deep Sea Drilling Project it became clear that marine microfossils in deep ocean basins were the same morphospecies as those recognized in marine sediments studied from outcrop, allowing global recognition of biostratigraphic schemes. Applying an age to an evolutionary or extinction events of marine microfossils relies upon sediments with continuous sedimentation and a clearly defined magnetostratigraphy or cyclostratigraphy—ocean cores do just that. In a recent paper published in *Earth Science Reviews*, Wade et al., (2011) bring together 187 tropical and subtropical planktonic foraminiferal biostratigraphic events for the Cenozoic. Such a compilation has not been attempted since 1995, however, the *JOIDES Resolution* began renewed ocean drilling operations in 2009, following a major refit, which acted as a catalyst to reassess the existing bioevents. Major advances by ODP and IODP in improved drilling recovery, multiple coring and high-resolution sampling, has allowed many biostratigraphic events to be refined. For example, detailed biostratigraphic investigations from Ocean Drilling Program Leg 154 (Ceara Rise; Chaisson and Pearson, 1997; Pearson and Chaisson, 1997; Turco et al., 2002), Leg 199 (Equatorial Pacific; Wade et al., 2007), as well as outcrop sections (Payros et al., 2007, 2009) have resulted in revision of the calibrations of numerous bioevents. The compilation by Wade et al., (2011) includes a series of convenient "look-up" tables against multiple geomagnetic time scales. The revised and recalibrated data provide a major advance in biochronologic resolution and a template for future progress to the Cenozoic time scale. This is one step towards the development of an integrated bio-magneto-astrochronology for the Cenozoic. The new cores drilled during IODP on cruises such as Expedition 320/321 in the equatorial Pacific Ocean (Pälike et al., 2010) will allow further refinements.

References

Chaisson, W. P, and Pearson, P. N. 1997. Planktonic foraminifer biostratigraphy at Site 925: Middle Miocene – Pleistocene. *In* Shackleton, N. J., Curry, W. B., Richter, C., Bralower, T. J. (Eds.), *Proc. ODP, Sci. Results* 154: College Station, TX (Ocean Drilling Program), 3–31.

Pälike, H., Nishi, H., Lyle, M., Raffi, I., Gamage, K., Klaus, A., and the Expedition 320/321 Scientists, 2010. Pacific Equatorial Age Transect. *Proc. IODP*, 320/321: Tokyo (Integrated Ocean Drilling Program Management International, Inc.). doi:10.2204/iodp.proc.320321.2010.

Payros, A., Bernaola, G., Orue-Etxebarria, X., Dinares-Turell, J., Tosquella, J., and Apellaniz, E., 2007. Reassessment of the Early-Middle Eocene biomagnetochronology based on evidence from the Gorrondatxe section (Basque Country, western Pyrenees). *Lethaia* 40:183–195.

Payros, A., Orue-Etxebarria, X., Bernaola, G., Apellaniz, E., Dinarès-Turell, J., Tosquella, J., and Caballero, F., 2009. Characterization and astronomically calibrated age of the first occurrence of Turborotalia frontosa in the Gorrondatxe section, a prospective Lutetian GSSP: implications for the Eocene time scale. *Lethaia* 42:255–264.

Pearson, P. N., and Chaisson, W. P. 1997. Late Paleocene to middle Miocene planktonic foraminifer biostratigraphy of the Ceara Rise. *In* Shackleton, N. J., Curry, W. B., Richter, C., Bralower, T. J. (Eds.), *Proc. ODP, Sci. Results* 154: College Station, TX (Ocean Drilling Program), 33–68.

Turco, E., Bambini, A.M., Foresi, L.M., Iaccarino, S., Lirer, F., Mazzei, R., and Salvatorini, G., 2002. Middle Miocene high-resolution calcareous plankton biostratigraphy at Site 926 (Leg 154, equatorial Atlantic Ocean): paleoecological and paleobiogeographical implications. *Geobios* 35:257–276.

Wade, B.S., Berggren, W.A., and Olsson, R.K., 2007. The biostratigraphy and paleobiology of Oligocene planktonic foraminifera from the equatorial Pacific Ocean (ODP Site 1218). *Mar. Micropaleontology,* 62:167–179.

Wade, B.S., Pearson, P.N., Berggren, W.A., and Pälike, H., 2011. Review and revision of Cenozoic tropical planktonic foraminiferal biostratigraphy and calibration to the geomagnetic polarity and astronomical time scale. *Earth Science Rev.* 104:111–142.

Bridget Wade, School of Earth and Environment, University of Leeds, Woodhouse Lane, Leeds, LS2 9JT, U.K., e-mail: B.Wade@leeds.ac.uk

25

IODP Expedition 319, NanTroSEIZE Stage 2: First IODP Riser Drilling Operations and Observatory Installation Towards Understanding Subduction Zone Seismogenesis

by Lisa McNeill, Demian Saffer, Tim Byrne, Eiichiro Araki, Sean Toczko, Nobu Eguchi, Kyoma Takahashi, and IODP Expedition 319 Scientists

Abstract

The Nankai Trough Seismogenic Zone Experiment (NanTroSEIZE) is a major drilling project designed to investigate fault mechanics and the seismogenic behavior of subduction zone plate boundaries. Expedition 319 is the first riser drilling operation within scientific ocean drilling. Operations included riser drilling at Site C0009 in the forearc basin above the plate boundary fault, non-riser drilling at Site C0010 across the shallow part of the megasplay fault system—which may slip during plate boundary earthquakes—and initial drilling at Site C0011 (incoming oceanic plate) for Expedition 322. At Site C0009, new methods were tested, including analysis of drill mud cuttings and gas, and *in situ* measurements of stress, pore pressure, and permeability. These results, in conjunction with earlier drilling, will provide a) the history of forearc basin development (including links to growth of the megasplay fault system and modern prism), b) the first *in situ* hydrological measurements of the plate boundary hanging wall, and c) integration of *in situ* stress measurements (orientation and magnitude) across the forearc and with depth. A vertical seismic profile (VSP) experiment provides improved constraints on the deeper structure of the subduction zone. At Site C0010, logging-while-drilling measurements indicate significant changes in fault zone and hanging wall properties over short (<5 km) along-strike distances, suggesting different burial and/or uplift history. The first borehole observatory instruments were installed at Site C0010 to monitor pressure and temperature within the megasplay fault zone, and methods of deployment of more complex observatory instruments were tested for future operations.

Introduction

Subduction zones account for 90% of global seismic moment release, generating damaging earthquakes and tsunamis with potentially disastrous effects as exemplified by recent earthquakes in Indonesia and Chile. Understanding the processes that govern the strength and nature of slip along these plate boundary fault systems by direct sampling and measurement of *in situ* conditions is a crucial step toward evaluating earthquake and tsunami hazards. To this end, the Integrated Ocean Drilling Program (IODP) Nankai Trough Seismogenic Zone Experiment (NanTroSEIZE) project (Tobin and Kinoshita, 2006) has been implemented, complementing other fault drilling projects worldwide (e.g., the San Andreas Fault Observatory at Depth (SAFOD) and the Taiwan-Chelungpu Drilling Project). NanTroSEIZE is a multistage program focused on understanding the mechanics of seismogenesis and rupture propagation along subduction plate boundary faults, targeting a transect of the Nankai margin offshore the Kii Peninsula, the location of the 1944 M 8.2 Tonankai earthquake (Fig. 1). The drilling program is a coordinated effort over a period of years to characterize, sample, and instrument the plate boundary system at several locations, culminating in drilling, sampling, and instrumenting the plate boundary fault near the updip limit of inferred coseismic slip, at ~6–7 km below sea-

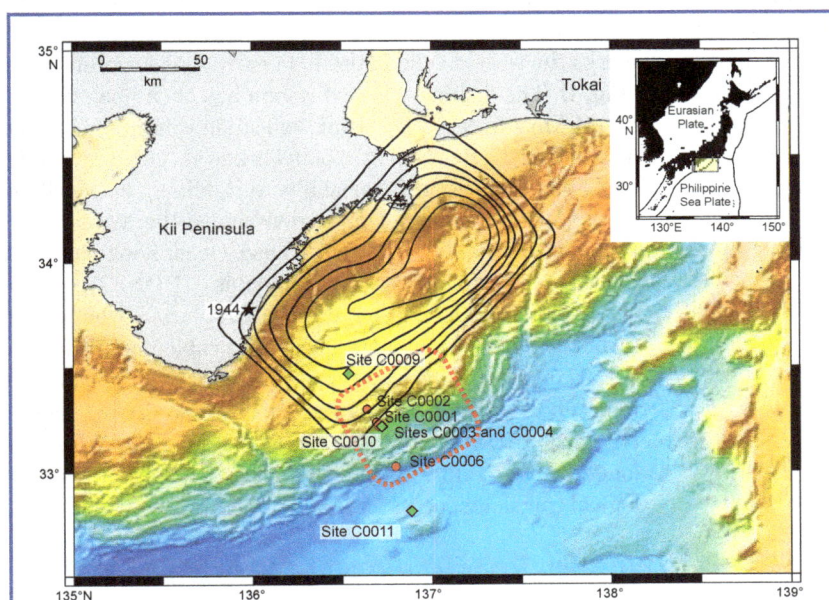

Figure 1. Map of the Nankai margin study area showing drill sites (from Stage 1 and from this expedition, including Site C0011 of Expedition 322; Underwood et al., 2010), details of 1944 plate boundary earthquake slip, and location of coseismic VLFs. See Fig. 6 and Saffer et al. (2009, 2010) for further geographic information. Contours = estimated slip from 1944 event (0.5 m intervals, increasing inwards; Kikuchi et al., 2003), red box = region of recorded VLFs (Obara and Ito, 2005). Inset shows tectonic setting.

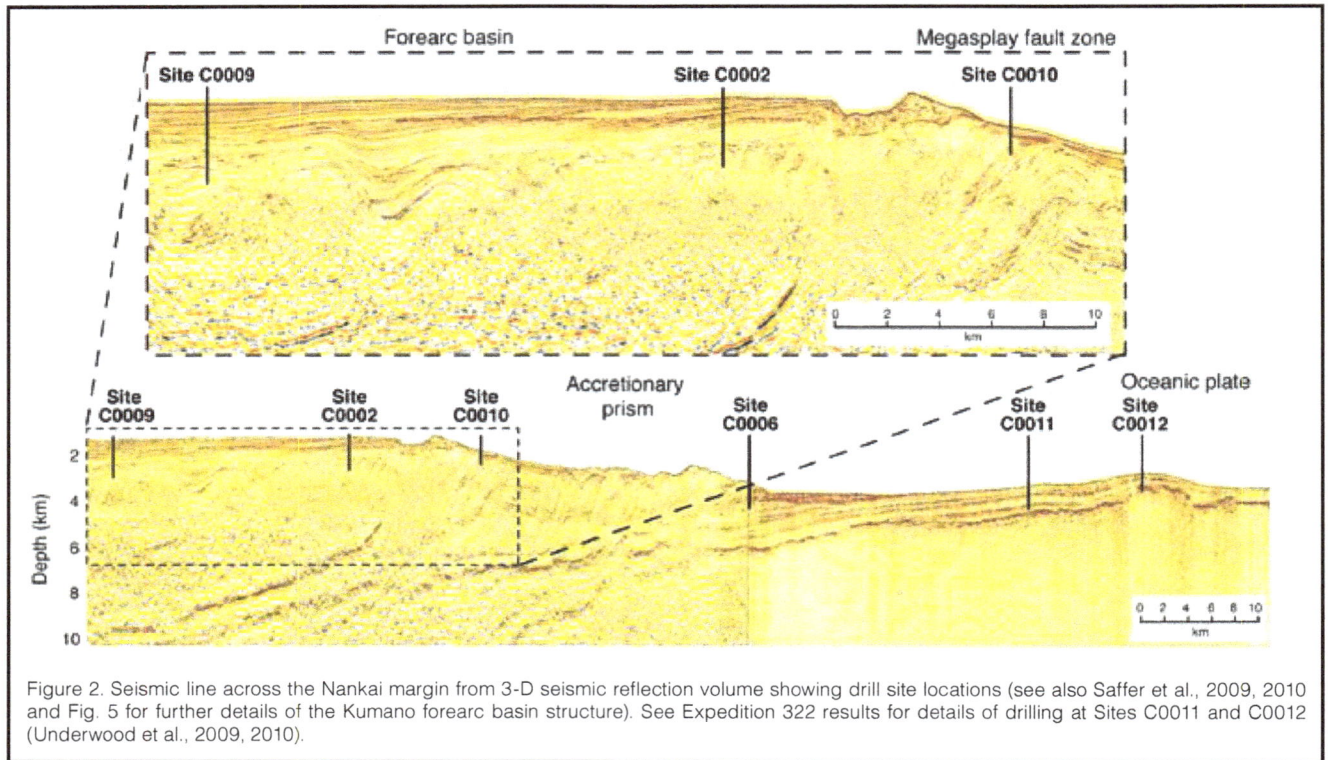

Figure 2. Seismic line across the Nankai margin from 3-D seismic reflection volume showing drill site locations (see also Saffer et al., 2009, 2010 and Fig. 5 for further details of the Kumano forearc basin structure). See Expedition 322 results for details of drilling at Sites C0011 and C0012 (Underwood et al., 2009, 2010).

floor (Tobin and Kinoshita, 2006). The main project objectives are to understand the following:

- *In situ* physical conditions and the state of stress within different parts of the subduction system during an earthquake cycle
- The mechanisms controlling the updip aseismic-seismic transition along the plate boundary fault system
- Processes of earthquake and tsunami generation and strain accumulation and release
- The mechanical strength of the plate boundary fault
- The potential role of a major fault system (termed the "megasplay" fault, Fig. 2) in accommodating earthquake slip and hence influencing tsunami generation

These objectives are being addressed through a combined program of non-riser and riser drilling and integration of linked long-term observatory, geophysical, laboratory, and numerical modeling efforts.

The Nankai Margin

The Nankai subduction zone forms as a result of subduction of the Philippine Sea Plate beneath the Eurasian Plate at ~40–65 mm yr^{-1} (Seno et al., 1993; Miyazaki and Heki, 2001; Fig. 1) and has generated M8 tsunamigenic earthquakes on identifiable segments of the margin regularly (~150–200 years) over at least the past ~1000–1500 years (Ando, 1975). The most recent earthquakes occurred in 1944 (Tonankai M 8.2) and 1946 (Nankaido M 8.3). The margin has been extensively studied using marine geophysics (including two 3-D multichannel seismic reflection volumes),

ocean drilling (across three transects along the margin), passive seismology, and geodesy. The position and effects of earthquake rupture are therefore relatively well understood, and the structure of the margin is well imaged. More recently, very low frequency (VLF) earthquakes and slow slip events have also been recorded, in the shallowest and deeper parts of the subduction system, respectively (Obara and Ito, 2005; Ito et al., 2007). The region offshore the Kii peninsula is the focus of the NanTroSEIZE experiment drilling transect and marks the western end of the 1944 earthquake rupture zone (Fig. 1). The drilling transect region has been imaged with an ~11 km x 55 km 3-D seismic reflection volume (Moore et al., 2007, 2009; Bangs et al., 2009; Park et al., 2010) revealing along-strike structural variability in addition to excellent across-strike imagery. This region is also the focus of the seafloor observatory project "Dense Oceanfloor Network System for Earthquakes and Tsunamis" (DONET; http://www.jamstec.go.jp/jamstec-e/maritec/donet/) which will install seafloor instruments as part of a cabled network connected to shore. This observatory network will also connect to borehole observatory instruments installed as part of the NanTroSEIZE project (Tobin and Kinoshita, 2006) to enable longer term monitoring of seismicity, deformation, hydrological transients, pressure, and temperature.

The NanTroSEIZE Project and Expedition 319

Shallow drilling of a series of holes across the forearc was conducted successfully during Stage 1 of the NanTroSEIZE project in late 2007 through early 2008 onboard the D/V *Chikyu* (Expeditions 314, 315, 316; Ashi et al., 2009; Kimura et al., 2009; Kinoshita et al., 2009a; Fig. 1). These expeditions collected logging-while-drilling (LWD) data (providing *in situ* measurements of physical properties and stress state) and core samples in the shallower aseismic regions of the frontal thrust and the major megasplay fault, and within the forearc basin. Stage 2 consisted of Expedition 319 (the subject of this report) and Expedition 322, conducted from May to October 2009. Expedition 322 drilled, logged, and sampled the input sedimentary section to characterize the properties of sediment that ultimately influence fault development and seismogenic behavior (Underwood et al., 2009, 2010).

Expedition 319 drilled two main sites, C0009 and C0010 (Figs. 1 and 2). Drilling operations were also started at Site C0011, contributing to Expedition 322. Site C0009 is located within the central Kumano forearc basin in the hanging wall of the locked seismogenically active plate boundary (Figs. 1 and 2) and is the site for a future long-term observatory site. Objectives were to drill, sample (cuttings and cores), and log (wireline log data) the hole, conduct downhole *in situ* measurements and a vertical seismic profile (VSP) experiment, and finally case and cement the hole for future observatory installation. The scientific objectives at Site C0009 were as follows: (1) document the lithology, structure, *in situ* properties and development of the central Kumano forearc basin (lying within the hanging wall of the seismogenic plate boundary fault); (2) collect core samples at the

depth of a future potential observatory for geotechnical analysis in advance of observatory instrument installation; and (3) constrain the seismic properties of the deeper subduction zone including the plate boundary below the drill site using a two-ship VSP experiment.

Site C0010 is located ~3.5 km along strike from Site C0004 (Fig. 1, see also Fig. 6 and Saffer et al., 2010) which was drilled during Expeditions 314 and 316 (Kinoshita et al., 2009a; Kimura et al., 2009). Objectives were to drill and log the hole, install an instrument package across the fault zone (designed to collect data for a few years), and prepare for the future installation of an observatory by conducting a "'dummy'" run of an instrument package containing a strainmeter and seismometer. Logging data at this site included measurement-while-drilling (MWD) and LWD (gamma ray, resistivity-at-bit, and resistivity image data); these data were used to constrain the lithology, structural deformation, and *in situ* stress within and across the shallow section of the megasplay fault zone (Figs. 1 and 2). This site provides the opportunity to document and quantify the degree of along-strike structural variability within the megasplay fault system through comparison with Site C0004.

Expedition 319 was noteworthy because it included the first riser drilling operations in scientific ocean drilling history, as well as the first observatory installations conducted

Figure 3. Drilling tools and operations: [A] riser drilling, [B] Modular Formation Dynamics Tester wireline logging tool (single probe at top, dual packer below), [C] observatory dummy run instruments.

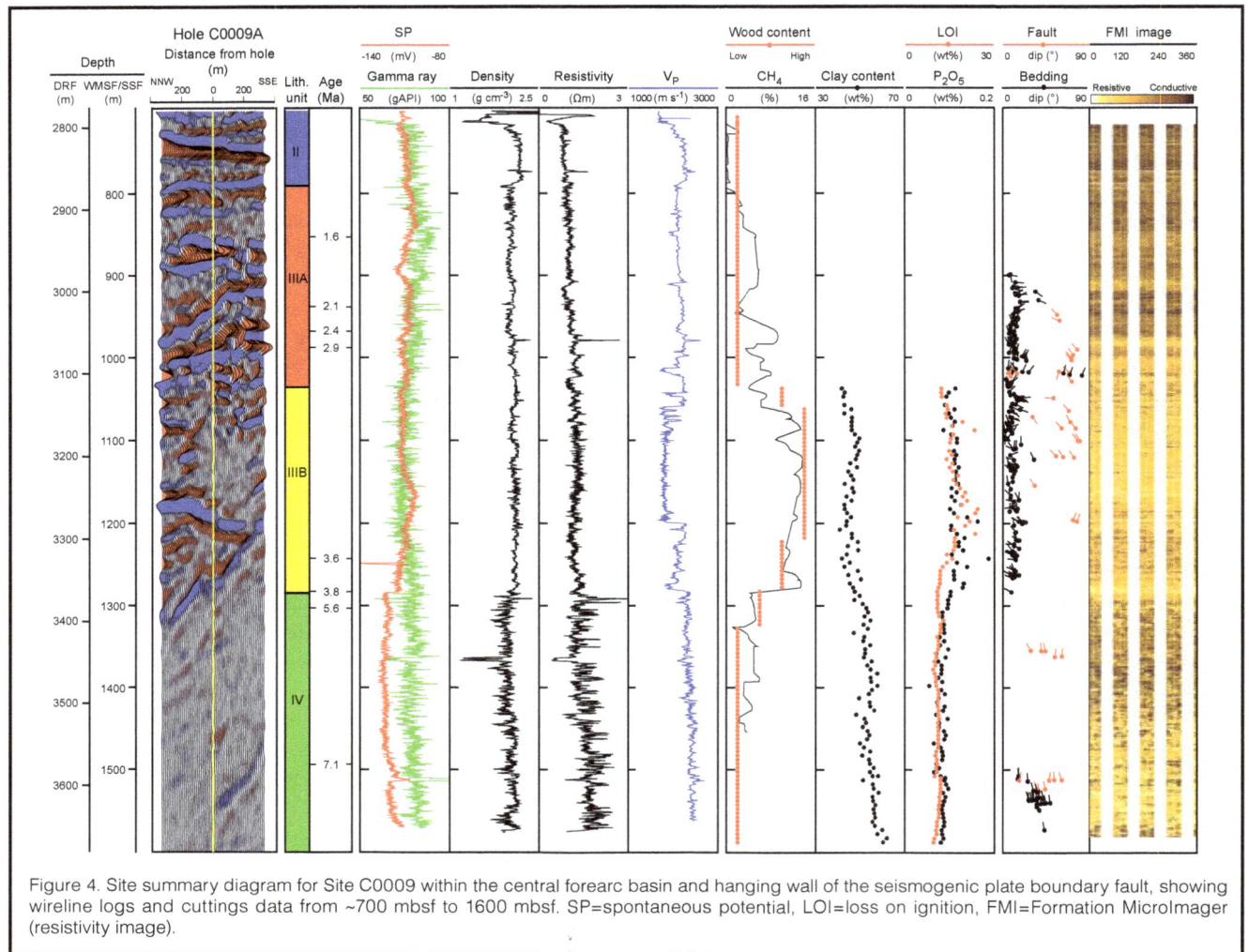

Figure 4. Site summary diagram for Site C0009 within the central forearc basin and hanging wall of the seismogenic plate boundary fault, showing wireline logs and cuttings data from ~700 mbsf to 1600 mbsf. SP=spontaneous potential, LOI=loss on ignition, FMI=Formation MicroImager (resistivity image).

by the D/V *Chikyu* (Saffer et al., 2009). Riser drilling (Fig. 3A) enabled several significant measurements new to IODP, including measurement of *in situ* stress magnitude and pore pressure using the Modular Formation Dynamics Tester (MDT) wireline tool (Fig. 3B), real-time mud gas analysis, and the analysis of drill cuttings. These types of measurements are critical for addressing the scientific objectives of the NanTroSEIZE project and will be utilized in future riser drilling operations targeting the deeper seismogenic portions of the plate boundary and megasplay fault systems. Data collected from the long offset 'walkaway' VSP experiment will allow remote imaging and resolution of properties of the deeper plate boundary fault system. The first observatory instruments of the NanTroSEIZE experiment were installed during Expedition 319, in the form of a simple instrument package (termed a "smart plug") at Site C0010, where it was designed to monitor pressure and temperature at the shallow megasplay fault (Araki et al., 2009; Saffer et al., 2009).

Site C0009

All of the primary planned scientific objectives were achieved through successful drilling operations at this site. Riserless drilling, with gamma ray logging data, and casing to 703.9 mbsf were followed by preparations for riser drilling of the deeper hole. Prior to riser drilling, a Leak-Off Test (LOT) was conducted for operational reasons, and it provided a measurement of minimum *in situ* stress magnitude. Riser drilling continued from 703.9 to a Total Depth (TD) of 1604 mbsf. Throughout riser operations, cuttings and mud gas were collected for analysis. From a depth of 1017 mbsf, cuttings were sufficiently cohesive to allow analyses of detailed lithology, bulk composition, and some structural and physical properties, in addition to lithology and biostratigraphic analysis of incohesive cuttings at shallower depths. Limited coring at the base of the hole (1510–1594 mbsf) allowed calibration of cutting measurements and log data as well as analysis of detailed lithology, structure, physical properties, and geochemistry. Cores will also provide material for shore-based geotechnical testing that will help in the design and engineering of the planned future borehole observatory at this site. Following riser drilling, three wireline runs from ~705 mbsf to TD collected a range of datasets within the riser hole, including density, resistivity image, caliper, and gamma ray logs. Cuttings and core samples and logs together provide information about stratigraphy, age, composition, physical properties, and structure of the forearc basin sediments and underlying material (Fig. 4). The third

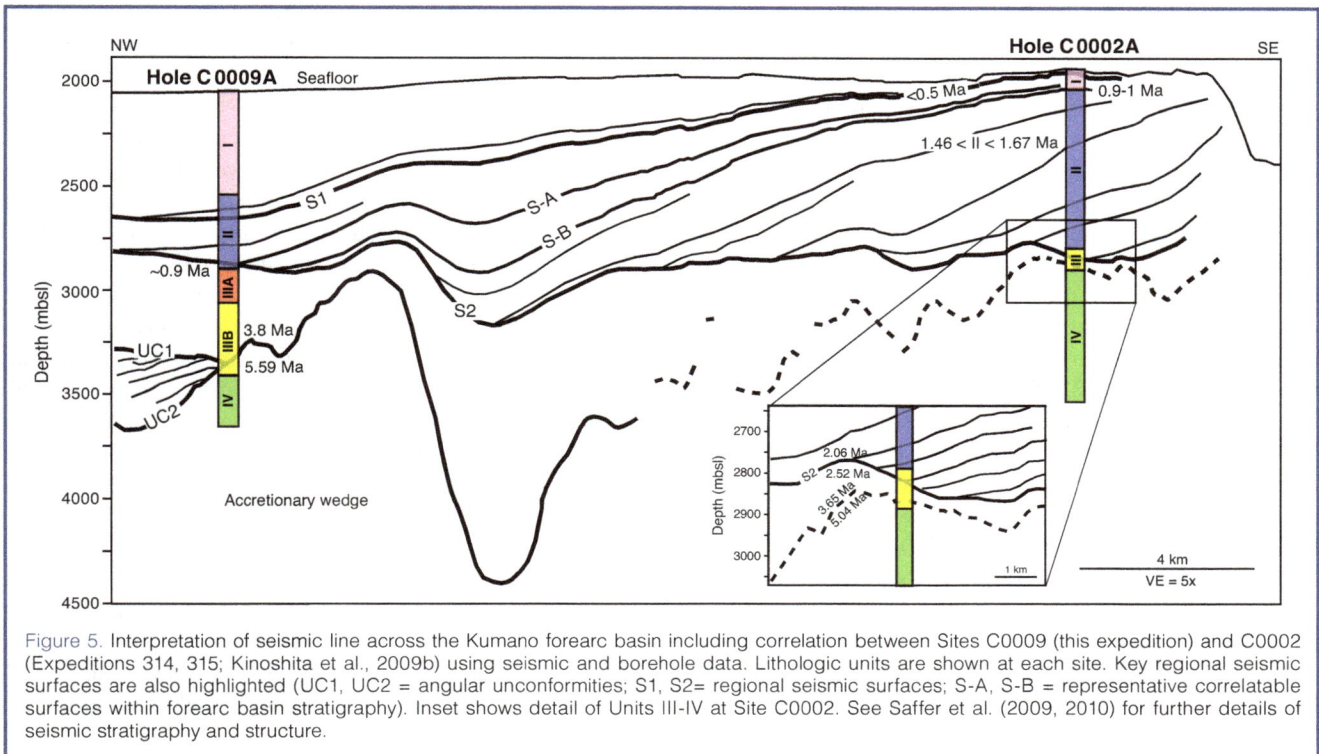

Figure 5. Interpretation of seismic line across the Kumano forearc basin including correlation between Sites C0009 (this expedition) and C0002 (Expeditions 314, 315; Kinoshita et al., 2009b) using seismic and borehole data. Lithologic units are shown at each site. Key regional seismic surfaces are also highlighted (UC1, UC2 = angular unconformities; S1, S2= regional seismic surfaces; S-A, S-B = representative correlatable surfaces within forearc basin stratigraphy). Inset shows detail of Units III-IV at Site C0002. See Saffer et al. (2009, 2010) for further details of seismic stratigraphy and structure.

wireline logging run utilized the MDT tool (Fig. 3B) to measure *in situ* stress magnitude, pore pressure, and permeability at multiple positions downhole between ~700 mbsf and 1600 mbsf. Following riser operations, the hole was cased and cemented in preparation for future observatory installation. The final operation at Site C0009 was a walkaway VSP experiment, conducted using the JAMSTEC vessel R/V *Kairei*, followed by a zero-offset VSP. For the walkaway VSP, the R/V *Kairei* fired airguns along a transect perpendicular to the margin (maximum offset 30 km) and along a circular trackline around the vessel to a wireline array of seismometers within the borehole. Airguns onboard D/V *Chikyu* were then fired and recorded on the borehole seismometer array for the zero offset VSP. These experiments together provide improved definition of velocity (and hence depth) and structure, including anisotropy, around the borehole and of the underlying plate boundary.

Preliminary Results, Site C0009

Combining cuttings, core data, and log data, four stratigraphic units were defined (Fig. 4). Units I and II (0–791 mbsf) are Quaternary forearc basin deposits characterized by mud interbedded with silt and sand, with shallower Unit I being relatively more sand rich. Plio-Pleistocene Unit III (silty mudstone with rare silty sand interbeds) is notable for its high wood/lignite and methane content from cuttings and from analysis of formation gases released and collected in the drill mud. These are particularly concentrated in the lower subunit (IIIB), which is defined by increased organic material (from wood chips, total organic carbon (TOC) and loss on ignition (LOI) concentrations), methane, and glauco-

nite (Fig. 4). The molecular composition of methane suggests a microbial source, consistent with estimated temperature at the bottom of the hole (~50°C) and with an interpretation of gas generated *in situ* from terrestrially sourced organic matter. A major angular unconformity documented in both cuttings and log datasets and representing a hiatus of ~1.8 Ma is crossed at 1285 mbsf and marks the Unit III-IV boundary (Fig. 4). This unconformity can be traced across the Kumano forearc basin (Fig. 5). The underlying unit (Unit IV) is composed of late Miocene silty mudstone with minor silt and vitric tuff turbiditic interbeds. All four units were deposited above the paleo CCD (at ~4000 m depth today). The stratigraphic succession is interpreted as a series of relatively fine-grained forearc basin-filling mudstones and thin turbidites (Units I, II, and/or III) underlain by older forearc basin, slope deposits or accreted prism sediments (Units IV and/or III) (see Saffer et al., 2010 for further discussion of the origin of Unit IV). As part of future research, the drilling results from Site C0009, together with results from Stage 1, can be integrated with 3-D seismic reflection data to better understand the history of forearc basin development and its potential relationship to activity on the megasplay fault.

Structural interpretations of the forearc basin and underlying slope basin-prism units were derived primarily from log and cuttings data and by comparison with cores at the base of the hole. Many types of minor structures, including faults with measurable displacement, are identified in cores. However, deformation is markedly more subdued than within the accretionary prism sediments (Unit IV) at the base of Site C0002, close to the seaward edge of the forearc basin.

Structures can also be identified and categorized in cuttings of ~2 cm or larger diameter. This technique allowed the distribution of vein structures, associated with probable tectonic-induced dewatering, to be determined downhole. Wireline log Formation MicroImager (FMI) resistivity images and caliper data were used to identify the orientation of borehole enlargement, indicating the minimum horizontal stress orientation ("borehole breakouts"). These results allowed the *in situ* stress orientation to be determined (Lin et al., 2010), complementing measurements of *in situ* stress magnitude from other tools (see below) and *in situ* stress measurements at other sites across the margin (McNeill et al., 2004; Ienaga et al., 2006; Byrne et al., 2009). Minimum horizontal stress consistently trends NE-SW downhole (~700–1600 mbsf); therefore, the maximum horizontal stress trends NW-SE (Fig. 6). This is similar to that observed in other boreholes across the accretionary prism and megasplay fault (TDs of ~400-1000 mbsf for NanTroSEIZE Stage 1 boreholes; see Kinoshita et al., 2009b), and is perpendicular to the margin and roughly parallel to the plate convergence direction (Fig. 6). This orientation, however, contrasts with that in the outer forearc basin, at Site C0002, where maximum horizontal stress is NE-SW. This Site C0002 orientation is consistent with multiple lines of evidence for margin-normal extension, potentially driven by uplift and tilting of the seaward forearc basin.

Measurements of the physical properties of sediments and rocks (bulk density, P- and S-wave velocity, resistivity, porosity, magnetic susceptibility, and thermal conductivity) were derived from wireline logs and core and cuttings materials. Cuttings materials are likely to be affected by the drilling process and the time of exposure to drilling mud; therefore, physical properties and some geochemical measurements are likely to be compromised. In particular, porosity is overestimated and bulk density underestimated. Relative bulk compositions are subject to errors due to artifacts in carbonate content associated with the interaction between cuttings and the drilling mud. However, relative downhole trends may be valid. Log-derived P-wave velocity and Poisson's ratio are markedly reduced (Fig. 4) where methane gas concentrations are high (primarily in Unit IIIB), and preliminary calculations suggest a gas saturation of ~10%. Corrected velocities from these sonic logs and from the later VSP experiment (see below) were applied to the 3-D seismic reflection data at the borehole to allow integration of borehole and seismic datasets.

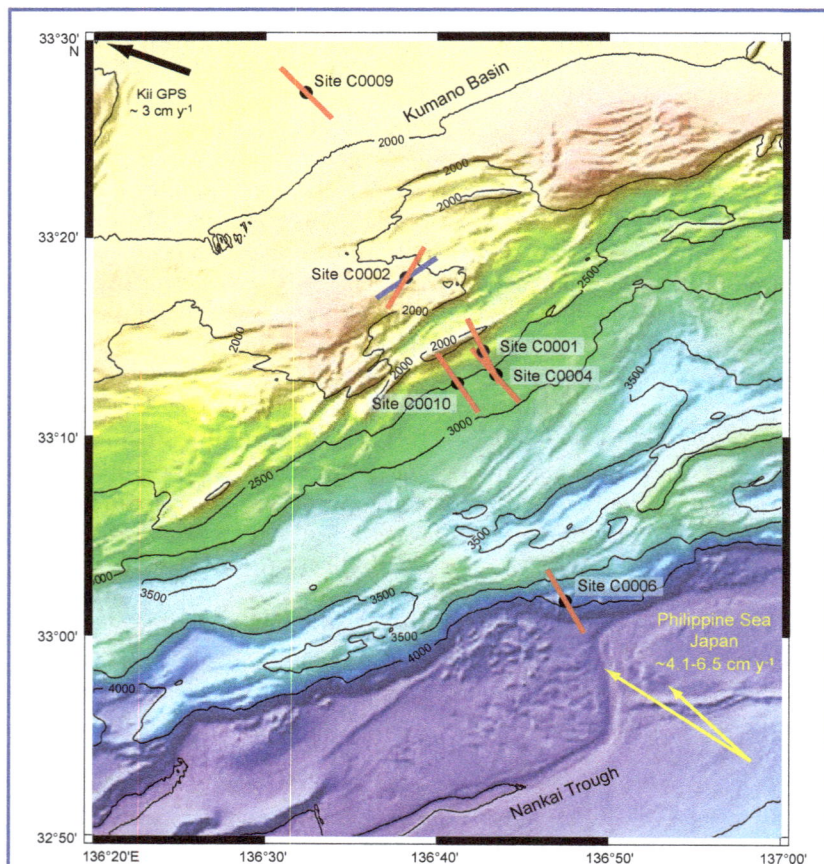

Figure 6. Map showing orientations of maximum horizontal stress inferred from borehole breakouts (see also Kinoshita et al., 2009a). At Site C0002, red line = orientation in forearc basin sediments, blue line = orientation in underlying accretionary prism. Yellow arrows indicate range of suggested convergence rates between the two plates and black arrow indicates GPS-constrained displacement on the Kii peninsula (Seno et al., 1993; Miyazaki and Heki, 2001; Heki, 2007).

A series of new downhole measurements of least principal stress magnitude (σ_3), pore pressure, and permeability were made using the MDT wireline logging tool (Fig. 3B) within the riser drilled section of the forearc basin and underlying sediments (~700–1600 mbsf). This was the first time this tool was used in ocean drilling (its diameter prevents usage in IODP non-riser holes). The tool has two components: a single probe which makes discrete measurements of pore pressure and fluid mobility; and the dual packer which isolates an interval of the borehole (set at 1 m for Expedition 319) to measure pore pressure and fluid mobility during a drawdown test and stress magnitude by hydraulic fracturing. Nine single-probe measurements, one dual packer drawdown test, and two dual packer *in situ* stress magnitude tests were conducted at Site C0009. The pore pressure measurements indicate that formation pore pressure is hydrostatic or very slightly elevated to depths of at least 1460 mbsf. Permeabilities from the single probe range from ~10^{-16} m^2 to 10^{-14} m^2, with variations that are generally consistent with lithology. Permeability from the drawdown test within the clay-rich Unit IV yielded slightly lower permeability of 1.3×10^{-17} m^2. However, the pore pressure and permeability measurements should be viewed with some caution, as the MDT tool is typically used in more permeable formations, and a long pressure recovery time is needed in the low permeability formations drilled here. Hydraulic fracturing tests were conducted at ~870 mbsf within the forearc basin sediments, and at ~1460 mbsf near the bottom of the borehole within older forearc basin, slope deposits, or accreted sediments of the prism. The shallower test is thought to be reliable and can be compared with the leak-off test at a comparable depth (~710 mbsf). Both tests suggest that σ_3 is ~30–35 MPa and horizontal (therefore the minimum horizontal stress).

The vertical seismic profiling experiment was conducted successfully and included a walkaway and zero offset component. For the walkaway experiment, a single 53.4-km line perpendicular to the margin (880 shots) and a circular path of 3.5 km radius around the borehole (275 shots) were shot by the R/V *Kairei*. During the walkaway experiment, direct wave arrivals, refractions from the accretionary prism, and reflections from prism, megasplay fault, and plate boundary interfaces were recorded. These will provide information on seismic velocity (enabling deeper drilling targets to be determined), seismic properties, and structure of the deeper subduction zone. Anisotropy was observed during the circular transect, compatible with the *in situ* stress orientation measurements from logging results (Hino et al., 2009). The zero-offset experiment provided improved seismic velocity

measurements around and immediately below the borehole, thus allowing the results from cores, cuttings, logs, and seismic data to be depth calibrated and integrated with confidence.

Site C0010

Operations at Site C0010 included running a minimal array of MWD/LWD logging tools (gamma ray, resistivity, including resistivity image) across the shallow megasplay fault system to a TD of 555 mbsf (Fig. 7), followed by casing of the hole, an observatory dummy run with a strainmeter and seismometer to test the impact of deployment on the instruments (Fig. 3C), and installation of a simple short-term observatory package ('smart plug') to measure temperature and pressure over a period of a few years (Fig. 8), which is a crucial component of the NanTroSEIZE experiment. The MWD/LWD data allow definition of the major lithologic units and identification of the megasplay fault zone and its properties. Comparison with Site C0004 (Kinoshita et al., 2009b) reveals considerable differences in both hanging wall and fault zone properties over only ~3.5 km along strike.

Preliminary Results, Site C0010

Three distinct lithologic units are defined at Site C0010 (Fig. 7) based on logging data and through comparison with

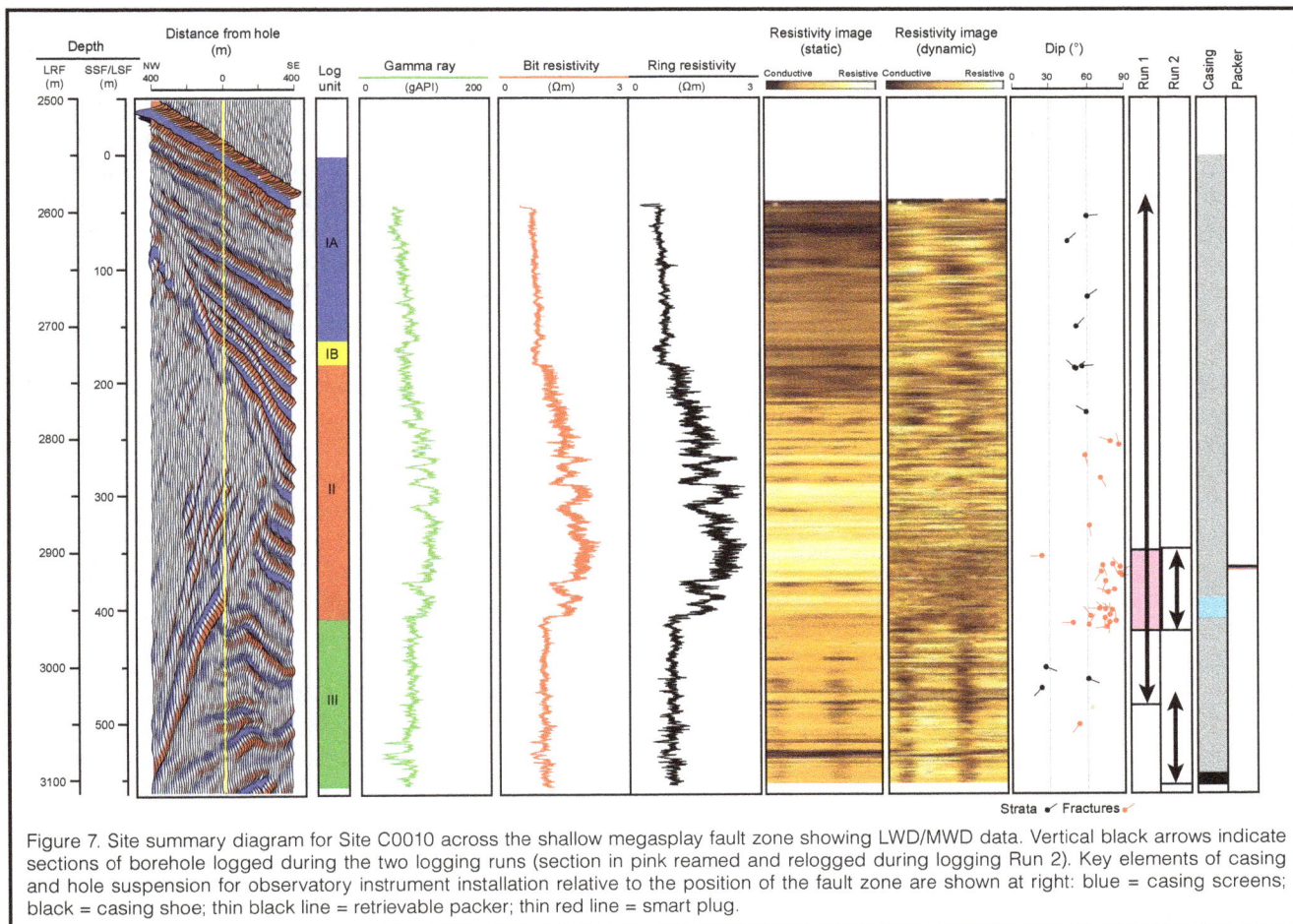

Figure 7. Site summary diagram for Site C0010 across the shallow megasplay fault zone showing LWD/MWD data. Vertical black arrows indicate sections of borehole logged during the two logging runs (section in pink reamed and relogged during logging Run 2). Key elements of casing and hole suspension for observatory instrument installation relative to the position of the fault zone are shown at right: blue = casing screens; black = casing shoe; thin black line = retrievable packer; thin red line = smart plug.

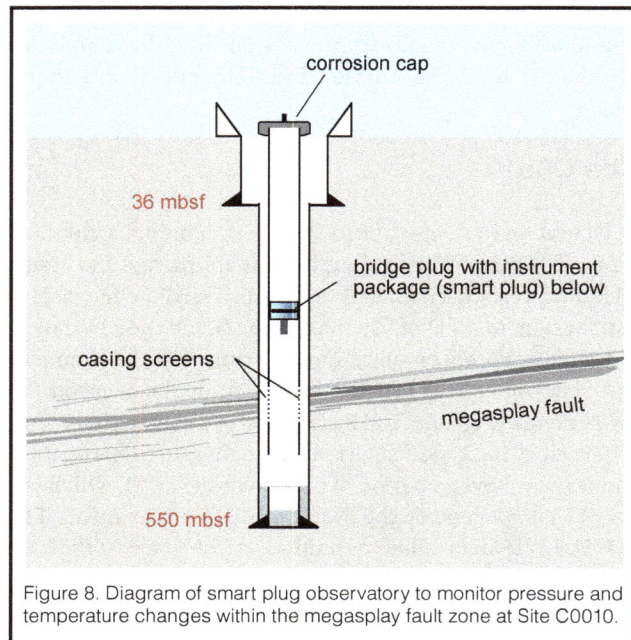

Figure 8. Diagram of smart plug observatory to monitor pressure and temperature changes within the megasplay fault zone at Site C0010.

Site C0004 (Kinoshita et al., 2009b)—slope deposits (Unit I, 0–183 mbsf); thrust wedge/hanging wall of the megasplay fault zone (Unit II, 183–407 mbsf); and overridden slope deposits/footwall of the megasplay fault zone (Unit III, 407 mbsf to TD). At Site C0010, the thrust wedge has lower gamma ray and higher resistivity values than its equivalent at Site C0004. These values suggest higher clay content and potentially increased compaction in the Site C0010 thrust wedge. Porosity estimated from resistivity log values indicates reduced porosity within the C0010 thrust wedge, although low resistivity may in part result from high clay content. Marked reductions in resistivity across the megasplay fault zone correspond to a negative (or inverted) polarity seismic reflector, suggesting reductions in velocity and density into the underthrust/overridden slope deposits of the footwall. At nearby Site C0004, the equivalent reflector is positive polarity, emphasizing that differences in properties, primarily of the hanging wall thrust wedge, can occur over a very short distance within the forearc. These differences may originate from contrasts in original composition or in degree of exhumation along the thrust fault.

Ship heave during logging resulted in variable quality of resistivity image data for structural interpretation; however, analysis of orientations of borehole breakouts revealed an orientation of horizontal maximum stress of NW-SE (Fig. 6). This orientation is similar to that measured at other sites across the prism during NanTroSEIZE Stage 1 and similar to the orientation at Site C0009 in the central forearc basin. An abrupt downhole change in breakout orientation across the megasplay fault zone at Site C0010 is consistent with a sharp mechanical discontinuity at the fault zone (Barton and Zoback, 1994); such a change is not observed at nearby Site C0004, where the megasplay is defined as a broad ~50-m-thick fault-bounded package. Minor faults are concentrated around the thrust zone, as might be expected.

Two sensor dummy runs were conducted using a strainmeter, seismometer, temperature loggers, and an accelerometer-tiltmeter (Fig. 3C) to test the degree of vibration and shock associated with running the instrument package through the water column and reentering the borehole. Unfortunately, during the first run the seismometer and strainmeter became detached and lost due to strong vibrations of the drill pipe in a high current velocity area; however, acceleration, tilt, and temperature data were recorded within the water column. The second run (including a dummy strainmeter with identical dimensions) attempted to test reentry conditions at the wellhead. During both runs, vibrations in the water column resulting from high current velocity were significantly greater than expected, and these results will be critical for modifying future installations of observatory instruments. On a more positive note, a temporary single observatory "smart plug" was successfully installed in the borehole. Screened casing intervals and a retrievable packer will isolate the megasplay fault zone (Figs. 7 and 8) and allow measurements of pressure (referenced to hydrostatic pressure) and temperature to be taken regularly at one-minute intervals over a period of a few years before the instrument package is recovered during future NanTroSEIZE operations.

Key Scientific and Technical Results and Future Work

Data from Expedition 319 and previous NanTroSEIZE drill sites can now be integrated to provide constraints on present-day stress orientation and magnitude across the forearc; they can also be compared with past records of deformation at a range of scales (from core to seismic), incorporating, for the first time, measurements of *in situ* stress magnitude. The emerging picture of stress conditions (Kinoshita et al., 2009a and 2009b; Tobin et al., 2009) is one in which maximum horizontal stress is slightly oblique to the plate convergence across the prism and the inner forearc basin, but deviates from this trend in the outer forearc basin where margin perpendicular extension dominates (Fig. 6). *In situ* stress magnitude at Site C0009 suggests that a normal or strike-slip faulting regime dominates today. Normal faults are observed in reflection data of nearby parts of the basin, but fault orientations from resistivity images are inconsistent with these present-day stress measurements and likely represent an earlier phase of deformation and evolution of stress regimes in the hanging wall of the plate boundary fault.

For the first time, *in situ* hydrological properties of sediments and rocks have been obtained for scientific analysis. These properties (e.g., formation pore pressure and permeability) are critical parameters for understanding the role of fluids in deformation of the forearc and will ultimately be important for determining the role of fluids in fault development and in seismogenic behavior. Properties and behavior can be inferred from core samples and logs, but only direct measurements at depth provide *in situ* properties where these processes are taking place.

Performing riser drilling for the first time in IODP presented a range of operational and analytical challenges. Experiences from this expedition will be valuable for future riser drilling operations within scientific ocean drilling. New methods of analyzing cuttings were developed, and the validity of specific measurements on cuttings was tested. Methods of integrating cores, cuttings data, and log data were also developed to provide the most scientifically realistic interpretation of downhole geological (including lithology, biostratigraphy, structure) and physical properties, particularly important for future deep boreholes where continuous coring will not be feasible. Existing methods for drill mud gas analysis established for continental drilling were also applied successfully to drilling in a marine environment.

Future work will focus on the following: a) results made possible by these new techniques, b) continued post-expedition shore-based laboratory and analytical study of samples, and c) integration of the drilling results from Expedition 319 with existing results across the broader forearc from NanTroSEIZE Stage 1 and with non-drilling datasets, such as 3-D seismic reflection data. Collectively, these will provide the context of regional forearc structure including that of the deeper seismogenic plate boundary.

Acknowledgements

We thank the crew of the D/V *Chikyu* and all drilling operations and related personnel, particularly the efforts of Marine Works Japan laboratory technicians and all Mantle Quest Japan onboard personnel who worked diligently to ensure the success of Expedition 319.

IODP Expedition 319 Scientists

David Boutt, David Buchs, Christophe Buret, Marianne Conin, Deniz Cukur, Mai-Linh Doan, Natalia Efimenko, Peter Flemings, Nicholas Hayman, Keika Horiguchi, Gary Huftile, Takatoshi Ito, Shijun Jiang, Koji Kameo, Yasuyuki Kano, Juniyo Kawabata, Kazuya Kitada, Achim Kopf, Weiren Lin, J. Casey Moore, Anja Schleicher, Roland von Huene, and Thomas Wiersberg

References

Ando, M., 1975. Source mechanisms and tectonic significance of historical earthquakes along the Nankai trough, Japan. *Tectonophysics*, 27:119–140.

Araki, E., Byrne, T., McNeill, L., Saffer, D., Eguchi, N., Takahashi, K., and Toczko, S., 2009. NanTroSEIZE Stage 2: NanTroSEIZE riser/riserless observatory. *IODP Sci. Prosp.*, 319. doi:10.2204/iodp.sp.319.2009.

Ashi, J., Lallemant, S., Masago, H., and the Expedition 315 Scientists, 2009. Expedition 315 summary. *In* Kinoshita, M., Tobin, H., Ashi, J., Kimura, G., Lallemant, S., Screaton, E.J., Curewitz, D., Masago, H., Moe, K.T., and the Expedition 314/315/316 Scientists, *Proc. IODP*, 314/315/316: Washington, DC (Integrated Ocean Drilling Program Management International, Inc.). doi:10.2204/iodp.proc.314315316.121. 2009

Bangs, N.L.B., Moore, G.F., Gulick, S.P.S., Pangborn, E.M., Tobin, H.J., Kuramoto, S., and Taira, A., 2009. Broad, weak regions of the Nankai megathrust and implications for shallow coseismic slip. *Earth Planet. Sci. Lett.*, 284:44–49.

Barton, C.E., and Zoback, M.D., 1994. Stress perturbations associated with active faults penetrated by boreholes: possible evidence for near-complete stress drop and a new technique for stress magnitude measurement. *J. Geophys. Res.*, 99:9379–9390, doi:10.1029/ 93JB03359.

Byrne, T., Lin, W., Tsutsumi, A., Yamamoto, Y., Lewis, J.C., Kanagawa, K., Kitamura, Y., Yamaguchi, A., and Kimura, G., 2009. Anelastic strain recovery reveals extension across SW Japan subduction zone. *Geophys. Res. Lett.*, 36:L23310, doi:10.1029/2009GL040749.

Heki, K., 2007. Secular, transient and seasonal crustal movements in Japan from a dense GPS array: implications for plate dynamics in convergent boundaries. *In* Dixon, T., and Moore, C. (Eds.), *The Seismogenic Zone of Subduction Thrust Faults*: New York (Columbia University Press), 512–539.

Hino, R., Kinoshita, M., Araki, E., Byrne, T.B., McNeill, L.C., Saffer, D.M., Eguchi, N.O., Takahashi, K., and Toczko, S., 2009. Vertical seismic profiling at riser drilling site in the rupture area of the 1944 Tonankai earthquake, Japan. *Eos Trans. AGU*, 90(52), Fall Meet. Suppl., Abstract T12A-04.

Ienaga, M., McNeill, L.C., Mikada, H., Saito, S., Goldberg, D., and Moore, J.C., 2006. Borehole image analysis of the Nankai accretionary wedge, ODP Leg 196: structural and stress studies. Tectonophysics, 426:207–220.

Ito, Y., Obara, K., Shiomi, K., Sekine, S., and Hirose, H., 2007. Slow earthquakes coincident with episodic tremors and slow slip events. *Science*, 315:503–506. doi:10.1126/science. 1134454.

Kikuchi, M., Nakamura, M., and Yoshikawa, K., 2003. Source rupture processes of the 1944 Tonankai earthquake and the 1945 Mikawa earthquake derived from low-gain seismograms. *Earth Planets Space*, 55:159–172.

Kimura, G., Screaton, E.J., Curewitz, D., and the Expedition 316 Scientists, 2009. Expedition 316 summary. *In* Kinoshita, M., Tobin, H., Ashi, J., Kimura, G., Lallemant, S., Screaton, E.J., Curewitz, D., Masago, H., Moe, K.T., and the Expedition 314/315/316 Scientists, *Proc. IODP*, 314/315/316: Washington, DC (Integrated Ocean Drilling Program Management International, Inc.). doi:10.2204/iodp. proc.314315316.131.2009

Kinoshita, M., Tobin, H., Moe, K.T., and the Expedition 314 Scientists, 2009a. Expedition 314 summary. *In* Kinoshita, M., Tobin, H., Ashi, J., Kimura, G., Lallemant, S., Screaton, E.J., Curewitz, D., Masago, H., Moe, K.T., and the Expedition 314/315/316 Scientists, *Proc. IODP*, 314/315/316: Washington, DC (Integrated Ocean Drilling Program Management International, Inc.). doi:10.2204/iodp.proc.

314315316.111.2009

Kinoshita, M., Tobin, H., Ashi, J., Kimura, G., Lallemant, S., Screaton, E.J., Curewitz, D., Masago, H., Moe, K.T., and the Expedition 314/315/316 Scientists, 2009b. NantroSEIZE Stage 1: investigations of seismogenesis, Nankai Trough, Japan, *Proc. IODP*, 314/315/316: Washington, DC (Integrated Ocean Drilling Program Management International, Inc.), doi: 10.2204/iodp.proc.314315316.111.2009.

Lin, W., Doan, M.-L., Moore, J.C., McNeill, L.C.,et al., 2010. Present-day principal horizontal stress orientations in the Kumano forearc basin of the southwest Japan subduction zone determined from IODP NanTroSEIZE drilling Site C0009. *Geophys. Res. Lett.*, 37, doi:10.1029/2010GL043158.

McNeill, L.C., Ienaga, M., Tobin, H., Saito, S., Goldberg, D., Moore, J.C., and Mikada, H., 2004. Deformation and in situ stress in the Nankai accretionary prism from resistivity-at-bit images, ODP Leg 196. *Geophys. Res. Lett.*, 31:L02602, doi:10.1029/2003GL018799.

Miyazaki, S., and Heki, K., 2001. Crustal velocity field of southwest Japan: subduction and arc-arc collision. *J. Geophys. Res.*, 106:4305–4326, doi:10.1029/2000JB900312.

Moore, G.F., Bangs, N.L., Taira, A., Kuramoto, S., Pangborn, E., and Tobin, H.J., 2007. Three-dimensional splay fault geometry and implications for tsunami generation. *Science*, 318:1128–1131. doi:10.1126/science.1147195.

Moore, G.F., Park, J.O., Bangs, N.L., Gulick, S.P., Tobin, H.J., Nakamura, Y., Sato, S., Tsuji, T., Yoro, T., Tanaka, H., Uraki, S., Kido, Y., Sanada, Y., Kuramoto, S., and Taira, A., 2009. *In* Kinoshita, M., Tobin, H., Ashi, J., Kimura, G., Lallement, S., Screaton, E.J., Curewitz, D., Masago, H., Moe, K.T., and the Expedition 314/315/316 Scientists. Structural and seismic sstratigraphic framework of the NanTroSEIZE Stage 1 transect, *Proc. IODP*, 314/315/316, Washington, DC (Integrated Ocean Drilling Program Management International, Inc.). doi:10.2204/ iodp.proc.314315316.102.2009.

Obara, K., and Ito, Y., 2005. Very low frequency earthquakes excited by the 2004 off the Kii peninsula earthquakes: a dynamic deformation process in the large accretionary prism. *Earth Planets Space*, 57:321–326.

Park, J.-O., Fujie, G., Wijerathne, L., Hori, T., Kodaira, S., Fukao, Y., Moore, G.F., Bangs, N.L., Kuramoto, S., and Taira, A., 2010. A low-velocity zone with weak reflectivity along the Nankai subduction zone. *Geology*, 38:283–286.

Saffer, D., McNeill, L., Araki, E., Byrne, T., Eguchi, N., Toczko, S., Takahashi, K., and the Expedition 319 Scientists, 2009. NanTroSEIZE Stage 2: NanTroSEIZE riser/riserless observatory. *IODP Prel. Rept.*, 319. doi:10.2204/iodp.pr.319.2009

Saffer, D., McNeill, L., Araki, E., Byrne, T., Eguchi, N., Toczko, S., Takahashi, K., and the Expedition 319 Scientists, 2010. *Proc. IODP*, 319: Washington, DC (Integrated Ocean Drilling Program Management International, Inc.).

Seno, T., Stein, S., and Gripp, A.E., 1993. A model for the motion of the Philippine Sea Plate consistent with NUVEL-1 and geological data. *J. Geophys. Res.*, 98(B10):17,941–17,948. doi:10.1029/93JB00782.

Tobin, H.J., and Kinoshita, M., 2006. NanTroSEIZE: the IODP Nankai Trough Seismogenic Zone Experiment. *Sci. Drill.*, 2:23–27. doi:10.2204/iodp.sd.2.06.2006.

Tobin, H., Kinoshita, M., Ashi, J., Lallemant, S., Kimura, G., Screaton, E.J., Moe, K.T., Masago, H., Curewitz, D., and the Expedition 314/315/316 Scientists, 2009. NanTroSEIZE Stage 1 expeditions: introduction and synthesis of key results. *In* Kinoshita, M., Tobin, H., Ashi, J., Kimura, G., Lallemant, S., Screaton, E.J., Curewitz, D., Masago, H., Moe, K.T., and the Expedition 314/315/316 Scientists, *Proc. IODP*, 314/315/316: Washington, DC (Integrated Ocean Drilling Program Management International, Inc.). doi:10.2204/iodp.proc.314315316.111.2009

Underwood, M.B., Saito, S., Kubo, Y., and the Expedition 322 Scientists, 2009. NanTroSEIZE Stage 2: subduction inputs. *IODP Prel. Rept.*, 322. doi:10.2204/iodp.pr.322.2009.

Underwood, M.B., Saito, S., Kubo, Y., and the IODP Expedition 322 Scientists, 2010. IODP Expedition 322 drills two sites to document inputs to the Nankai Trough Subduction Zone, *Sci. Drill.* 10:14–25, doi: 10.2204/iodp.sd.10.02.2010.

Authors

Lisa McNeill, School of Ocean and Earth Science, National Oceanography Centre, Southampton, University of Southampton, Southampton, SO14 3ZH, UK, lcmn@noc.soton.ac.uk.

Demian Saffer, The Pennsylvania State University, University Park, PA 16802, U.S.A.

Tim Byrne, Center for Integrative Geosciences, University of Connecticut, Storrs, CT 06269, U.S.A.

Eiichiro Araki, Earthquake and Tsunami Research Project for Disaster Prevention, Japan Agency for Marine-Earth Science and Technology (JAMSTEC), Kanagawa 237-0061, Japan.

Sean Toczko, Nobu Eguchi, and **Kyoma Takahashi**, Center for Deep Earth Exploration (CDEX), Japan Agency for Marine-Earth Science and Technology (JAMSTEC), Kanagawa 237-0061, Japan.

and the IODP Expedition 319 Scientists

Related Web Link

http://www.jamstec.go.jp/jamstec-e/maritec/donet/

Photo Credits

Fig. 3 A. by Lisa McNeill, University of Southhampton
Fig.3 B and C. by CDEX-Jamstec

26

IODP Expedition 322 Drills Two Sites to Document Inputs to The Nankai Trough Subduction Zone

by Michael B. Underwood, Saneatsu Saito, Yu'suke Kubo, and the IODP Expedition 322 Scientists

Abstract

The primary goals during Expedition 322 of the Integrated Ocean Drilling Program were to sample and log the incoming sedimentary strata and uppermost igneous basement of the Shikoku Basin, seaward of the Nankai Trough (southwestern Japan). Characterization of these subduction inputs is one piece of the overall science plan for the Nankai Trough Seismogenic Zone Experiment. Before we can assess how various material properties evolve down the dip of the plate interface, and potentially change the fault's behavior from stable sliding to seismogenic slip, we must determine the initial pre-subduction conditions. Two sites were drilled seaward of the trench to demonstrate how facies character and sedimentation rates responded to bathymetric architecture. Site C0011 is located on the northwest flank of a prominent basement high (Kashinosaki Knoll), and Site C0012 is located near the crest of the seamount. Even though significant gaps remain in the coring record, and attempts to recover wireline logs at Site C0012 failed, correlations can be made between stratigraphic units at the two sites. Sedimentation rates slowed down throughout the condensed section above the basement high, but the seafloor relief was never high enough during the basin's evolution to prevent the accumulation of sandy turbidites near the crest of the seamount. We discovered a new stratigraphic unit, the middle Shikoku Basin facies, which is typified by late Miocene volcaniclastic turbidites. The sediment-basalt contact was recovered intact at Site C0012, giving a minimum basement age of 18.9 Ma. Samples of interstitial water show a familiar freshening trend with depth at Site C0011, but chlorinity values at Site C0012 increase above the values for seawater toward the basement contact. The geochemical trends at Site C0012 are probably a response to hydration reactions in the volcaniclastic sediment and diffusional

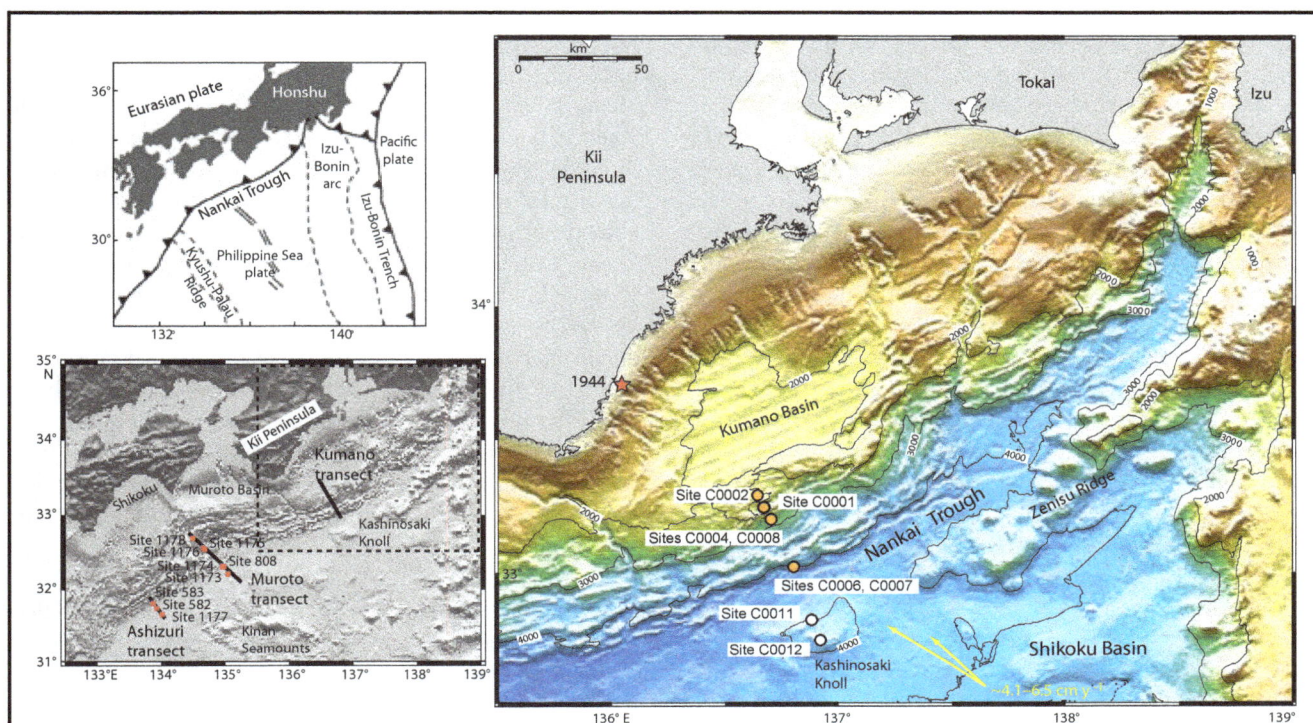

Figure 1. Maps of Nankai Trough and Shikoku Basin showing locations of previous DSDP and ODP drill sites (Ashizuri and Muroto transects), Kumano transect area, Stage 1 NanTroSEIZE coring sites (C0001, C0002, C0004, C0006, C0007, C0008), and Expedition 322 drill sites (C0011 and C0012). Yellow arrows show convergence vector between Philippine Sea plate and Japanese Islands (Eurasian plate). Red star gives epicenter for nucleation of large subduction earthquake in 1944.

exchange with seawater-like fluid in the upper igneous basement. These data are important because they finally establish an authentic geochemical reference site for Nankai Trough, unaffected by dehydration reactions, and they provide evidence for active fluid flow within the upper igneous crust. Having two sets of geochemical profiles also shows a lack of hydrogeological connectivity between the flank and the crest of the Kashinosaki Knoll.

Introduction and Goals

Subduction megathrusts are responsible for some of the world's deadliest earthquakes and tsunamis. To improve our understanding of these hazards, an ambitious project known as the Nankai Trough Seismogenic Zone Experiment (NanTroSEIZE) was initiated along the subduction boundary of southwestern Japan, with the overarching goal of creating a distributed observatory spanning the up-dip limit of seismogenic and tsunamigenic behavior (Tobin and Kinoshita, 2006a, 2006b). Using an array of boreholes across the Kumano transect area (Fig. 1), scientists hope to monitor *in situ* conditions near a major out-of-sequence thrust (megasplay) and the subduction megathrust (plate interface). This multi-stage project is in the process of documenting several key components of the subduction margin, starting with the pre-subduction inputs of sediment and oceanic basement (Underwood et al., 2009), moving landward into the shallow plate interface (Tobin et al., 2009), and finally drilling to depths of 6–7 km where earthquakes occur.

Expedition 322 of the Integrated Ocean Drilling Program (IODP) was organized to sample and log incoming sedimentary strata and igneous basement of the Shikoku Basin, prior to their arrival and burial at the Nankai subduction front (Saito et al., 2009). It is only through such sampling that we can pinpoint how various geologic properties and diagenetic transformations (e.g., clay composition, cementation, microfabric, fluid production, pore pressure, friction, thermal state) change in 3-D space and through time. When viewed in the broader context of NanTroSEIZE, it is particularly important to understand pre-subduction conditions, because the down-dip evolution of those initial properties is what ultimately changes slip behavior along the plate interface from aseismic to seismic (Vrolijk, 1990; Hyndman et al., 1997; Moore and Saffer, 2001). Drilling was therefore conducted at two sites seaward of the trench (Fig. 1). Site C0011 is located on the northwest flank of a prominent bathymetric high (the Kashinosaki Knoll), whereas Site C0012 is located near the crest of the seamount.

Geological Setting and Earlier Work

The Shikoku Basin formed as part of the Philippine Sea plate during the early to middle Miocene by rifting and seafloor spreading along the backarc side of the Izu-Bonin volcanic chain (Okino et al., 1994; Kobayashi et al., 1995). Prominent basement highs within the Shikoku Basin include the Kinan seamount chain (Fig. 1), which grew along the axis of the extinct backarc spreading center, and isolated seamounts such as the Kashinosaki Knoll (Ike et al., 2008a). The subducting plate is currently moving toward the northwest beneath the Eurasian plate at a rate of ~4 to 6 cm yr^{-1} (Seno et al., 1993; Miyazaki and Heki, 2001), roughly orthogonal to the axis of the Nankai Trough. Shikoku Basin deposits, together with the overlying Quaternary trench wedge, are actively accreting at the deformation front, as demonstrated within the Kumano transect area by IODP Expeditions 314, 315, and 316 (Tobin et al., 2009).

As summarized by Underwood (2007), our knowledge of inputs to the Nankai subduction zone is rooted in pioneering drilling discoveries from the Muroto and Ashizuri transects (Deep Sea Drilling Project [DSDP] Legs 31 and 87 and Ocean Drilling Program [ODP] Legs 131, 190, and 196) (Karig et al., 1975; Kagami et al., 1986; Taira et al., 1991; Moore et al., 2001a; Mikada et al., 2002). Those studies demonstrated, among other things, that the plate-boundary fault (décollement) propagates through Miocene strata of the lower Shikoku Basin facies, at least near the toe of the accretionary prism (Taira et al., 1992; Moore et al., 2001b). One of the primary objectives of NanTroSEIZE is to track physical/chemical changes down the plate interface from shallow depths toward seismogenic depths, so the highest-priority sampling targets lie within the lower Shikoku Basin.

Seismic reflection data from across the width of the Shikoku Basin reveal a large amount of heterogeneity in terms of acoustic character and stratigraphic thickness (Ike et al., 2008a, 2008b). Seafloor relief was created during construction of the underlying igneous basement, and that relief strongly influenced the basin's early depositional history (Moore et al., 2001b; Underwood, 2007). As an example of such influence, elevation of the seafloor along the Kinan seamount chain inhibited transport and deposition of sand by gravity flows, so Miocene–Pliocene sediments above the extinct ridge consist almost entirely of hemipelagic mudstone. In contrast, coeval Miocene strata on the flanks of the Kinan basement high consist largely of sand-rich turbidites (Moore et al., 2001b). There are also important differences across the width of the basin (from SW to NE) in values of heat flow, clay mineral assemblages, and the progress of clay-mineral diagenesis (Yamano et al., 2003; Underwood, 2007; Saffer et al., 2008).

Within the Kumano transect area (Fig. 1), seismic reflection data show that the décollement is hosted by lower Shikoku Basin strata to a distance of at least 25–35 km landward of the trench (Moore et al., 2009). Farther landward, the plate-boundary fault steps down section to a position at or near the interface between sedimentary rock and igneous basement (Park et al., 2002). To learn more about the stratigraphic architecture of the Shikoku Basin, Expedition 322 drilled Site C0012 at the crest of Kashinosaki Knoll and Site

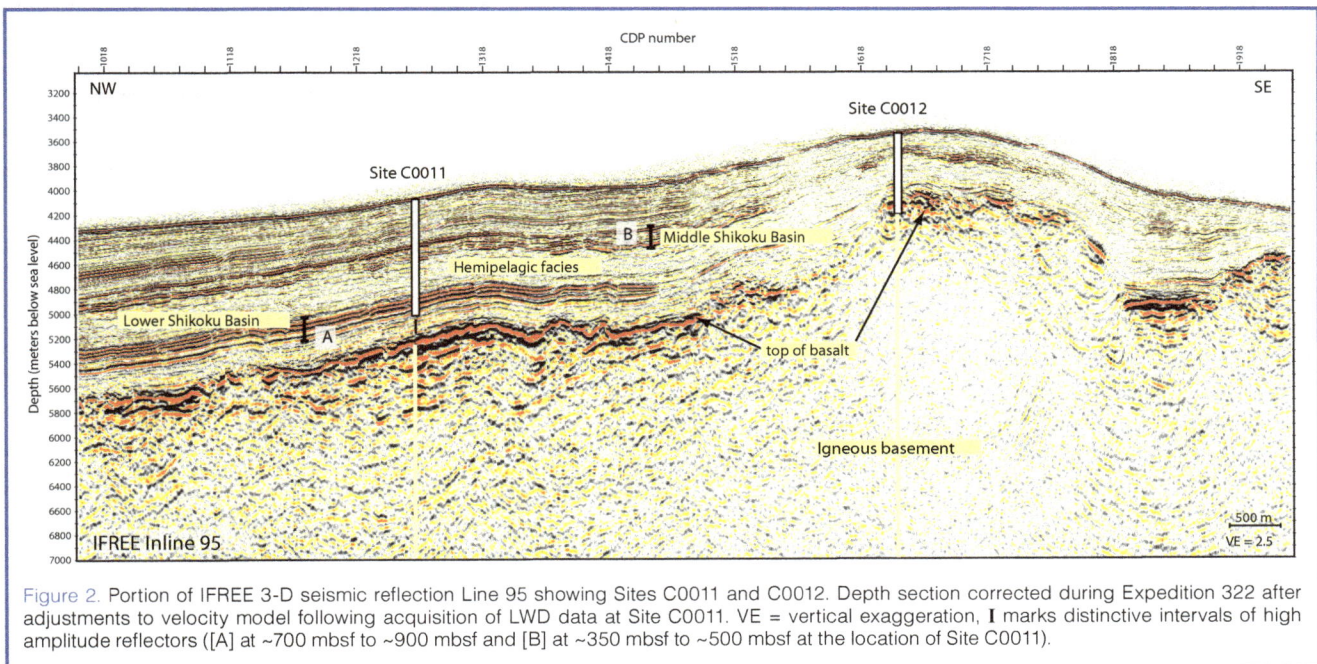

Figure 2. Portion of IFREE 3-D seismic reflection Line 95 showing Sites C0011 and C0012. Depth section corrected during Expedition 322 after adjustments to velocity model following acquisition of LWD data at Site C0011. VE = vertical exaggeration, I marks distinctive intervals of high amplitude reflectors ([A] at ~700 mbsf to ~900 mbsf and [B] at ~350 mbsf to ~500 mbsf at the location of Site C0011).

C0011 along its northwest flank (Fig. 2). In the vicinity of Site C0011, seismic profiles include two particularly distinctive intervals of high-amplitude reflectors: the first from ~350 meters below seafloor (mbsf) to ~500 mbsf and a second with better lateral continuity of reflectors beginning at ~700 mbsf (Fig. 2). Both of these packets of reflectors change thickness but can be traced up and over the crest of Kashinosaki Knoll. These particular depth intervals, plus the sediment-basalt interface, represented the primary coring targets for the expedition (Saito et al., 2009).

Principal Results from Site C0011

Near the end of Expedition 319, measurement-while-drilling and logging-while-drilling (LWD) data were collected at Hole C0011A (Saffer et al., 2009). Generally, we were able to make confident correlations between the logs and subsequent data from rotary core barrel (RCB) cores, with a vertical offset of ~4 m between the coring hole and the logging hole. Five logging units were defined on the basis of visual inspection of the gamma ray and ring resistivity responses (Underwood et al., 2009). These subdivisions also correlate reasonably well with the seismic stratigraphy. For example, the two intervals with high amplitude reflectors (Fig. 2) match up with log responses that are indicative of sand (stone) and volcanic ash/tuff beds. Structural analysis of the borehole resistivity images showed that bedding dips <20° toward the north, which is consistent with the relatively gentle dip observed in the seismic profiles down the landward-facing slope of Kashinosaki Knoll. Analysis of borehole breakouts indicates that the maximum horizontal stress field (SHmax) is orientated north-northeast to south-southwest, roughly perpendicular to the convergence direction of the Philippine Sea plate.

Because of contingency time restrictions (including anticipation of time lost for typhoon evacuations), coring at Hole C0011B began at 340 mbsf rather than the mudline (Fig. 3). Lithologic Unit I was not cored, so its character is inferred from LWD data (Underwood et al., 2009) and by analogy with the upper part of the Shikoku Basin at ODP Sites 808, 1173, 1174, and 1177 (Taira et al., 1992; Moore et al., 2001b). The dominant lithology of this upper Shikoku Basin facies is hemipelagic mud (silty clay to clayey silt) with thin interbeds of volcanic ash (mostly air-fall tephra). Below 340 mbsf, we used biostratigraphic data to merge certain parts of the magnetic polarity interval with the geomagnetic polar-ity reversal time-scale; this resulted in an integrated age-depth model that extends from ~7.6 Ma to ~14.0 Ma (Fig. 4). The composite model yields average rates of sedimentation (uncorrected for either compaction or rapid event deposition by gravity flows) ranging from 4.0 cm k.y.$^{-1}$ to 9.5 cm k.y.$^{-1}$.

Lithologic Unit II is late Miocene (~7.6–9.1 Ma) in age and extends in depth from 340 to 479 mbsf (Fig. 3). We named this unit the middle Shikoku Basin facies; it consists of moder-ately lithified bioturbated silty claystone with inter-beds of tuffaceous sandstone, volcaniclastic sandstone, dark gray clayey siltstone without appreciable bioturbation (mud turbidites), and a chaotic interval of intermixed volcanic-lastic sandstone and bioturbated silty claystone (mass trans-port deposit). Channel-like sand-body geometry is evident in both LWD data and seismic character. The volcanic-rich sandstones contain mixtures of primary eruptive products (e.g., fresh volcanic glass shards) and reworked fragments of pyroclastic and sedimentary deposits. The closest volcanic terrain at the time was probably the Izu-Bonin arc, located along the northeast margin of the Shikoku Basin (Taylor, 1992; Cambray et al., 1995); we interpret Izu-Bonin to be the main sediment source for SW-directed, channelized turbidity currents. No such deposits are known to exist within the western half of the Shikoku Basin.

Lithologic Unit III is middle to late Miocene (~9.1–~12.2 Ma) in age and extends from 479mbsf to 674 mbsf (Fig. 3). Its dominant lithology is bioturbated silty claystone, typical of the hemipelagic deposits throughout the Shikoku Basin. Secondary lithologies include sporadic dark gray silty claystone, lime mudstone, and very thin beds of ochre calcareous claystone. The most unusual aspect of this unit is the deceleration in the rate of hemipelagic sedimentation at ~10.8 Ma (Fig. 4).

Unit IV is middle Miocene (~12.2–14.0 Ma) in age and extends from 674 mbsf to 850 mbsf (Fig. 3). Core recovery within this interval was particularly poor, and our interpretations were further hampered by poor core quality and the decision to wash down without coring from 782mbsf to 844 mbsf. The dominant lithology is bioturbated silty claystone with abundant interbeds of dark gray clayey siltstone (deposited by muddy turbidity currents) and fine-grained siliciclastic sandstone (deposited by sandy turbidity currents). We suggest that the most likely terrigenous sources for this sandy detritus were rock units now exposed across the Outer Zone of southwest Japan, including the Shimanto Belt (Taira et al., 1989; Nakajima, 1997). Superficially similar sand deposits, with overlapping ages, have been documented on the southwest side of the Shikoku Basin at ODP Site 1177 and DSDP Site 297 offshore the Ashizuri Peninsula of Shikoku (Marsaglia et al., 1992; Fergusson, 2003; Underwood and Fergusson, 2005).

The age of Unit V is poorly constrained within the range of middle Miocene (~14.0 Ma). It extends from 850 mbsf to 876 mbsf, but our ability to characterize these strata was prevented by poor core recovery. In fact, the unit's lower boundary coin-cides with destruction of the drill bit, which forced us to abandon the hole. The dominant lithologies are tuffaceous silty claystone and light gray tuff with minor occurrences of tuffaceous sandy siltstone. X-ray diffraction data show an abundance of smectite and zeolites within this unit as alteration products of volcanic glass. These deposits probably correlate with the thick rhyolitic tuffs at ODP Site 808 (Muroto transect of the Nankai Trough), which yielded an age of ~13.6 Ma (Taira et al., 1991).

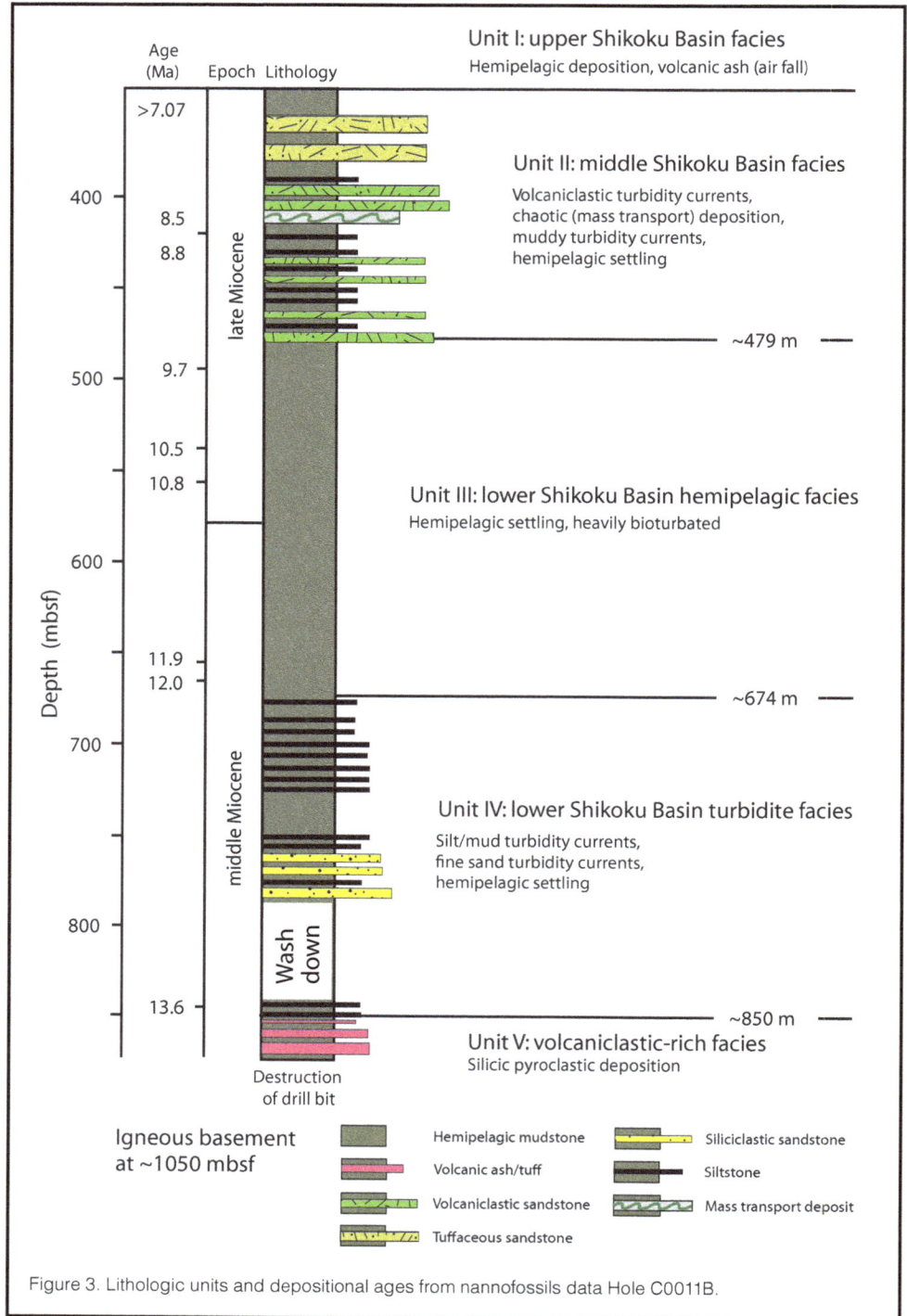

Figure 3. Lithologic units and depositional ages from nannofossils data Hole C0011B.

Structural features at Site C0011 include sub-horizontal to gently dipping bedding planes and small faults. Synsedimentary creep and layer-parallel faults developed in Units II and III, whereas a high-angle normal fault/fracture system is pervasive in Units IV and V. Deformation-fluid interactions were also deduced from mineral-filled veins precipitated along faults. Poles to these structures are distributed along a north-northwest to south-southeast trend, perpendicular to the present trench axis, and the orientations correlate nicely with the LWD-based measurements.

Although the data are adversely affected by widespread damage to cores, physical properties show downhole increases in bulk density and decreases in porosity indicative of sediment consolidation. These trends coincide with increases in P-wave velocity and electrical resistivity. The velocity-porosity relation is consistent with previous observations from Shikoku Basin sediments

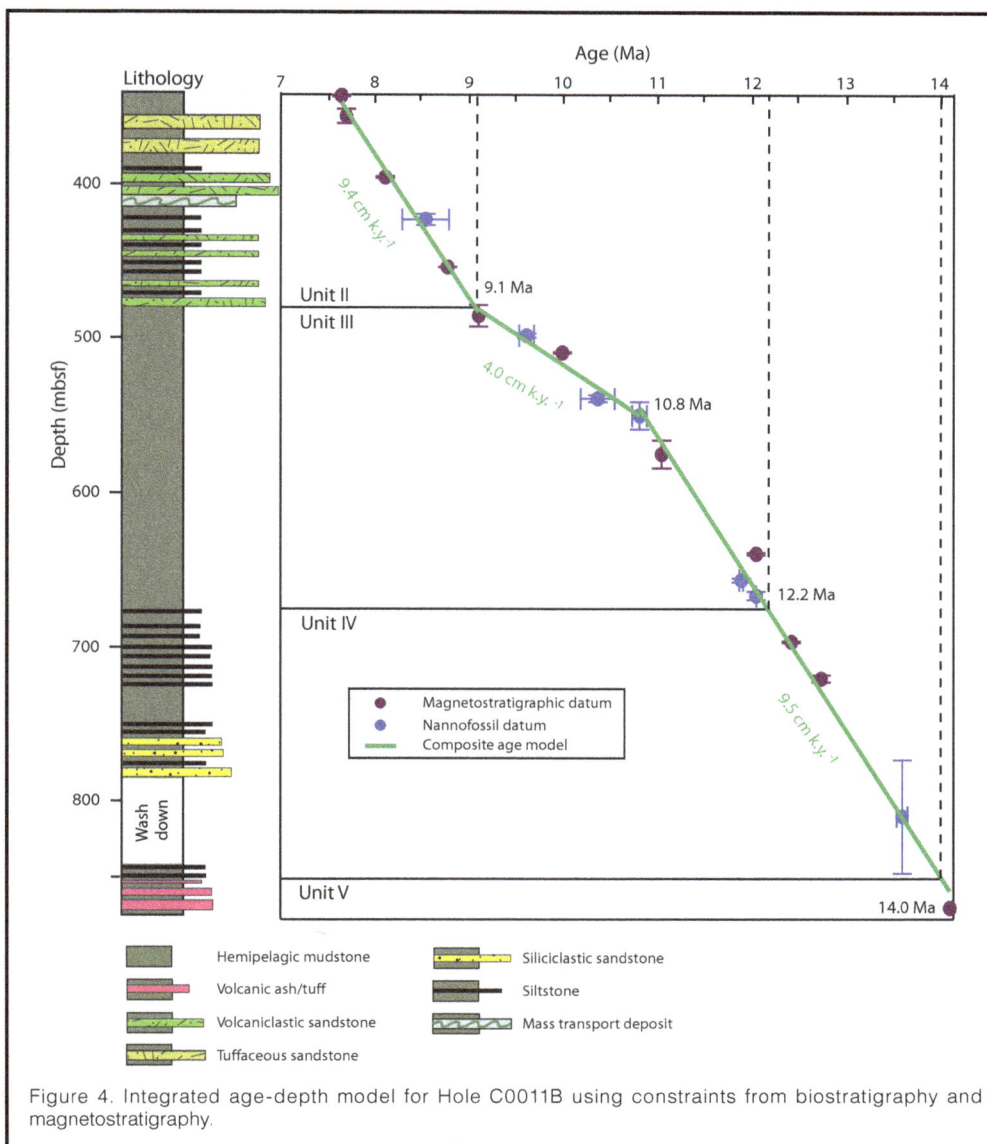

Figure 4. Integrated age-depth model for Hole C0011B using constraints from biostratigraphy and magnetostratigraphy.

(Hoffman and Tobin, 2004). Velocity anisotropy changes from isotropic to anisotropic (i.e., horizontal velocity faster than vertical velocity) near 440 mbsf, which we attribute to compaction-induced alignment of mineral grains. We also detected a shift in magnetic susceptibility near 575 mbsf, which correlates with the change in sedimentation rate at 10.8 Ma (Fig. 4).

Good quality samples were difficult to collect for interstitial water analysis because of persistent core disturbance and low water content. Contamination by seawater (drilling fluid) is evident in all data profiles, and corrections had to be made on the basis of sulfide concentrations. The top of the sampled sediment section (340 mbsf) lies beneath the sulfate-methane interface; thus, we have no information on shallow processes associated with organic carbon diagenesis. Chlorinity in the sampled fluids decreases from ~560 mM to ~510 mM, which is ~9% less than the typical value of 558 mM for seawater (Fig. 5). This freshening trend

is superficially consistent with the pattern observed at ODP Site 1177, Ashizuri transect of the Nankai Trough (Moore et al., 2001a). Judging from the similarity of chlorinity profiles, we tentatively suggest that the sampled fluids were altered at greater depths by clay dehydration reactions. If this interpretation is correct, then fluid migration toward Site C0011—from zones of diagenesis in the frontal accretionary prism and/or beneath the trench wedge—probably occurred along permeable conduits of turbidite sandstone. The lack of borehole temperature measurements at Site C0011 hinders a more definitive interpretation (i.e., in situ dehydration versus deeper seated dehydration). The distributions of major and minor cations document extensive alteration of volcanogenic sediments and oceanic basement, including the formation of zeolites and smectite-group clay minerals. These reactions lead to consumption of silica, potassium, and magnesium and the production of calcium. The very high calcium concentrations (>50 mM) favor authigenic carbonate formation, even at alkalinity of <2 mM.

In spite of the sediment's low content of organic carbon (average 0.31 ± 0.17 wt%), the dissolved hydrocarbon gas concentrations in interstitial water increase with depth. Methane is present as a dis-solved phase in all samples. Ethane was detected in all but one core taken from depths >422 mbsf. Dissolved propane was first observed at 568 mbsf and is present in almost all deeper cores. Butane occurs sporadically below 678 mbsf. The widespread occurrence of ethane results in low C1/C2 ratios ($\sim277 \pm 75$), which are unusual for sediments with organic carbon contents of <0.5 wt%. Without better con-straints on temperature at depth, it is difficult to resolve the potential contributions of heavier hydrocarbons from in situ production versus a deeper/hotter source coupled with up-dip migration along sandy intervals with higher permeability.

Principal Results from Site C0012

Attempts to acquire wireline logs failed at Site C0012, so our results are based on RCB coring. During the RCB jet-in, lithologic Unit I (upper Shikoku Basin facies) was not cored between 0.81 mbsf and 60 mbsf. The unit extends from the seafloor to ~151 mbsf, below which we recovered the first volcaniclastic sandstone of middle Shikoku Basin facies (Fig. 6). The dominant lithology of Unit I is green-gray intensely bioturbated silty clay(stone), and thin layers of volcanic ash are scattered throughout. Modest amounts of

biogenic calcite are compatible with hemipelagic settling on top of Kashinosaki Knoll at a water depth close to (but above) the calcite compensation depth. The integrated age-depth model (Fig. 7) places the lower boundary of Unit I at ~7.8 Ma and shows sedimentation rates decelerating from 4.3 cm k.y.$^{-1}$ to 1.2 cm k.y.$^{-1}$ at ~7.1 Ma.

Lithologic Unit II (middle Shikoku Basin facies) is late Miocene (7.8 Ma to ~9.4 Ma) in age and extends from 151 mbsf to 220 mbsf (Fig. 6). The dominant lithology recovered is green-gray silty claystone, alternating with medium- to thick-bedded tuffaceous/volcaniclastic sandstone and dark gray clayey siltstone. Unit II also contains two chaotic deposits with disaggregated pieces of volcanic-rich sandstone and mudstone that show folding, thinning, and attenuation of primary bedd-ing, probably deformed by gravitational sliding on the north-facing slope of the Kashinosaki Knoll. The volcanic sandstones were not expected at the top of the knoll, and they probably shared a common Izu-Bonin source with Unit II at Site C0011. Their existence seemingly requires upslope transfer by turbidity currents (Muck and Underwood, 1990) and/or post-depositional uplift of the basement high. Seismic data show that the facies thins toward the basement high but drapes over the crest and continues onto the seaward-facing flank (Fig. 2).

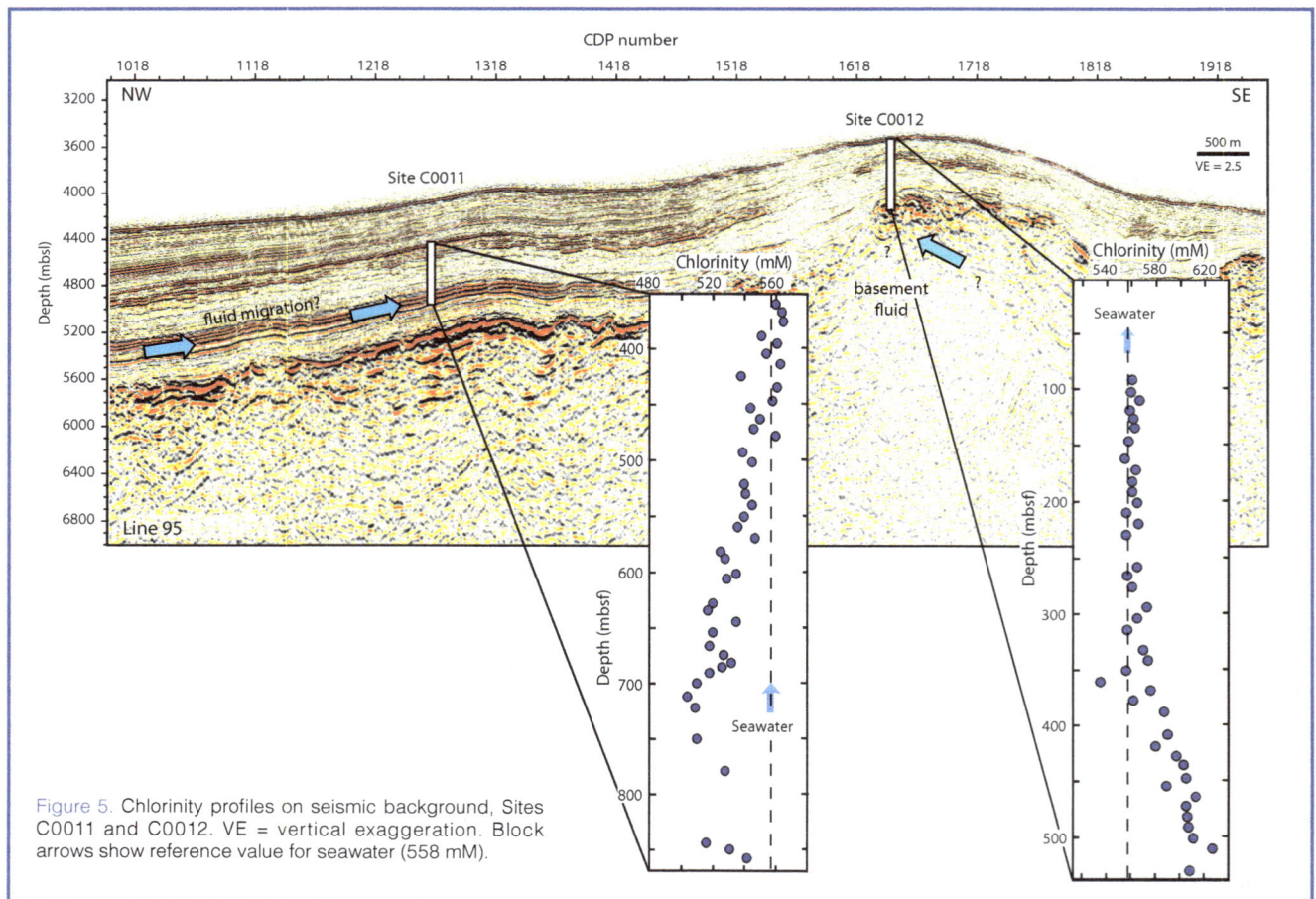

Figure 5. Chlorinity profiles on seismic background, Sites C0011 and C0012. VE = vertical exaggeration. Block arrows show reference value for seawater (558 mM).

Unit III (lower Shikoku Basin hemipelagic facies) is middle Miocene (9.4 Ma to ~12.7 Ma) in age and extends from 220 mbsf to 332 mbsf (Fig. 6). The lithology is dominated by bioturbated silty claystone, with scattered carbonate beds and clay-rich layers (possible bentonites). The upper portion of Unit III contains an interval with steeply inclined (40°–45°) bedding. From seismic data, this zone of disruption appears to be associated with rotational normal faulting. We also see evidence from nannofossils of a hiatus near the top of this deformed interval (Fig. 7). Rates of hemipelagic sedimenta-tion changed at ~12 Ma, although the magnitude of this shift is far more subtle than the effect at Site C0011 (Fig. 4).

Lithologic Unit IV (lower Shikoku Basin turbidite facies) is middle Miocene (12.7–13.5 Ma) in age and extends from 332 mbsf to 416 mbsf (Fig. 6). This interval is characterized by alternations of silty claystone, clayey siltstone, and thin, normally graded siltstone turbi-dites. Prior to drilling, we had expected this turbidite facies to pinch out against the basement high (Saito et al., 2009). The transition down-section into Unit V is not sharp (based on coarser-grained, volcani-clastic sandstone). These volcaniclastic-rich deposits range in age from early to middle Miocene (13.5 Ma to ≥18.9 Ma) and extend in depth from 416 mbsf to 528.5 mbsf. Unit V also includes beds of siltstone, siliciclastic sandstone, and tuff. Some of the sandstones display spectacular cross-laminae, plane-parallel laminae, convolute laminae, and soft-sediment sheath folds. Two detrital sources are evident for the sand grains: a volcanic terrain with fresh volcanic glass together with relatively large amounts of feldspar, and a siliciclastic source enriched in sedimentary lithic grains, quartz, and heavy minerals (including pyroxene, zircon, and amphibole). This compositional heterogeneity is reminiscent of the volcaniclastic-rich facies at ODP Site 1177 (Moore et al., 2001a). Nannofossils also provide evidence for a significant unconformity within Unit V at ~510 mbsf; the associated hiatus spans approximately 4 m.y. (Fig. 7). This unconformity may have been caused by mass wasting during early stages of turbidite sedimentation on Kashinosaki Knoll. Additional support for this interpretation comes from the chaotic and discontinuous seismic reflectors just

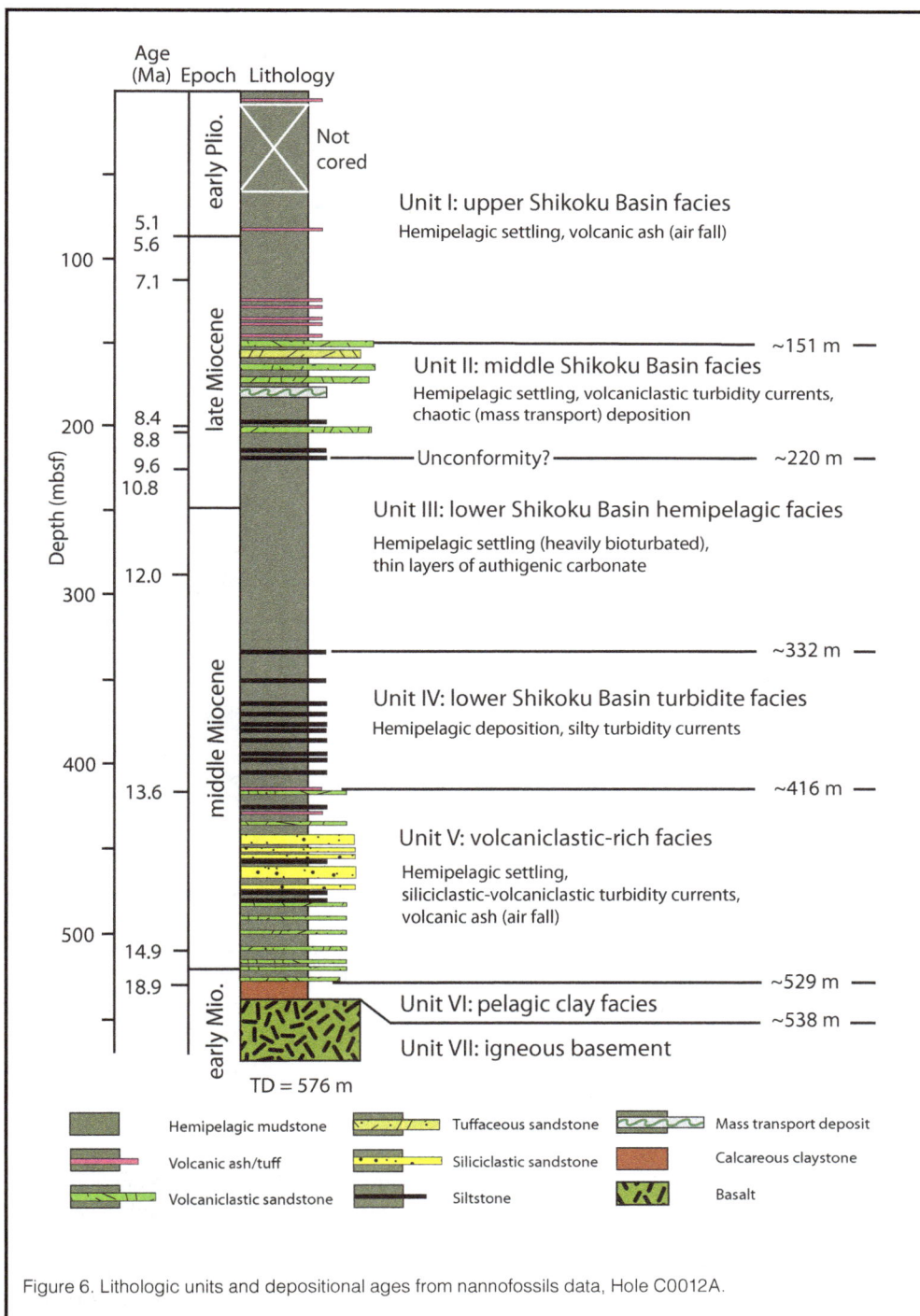

Figure 6. Lithologic units and depositional ages from nannofossils data, Hole C0012A.

below the crest of the basement high on its landward-facing flank (Fig. 2).

We succeeded in recovering the depositional contact between pelagic claystone (Unit VI) and igneous basement (Unit VII) at a depth of 537.8 mbsf (Fig. 6). The age of Unit VI is early Miocene (>18.921 Ma, with a lower limit of 20.393 Ma for the age-constraining nannofossil zone). These thin pelagic deposits are only 9.3 m thick, and they include variegated red, reddish brown, and green calcareous claystone, rich in nannofossils, with minor amounts of radiolarian spines. Carbonate content is ~20 wt%. The cored interval for the underlying basalt extends to 560.74 mbsf. Four types of lava morphology were distinguished: pillow lava, 'massive' basalt, breccia, and mixed rubble pieces caused by drilling disturbance. The basalts are aphanitic to porphyritic, and abundance of phenocrysts is highly variable, from slightly to highly phyric textures. Vesicularity is highly heterogeneous, and alteration of the basalt ranges from moderate to very high. Secondary minerals include saponite, zeolite, celadonite, pyrite, iddingsite, quartz, and calcite.

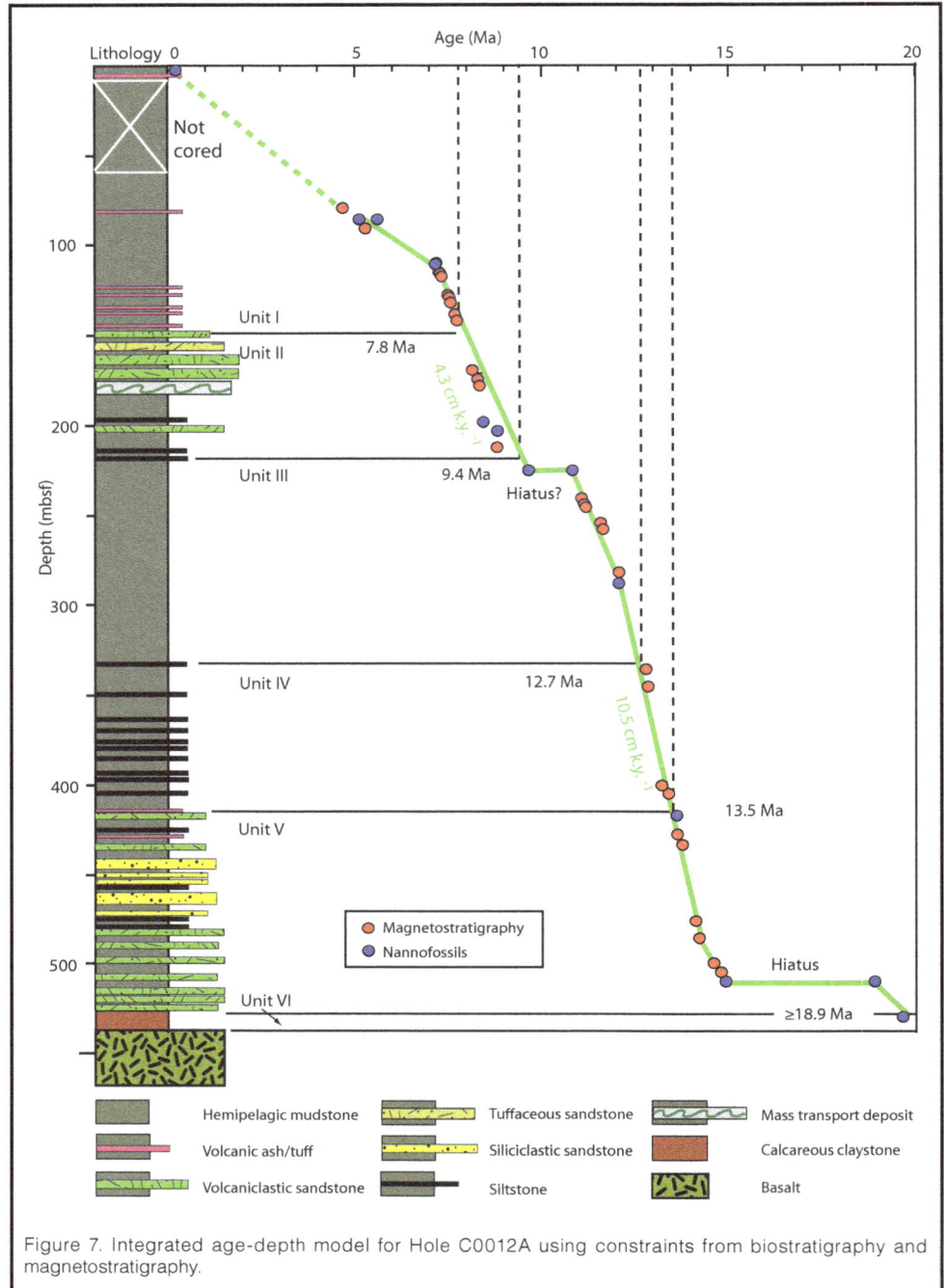

Figure 7. Integrated age-depth model for Hole C0012A using constraints from biostratigraphy and magnetostratigraphy.

Structures in the cores from Site C0012 consist of bedding planes, minor faults, and fractures. Most of the beds dip gently to the north, and seismic data show reflectors rolling over the crest of the knoll to the south-dipping flank just seaward of the drill site (Fig. 2). Steeper bedding dips (>25°) occur only in the upper part of Unit III, within the interval of inferred block sliding.

Physical properties data reveal depth trends similar to those documented at Site C0011, with gradual downhole increases in bulk density, electrical resistivity, thermal conductivity, and P-wave velocity, and a downhole decrease in porosity. Overall, these trends are consistent with normal consolidation, although multiple forms of drilling distur-

bance adversely affected data quality. Mudstone compressibility at Site C0012 is twice that interpreted for Site C0011, and there are several zones of anomalous porosity, including an interval from ~100 mbsf to 136 mbsf where changes in porosity are subdued. Porosity values are also abnormally high near the top of Unit V, these may help facilitate fluid migration at depth.

The results of interstitial water analyses from Site C0012 come as close as we can get to a true geochemical reference site for the Nankai Trough, and the problems encountered with contamination at Site C0011 are far less severe. The profile for dissolved sulfate shows quite a bit of structure, which is consistent with biogeochemical processes. The sulfate reduction zone is significantly deeper (~300 mbsf) than those detected at other sites along the Nankai margin. We suspect that subdued microbial activity in the upper sections of Hole C0012A is due to lower sedimentation rates above the basement high (Fig. 7), as compared to the flanks of the Kashinosaki Knoll where terrigenous organic matter is more voluminous (Fig. 4). Dissolved hydrocarbon gases were not detected in the upper 189 m of sediment, but increases in dissolved methane and ethane concentrations then begin to coincide with sulfate depletion at ~300 mbsf. The hydrocarbons at Site C0012 occur in significantly lower concentrations than at Site C0011, and gases heavier than ethane were not detected. One explanation for the presence of methane and ethane is up-dip migration of gas in solution from deeper thermogenic sources. Another possibility is *in situ* biogenic formation from terrigenous organic matter. Regardless of their origin, sulfate concentrations at Site C0012 are probably modulated by anaerobic methane oxidation (AMO), thereby leading to the production of hydrogen sulfide. In support of this idea, we observed a marked increase in hydrogen sulfide concentration concomitant with the peak in methane concentration, and pyrite is common in the sediments over a comparable depth range.

Unlike Site C0011, values of chlorinity at Site C0012 increase by about 12% relative to seawater. Values within Units I and II are uniform (~560 mM) and begin to increase below ~300 mbsf to the maximum value of 627 mM at 509 mbsf (Fig. 5). We suggest that this steady increase in chlorinity is a response to hydration reactions affecting dispersed volcanic ash and volcanic rock fragments (i.e., volcanic glass to smectite and zeolites) within Units IV and V. The trend of increasing chlorinity is unique with respect to other drilling sites on the Nankai trench floor and in the Shikoku Basin (including Site C0011), all of which show freshening at depth (Taira et al., 1992; Moore et al., 2001b) (Fig. 5). Regardless of the cause of freshening elsewhere, the absence of freshening at Site C0012 is important because it shows a lack of hydrogeologic connectivity between the northwest flank and the crest of the basement high. In addition, temperature-dependent alteration of volcanic ash and volcaniclastic sandstone probably controls changes in silica, potassium, and magnesium concentrations within the middle range of the sedimentary section. We attribute the documented depletion of dissolved sodium to formation of zeolites. Increases in dissolved calcium, which begin in Unit I, are probably overprinted at greater depths by deep-seated reactions in the basal pelagic claystone. The high concentrations of dissolved calcium help explain the precipitation of $CaCO_3$ as thin layers and nodules, even at very low alkalinity (<2 mM). In addition, all of the profiles of major cations and sulfate show reversals toward more seawater-like values within the lower half of lithologic Unit V. We tentatively attribute this shift to the presence of a seawater-like fluid migrating through the upper basaltic crust and diffusive exchange through the turbidites of lithol-

ogic Unit V. The hydrology responsible for this pattern, including potential recharge and discharge zones for fluids within upper igneous basement, remains unidentified.

History of Sedimentation and Fluids Around Kashinosaki Knoll

Figure 8 shows the provisional correlation of units and unit boundaries at Sites C0011 and C0012. Recovery of the basal pelagic deposits in contact with pillow basalt at Site C0012 constitutes a major achievement. From this we know that the age of the basement is older than ~18.9 Ma, but radiometric age dating will be needed to establish the eruptive age. After a brief period of pelagic settling, a long interval of mixed volcaniclastic/siliciclastic sedimentation began. The turbidite section was interrupted near the crest of Kashinosaki Knoll by a hiatus of ~4 m.y., but this is probably a consequence of local mass wasting. Deposition of the sandy to silty turbidites in the lower Shikoku Basin seems to match up with broadly coeval siliciclastic turbidites at Site 1177 (Ashizuri transect of Nankai Trough). The subsequent transition into a long period of hemipelagic sedimentation is reminiscent of a similar lithologic transition at other sites in the Shikoku Basin, but the ages are different: ~7.0 Ma to ~2.5 Ma (Site 1177) versus ~12.7 Ma to ~9.1 Ma (Sites C0011 and C0012). The middle Shikoku Basin facies (Unit II) is unique to the Kumano transect area based on its age (late Miocene) and volcanic sand content. The closest volcanic terrain at that time was probably the Izu-Bonin arc, which we interpret to be the primary source for the volcaniclastic turbidites. Sedimentation decelerated again to a regime of hemipelagic settling and air-falls of volcanic ash (upper Shikoku Basin facies) beginning at ~7.8 Ma and continuing through the Quaternary.

When viewed as a pair of sites, it is clear that the condensed section at Site C0012 displays significant reductions in unit thickness and average sedimentation rate for all parts of the stratigraphic column, relative to the expanded section at Site C0011 (Fig. 8). The basement architecture clearly modulated sedimentation rates throughout the history of the Shikoku Basin, but relief on Kashinosaki Knoll was never high enough to completely prevent the transport and deposition of sandy detritus atop the crest. This comes as something of a surprise, although comparable deposits from thick turbidity currents and/or upslope flow of gravity flows have been documented elsewhere (Muck and Underwood, 1990). This discovery is important because the basement highs of the Shikoku Basin could act as asperities once they reach seismogenic depths along the plate interface (Cloos, 1992; Bilek, 2007). The thickness, texture, and mineral composition of sedimentary strata above such subducting seamounts probably contribute to variations in friction along the fault plane.

The interactions among sedimentation, basement topography, and diagenesis set up an intriguing possibility of multiple fluid regimes within the Shikoku Basin (Fig. 5). We need *in situ* temperature constraints and refined shore-based analyses to interpret the geochemistry with greater confidence, but one regime is modulated by compaction and mineral dehydration reactions. Fluid freshening and generation of heavier hydrocarbons (ethane, propane, butane) may be occurring at greater depths (i.e., below the trench wedge or frontal accretionary prism) in concert with up-dip migration toward the Shikoku Basin through high-permeability sand-rich facies. No such freshening, however, is observed in the condensed section at Site C0012 (Fig. 5). Unlike its flanks, the crest of Kashinosaki Knoll reveals a separate pore water regime driven by *in situ* hydration reactions and diffusional exchange with a higher-chlorinity, more-seawater-like fluid that is migrating through the underlying igneous basement. In essence, that site is a bona fide pre-subduction geochemical reference site for the Nankai subduction zone, with pore fluid chemistry unaffected by diagenesis and/or focused flow closer to the prism toe. The observed increase in sulfate below 490 mbsf is especially noteworthy; it cannot be supplied by the fluids with high concentrations of methane that exist on the deeper landward side of the seamount. Furthermore, we see an increase in hydrogen sulfide produced by AMO in the sediments above 490 mbsf, which argues for a sustained presence of sulfate below 490 mbsf. The sulfate must be replenished by active flow of methane-impoverished fluids within the highly permeable basalt below. This discovery is also very significant because it sets up the possibility of having hydrothermal circulation in the upper igneous crust (Spinelli and Wang, 2008; Fisher and Wheat, 2010) continue to modulate the hydrogeology and fluid chemistry of a subduction margin even after seamounts on the downgoing plate are subducted. This scenario of basement-hosted fluids is particularly intriguing at the seismogenic depths of the Nankai Trough, because the plate interface there is positioned at or near the top of igneous basement (Park et al., 2002).

Plans for Future Drilling

By drilling and coring two sites on the incoming Philippine Sea plate, we captured most of the fundamental compositional, geotechnical, and fluid properties of the Shikoku Basin that are likely to change down dip along the key stratigraphic intervals through which the Nankai plate boundary passes. In the future, IODP Expedition 333 will need to fill in some of the disconcerting gaps in coring and logging that remain from washed-down intervals and operational difficulties, and we must complete some much-needed measurements of borehole temperature to assess thermal history and the extent of in situ diagenesis. In the meantime, Expedition 322 will segue into a broad range of shore-based laboratory projects aimed at evaluating the many

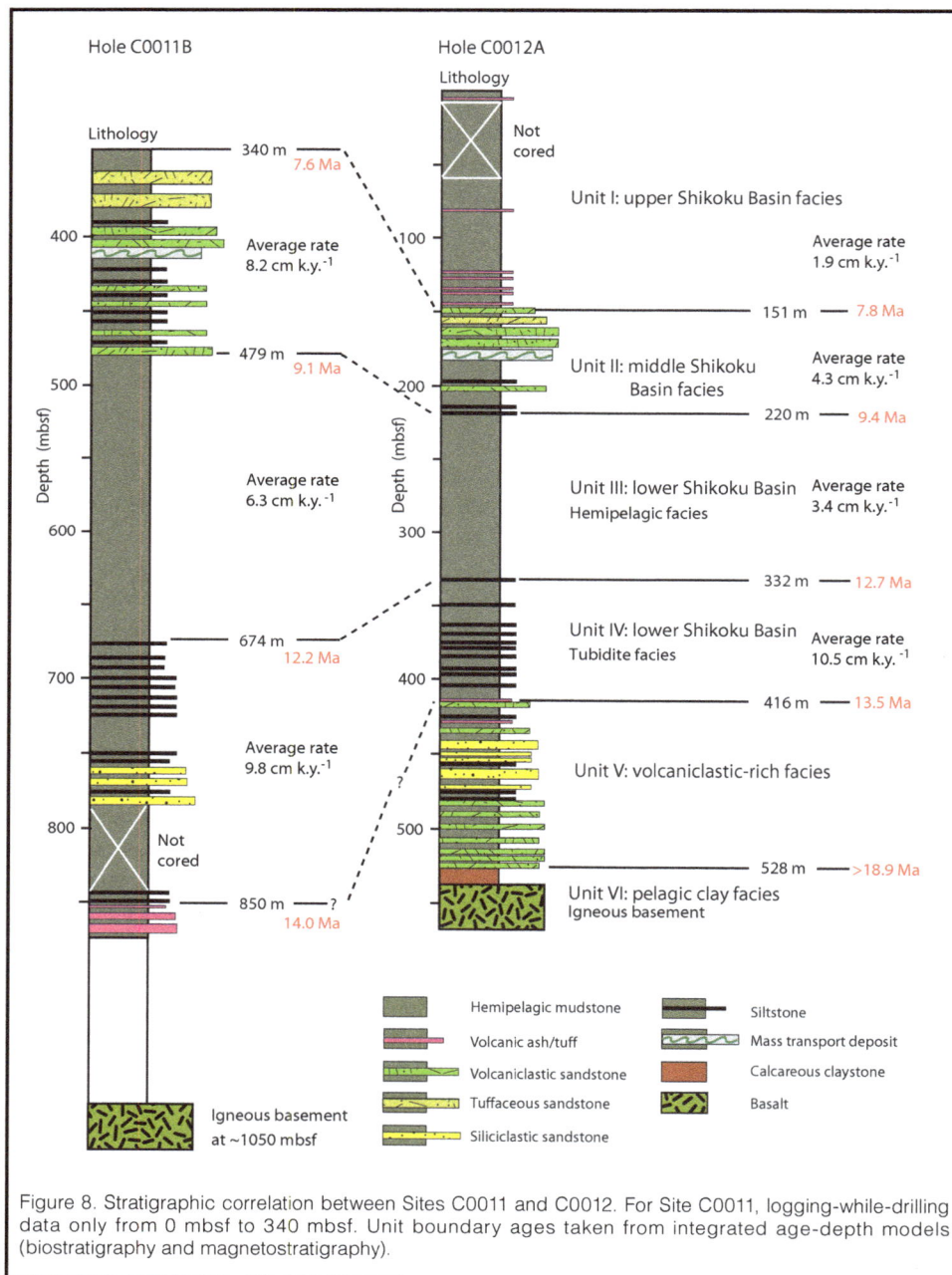

Figure 8. Stratigraphic correlation between Sites C0011 and C0012. For Site C0011, logging-while-drilling data only from 0 mbsf to 340 mbsf. Unit boundary ages taken from integrated age-depth models (biostratigraphy and magnetostratigraphy).

interwoven factors that collectively govern the initial, pre-subduction conditions. By expanding our collective knowledge of the subduction inputs, the NanTroSEIZE project will eventually be able to refine the observational and theoretical context for deep and ultra-deep riser drilling (Tobin and Kinoshita, 2006b), with the ultimate aim of understanding why transitions in fault behavior change from stable sliding to seismogenic slip (Moore and Saffer, 2001).

Acknowledgements

We are indebted to the captains, operations superintendents, offshore installation managers, shipboard personnel, laboratory officers, curators, and laboratory technicians who sailed during IODP Expedition 322 for their dedication and assistance with all aspects of logging, coring, sampling, and shipboard laboratory measurements. We also thank the Project Management Team and specialty coordinators of NanTroSEIZE for their organizational know-how and guidance with some of the scientific interpretations. Michael Strasser and two anonymous reviewers provided helpful comments to improve the manuscript.

The IODP Expedition 322 Scientists

M. Underwood (Co-Chief Scientist), S. Saito (Co-Chief Scientist), Y. Kubo (Expedition Project Manager), Y. Sanada (Logging Staff Scientist), S. Chiyonobu, C. Destrigneville, B. Dugan, P. Govil, Y. Hamada, V. Heuer, A. Hupers, M. Ikari, Y. Kitamura, S. Kutterolf, S. Labanieh, J. Moreau, H. Naruse, H. Oda, J-O. Park, K. Pickering, R. Scudder, A. Slagle, G. Spinelli, M. Torres, J. Tudge, H. Wu, T. Yamamoto, Y. Yamamoto, and X. Zhao.

References

Bilek, S., 2007. Influence of subducting topography on earthquake rupture. *In* Dixon, T., and Moore, J.C. (Eds.), *The Seismogenic Zone of Subduction Thrust Faults*: New York (Columbia Univ. Press), 123–146.

Cambray, H., Pubellier, M., Jolivet, L., and Pouclet, A., 1995. Volcanic activity recorded in deep-sea sediments and the geodynamic evolution of western Pacific island arcs. *In* Taylor, B., and Natland, J. (Eds.), *Active Margins and Marginal Basins of the Western Pacific. AGU Geophys. Monogr.*, 88:97–124.

Cloos, M., 1992. Thrust-type subduction-zone earthquakes and seamount asperities: a physical model for seismic rupture. *Geology*, 20:601–604.

Fergusson, C.L., 2003. Provenance of Miocene-Pleistocene turbidite sands and sandstones, Nankai Trough, Ocean Drilling Program Leg 190. *In* Mikada, H., Moore, G.F., Taira, A., Becker, K., Moore, J.C., and Klaus, A. (Eds.), *Proc. ODP, Sci. Results*, 190/196: College Station, TX (Ocean Drilling Program).

Fisher, A.T., and Wheat, C.G., 2010. Seamounts as conduits for massive fluid, heat, and solute fluxes on ridge flanks. *Oceanography*, 23(1):74–87.

Hoffman, N.W., and Tobin, H.J., 2004. An empirical relationship between velocity and porosity for underthrust sediments in the Nankai Trough accretionary prism. *In* Mikada, H., Moore, G.F., Taira, A., Becker, K., Moore, J.C., and Klaus, A. (Eds.), *Proc. ODP, Sci. Results*, 190/196: College Station, TX (Ocean Drilling Program).

Hyndman, R.D., Yamano, M., and Oleskevich, D.A., 1997. The seismogenic zone of subduction thrust faults. *Island. Arc*, 6:244–260.

Ike, T., Moore, G.F., Kuramoto, S., Park, J-O., Kaneda, Y., and Taira, A., 2008a. Tectonics and sedimentation around Kashinosaki Knoll: a subducting basement high in the eastern Nankai Trough. *Island Arc*, 17(3):358–375. doi:10.1111/j.1440-1738.2008.00625.x.

Ike, T., Moore, G.F., Kuramoto, S., Park, J.-O., Kaneda, Y., and Taira, A., 2008b. Variations in sediment thickness and type along the northern Philippine Sea plate at the Nankai Trough. *Island Arc*, 17(3):342–357. doi:10.1111/j.1440-1738.2008.00624.x.

Kagami, H., Karig, D.E., Coulbourn, W.T., et al., 1986. *Init. Repts. DSDP*, 87: Washington, DC (U.S. Govt. Printing Office). doi:10.2973/dsdp.proc.87.1986.

Karig, D.E., Ingle, J.C., Jr., et al., 1975. *Init. Repts. DSDP*, 31: Washington, DC (U.S. Govt. Printing Office). doi:10.2973/dsdp.proc.31.1975.

Kobayashi, K., Kasuga, S., and Okino, K., 1995. Shikoku Basin and its margins. *In* Taylor, B. (Ed.), *Backarc Basins: Tectonics and Magmatism*: New York (Plenum), 381–405.

Marsaglia, K.M., Ingersoll, R.V., and Packer, B.M., 1992. Tectonic evolution of the Japanese Islands as reflected in modal compositions of Cenozoic forearc and backarc sand and sandstone. *Tectonics*, 11(5):1028–1044. doi:10.1029/91TC03183.

Mikada, H., Becker, K., Moore, J.C., Klaus, A., et al., 2002. *Proc. ODP, Init. Repts.*, 196: College Station, TX (Ocean Drilling Program). doi:10.2973/odp.proc.ir.196.2002.

Miyazaki, S., and Heki, K., 2001. Crustal velocity field of southwest Japan: subduction and arc-arc collision. *J. Geophys. Res.*, 106(B3):4305–4326. doi:10.1029/2000JB900312.

Moore, G.F., Park, J.-O., Bangs, N.L., Gulick, S.P., Tobin, H.J., Nakamura, Y., Sato, S., Tsuji, T., Yoro, T., Tanaka, H., Uraki, S., Kido, Y., Sanada, Y., Kuramoto, S., and Taira, A., 2009. Structural and seismic stratigraphic framework of the NanTroSEIZE Stage 1 transect. *In* Kinoshita, M., Tobin, H., Ashi, J., Kimura, G., Lallemant, S., Screaton, E.J., Curewitz, D., Masago, H., Moe, K.T., and the Expedition 314/315/316 Scientists, *Proc. IODP*, 314/315/316: Washington, DC (Integrated Ocean Drilling Program Management International, Inc.). doi 10.2204/iodp.proc.314315316.102.2009.

Moore, G.F., Taira, A., Klaus, A., et al., 2001a. *Proc. ODP, Init. Repts.*, 190: College Station, TX (Ocean Drilling Program). doi:10.2973/odp.proc.ir.190.2001.

Moore, G.F., Taira, A., Klaus, A., Becker, L., Boeckel, B., Cragg, B.A., Dean, A., Fergusson, C.L., Henry, P., Hirano, S., Hisamitsu, T., Hunze, S., Kastner, M., Maltman, A.J., Morgan, J.K., Murakami, Y., Saffer, D.M., Sánchez-Gómez, M., Screaton, E.J., Smith, D.C., Spivack, A.J., Steurer, J., Tobin, H.J., Ujiie,

K., Underwood, M.B., and Wilson, M., 2001b. New insights into deformation and fluid flow processes in the Nankai Trough accretionary prism: results of Ocean Drilling Program Leg 190. *Geochem., Geophys., Geosyst.*, 2(10):1058. doi:10.1029/2001GC000166.

Moore, J.C., and Saffer, D., 2001. Updip limit of the seismogenic zone beneath the accretionary prism of southwest Japan: an effect of diagenetic to low-grade metamorphic processes and increasing effective stress. *Geology*, 29(2):183–186. doi:10.1130/0091-7613(2001)029<0183:ULOTSZ>2.0.CO;2.

Muck, M.T., and Underwood, M.B., 1990. Upslope flow of turbidity currents: a comparison among field observations, theory, and laboratory models. *Geology*, 18:54–57.

Nakajima, T., 1997. Regional metamorphic belts of the Japanese Islands. *Island Arc*, 6:69–90.

Okino, K., Shimakawa, Y., and Nagaoka, S., 1994. Evolution of the Shikoku Basin. *J. Geomagn. Geoelectr.*, 46:463–479.

Park, J.-O., Tsuru, T., Kodaira, S., Cummins, P.R., and Kaneda, Y., 2002. Splay fault branching along the Nankai subduction zone. *Science*, 297(5584):1157–1160. doi:10.1126/science.1074111.

Saffer, D., McNeill, L., Araki, E., Byrne, T., Eguchi, N., Toczko, S., Takahashi, K., and the Expedition 319 Scientists, 2009. NanTroSEIZE Stage 2: NanTroSEIZE riser/riserless observatory. *IODP Prel. Rept.*, 319. doi:10.2204/iodp.pr.319.2009.

Saffer, D.M., Underwood, M.B., and McKiernan, A.W., 2008. Evaluation of factors controlling smectite transformation and fluid production in subduction zones: application to the Nankai Trough. *Island Arc*, 17(2):208–230. doi:10.1111/j.1440-1738.2008.00614.x.

Saito, S., Underwood, M.B., and Kubo, Y., 2009. NanTroSEIZE Stage 2: subduction inputs. *IODP Sci. Prosp.*, 322. doi:10.2204/iodp.sp.322.2009.

Seno, T., Stein, S., and Gripp, A.E., 1993. A model for the motion of the Philippine Sea Plate consistent with NUVEL-1 and geological data. *J. Geophys. Res.*, 98(B10):17941–17948. doi:10.1029/93JB00782.

Spinelli, G.A., and Wang, K., 2008. Effects of fluid circulation in subducting crust on Nankai margin seismogenic zone temperatures. *Geology*, 36(11):887–890. doi:10.1130/G25145A.1.

Taira, A., Hill, I., Firth, J.V., et al., 1991. *Proc. ODP, Init. Repts.*, 131: College Station, TX (Ocean Drilling Program). doi:10.2973/odp.proc.ir.131.1991.

Taira, A., Hill, I., Firth, J., Berner, U., Brückmann, W., Byrne, T., Chabernaud, T., Fisher, A., Foucher, J.-P., Gamo, T., Gieskes, J., Hyndman, R., Karig, D., Kastner, M., Kato, Y., Lallement, S., Lu, R., Maltman, A., Moore, G., Moran, K., Olaffson, G., Owens, W., Pickering, K., Siena, F., Taylor, E., Underwood, M., Wilkinson, C., Yamano, M., and Zhang, J., 1992. Sediment deformation and hydrogeology of the Nankai Trough accretionary prism: synthesis of shipboard results of ODP Leg 131. *Earth Planet. Sci. Lett.*, 109(3–4):431–450. doi:10.1016/0012821X(92)901044.

Taira, A., Tokuyama, H., and Soh, W., 1989. Accretion tectonics and evolution of Japan. *In* Ben-Avraham, Z. (Ed.), *The Evolution of the Pacific Ocean Margins*: New York (Oxford University Press), 100–123.

Taylor, B., 1992. Rifting and the volcanic-tectonic evolution of the Izu-Bonin-Mariana arc. *In* Taylor, B., Fujioka, K., et al., *Proc. ODP, Sci. Results*, 126: College Station, TX (Ocean Drilling Program), 627–651.

Tobin, H.J., and Kinoshita, M., 2006a. Investigations of seismogenesis at the Nankai Trough, Japan. *IODP Sci. Prosp.*, NanTroSEIZE Stage 1. doi:10.2204/iodp.sp.nantroseize1.2006.

Tobin, H.J., and Kinoshita, M., 2006b. NanTroSEIZE: the IODP Nankai Trough Seismogenic Zone Experiment. *Sci. Drill.*, 2:23–27. doi:10.2204/iodp.sd.2.06.2006.

Tobin, H.J., Kinoshita, M., Ashi, J., Lallemant, S., Kimura, G., Screaton, E., Thu, M.K., Masago, H., Curewitz, D., and the Expedition 314/315/316 Scientists, 2009. NanTroSEIZE Stage 1 expeditions: introduction and synthesis of key results. *Proc. IODP*, 314/315/316: Washington, DC (Integrated Ocean Drilling Program Management International, Inc.). doi:10.2204/iodp.proc314315316.101.2009.

Underwood, M.B., 2007. Sediment inputs to subduction zones: why lithostratigraphy and clay mineralogy matter. *In* Dixon, T., and Moore, J.C. (Eds.), *The Seismogenic Zone of Subduction Thrust Faults*: New York (Columbia University Press), 42–85.

Underwood, M.B., and Fergusson, C.L., 2005. Late Cenozoic evolution of the Nankai trench-slope system: evidence from sand petrography and clay mineralogy. *In* Hodgson, D., and Flint, S. (Eds.), *Submarine Slope Systems: Processes, Products and Prediction. Geol. Soc. Spec. Publ.*, 244(1):113–129. doi:10.1144/GSL.SP.2005.244.01.07.

Underwood, M.B., Saito, S., Kubo, Y., and the Expedition 322 Scientists, 2009. NanTroSEIZE Stage 2: subduction inputs. *IODP Prel. Rept.*, 322. doi:10.2204/iodp.pr.322.2009.

Vrolijk, P., 1990. On the mechanical role of smectite in subduction zones. *Geology*, 18:703–707.

Yamano, M., Kinoshita, M., Goto, S., and Matsubayashi, O., 2003. Extremely high heat flow anomaly in the middle part of the Nankai Trough. *Phys. Chem. Earth*, 28(9–11):487–497. doi:10.1016/S1474-7065(03)00068-8.

Authors

Michael B. Underwood, Department of Geological Sciences, University of Missouri, Columbia, MO 65203, U.S.A., E-mail: UnderwoodM@missouri.edu.

Saneatsu Saito, Institute for Research on Earth Evolution, Japan Agency for Marine-Earth Science and Technology, 2-15 Natsushima-cho, Yokosuka 237-0061, Japan.

Yu'suke Kubo, Center for Deep Earth Exploration, Japan Agency for Marine-Earth Science and Technology, 3173-25 Showa-machi, Kanazawa-ku, Yokohama 236-0001, Japan.

and the IODP Expedition 322 Scientists

27

The New Jersey Margin Scientific Drilling Project (IODP Expedition 313): Untangling the Record of Global and Local Sea-Level Changes

Gregory Mountain, Jean-Noël Proust and the Expedition 313 Science Party

Introduction

Much of the world is currently experiencing shoreline retreat due to global sea level rising at the rate of 3–4 mm yr^{-1}. This rate will likely increase and result in a net rise to roughly 1 m above present sea-level by the year 2100 (e.g., Rahmstorf, 2007; Solomon et al., 2007), with significant consequences for coastal populations, infrastructures, and ecosystems. Preparing for this future scenario calls for careful study of past changes in sea level and a solid understanding of processes that govern the shoreline response to these changes. One of the best ways to assemble this knowledge is to examine the geologic records of previous global sea-level changes. Integrated Ocean Drilling Program (IODP) Expedition 313 set out to do this by recovering a record of global and local sea-level change in sediments deposited along the coast of eastern North America during the Icehouse world of the past 35 m.y. What we learn from this record—the factors driving sea-level changes, and the impact of this change on nearshore environments—will help us understand what lies ahead in a warming world.

Eustatic history can be derived from three archives: corals, oxygen isotopes, and shallow-water marine sediments. Corals provide the most direct and detailed record (millennial-scale resolution or better), but it can be traced back no farther than latest Pleistocene (Fairbanks, 1989; Bard et al., 1996; Camoin et al., 2007). Oxygen isotopic ratios in carbonate-secreting organisms yield a glacio-eustatic proxy, but uncertainties arise further back in time because past water temperatures, which affect the oxygen isotopic ratio of seawater, are not known with sufficient accuracy. Furthermore, changes in the total volume of ocean water as inferred by oxygen isotopes are not the only eustatic drivers; the total volume of the world's ocean basins can change as well and impart eustatic change. For example, variations in the rate of seafloor spreading and in the amount of sediment deposited on the ocean floor affect basin volume and cause long-term (~10 m.y.) eustatic changes on the scale of tens of meters (Hays and Pitman, 1973; Kominz, 1984; Harrison, 1990) that are not accounted for by oxygen isotopic measurements.

The spatial and temporal arrangement of shallow-water marine sediments is a third archive of eustatic history. Their analysis is not a direct measure like corals, and it is rarely able to track changes at the Milankovitch scale that is possible with oxygen isotopes; however, the shallow marine sedimentary record can detect changes in water depth throughout the Phanerozoic and provide the sum of all processes that contribute to these changes and to the lateral migration of the shoreline. Therein lies the difficulty. Besides responding to eustatic change, the position of the shoreline and the accompanying change in facies respond to changes in sediment supplied to the coastal zone, compaction of deposited sediments, isostatic and/

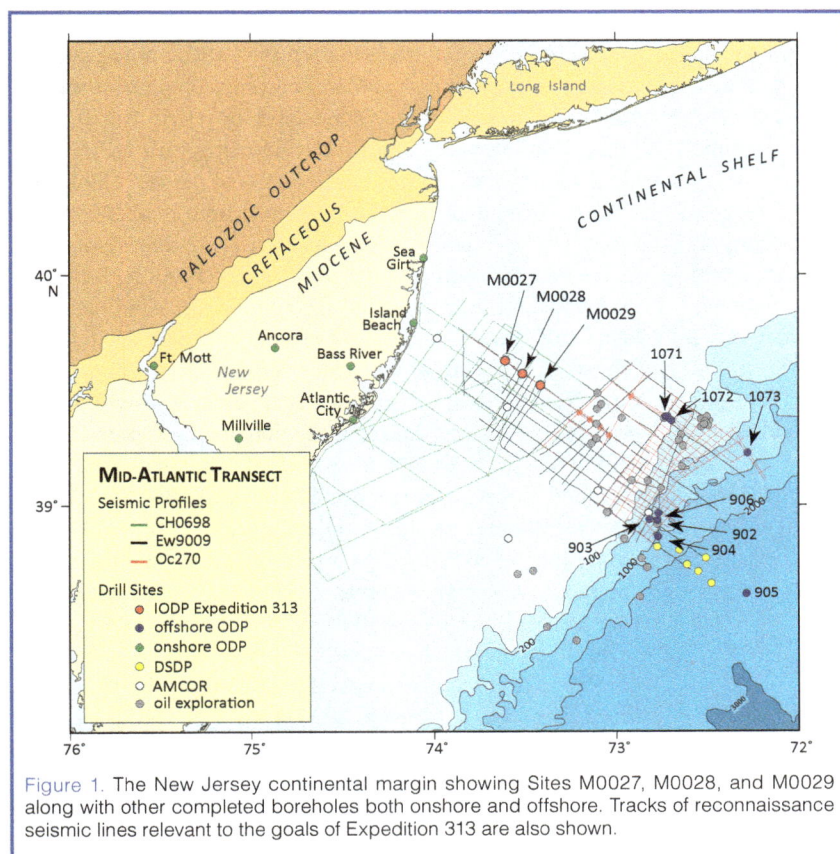

Figure 1. The New Jersey continental margin showing Sites M0027, M0028, and M0029 along with other completed boreholes both onshore and offshore. Tracks of reconnaissance seismic lines relevant to the goals of Expedition 313 are also shown.

or flexural loading of the crust, thermal subsidence of the lithosphere, and any other vertical tectonic motions of the basin. Furthermore, the magnitude of eustatic change is often considerably smaller than that of these other processes (Watts and Steckler, 1979), making the accurate measurement of this record a very challenging procedure.

Sloss (1963) cited eustasy as a possible cause of continent-wide unconformities dividing successions of shallow-water sediment across North America and Eurasia. His analysis suggested global sea level rose and fell in ~100-m.y. cycles throughout the Paleozoic. Other researchers reported that seismic profiles from sedimentary basins and continental margins revealed additional, globally correlated, unconformity-bound packages that indicated a higher-order cyclicity of eustatic change (~1.5 m.y.; Vail et al., 1977). By measuring variations in the elevation of what they termed coastal onlap seen in seismic profiles, they calculated that many eustatic oscillations were 100 m or more. Subsequent research showed, however, that the likelihood of shallow-water sediment accumulating along passive margins depends on the rate of eustatic change in relation to rates of change in sediment supply, basement subsidence, and compaction (Pitman and Golovchenko, 1983). This means that in the absence of detailed age control, compaction history, and paleo-water depth estimates, eustatic magnitudes cannot be derived directly from the architecture of stratal boundaries revealed by seismic profiles (Watts and Thorne, 1984).

Despite an updated eustatic history based, as before, on patterns of coastal onlap, the underlying data was still proprietary (Haq et al., 1987), and researchers saw the need for passive margin records open to public scrutiny (Imbrie et al., 1987; Watkins and Mountain, 1990.) After many locations were evaluated, a transect across the New Jersey (NJ) coastal plain, shelf, and slope was chosen because of its relatively thick and continuous mid- to late-Tertiary section, lack of tectonic disturbance, wealth of background information, and mid-latitude setting that suggested a strong likelihood of yielding excellent geochronology and paleobathymetric control.

IODP Expedition 313 Background

Ocean Drilling Program (ODP) Leg 150 (Mountain et al., 1994), benfited from earlier drilling (Deep Sea Drilling Project (DSDP) Legs 11 and 95; Hollister et al., 1972; Poag et al., 1987), and was the first step in the multi-leg New Jersey Transect designed to recover a record of eustatic history

Figure 2. L/B *Kayd* (Montco Offshore, Inc.) outfitted with a commercial drill rig using coring tools developed and operated by DOSECC, Inc., during Expedition 313 while under direction from ESO. It stands in 35 m of water 45 km off the New Jersey coast.

(Fig. 1). Leg 150 sampling was limited to the continental slope and rise, but it proved that sequence boundaries, defined by facies successions identified in cores and tied to seismic unconformities, coincided with increases in $\delta^{18}O$ that formed during times of glacio-eustatic lowering. Due to the distance from paleo-shorelines, these drill cores provided no information about magnitudes of eustatic change. ODP Leg 174A on the outer shelf attempted with limited success to sample more proximal Miocene sediments (Austin et al., 1998). Onshore drilling, sponsored by ODP, the International Continental Scientific Drilling Program (ICDP), the U.S. Geological Survey, and the New Jersey Geological Survey, cored equivalent sediments (Miller et al., 1998) and recovered vertical facies associations consistent with glacio-eustatic control. But due to their updip locations, each was stratigraphically incomplete, missing sea-level lowstands and lacking in seismic profiles that would otherwise place all in a broader context of stratal architecture. Nonetheless, calculations that removed the imprint of processes affecting the accumulation of Oligocene-Miocene sediments left 30–50 m sea-level changes that were assumed to be eustatic (Van Sickel et al., 2004). Offshore high-resolution profiles collected in 1995 and 1998 (Mountain et al., 2007; Monteverde et al., 2008) located clinoform topsets, foresets, and toesets of presumed Oligocene-Miocene age that, if sampled, would capture several complete Icehouse sea-level cycles. With the development of mission-specific operations in the IODP, three sites were selected in 2005 and placed on the schedule for drilling from a jack-up platform. It took until 2009 to secure a lift boat (L/B), drill rig, and crew under contract to the European Consortium for Ocean Research Drilling (ECORD) Science Operator (ESO) (Fig. 2).

Scientific Objectives

Encouraged by the relatively well-known geologic setting of the NJ transect, and equipped with a platform immune to vertical and lateral motion and a drill rig well-suited for coring in sand-prone formations, Expedition 313 set out to overcome these challenges with the following objectives:

- Compare the age of Oligocene-Miocene Icehouse base-level changes with the age of sea-level lowerings predicted by the global $\delta^{18}O$ glacio-eustatic proxy
- Estimate amplitudes and rates and infer mechanisms of eustatic sea-level changes
- Evaluate models that predict lithofacies successions, depositional environments, and the arrangement of seismic reflections in response to eustatic change
- Provide a database to compare to sea-level studies on other margins

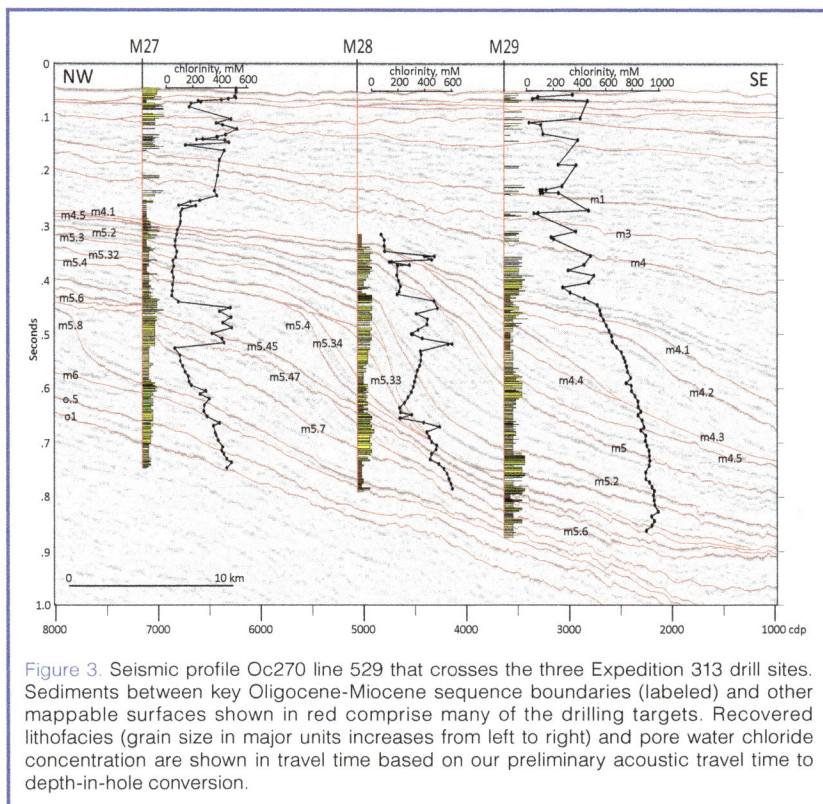

Figure 3. Seismic profile Oc270 line 529 that crosses the three Expedition 313 drill sites. Sediments between key Oligocene-Miocene sequence boundaries (labeled) and other mappable surfaces shown in red comprise many of the drilling targets. Recovered lithofacies (grain size in major units increases from left to right) and pore water chloride concentration are shown in travel time based on our preliminary acoustic travel time to depth-in-hole conversion.

Operations

The L/B *Kayd* (owned and operated by Montco Offshore, Inc.) sailed north from the Gulf of Mexico and arrived in Atlantic City where offshore operations mobilized under contract to Drilling, Observation and Sampling of the Earth's Continental Crust (DOSECC). Operations began 45 km offshore on 30 April 2009 and continued until the return and demobilization at Atlantic City on 19 July 2009 (Table 1).

At each site conductor pipe was run 12–25 m into the seabed to stabilize the top of the hole and provide re-entry into the seabed for subsequent operations. Various biodegradable drilling fluids were used to condition the hole, cool the drill bit, and lift cuttings to perforations in the casing in the water column from which they settled out onto the seabed. Drilling and coring were conducted with 114.3-mm-diameter PHD drill pipe and top drive assembly commonly used in onshore mining operations. Hydraulic piston and extended-nose rotary coring similar to operations on the *JOIDES Resolution* were employed, but it was drilling with the extended, rotating coring "Alien" bit developed by DOSECC that proved most successful. Although the drill

pipe became stuck numerous times, we had to abandon hope of freeing it only once (at 404 meters below seafloor [mbsf] in Hole 28A). In this case, narrower diameter (96-mm) HQ pipe and extended-nose rotary drilling continued to total depth through the center of the fixed PHD pipe. As anticipated in pre-expedition planning, to maximize operational time in the intervals of highest expedition priorities, we chose to spot core the top ~180–220 mbsf at each site (Fig. 3). Five types of wireline electric and imaging logs plus a Vertical Seismic Profile (VSP) were collected in segments at each site (Table 2, Fig. 4).

The onboard complement typically comprised the following personnel: 10 *Kayd* crew, 7–9 DOSECC drillers, 13–15 ESO staff, and 4–5 expedition science party members. Basic core curation, through-liner descriptions, measurements of ephemeral properties on un-split cores using the Multi-Sensor Core Logger (MSCL), and sampling at liner boundaries were conducted around the clock for the duration of offshore operations (Table 2).

Table 1. Expedition 313 Operations Summary (mbsl: meters below sea-level; mbsf: meters below seafloor)

Hole	Latitude (N)	Longitude (W)	Water Depth (mbsl)	No. Cores	Total Hole Drilled (mbsf)	Total Core Attempted (m)	Total Core Recovered (m)	Total Core Recovered (%)	Total Hole Recovered (%)	Time on Site (days)
M0027A	39°38.046067'	73°37.301460'	34	224	631.01	547.01	471.59	86.21	74.74	22
M0028A	39°33.942790'	73°29.834810'	35	171	668.66	476.97	385.5	80.82	57.65	28.7
M0029A	39°31.170500'	73°24.792500'	36	217	754.55	609.44	454.31	74.55	60.04	26.3
Totals				612	2054.22	1633.42	1311.4	80.29	63.77	77

Table 2. Expedition 313 Measurements Summary

	L/B *Kayd*, Offshore New Jersey	Onshore Science Party, Bremen	Moratorium Studies
Lithostratigraphy	Core catcher description Smear slide identification Core catcher photography	Split-core visual description archived according to IODP protocol Smear slides, thin sections, CT scans Observations incorporated into facies model Full core & close-up photography	Lithofacies (petrography, sedimentary structures, etc.), depositional environment & sequence stratigraphic analysis Ichnofacies & benthic macrofauna Diagenetic alteration & cementation Compaction & dewatering CT scan analysis of sediment structure & texture Semiquantitative petrography of coarse sediment Quantitative clay mineralogy Mapping Pleistocene valleys
Biostratigraphy & Micropaleontology	Benthic forams paleobathymetry Planktonic forams biostratigraphy Many samples (forams, palynomorphs, nannos) were taken from core catchers & some were analyzed before OSP	Additional sampling & sample preparation (primarily nannos, some forams & palynomorphs samples) Biostratigraphic analysis of nannos, planktonic forams & dinocysts Benthic forams paleobathymetry Palynology paleoenvironments	Semiquantitative nannofossil biostratigraphy & paleoecology Controls on planktonic foram abundance & diversity Palynomorph taphonomy Climate & sea-level controls on ecosystem evolution based on pollen (Eocene-Miocene) Diatom & silicoflagellate biostratigraphy (mid-upper Miocene) Amino acid racemization age dating (Pleistocene molluscs) Radiolarian biostratigraphy
Paleo-magnetism		Discrete measurements in fine-grained sediment U-channel sampling	U-channel sample measurement & analysis Detailed magnetic stratigraphy of selected sections (O/M transition, clinoform tops, etc.)
Sr Isotopes	300 core catcher samples with shell material & benthic forams dated (using Sr) before OSP	Additional 900 samples taken	$^{87}Sr/^{86}Sr$ age dating of 900 samples of mollusks & benthic forams
Downhole Logging	Wireline logs, through pipe: Spectral gamma ray (98%) VSP (71%) Wireline logs, open hole: Spectral gamma ray (35%) Induction resistivity (46%) Magnetic susceptibility (47%) Full waveform sonic (34%) Acoustic imaging & caliper (34%)		Refine VSP data to use up- & down-going energy for core-seismic correlation Integration of acoustic images, depositional fabric & CT scans for core-log correlation Synthetic seismograms & core-log-seismic correlation Statistical analysis of log character & ties to lithofacies
Core Logging	Whole-core multi-sensor logging: Density P-wave velocity Magnetic susceptibility Electrical resistivity	Whole-core logging: Natural gamma ray Thermal conductivity Split-core logging: Color reflectance of split-core surface at discrete points Continuous digital line-scanning of split core CT-scanning (selected cores)	Lateral changes in physical properties Core-log correlation Comparison of core quality with MSCL measurements
Petrophysics		Discrete sample index properties: Compressional P-wave velocity Bulk, dry & grain density Water content Porosity & void ratio	Changes of permeability, porosity & thermal conductivity with depth Magnetic mineralogy & links to MSCL data Cross plots of petrophysical data
Geochemistry, organic & inorganic	Pore water extraction with rhizome or hydraulic press: Ephemeral pH by ion-specific electrode Alkalinity by single-point titration to pH Ammonium by conductivity Chlorinity by automated electrochemical titration Headspace samples for methane & stable carbon isotopes	IW analysis by ICP-AES & ICP-MS for 24 major & trace elements Sediment TOC, TC, & TS by LECO (carbon-sulfur analyzer) Sediment mineralogy by XRD (28 samples)	Relationship of sea-level change to phosphorus & organic carbon burial Stable isotope (C, O, S) geochemistry of sediments Nd & Os in sediment pore fluids Quantitative assessment & carbon isotope stratigraphy of organic material (phytoclasts) Iron-rich chlorite precursors in ichnofabrics Full suite of elemental analyses & C isotopes of pore waters Carbonates & other authigenic minerals 37Cl & origin of sediment porewater
Seismic Stratigraphy	Travel-time depth below sea floor relationship based on stacking velocities used as preliminary core-log-seismic correlation	Refined seismic-core correlations to improve agreement with lithostratigraphy & physical properties	Backstripping NJ transect compaction & subsidence history to estimate eustasy Mapping seismic sequences on the NJ margin with ties to boreholes
Microbiology	Sampling & preparation of sediment samples		Identification & quantification of phylogenetic groups of microorganisms (microscopic & molecular techniques) living in the subsurface

Note: L/B, lift boat; CT, computed tomography; OSP, Onshore Science Party; O/M transition, Oligocene-Miocene transition; Sr, strontium; VSP, Vertical Seismic Profile; MSCL, Multi Sensor Core Logger; IW, interstitial water; ICP-AES & MS, Inductively Coupled Plasma Atomic Emission Spectroscopy & Mass Spectrometry; TOC, Total Organic Carbon; TC, Total Carbon; TS, Total Sulfur; XRD, X-Ray Diffraction; C, O. S, carbon, oxygen, sulfur; Nd, Os, Cl, neodymium, osmium, chloride. Numbers in parentheses equal percentage of total drilled section that was logged.

All cores and data were transferred to the Bremen Core Repository (BCR) at the end of the offshore phase. Additional measurements of natural gamma radiation, thermal conductivity, and computed tomography (CT) scans on selected cores were performed prior to splitting at the Onshore Science Party (OSP). The entire 28-member science party plus 37 others from the ESO and BCR plus student helpers met at the BCR from 6 November to 4 December 2009 to split, sample, and analyze the 612 cores and logs collected offshore (Table 2).

Preliminary Results

Drilled-through and recovered sands. Interpreting the results of Exp. 313 began aboard the L/B *Kayd* by correlating the driller's reports of subsurface conditions with through-liner core descriptions, core-catcher samples, whole-round MCSL measurements, and inferences derived from seismic profiles (Fig. 3). The first strong reflector within a few tens of milliseconds below the seabed at each site indicated fine-grained, relatively firm sediment

that provided a stable base for the conductor pipe designed to keep loose surficial sediment from caving into the hole.

The uppermost ~168–220 mbsf at all three sites contained few strong and continuous reflectors that would indicate regional stratal boundaries of interest and primary coring targets. Although we tried to core continuously in this shallowest interval at Site M27, unconsolidated and coarse-grained sediments led to slow difficult drilling and poor core recovery. This forced the eventual decision to drill without coring until we calculated we were approaching high-priority objectives. The equivalent stratigraphic units (based on seismic correlation) were drilled without coring at Site M28 and were only spot cored at Site M29; all information indicates these were upper Miocene to upper Pleistocene sands and gravels (we identified no Pliocene at any site) deposited in a range of shoreface, estuarine, fluvial, incised valley, and coastal plain environments. A possible paleosol was recovered in this spot-cored interval at Sites M27 and M29 at the depth calculated at both sites to correspond to seismic reflector m1.

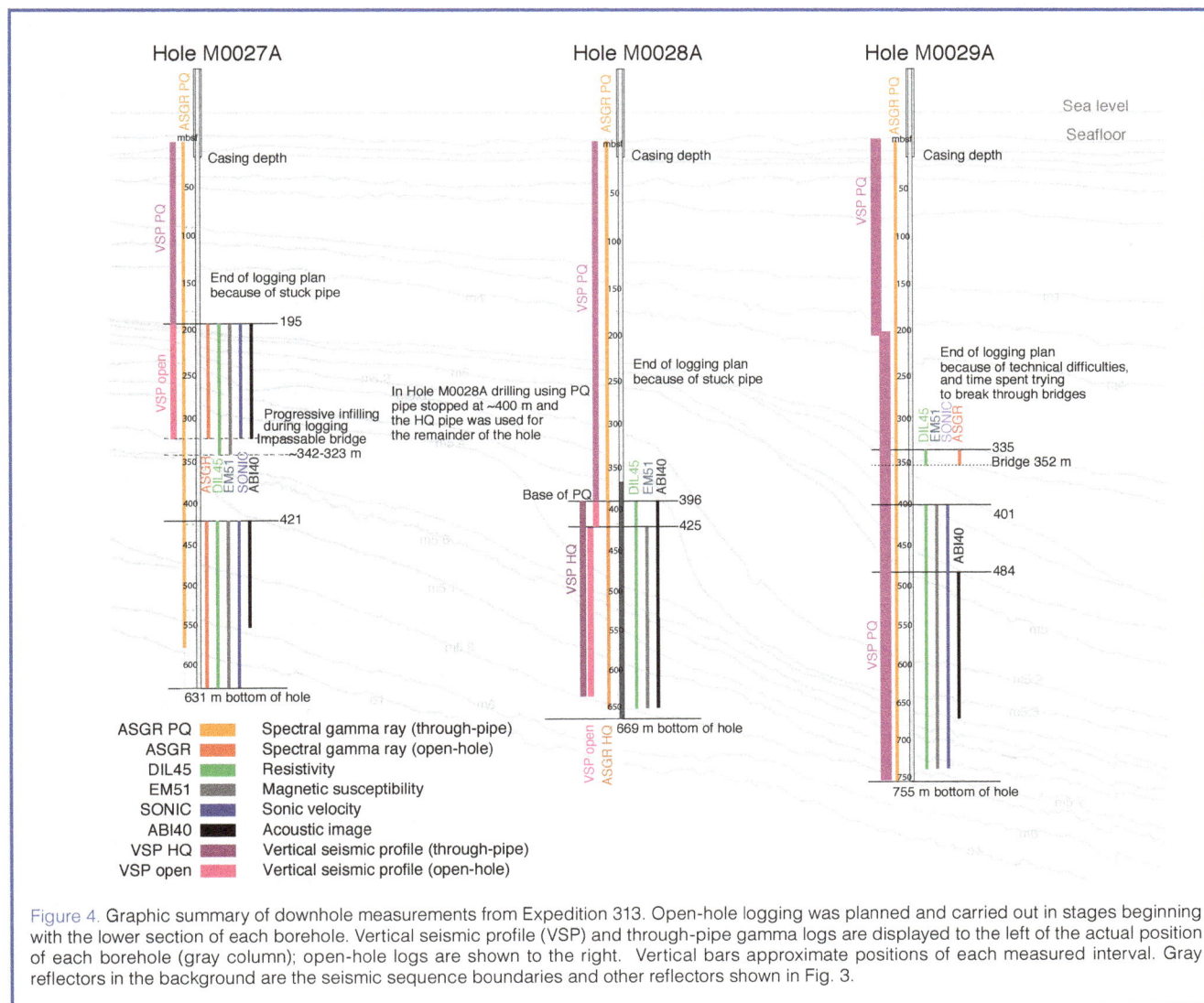

Figure 4. Graphic summary of downhole measurements from Expedition 313. Open-hole logging was planned and carried out in stages beginning with the lower section of each borehole. Vertical seismic profile (VSP) and through-pipe gamma logs are displayed to the left of the actual position of each borehole (gray column); open-hole logs are shown to the right. Vertical bars approximate positions of each measured interval. Gray reflectors in the background are the seismic sequence boundaries and other reflectors shown in Fig. 3.

Figure 5. Close-up photographs of 3 split cores showing (from left to right): laminated silty clay from toeset beds seaward of a clinoform (Hole 28 core 166 sec 2); sharp-based sandy storm bed in sandy silt from topset beds (Hole 28 core 92 sec 1); and well-preserved gastropod and mollusk shells in silty offshore flooding surface (Hole 27 core 89 sec 1).

Integrated seismic, log, and core information. Prior seismic stratigraphy studies (Greenlee et al., 1992; Monteverde et al., 2008) identified probable sequence boundaries on the basis of seismic onlap/offlap patterns and ties to drill cores on the continental slope (Mountain et al., 1994). As a first approximation, stacking velocities from the processing of seismic data surrounding and passing through Exp. 313 sites were used to derive an acoustic travel time to depth-below-seafloor conversion. Check-shot VSP measurements at each site, plus physical properties measurements and synthetic seismograms, will firm up these preliminary seismic-core correlations in subsequent shore-based study.

As many as fifteen regionally mapped reflectors were intersected by one or more Exp. 313 drill sites (Fig. 3). Calculations using our preliminary time-depth equation provided expectations of depths at which surfaces and/or facies changes would be encountered, and typically these came within 5–10 m of a probable match in the cores, wireline logs, and/or MSCL measurements. In cases where seismic reflectors were especially closely spaced (vertical separation of <5 m), there remains uncertainty concerning which reflector ties to which geologic feature. Future work planned by the Exp. 313 science party will improve the reliability of correlations.

Petrophysical, MSCL, and downhole log data provided additional lithofacies characterization and greatly aided intersite correlations (Fig. 4). By providing continuous data for intervals with poor core recovery, logs enabled us to assign with reasonable confidence major seismic reflectors

to depths where there was no core and hence no lithofacies feature to examine.

Described and interpreted the lithofacies. Sediments have been assigned to eight lithostratigraphic units that were deposited under two broadly defined conditions: (1) on a mixed wave- to river-dominated shelf where well-sorted silt and sand accumulated in offshore to shoreface environments and (2) during intervals of clinoform slope degradation that resulted in the interbedding of poorly sorted silts plus debris flow and turbidite sands with toe-of-slope silt and silty clays (Fig. 5). Deposits at all sites and all ages indicate a silt-rich sediment supply notably lacking in clay-sized components. Both *in situ* and reworked glauconite were common components of top-set and toe-set strata. The open shelf experienced frequent and sometimes cyclic periods of dysoxia. We found no evidence of subaerial exposure at the clinoform inflection point (depositional shelf break), but the periodic occurrences of shallow-water facies along clinoform slopes and of deepwater facies on the topset of the clinoforms suggest large-amplitude changes in relative sea level. Backstripping will provide estimates of the true eustatic component involved in the ~60-m changes in relative sea level observed on the shelf.

Developed geochonology of the sedimentary record. 1) Biostratigraphy: Roughly 300 samples were taken from core catchers on the *Kayd* and brought ashore for extraction of suitable mollusk shells and forams for Sr isotopic dating prior to the OSP in Bremen. Additional samples for palynomorphs, foraminifera, and calcareous nannofossils were also collected offshore and prepared prior to the OSP. Additional biostratigraphic control available at this preliminary stage comes largely from calcareous nannofossils sampled in split cores and analyzed during the OSP; as time allowed, a limited number of additional samples for planktonic foraminifera and palynomorphs were prepared and analyzed in Bremen. Additional biostratigraphic shore-based studies will be performed using diatoms and radiolaria. Amino acid racemization studies will be conducted on shells from the upper few tens of mbsf at Sites M27 and M29.

2) Magnetostratigraphy: Due to the paucity of fine-grained intervals and/or the lack of carrier minerals preserving a remnant signal, geomagnetic reversal chronology measured during the OSP was restricted to a few short intervals. More sensitive and time-consuming demagnetization techniques on discrete samples, as well as on U-channel samples, will be applied during post-OSP studies and may contribute further to Exp. 313 geochronology.

3) Sr isotopes: There is generally good agreement between the biohorizons of the different microfossil groups and the Sr isotope ages. One exception is within the thick Oligocene section at Site M27 where more shore-based analysis is warranted. The abundance and preservation of calcareous microfossils and dinocysts vary significantly, with barren intervals coinciding with coarse-grained sediments. Sr isotopic measurements provided ages that approached a precision of ±0.5 m.y. in some intervals and ±1 m.y. in others. A notable exception was the middle Miocene silty clay at Site M29 that showed substantial scatter. Microfossils are most abundant in this latter, most distal site, allowing for age refinements within the lower Miocene sections that proved barren of planktonic microfossils at the more proximal sites. Reworked Paleogene material occurs throughout the Miocene at all sites. A preliminary age-depth plot was developed in Bremen (Fig. 6) and will be refined when these additional shore-based studies are complete.

Detected sea-level changes and paleoclimate. Paleobathymetry and paleoenvironments determined from benthic foraminifers, dinocysts, and terrigenous palynomorphs track similar paleodepth variations at each site and in general agree well with paleobathymetric changes indicated by lithofacies. Values at each site range from inner neritic (0–50 m) to outer neritic (100–200 m). Paleodepth variations within individual unconformity-bound topset

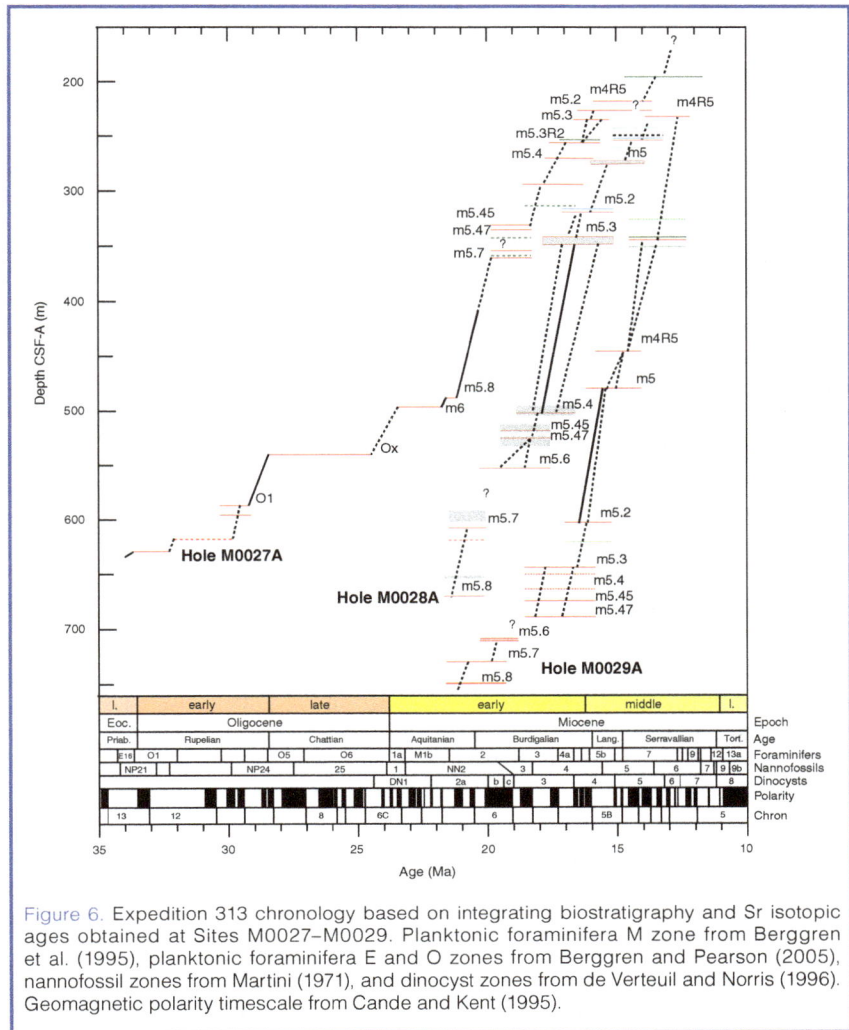

Figure 6. Expedition 313 chronology based on integrating biostratigraphy and Sr isotopic ages obtained at Sites M0027–M0029. Planktonic foraminifera M zone from Berggren et al. (1995), planktonic foraminifera E and O zones from Berggren and Pearson (2005), nannofossil zones from Martini (1971), and dinocyst zones from de Verteuil and Norris (1996). Geomagnetic polarity timescale from Cande and Kent (1995).

intervals suggest relative sea-level changes were as large as 60 m. While we detected mass failure of clinoform foresets, we found no evidence that sea level ever fell below the elevation of clinoform topsets. Water depth estimates based on benthic for-aminifers and palynological estimates of proximity to the shoreline show good to excellent agreement at all sites. Pollen studies identified a hemlock horizon across all three sites, indicating temperate forests and humid conditions on the Atlantic coastal plain during the early Miocene. Middle Miocene pollen assemblages record the expansion of grasses and sedges, indicating increasing aridity.

Analyzed pore water chemistry. Pore-water studies show that the upper several hundred meters of sediment at each site is dominated by freshwater interlayered with salt water of nearly seawater chlorinity. The abrupt boundaries between fresh and saline pore waters are especially remarkable (Fig. 3); whether they are maintained by dynamic flow/ recharge or by strongly impermeable boundaries is not yet recognized. The pore water in these layers may have chemistries sufficiently distinct to enable correlation of individual layers from site to site; more shore-based analysis is required. Chloride concentrations increase with depth below seafloor, reaching seawater values in Site M27 and even higher concentrations in the other two sites. In Site M29, brine was encountered toward the bottom of the hole, reaching a chlorinity value twice that of seawater.

Acknowledgements

We are grateful to Dennis Nielson, Chris Delahunty, Beau Marshall, and the rest of the DOSECC drilling team, to Captains Clem Darda and Farrel Charpentier plus the crew of the L/B *Kayd*, to Dan Evans, David Smith, Colin Graham, David McInroy, Carol Cotterill, and the rest of the very talented ESO support staff of the mission-specific operations for ECORD, and to Ursula Röhl, Holger Kuhlmann, and the BCR staff in Bremen. Without these individuals and the many others left unnamed, the New Jersey Shallow Shelf operations would not have been possible. Two decades of preparatory and related studies were supported by grants from the U.S. National Science Foundation's Division of Ocean Sciences as well as the Division of Earth Sciences, the U.S. Office of Naval Research, the U.S. Geological Survey, and the New Jersey State Geological Survey. IODP funds were augmented by a long-standing commitment from ICDP, for which we are grateful.

References

Austin, J.A., Jr., Christie-Blick, N., Malone, M.J., Berné, S., Borre, M.K., Claypool, G., Damuth, J., Delius, H., Dickens, G., Flemings, P., Fulthorpe, C., Hesselbo, S., Hoyanagi, K., Katz, M., Krawinkel, H., Major, C., McCarthy, F., McHugh, C., Mountain, G., Oda, H., Olson, H., Pirmez, C., Savrda, C., Smart, C., Sohl, L., Vanderaveroet, P., Wei, W., Whiting, B., 1998. *Proc. ODP, Init. Repts.,* 174A: College Station, TX (Ocean Drilling Program).

Bard, E., Hamelin, B., Arnold, M., Montaggioni, L., Cabioch, G., Faure, G., and Rougerie, F., 1996. Deglacial sea-level record from Tahiti corals and the timing of global meltwater discharge. *Nature,* 382:24–244; doi:10.1038/382241a0.

Berggren, W.A., and Pearson, P.N., 2005. A revised tropical to subtropical Paleogene planktonic foraminiferal zonation. *J. Foram. Res.,* 35(4):279–298. doi:10.2113/35.4.279.

Berggren, W.A., Kent, D.V., Swisher, C.C., III, and Aubry, M.-P., 1995. A revised Cenozoic geochronology and chronostratigraphy. *In* Berggren, W.A., Kent, D.V., Aubry, M.-P., and Hardenbol, J. (Eds.), *Geochronology, Time Scales and Global Stratigraphic Correlation.* Spec. Publ. - SEPM (Society for Sedimentary Geology), 54:129–212.

Camoin, G.F., Iryu, Y., McInroy, D., and the IODP Expedition 310 Scientists, 2007. IODP Expedition 310 reconstructs sea level, climatic, and environmental changes in the South Pacific during the last deglaciation. *Sci. Drill.,* 5:4–12; doi:10.2204/iodp.sd.5.01.2007.

Cande, S.C., and Kent, D.V., 1995. Revised calibration of the geomagnetic polarity timescale for the Late Cretaceous and Cenozoic. *J. Geophys. Res.,* 100(B4):6093–6095. doi:10.1029/94JB03098.

De Verteuil, L. and Norris, G., 1996. Miocene dinoflagellate stratigraphy and systematics of Maryland and Virginia. *Micropaleontology,* 42, supp., part 1:1–82.

Fairbanks, R.G., 1989. A 17,000-year glacio-eustatic sea level record: influence of glacial melting rates on the Younger Dryas event and deep-ocean circulation. *Nature,* 342(6250):637–642, doi:10.1038/342637a0.

Greenlee, S.M., Devlin, W.J., Miller, K.G., Mountain, G.S., and Flemings, P.B., 1992. Integrated sequence stratigraphy of Neogene deposits, New Jersey continental shelf and slope: comparison with the Exxon model. *Geol. Soc. Am. Bull.,* 104(11):1403–1411, doi:10.1130/0016-7606(1992)104<1403: ISSOND>2.3.CO;2.

Harrison, C.G.A., 1990. Long-term eustasy and epeirogeny in continents. *In* Revelle, R. (Ed.), *Sea-level Change*: Washington, DC (National Academy Press), 141–158.

Haq, B.U., Hardenbol, J., and Vail, P.R., 1987. Chronology of fluctuating sea levels since the Triassic. *Science,* 235(4793):1156–1167, doi:10.1126/science.235.4793.1156.

Hays, J.D., and Pitman, W.C., III, 1973. Lithospheric plate motion, sea level changes and climatic and ecological consequences. *Nature,* 246(5427):18–22, doi:10.1038/246018a0.

Hollister, C.D., Ewing, J.I., Hathaway, J.C., Lancelot, Y., Paulus, F.J., Habib, D., Luterbacher, H., Poag, C.W., Wilcoxon, J.A., Worstell, P., 1972. *Init. Repts. DSDP,* 11: Washington, DC (U.S. Govt. Printing Office).

Imbrie, J., Barron, E.J., Berger, W.H., Bornhold, B.D., Cita Sironi, M.B., Dieter-Haass, L., Elderfield, H., Fischer, A., Lancelot, Y., Prell, W.L., Togweiler, J.R., and Van Hinte, J., 1987. Scientific goals of an Ocean Drilling Program designed to investigate changes in the global environment. *In* Munsch, G.B. (Ed.), *Report of the Second Conference on Scientific Ocean Drilling (COSOD II):* Strasbourg (European Science Foundation), 15–46.

Kominz, M., 1984. Oceanic ridge volumes and sea-level change: an error analysis. *In* Schlee, J.S. (Ed.), *Interregional Unconformities and Hydrocarbon Accumulation, AAPG Mem.,* 36:109–127.

Martini, E., 1971. Standard Tertiary and Quaternary calcareous nannoplankton zonation. *In* Farinacci, A. (Ed.), *Proc. 2nd Int. Conf. Planktonic Microfossils Roma*: Rome (Ed. Tecnosci.), 2:739–785.

Miller, K.G., Mountain, G.S., Browning, J.V., Kominz, M., Sugarman, P.J., Christie-Blick, N., Katz, M.E., and Wright, J.D., 1998. Cenozoic global sea level, sequences, and the New Jersey transect: results from coastal plain and continental slope drilling. *Rev. Geophys.,* 36(4):569–602, doi:10.1029/98RG 01624.

Monteverde, D., Mountain, G., and Miller, K., 2008. Early Miocene sequence development across the New Jersey margin. *Basin Res.,* 20:249–267, doi:10.1111/j.1365-2117.2008. 00351.x.

Mountain, G.S., Burger, R.L., Delius, H., Fulthorpe, C.S., Austin, J.A., Goldberg, D.S., Steckler, M.S., McHugh, C.M., Miller, K.G., Monteverde, D.H., Orange, D.L., and Pratson, L.F., 2007. The long-term stratigraphic record on continental margins. *In* Nittrouer, C.A., Austin, J.A., Jr., Field, M.E., Kravitz, J.H.,

Syvitski, J.P.M., and Wiberg, P.L. (Eds.), *Continental Margin Sedimentation: From Sediment Transport to Sequence Stratigraphy, IAS Spec. Publ., 37*: Oxford (Blackwell Publishing Ltd.), 381–458.

Mountain, G.S., Miller, K.G., Blum, P., et al., 1994. *Proc. ODP, Init. Repts.*, 150: College Station, TX (Ocean Drilling Program).

Pitman, W.C., III, and Golovchenko, X., 1983. The effect of sea level change on the shelf edge and slope of passive margins. *Spec. Publ.—Soc. Econ. Paleontol. Mineral.*, 33:41–58.

Poag, C., Watts A.B., et al., 1987. *Init. Repts. DSDP*, 95: Washington, DC, (U.S. Govt. Printing Office), doi:10.2973/dsdp. proc.95.1987.

Rahmstorf, S., 2007. A semi-empirical approach to projecting future sea-level rise. *Science*, 315(5810):368–370, doi:10.1126/science.1135456.

Sloss, L.L., 1963. Sequences in the cratonic interior of North America. *GSA Bulletin*, 74(2):93–114; doi:10.1130/0016-7606(1963)74[93:SITCIO]2.0.CO;2.

Solomon, S., Qin, D., Manning, M., Chen, Z., Marquis, M., Averyt, K.B., Tignor, M., and Miller, H.L. (Eds.), 2007. *Contribution of Working Group I to the Fourth Assessment Report of the Intergovernmental Panel on Climate Change*: Cambridge, U.K. and New York, U.S. (Cambridge University Press).

Vail, P.R., Mitchum, R.M., Jr., and Thompson, S., III, 1977. Seismic stratigraphy and global changes of sea level, Part 2. The depositional sequence as a basic unit for stratigraphic analysis. *In* Payton, C.E. (Ed.), *Seismic Stratigraphy: Applications to Hydrocarbon Exploration, AAPG Mem.*, 26:53–62.

Van Sickel, W.A., Kominz, M.A., Miller, K.G., and Browning, J.V., 2004. Late Cretaceous and Cenozoic sea-level estimates: backstripping analysis of borehole data, onshore New Jersey. *Basin Res.*, 16:451–465, doi:10.1111/j.1365-2117.2004.00242.x.

Watkins, J.S., and Mountain, G.S., 1990. Role of ODP drilling in the investigation of global changes in sea level. Report of a JOI/USSAC Workshop, El Paso, Texas, 24–26 October 1988, 70 pp.

Watts, A.B., and Steckler, M.S., 1979. Subsidence and eustasy at the continental margin of eastern North America. *In* Talwani, M., Hay, W., and Ryan, W.B.F. (Eds.), *Deep Drilling Results in the Atlantic Ocean: Continental Margins and Paleoenvironment:* Washington, DC (American Geo-physical Union), 218–234.

Watts, A.B., and Thorne, J.A., 1984. Tectonics, global changes in sea-level and their relationship to stratigraphic sequences at the U.S. Atlantic continental margin. *Mar. Pet. Geology*, 1:319–339, doi:10.1016/0264-8172(84)90134-X.

Authors

Gregory Mountain (Co-chief Scientist), Department of Earth and Planetary Sciences, Rutgers University, 610 Taylor Road, Piscataway, NJ 08854, U.S.A., e-mail: gmtn@rci.rutgers.edu.

Dr. Jean-Noël Proust (Co-chief Scientist), Géosciences, CNRS, Université Rennes1, Campus de Beaulieu, 35042 Rennes, France, e-mail: jean-noel.proust@univ-rennes1.fr.

IODP Expedition 313 Science Party

Co-Chief Scientists: Gregory Mountain (U.S.), Jean-Noël Proust (France); ESO Staff Scientists: David McInroy (U.K.), Dayton Dove (U.K.); Sedimentologists: Hisao Ando (Japan), James Browning (U.S.), Stephen Hesselbo (U.K.), David Hodgson (U.K.), Marina Rabineau (France), Peter Sugarman (U.S.); Stratigraphic Correlators: Maria-Angela Bassetti (France), Kenneth Miller (U.S.), Donald Monteverde (U.S.); Paleontologists: Baoqi Huang (China), Miriam Katz (U.S.), Ulrich Kotthof (Germany), Denise Kulhanek (U.S.), Francine McCarthy (Canada); Petrophysicists: Jenny Inwood (U.K.), Johanna Lofi (France), Christophe Basile (France,) Christian Bjerrum (Denmark), Hironori Otsuka (Japan), Henna Valppu (Finland); Porewater Geochemists: Takeshi Hayashi (Japan,) Michael Mottl (U.S.); Paleomagneticists: Youn Soo Lee (Korea), Andres Nilsson (Sweden); Microbiologist: Susanne Stadler (Germany)

Related Web Links

http://www.montco.com
http://www.dosecc.org
http://www.eso.ecord.org/expeditions/313/313.php
http://publications.iodp.org/preliminary_report/313/313_t1.htm
http://publications.iodp.org/preliminary_report/313/313_t2.htm
http://publications.iodp.org/scientific_prospectus/313/313_t5.htm#1022416

Figures and Photo Credits

Figure 2. photo cortesy of ESO/IODP
Figure 5. by Gregory Mountain

28

Establishing Sampling Procedures in Lake Cores for Subsurface Biosphere Studies: Assessing *In Situ* Microbial Activity

by Aurèle Vuillemin, Daniel Ariztegui, Crisogono Vasconcelos, and the PASADO Scientific Drilling Party

Introduction

Sub-recent sediments in modern lakes are ideal to study early diagenetic processes with a combination of physical, chemical, and biological approaches. Current developments in the rapidly evolving field of geomicrobiology have allowed determining the role of microbes in these processes (Nealson and Stahl, 1997; Frankel and Bazylinski, 2003). Their distribution and diversity in marine sediments have been studied for some years (Parkes et al., 1994; D'Hondt et al., 2004; Teske, 2005). Comparable studies in the lacustrine realm, however, are quite scarce and mainly focused on the water column (Humayoun et al., 2003) and/or very shallow sediments (Spring et al., 2000; Zhao et al., 2007). Thus, there is a need to determine the presence of living microbes in older lacustrine sediments, their growth, and metabolic paths, as well as their phylogenies that seem to differ from already known isolates.

During the PASADO (Potrok Aike Maar Lake Sediment Archive Drilling Project) ICDP (International Continental Scientific Drilling Program) drilling, more than 500 meters of sedimentary cores were retrieved from this crater lake (Zolitschka et al., 2009). A 100-m-long core was dedicated to a detailed geomicrobiological study and sampled in order to fill the gap of knowledge in the lacustrine subsurface biosphere.

Here we report a complete *in situ* sampling procedure that aims to recover aseptic samples as well as determining active *in situ* biological activity. Preliminary results demonstrate that these procedures provide a very useful semi-quantitative index which immediately reveals whether there are biologically active zones within the sediments.

The PASADO Project

Laguna Potrok Aike is a 770-ka-old maar lake located at $51°58'$ S and $70°22'$ W in the Santa Cruz Province, Argentina, within the 3.8-Ma-old Pali Aike Volcanic Field (Fig. 1; Zolitschka et al., 2006). Although annual precipitation ranging between 200 mm and 300 mm gives a semi-arid character to the area, the lake is presently the only permanently water-filled lacustrine system in the southeastern Patagonian steppe. Today it has a maximum diameter of 3.5 km, a total surface of 7.74 km^2, and a maximum water depth of 100 m. The lake regime is polymictic, and the water-column is non-stratified with an anoxic sediment-water interphase.

A seismic study of this lacustrine basin showed a thick sedimentary sequence (Anselmetti et al., 2009; Gebhardt et al., in review) that was the target of the PASADO project. This international research initiative had a key objective: quantitative climatic and environmental reconstruction of this remote area through time. The multi-proxy study also provides unique material to initiate, for the first time in an ICDP project, a systematic study of the living lacustrine subsurface environment. From a total of 533 meters of sediment cores recovered at 100 m

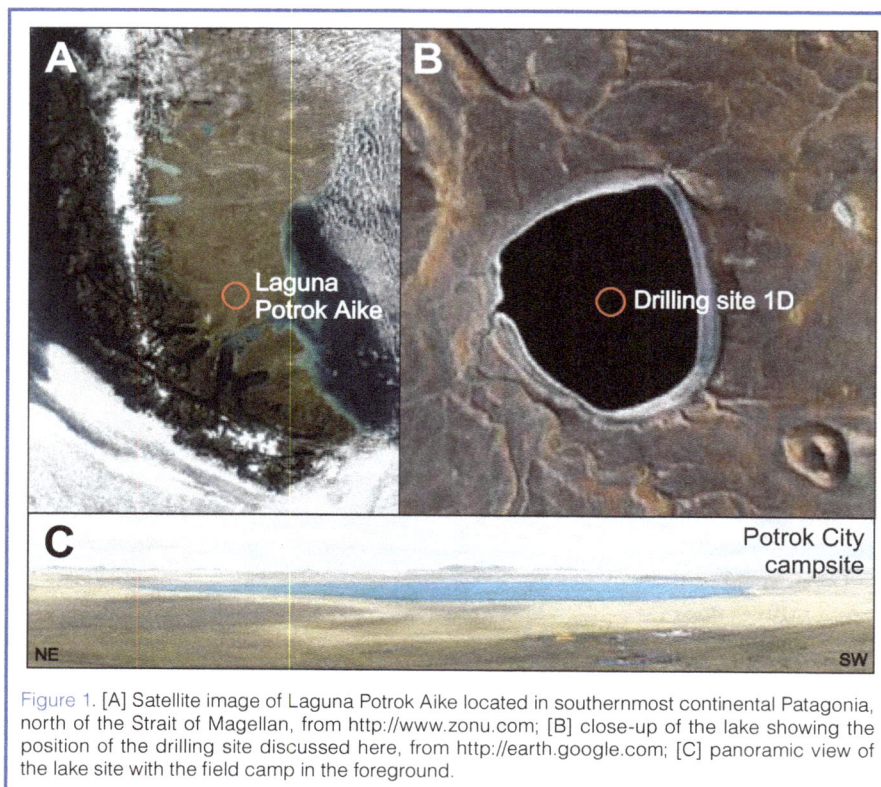

Figure 1. [A] Satellite image of Laguna Potrok Aike located in southernmost continental Patagonia, north of the Strait of Magellan, from http://www.zonu.com; [B] close-up of the lake showing the position of the drilling site discussed here, from http://earth.google.com; [C] panoramic view of the lake site with the field camp in the foreground.

water depth (Fig. 1), a one-meter-long gravity core PTA-1I and the 97-m-long hydraulic piston core PTA-1D were sampled following a newly established strategy to obtain aseptic samples for geomicrobiological studies.

Sampling Procedure

A procedure was designed to minimize contamination risks in the field and laboratory. The size and configuration of the drilling platform prevented the setting up of a sampling laboratory with maximum conditions of asepsis. Thus, the retrieved cores were transported every 90 min from the platform to a laboratory in the campsite where they were sampled (Fig. 2). The liners of hydraulic cores were first disinfected with isopropanol and then sprayed with fungicide. Thereafter, cut in the liner using a portable circular saw every one or two meters and at higher resolution for the upper 15 m (Fig.3). Conversely, in the gravity core twenty windows were cut at 5-cm spacing in the empty liner and sealed with strong adhesive tape prior to coring. This latter technique facilitated opening windows and allowed sampling quickly at a higher resolution. Samples from these windows were immediately chemically fixed and/or frozen, optimizing the preservation of their initial conditions for further analyses.

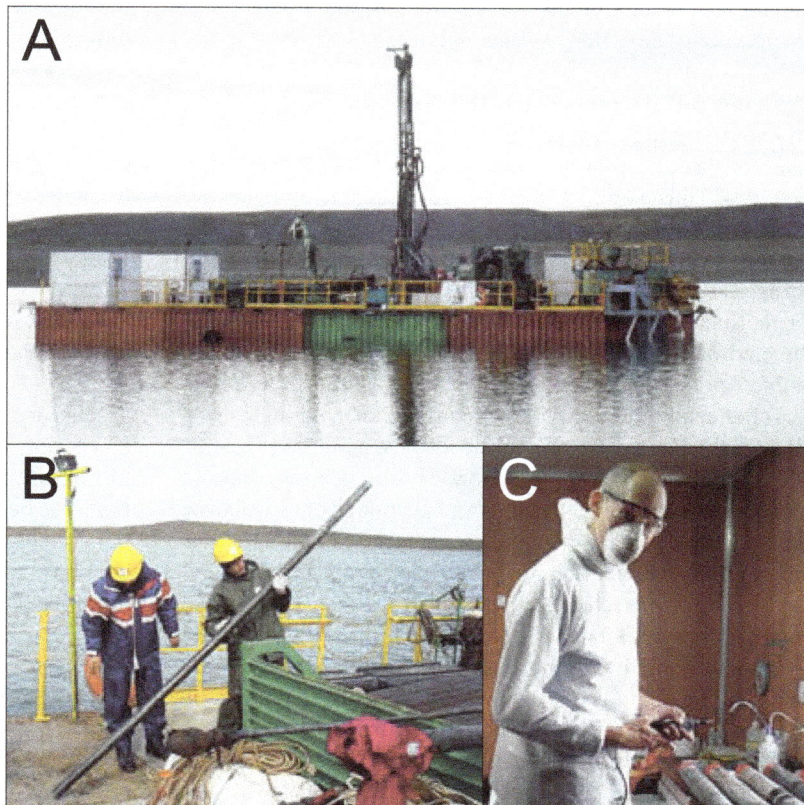

Figure 2. [A] Drilling platform GLAD 800. After retrieval [B], the cores were transported from the platform to the laboratory where they were sampled at once [C].

A rapid biological activity test, which is commercially available for industrial hygiene monitoring, was applied immediately after coring in order to test for microbial activity in the sediments. *In situ* adenosine-5'-triphosphate (ATP) measurements were taken as an indication of living organisms within the sediments. The presence of ATP is a marker molecule for metabolically active cells (Bird et al., 2001), since it is not known to form abiotically. ATP can be easily detected with high sensitivity and high specificity using an enzymatic assay (Lee et al., 2010).

ATP + luciferin + O_2 –> AMP + oxyluciferin + PPi + CO_2 + light

ATP is degraded to adenosine monophosphate (AMP) and pyrophosphate (PPi) while luciferin is oxidized. Light is emitted as a result of the reaction, and the light is detected by a photomultiplier. We used the Uni-Lite® NG Luminometer (Biotrace International Plc, Bridgend, U.K.), in combination with the "Clean-Trace" and "Aqua-Trace" swab kits (3M, U.S., Fig. 3E). The sensitivity of the test is on the order of 10^{-20} moles of ATP per mL of water, corresponding to a standard of 5 cells of *Escherichia coli* as expressed in RLU (relative luminescence units). This handheld device was previously tested at the Geomicrobiology Laboratory, ETH Zurich (Switzerland), where it was determined that this method could be applied on geological material such as rock surfaces and other environmental biofilms. It was also successfully used for fast and accurate measurements of life activity for freshly retrieved cores in lithified sediments of the IODP Expedition 310 in Tahiti (Camoin et al., 2007). The performance of this instrument in fresh sediments was uncertain, however, and to our knowledge this is the first time that it was successfully applied to lacustrine sediments. Additionally, the application of this test to water samples can aid in the evaluation of

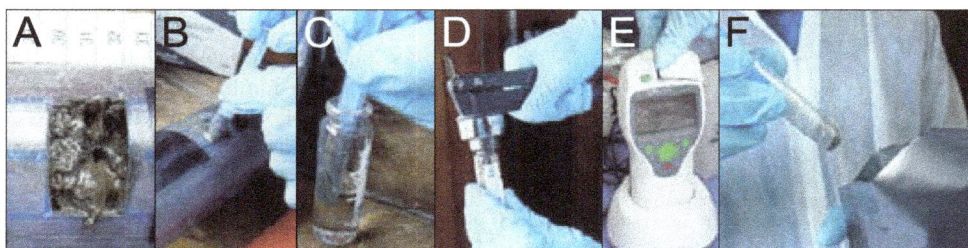

Figure 3. [A] Window cut for sampling; [B-D] sampling for methane headspace determinations; [E] preparation of the sample for *in situ* ATP measurements: sample is mixed with deionized water prior to centrifugation, then tested with the Uni-Lite® NG water tester (shown); and [F] storage of the remaining sediment for cell culture. Refer to text for details.

the degree of contamination of the drilling water which percolates along the inside of the core liner.

Figure 3A-3F summarizes the sequence and sampling procedures established in this project. Part of the sampling required precise volumes that were obtained using sterile syringes. Thus, samples of 3 mL and 5 mL of sediment were extracted from freshly opened windows using these syringes whose narrow tips were cut off in order to collect "minicores" (Fig. 3B). The first extracted sample was designated for methane analyses because of its immediate release into the environment due to volume expansion when exposed to ambient pressure. Hence, a portion (3 mL) of this first sample was chemically stabilized using 10 mL of 2.5% sodium hydroxide, and then sealed in vials for headspace analysis (Figs. 3C and 3D). The sediments were further sampled for different techniques using 5-mL syringes and portioned out as follows: the first 1-mL portion of sample was placed in an Eppendorf tube and kept frozen for further DNA extraction; a second 1-mL portion was chemically fixed in formaldehyde (final concentration, 2%) for DAPI (4',6-diamidino-2-phenylindole) cell count; a third 1-mL portion of the sediment was mixed with 1-mL of deionized water in an Eppendorf tube and centrifuged for five minutes. Commercially available water testers (Biotrace International) were carefully submerged in the supernatant, and ATP content was measured with the Uni-Lite® NG luminometer as an index of *in situ* microbial activity (Fig. 3E). The remaining sediment in the syringe was coated with plastic foil and hermetically sealed into alu-

minum foil bags (Fig. 3F). These bags were flushed with nitrogen (to prevent oxidation) prior to sealing with a heating device. These samples can be further used for microbial culture experiments back at the home laboratory. Once the sampling was accomplished, the windows were sealed with strong adhesive tape. This sampling procedure was carried out non-stop over a 48-hour period. A comparable sampling procedure for marine sediments can be found in Bird et al. (2001).

Assessing *In Situ* Microbial Activity in Sediments

The presence of nutrients as energy sources is critical, promoting an active behavior of the inner microbial communities within sediments. When certain nutrient concentrations are below a threshold, microbial metabolism and population density are lowered progressively as these microbial communities enter in dormant state. Thus, microbial communities in deep sediments can be considered as mainly oligotrophic and dormant.

The 97-m-long sediment core retrieved from Laguna Potrok Aike provided us the opportunity to identify a transition from a weak but active to a dormant state of microbial communities as reflected by *in situ* ATP measurements (Fig. 4A). These results were further compared with those from DAPI counting on the fixed samples carried out several months later in the laboratory (Fig. 4B). The DAPI fluorochrome dyes DNA without distinction —active, dormant, and dead cells, either eukaryote or prokaryote—and it is considered as a semi-quantitative index of cell density within the sediment. ATP and DAPI datasets, however, show an increasing trend from the sediment surface to ~6-m depth within sediments mainly composed of black mud and subject to gas expansion. The DAPI and ATP trends throughout depth suggest an exponential decrease in microbial activity that is most probably linked to a progressive compaction and gradual nutrient depletion within the sediments. There is, however, detectable microbial activity down to 40–50 m and recoverable DNA down to 60 m sediment depth.

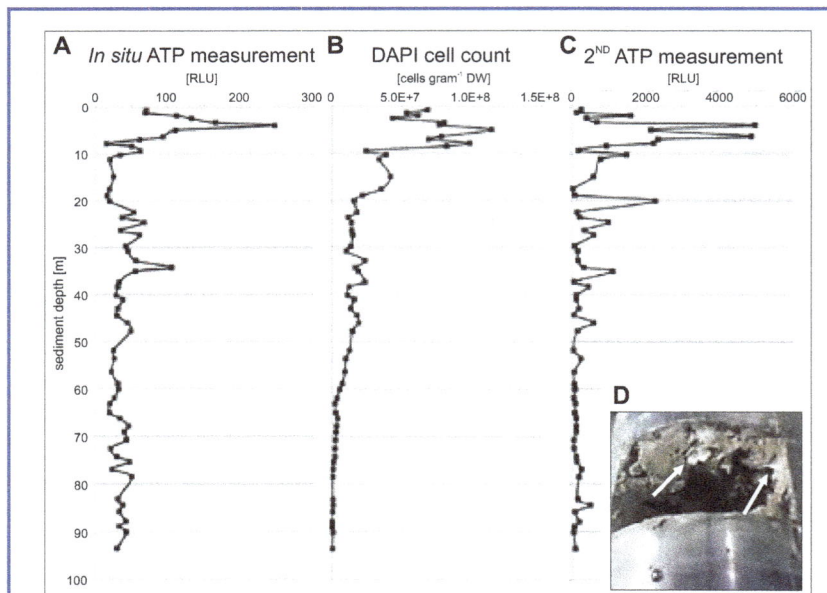

Figure 4. [A] The first ATP measurements were taken in an average of an hour and a half after each core recovery. They are considered as excellent indicators of *in situ* microbial activity. Noise was measured around 30 RLU (relative luminescence unit); [B] DAPI cell count provides a quantification of DNA present in the same samples; [C] second ATP measurements performed ten months later to test for eventual shifts in microbial activity. Although ATP indexes of active layers increased up to 20-fold, the originally nutrient-depleted layers remained inactive. Insert [D] shows a picture of mold (white arrows) which developed after exposure of the sediments to oxygen and pressure temperature (PT) ambient conditions. This partially caused the increased ATP values for the second run of measurements.

The sediments recovered from Laguna Potrok Aike are dominantly argillaceous but are occasionally interrupted by coarser sandy layers associated to slumps triggered by erosional and/or volcanic activities (Zolitschka et al., 2009). The latter are very important since allochthonous organic matter is harder to degrade, and microbial pres-ervation is highly dependent on grain size. Different sediment features further constrain microbial activity, as they provide colonization niches. Although microbial communities may adapt to trophic changes by shifting either their activity and/or dominant species, they are still highly representative of the lake catchment and their dominating climate. Ongoing multiproxy analyses of these cores will allow char-acterizing the sedimentary sequence and provide the critical grounds to interpret the results of the observed microbial behavior.

Validating *In Situ* ATP Measurements

Metabolic microbial activity can change drastically when samples are exposed to ambient temperature and pressure, light, and oxygen. In order to identify and possibly quantify the magnitude of these metabolic changes, a second set of ATP measurements was produced ten months after cores were retrieved (Fig. 4C). Both results indicate very similar distributions of microbial activity displaying the highest values at the same depths. In spite of the liner disinfection and the sealing of the sampling windows, mold had grown superficially on some windows, as shown in Figure 4D. The development of mesophilic aerobic microorganisms explains the comparatively higher ATP index of this second data set. These measurements warn about the omnipresent risks of contamination during sampling and further storage of the samples. They secondarily provide information about the nutrient resources of the sediments and their accessibility and use by microbes. Thus, this comparison between *in situ* and later ATP measurements highlights the relevance of the immediate measurement of microbiological living activity in the field. The comparison presented here between ATP values quickly obtained with a handset device further validates those *in situ* results produced by more established and tedious analyses such as DAPI cell counting of microbial cells.

Future Improvements in Detecting the Living Biosphere in Lake Sediments

Lacustrine systems gather widely diverse water types such as brackish (Banning et al., 2005), acidic (Chan et al., 2002), hypersaline (Cytryn et al., 2000), or alkaline (Jones et al., 1998), among others. Each of them contains very different sediment and associated microbial assemblages. Understanding trophic states within the water columns and the sediments is essential to reconstructing past climates (Nelson et al., 2007) as well as to managing anthropogenic impact on modern lakes (Ye et al., 2009).

The assessment of microbial activity presented here provides information on various ongoing organic matter mineralization processes in the sediments and helps to understand the influence of microbes during early diagenesis. Our procedure can be easily applied as routine, adding valuable microbiological information that is complementary and relevant to several standard lacustrine proxies such as the stable isotope composition of authigenic carbonates and

organic matter. Thus, the Uni-Lite® NG ATP tester is an excellent alternative to previously proposed complex ATP extractions (Stoeck et al., 2000; Bird et al., 2001; Nakamura and Takaya, 2003).

We are confident that the sampling protocol proposed here will allow scientists to sample cores in other ICDP projects with minimal contamination risks. It further points towards new research avenues and technical developments to better detect microbial activity and metabolic functions of the subsurface lacustrine biosphere.

Acknowledgements

We are indebted to S. Templer (MIT, Boston, U.S.) for productive discussions and introducing us to geomicrobiological sampling techniques. C. Recasens, R. Farah (University of Geneva, Switzerland) and C. Mayr (University of Erlangen, Germany) are kindly acknowledged for their help during field sampling. We thank the PASADO Scientific Drilling Party for fruitful discussions and help during drilling operations. B. Zolitschka's comments on an earlier version of the manuscript are specially acknowledged.

Funding for drilling was provided by the ICDP, the German Science Foundation (DFG), the Swiss National Funds (SNF), the Natural Sciences and Engineering Research Council of Canada (NSERC), the Swedish Vetenskapsradet (VR), and the University of Bremen. We are also grateful to the Swiss National Science Foundation (Grant 200020-119931/2 to D. Ariztegui) and the University of Geneva, Switzerland.

References

Anselmetti, F.S., Ariztegui, D., De Batist, M., Gebhardt, C., Haberzettl, T., Niessen, F., Ohlendorf, C., and Zolitschka, B., 2009. Environmental history of southern Patagonia unraveled by the seismic stratigraphy of Laguna Potrok Aike. *Sedimentology* 56/4:873–892, doi:10.1111/j.1365-3091.2008.01002.x.

Banning, N., Brock, F., Fry, J.C., Parkes, R.J., Hornibrook, E.R.C., and Weightman, A.J., 2005. Investigation of the methanogen population structure and activity in a brackish lake sediment. *Environ. Microbiol.*, 7:947–960, doi:10.1111/j.1462-2920.2004.00766.x.

Bird, D.F., Juniper, S.K., Ricciardi-Rigault, M., Martineu, P., Prairie, Y.T., and Calvert, S.E., 2001. Subsurface viruses and bacteria in Holocene/Late Pleistocene sediments of Saanich Inlet, BC: ODP Holes 1033B and 1034B, Leg 169S. *Mar. Geol.*, 174:227–239, doi:10.1016/S0025-3227(00)00152-3.

Camoin, G.F., Iryu, Y., McInroy, D.B., and Expedition 310 Scientists, 2007. *Proc. IODP*, 310: College Station, TX (Integrated Ocean Drilling Program Management International, Inc.).

Chan, O.C., Wolf, M., Hepperle, D., and Casper, P., 2002. Methanogenic archaeal community in the sediment of an artificially partitioned acidic bog lake. *FEMS Microbiol. Ecol.*, 42:119–129, doi:10.1111/j.1574-6941.2002.tb01001.x.

Cytryn, E., Minz, D., Oremland, R.S., and Cohen, Y., 2000. Distribution and diversity of Archaea corresponding to the limnological cycle of a hypersaline stratified lake (Solar Lake, Sinai, Egypt). *Appl. Environ. Microbiol.,* 66:3269–3276, doi:10.1128/AEM.66.8.3269-3276.2000.

D'Hondt, S., Jorgensen, B.B., Millet, D.J., Batzke, A., Blake, R., Cragg, B.A., Cypionka, H., Dickens, G.R., Ferdelman, T., Hinrichs, K.-U., Holm, N.G., Mitterer, R., Spivack, A., Wang, G., Bekins, B., Engelen, B., Ford, K., Gettemy, G., Rutherford, S.D., Sass, H., Skilbeck, C.G., Aiello, I.W., Guèrin, G., House, C.H., Inagaki, F., Meister, P., Naehr, T., Niitsuma, S., Parkes, R.J., Schippers, A., Smith, D.C., Teske, A., Wiegel, J., Padilla, C.N., and Acosta, J.L.S., 2004. Distributions of microbial activities in deep subseafloor sediments. *Science,* 306: 2216–2221, doi:10.1126/science.1101155.

Frankel, R.B., and Bazylinski, D.A., 2003. Biologically induced mineralization by bacteria. *Biomineralization,* 54:95–114.

Gebhardt, C.A., De Batist, M., Niessen, F., Anselmetti, F.S., Ariztegui, D., Kopsch, C., Ohlendorf, C., and Zolitschka, B., in review. Origin and evolution of Laguna Potrok Aike maar (Southern Patagonia, Argentina) as revealed by seismic refraction and reflection data. *Geophys. J. Intl.*

Humayoun, S.B., Bano, N., and Hollibaugh, J.T., 2003. Depth distribution of microbial diversity in Mono Lake, a meromictic soda lake in California. *Appl. Environ. Microbiol.,* 69:1030–1042, doi:10.1128/AEM.69.2.1030-1042.2003.

Jones, B.E., Grant, W.D., Duckworth, A.W., and Owenson, G.G., 1998. Microbial diversity of soda lakes. *Extremophiles,* 2:191–200, doi:10.1007/s007920050060.

Lee, H.J., Ho, M.R., Bhuwan, M., Hsu, C.Y., Huang M.S., Peng H.L., and Chang H.Y., 2010. Enhancing ATP-based bacteria and biofilm detection by enzymatic pyrophosphate regeneration. *Analytical Biochemistry,* 399:168-173, doi:10.1016/j.ab.2009.12.032.

Nakamura, K.-I., and Takaya, C., 2003. Assay of phosphatase activity and ATP biomass in tideland sediments and classification of the intertidal area using chemical values. *Mar. Poll. Bull.,* 47:5–9, doi:10.1016/S0025-326X(02)00471-X.

Nealson, K.H., and Stahl, D.A., 1997. Microorganisms and biogeochemical cycles: what can we learn from layered microbial communities? *Rev. Mineral. Geochem.,* 35:5–34.

Nelson, D.M., Ohene-Adjei, S., Hu, F.S., Cann, I.K.O., and Mackie, R.I., 2007. Bacterial diversity and distribution in the Holocene sediments of a northern temperate lake. *Microb. Ecol.,* 54:252–263, doi:10.1007/s00248-006-9195-9.

Parkes, R.J., Cragg, B.A., Bale, S.J., Getliff, J.M., Goodmann, K., Rochelle, P.A., Fry, J.C., Weightman, A.J., and Harvey, S.M., 1994. Deep bacterial biosphere in Pacific Ocean sediments. *Nature,* 371:410–413, doi:10.1038/371410a0.

Spring, S., Schulze, R., Overmann, J., and Schleifer, K.-H., 2000. Identification and characterization of ecologically significant prokaryotes in the sediment of freshwater lakes: molecular and cultivation studies. *FEMS Microbiol. Rev.,* 24:573–590, doi:10.1111/j.1574-6976.2000.tb00559.x.

Stoeck, T., Duineveld, G.C.A., Kok, A., and Albers, B.P., 2000. Nucleic acids and ATP to assess microbial biomass and activity in a marine biosedimentary system. *Mar. Biol.,* 137:1111–112, doi:10.1007/s002270000395.

Teske, A.P., 2005. The deep subsurface biosphere is alive and well.

Trends Microbiol., 13(9):402–404, doi:10.1016/j.tim.2005.07.004.

Ye, W., Liu, X., Lin, S., Tan, J., Pan, J., Li, D., and Yang, H., 2009. The vertical distribution of bacterial and archaeal communities in the water and sediment of Lake Taihu. *FEMS Microbiol. Ecol.,* 70:263–276, doi:10.1111/j.1574-6941.2009.00761.x.

Zhao, X., Yang, L., Yu, Z., Peng, N., Xiao, L., Yin, D., and Qin, B., 2007. Characterization of depth-related microbial communities in lake sediments by denaturing gradient gel electrophoresis of amplified 16S rRNA fragments. *J. Environ. Sci.,* 20:224–230, doi:10.1016/S1001-0742(08)60035-2.

Zolitschka, B., Anselmetti, F., Ariztegui, D., Corbella, H., Francus, P., Ohlendorf, C., Schäbitz, F., and the PASADO Scientific Drilling Team, 2009. The Laguna Potrok Aike Scientific Drilling Project PASADO (ICDP Expedition 5022). *Sci. Drill.,* 8:29–34.

Zolitschka, B., Schäbitz, F., Lücke, A., Clifton, G., Corbella, H., Ercolano, B., Haberzettl, T., Maidana, N., Mayr, C., Ohlendorf, C., Oliva, G., Paez, M.M., Schleser, G.H., Soto, J., Tiberi, P., and Wille, M., 2006. Crater lakes of the Pali-Aike Volcanic Field as key sites of paleoclimatic and paleoecological reconstructions in southern Patagonia, Argentina. *J. S. Am. Earth Sci.,* 21:294–309, doi:10.1016/j.jsames.2006.04.001

Authors

Aurèle Vuillemin and Daniel Ariztegui, Section of Earth & Environmental Sciences, University of Geneva, Rue des Maraîchers 13, CH-1205 Geneva, Switzerland, e-mail: aurele. vuillemin@unige.ch, daniel.ariztegui@unige.ch
Crisogono Vasconcelos, Geological Institute, ETH Zürich, Sonneggstr. 5, 8092 Zürich, Switzerland, e-mail: cris.vasconcelos@erdw.ethz.ch
and the PASADO Scientific Drilling Party

Photo Credits

Figures 1C, 2A–C, 3A–F, and 4D by Aurèle Vuillemin

Related Web Links

http://www.icdp-online.org/
http://www.pasado.uni-bremen.de
http://www.biotraces.com
http://earth.eo.esa.int/satelliteimages/
http://www.zonu.com
http://earth.google.com

29

Design, Manufacture, and Operation of a Core Barrel for the Iceland Deep Drilling Project (IDDP)

Alexander C. Skinner, Paul Bowers, Sverrir Þórhallsson, Guðmundur Ómar Friðleifsson, and Hermann Guðmundsson

Abstract

The science program of the Iceland Deep Drilling Project (IDDP) requires as much core as possible in the transition zone to supercritical and inside the supercritical zone (>374°C), in the depth interval 2400–4500 m. The spot coring system selected has a 7 ¼" (184.15 mm) OD at 10 m length and collects a 4" (101.6 mm) diameter core using an 8 ½" (215.9 mm) OD core bit. It incorporates design characteristics, materials, clearances and bearings compatible with operation of the core barrel at temperatures as high as 600°C. Special attention was given to the volume of flushing which could be applied to the core barrel and through the bit while running in and out of the borehole and while coring. In November 2008 a successful spot coring test using the new core barrel was performed at 2800 m depth in the production well RN-17 B at Reykjanes, Iceland, where the formation temperature is 322°C. A 9.3-m hydrothermally altered hyaloclastite breccia was cored with 100% core recovery, in spite of it being highly fractured. A core tube data logger was also designed and placed inside the inner barrel to monitor the effectiveness of cooling. The temperature could be maintained at 100°C while coring, but it reached 170°C for a very short period while tripping in. The effective cooling is attributed to the high flush design and a top drive being employed, which allows circulation while tripping in or out, except for the very short time when a new drill pipe connection is being made.

Introduction

In late 2003, a member of the IDDP consortium had offered one of its planned exploratory wells—RN-17, located on the Reykjanes peninsula—for deepening by the IDDP (Elders and Friðleifsson, 2005). It was drilled to 3.1 km depth, where it was planned to deepen it further some 2 km, partly by continuous wireline coring. The 3.1-km-deep well was flow tested in November 2005, and it collapsed during that test. In February 2006 the well had to be abandoned after several failed attempts at reconditioning. In 2008, the field operator decided to sidetrack this well southwards. During that operation the request by IDDP came to allow test-coring in this well with the new IDDP coring equipment.

Prior to the RN-17 borehole being occupied for the test spot coring, IDDP had determined that a deep borehole would be scientifically examined by the collection of core and logs. Cost constraints determined that continuous coring would not be possible and that a series of spot cores at strategic intervals would have to be collected. Further cost analyses indicated that it would be cost efficient to purchase a core barrel for the project and that the design could incorporate specific requirements for operation in deep, hot boreholes.

Core Barrel Specifications and Manufacturing

The specification drawn up for international tender can be summarized as follows. A conventional core barrel is required for spot coring work in a deep scientific geothermal borehole. Temperatures in the borehole are expected to exceed 500°C, but with cooling and other measures currently employed in drilling shallower geothermal boreholes, it is hoped to keep coring temperatures below 200°C (Fig. 1). However, they could be as high as 250°C. The temperatures shown in Figure 1 are extrapolated from real data. They indicate the hole temperature over time when continually flushing while drilling or coring, and are the typical temperatures the core barrel could experience at the drilled depths shown. Data is verifiable to 2500 m from current drilling.

The outer core barrel is double-walled with (7" [177.8 mm] OD) with API thread and bit connections, and it has an overall length of 10 m. This overall length should be broken down to allow for efficient transportation and assembly and to incorporate top and bottom stabilizers, possibly a central stabilizer, and the opportunity to run a 5-m or 10-m assembly.

Rig configuration limits tubular handling operations and thus the maximum possible core barrel length to an effective 10-m core run. Stabilizers on the core barrel need to be compatible with coring operations with an 8½" core bit. Core barrel head or crossover sub-dimensions will be finalized when all rig tubular details are also known.

The inner core barrel produces a core of 4" (101.2 mm) diameter. The bearing assembly is water cooled and has no components susceptible to failure up to 250°C. During operation the barrel adjustments and tolerances must make allowance for core barrel heating, which can only be limited by the

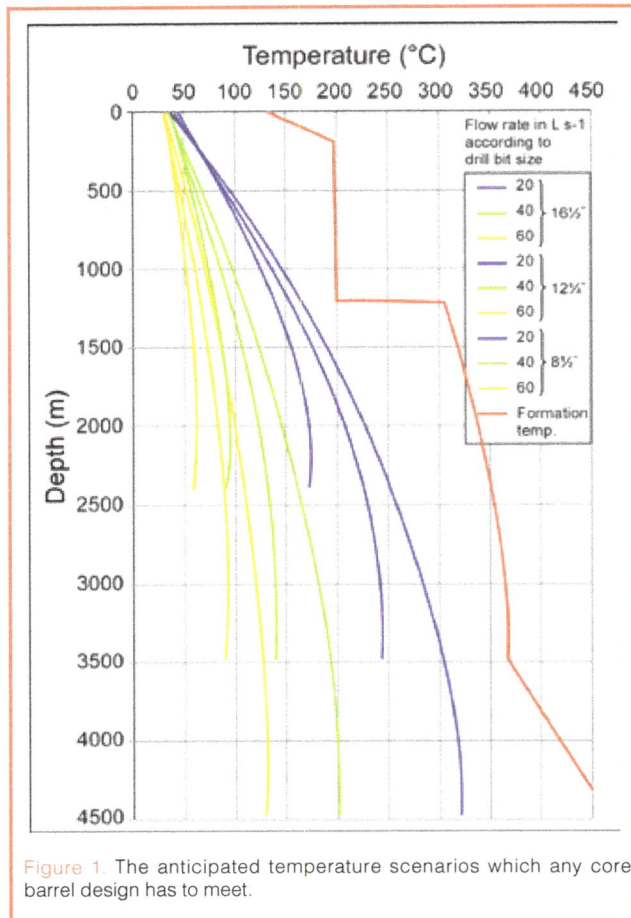

Figure 1. The anticipated temperature scenarios which any core barrel design has to meet.

water flushing. Figure 1 suggests that a minimum flushing capacity of 30–40 L s^{-1} will have to be maintained for all of the tripping time and preferably for at least some of the coring time.

A conventional core spring type catcher and screw on shoe is essential. Other catcher options can also be considered as additions. Consideration may also be needed to keep the catcher shoe cool so it does not expand and interfere with core passage through it. Single or limited re-use inner core barrels will be considered if this is thought beneficial for annular clearances for flow requirements and if the wall thickness is too weak for robust use but sufficient for core collection and stability.

The core barrel should allow for a minimum of 30–40 L s^{-1} flush while running string and should be able to accept 30 L s^{-1} while coring (i.e., there must be sufficient annulus to allow bit cooling at this flush rate with no "lift-off" [hydraulicking]). Because of the danger of a blowout in the borehole, there will be check valves in the string and core barrel. These will preclude sending down steel balls to reduce flow to the core barrel prior to coring. In any event these would also preclude pumping at higher flushing rates should the string be held downhole or tripping delayed for longer than anticipated when the core barrel is attached.

Core bits to be designed for the core barrel have 8 ½" (215.9 mm) OD and (possibly) have peripheral and face discharge to accommodate the volumes of flush required without undue pressure or flushing loss at the bit cutting face. Only spot coring will be undertaken, and this is anticipated to be carried out only at drill bit changes or at other places where the geology may dictate, such as in a lost circulation zone where rock alteration properties would be investigated. Bit life but not robustness can be sacrificed for high penetration rates.

The formation is most likely to be basalt, dolerite, or gabbro as the borehole progresses. It may be fractured and have alteration zones in areas where there may be lost circulation. The rig top drive is well-instrumented and has good bit weight control. It is also capable of 140–200 rpm but using this speed may not be possible due to API string harmonics, and 70–100 rpm may be a more likely useable range. Weight on bit (WOB) can be well controlled, and an appropriate bottom hole assembly (BHA) to suit the bit can be made up, including uphole stabilizers (see below). Information gained from previous large-diameter coring suggests that surface set diamond bits performed well and polycrystalline diamond compact (PDC) bits less well due to breaking, although the penetration rate was good until this happened. Smaller diameter core bits using impregnated diamond bits perform very well and are robust. Provided the rig can accommodate the rotary requiremetnts of such bits, they should also be considered for this spot coring exercise.

Due to borehole depths (>3500 m) and slow drillstring trip times, it is not possible to clean the hole after drilling and before coring, unless there is significant metal junk in the hole, nor is it intended to drill a pilot hole to stabilize the coring bit at start. The core barrel and core bit should therefore be sturdy and aggressive enough to withstand some difficult conditions while establishing the coring regime. However, if the core bit can achieve this and make good penetration rate for 10 m of core, then this will suffice and the bit can be changed if necessary for the next spot core. In order for the project to be financially viable, a penetration rate (given suitable operating parameters) of 2.5 m h^{-1} while coring should be targeted as a minimum in the formations indicated above.

According to the specifications, the core barrels were fabricated by Rok-Max Drilling Tools Ltd., and the core bits were made by GEOGEM Ltd., both UK companies with good track records in making specialist coring equipment and core bits. The core barrel is of all-steel construction. Bearings are heat-treated and plated to withstand an extremely harsh in-hole environment together with high operating temperatures. No rubber or plastic seals are used in its construction, and special high-temperature grease is used to lubricate the bearings and threaded components. It comprises an 11.6-m non-wireline conventional double tube corebarrel, with non-rotating inner tube assembly, and it is specifically

Figure 2. The core bit and core catcher assembly.

designed to take spot cores of 4″ (101.6 mm) diameter x 10 m length and to give maximum flow through the core barrel of at least 40 L s⁻¹ flush for cooling. Special attention was given to the volume of flushing which could be applied to the core barrel and through the bit while running in and out and while coring. This in turn impacts on internal core barrel design of waterways and bearing configuration.

The core barrel bits for use with the system were designed with large waterways and a rounded profile crown to allow an element of hole cleaning when spudding in and maximum cooling when down hole. The composition is impregnated diamond with natural diamond and carbide gauging on both OD and ID. Matrix composition is designed for high-end temperatures and fast wear to allow clean fast cutting over the whole life of the bit. The design of the bit and matrix allows for a rotational speed of 70–160 rpm. The WOB should be 5.45–11.36 tonnes (12,000–25,000 lbs). Generally speaking, the higher rpm used, the lower the WOB, and *vice versa*, always within the given parameters. Flushing with rates as high as 40 L s⁻¹ do not hydraulically influence the bit performance; this was tested at the design stage by the manufacturer. A close-up of bit and core catcher is shown in Fig. 2.

The outer core barrel assembly comprises the following: bit, lower stabilizer section, lower outer barrel body, middle stabilizer section, upper outer barrel body, top stabilizer section, and core barrel head. Stabilizers and core bit are designed for 8½″ (215.9 mm) hole size. There are no landing, latching, or stabilizing rings for the inner core barrel interconnected with the outer assembly. All materials, manufacture, and threads on the core barrel are to API specifications, but some internal core barrel threads are modified. The box thread and diameter on the core barrel head section of the outer core barrel are directly compatible with the drill collars being used. Two types were manufactured to accommodate 6¾″ (171.5 mm) and 8″ (203.2 mm) drill collar types.

The inner core barrel comprises a lower shoe which contains the core catcher or core spring, upper shoe, lower inner barrel stabilizer, lower core barrel section, middle stabilizer section, upper core barrel section, and core barrel bearing assembly. The bearing assembly housing screws into the outer core barrel head, and it is the only fixed point of contact with the outer core barrel. The bearings allow the inner barrel assembly to rotate freely within the outer tube, and adjustment on the bearing shaft ensures that the correct inner to outer spacings are set. The stabilizer sections keep the inner barrel central to the outer tube and the lower shoe. When the inner core barrel assembly length is adjusted via the bearing shaft, the face of the lower shoe sits inside the core bit throat with sufficient clearance for flush but not too much to hamper core ingress to the barrel. The core catcher design is "conventional" wedge spring catchers for competent formations. Heat treatment took into account metal fatigue and operating temperatures.

A digital temperature probe was designed and fitted into a pressure housing at the top of the inner barrel core chamber. Also within the pressure housing, a selection of temperature recording wax "spots" were inserted to allow a backup temperature reading to be recorded. Because this probe projects into the core chamber, it is not possible to take a full 10-m core run, as it could be crushed if full core recovery was achieved. Figure 3 shows fitting of the temperature probe.

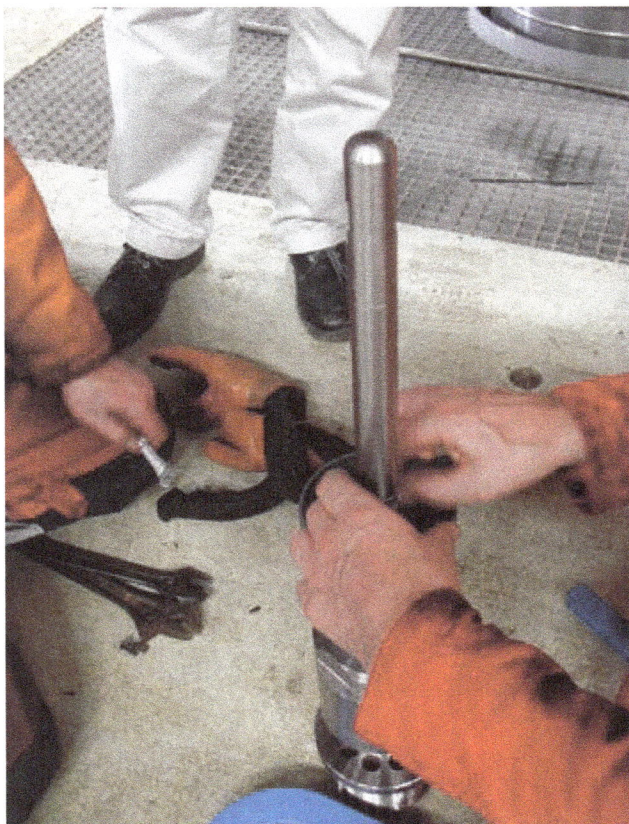

Figure 3. Temperature logger.

Spot Coring Operations

A spot core test was thought prudent as part of the operational planning for IDDP's planned drilling at Krafla, Iceland in order to try maximizing operational information and train the drilling crew in coring procedures. The main aims were as follows.

- Learn what may be the percentage core recovery, the core condition, information on core washing, bit performance and wear
- Approve the bit design based on grading of the used bit, so additional bits can be ordered
- Monitor the function of the core catcher and the bit cooling:

  o Inspect core barrel for adverse temperature effects
  o Read from temperature logger and temperature indicating strips/paint placed inside the core barrel
  o Determine how to optimize the bit cooling procedure with the top-drive while tripping in
  o Learn how much and how long to circulate at each stand

- Learn if any parts are missing or should be modified to speed up the handling of the core barrel
- Learn how to maximize the tripping speed
- Train the drillers and core hands in the proper coring procedures
- Collect information that can be put into a handbook for IDDP
- Collect information relative to 'risk assessment'

- Establish best parameters of rotation, WOB, and sensitivity of recording of parameters on the console to create a profile and add to procedures for spot coring on the full-scale IDDP-1 project in Krafla

The rig used for the spot coring is a Soilmech HH300 Drilling Rig with variable-speed top drive, automated pipe handling, and all safety features necessary for high pressure, high temperature geothermal well drilling. All rig tubulars are to API specification, and the rig crew is extremely competent in the handling and maintenance of all of the tools and machinery.

Data recording of drilling parameters was logged and made available for future analysis. A graphical output of drilling parameters is available to the driller during operation. Figures 4–6 have greatly assisted in allowing a full core barrel penetration to be achieved without any resistance to penetrate further.

Flush rate while coring was varied to observe any difference in core recovery results and recorded core barrel temperatures. The flush rates available while coring are very high compared to other core barrels, and high flush rates can affect both core recovery and penetration rates by washing away core and lifting the bit off cutting contact with the bottom, respectively. We used 40 L s⁻¹ while spudding in and for the first five minutes, then it was cut to 30 L s⁻¹ and after a few more minutes was reduced to 25 L s⁻¹ for the remainder of the coring run. No marked changes in bit weight or penetration rate were observed, so provided the formation is competent, there should be no problems in trying to core with full flush capability in boreholes which have a higher ambient temperature, such as that anticipated at Krafla. This suggests that the bit design meets requirements and allows for full flow through the waterways and then uphole without causing undue pressure build-up at the bit face or in the core barrel head ports.

It was not possible to reach and maintain high rotational speed (>100 rpm) on the drill string without excessive vibration and string movement. This could be due to a number of factors, but certainly the vertical to inclined borehole configuration would have something to do with this. However, rotation without excessive vibration or string movement was possible within the operating bit para-

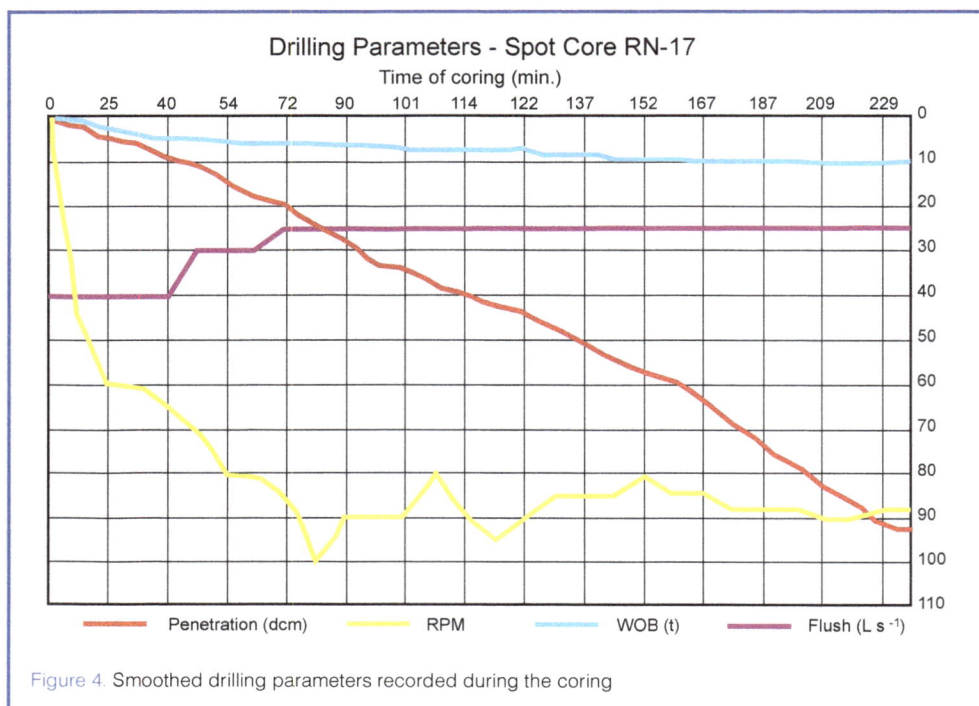

Figure 4. Smoothed drilling parameters recorded during the coring

meters. Generally if low-end rpm is used, then high-end bit weights have to be applied to maintain acceptable penetration and smooth operation.

Spot Coring Results

Indications while coring suggested that core was being collected, but it was not until the pull-off force at the base of the coring run increased markedly then dropped back that it could be said that there was core caught inside the core barrel (which had to be broken off). Then it was not until the string was tripped out, the core barrel dismantled, and the core pumped out of the inner tubes that the full measure of the success of the core run was established.

Core recovery was excellent despite the fact that the core barrel was being run in an inclined borehole and the rock was fractured with many wedge-shaped planar fractures which could have easily caused core jamming. Figures 7 and 8 show the nature of the core collected. The cored section consisted of a hyaloclastite breccia, thoroughly altered to greenschist facies mineralogy. Apparently, this breccia was deposited in shallow marine environment, despite the fact it is now at 2.500 m depth below the surface. The age of this breccia is one of the key questions to be unraveled to

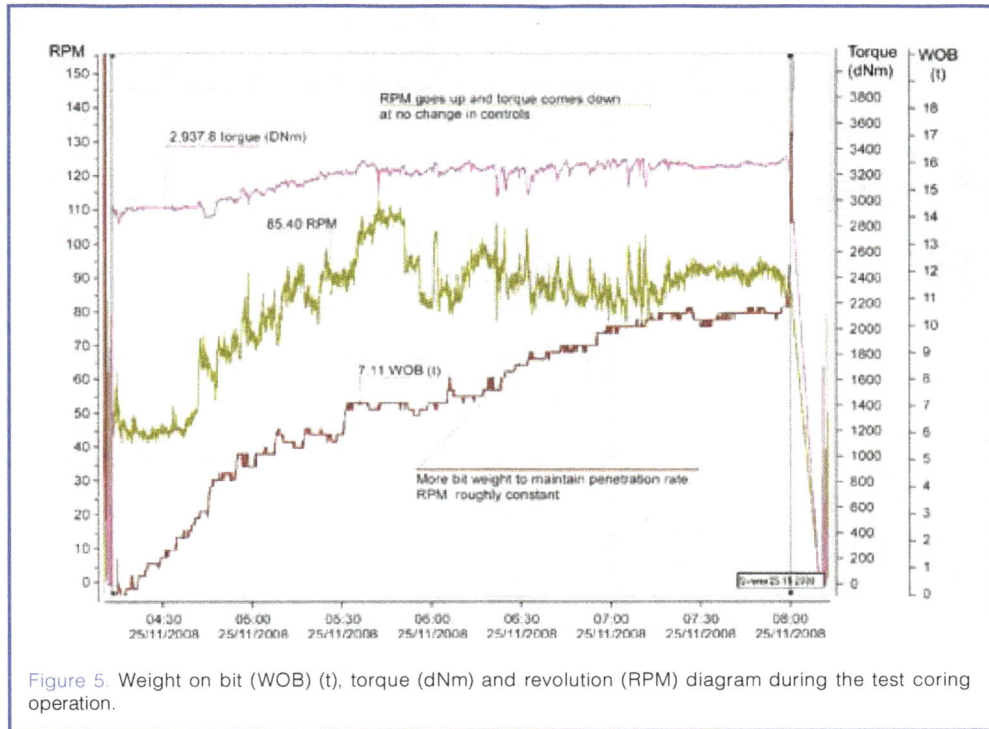

Figure 5. Weight on bit (WOB) (t), torque (dNm) and revolution (RPM) diagram during the test coring operation.

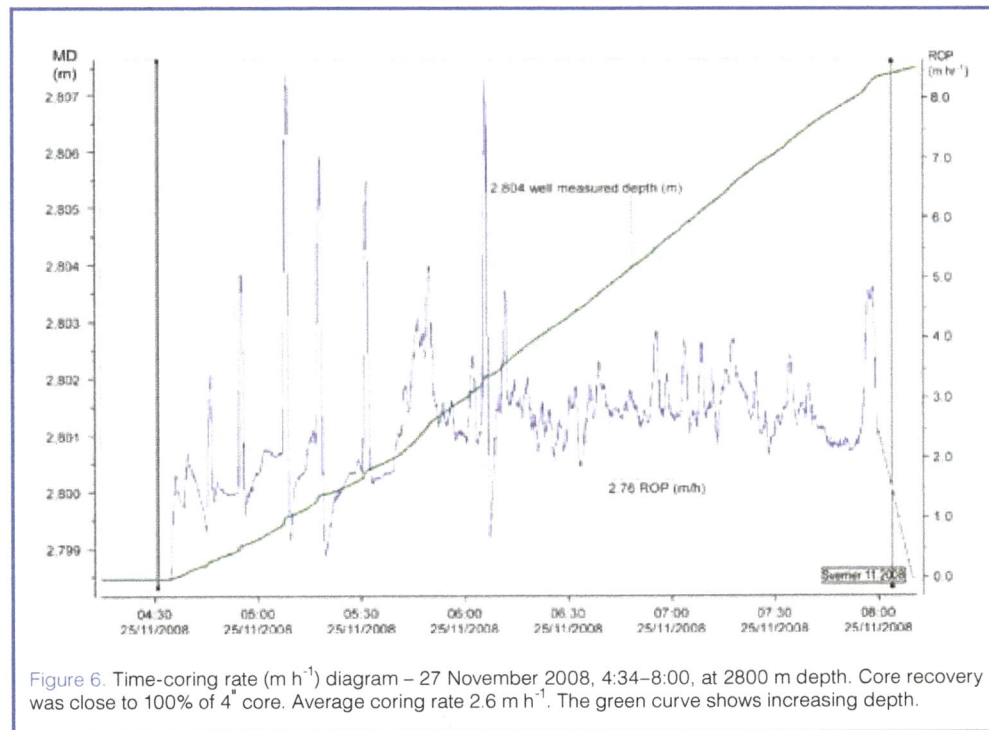

Figure 6. Time-coring rate (m h^{-1}) diagram – 27 November 2008, 4:34–8:00, at 2800 m depth. Core recovery was close to 100% of 4" core. Average coring rate 2.6 m h^{-1}. The green curve shows increasing depth.

allow estimation of the subsidence rate in the middle of the Reykjanes rift zone. The drill core is now being studied, and first results have been published by Friðleifsson and Richter (2010).

In the 35°-inclined RN-17 B test hole, the inner core barrel stabilizers could not operate properly as centralizers, since wear was always on the side "lying down". Although this did not materially affect the coring, it induced more refurbish-

ment of stabilizers than would otherwise occur, and it may be that for inclined coring (not routinely planned) a set of stabilizer rings may have to be incorporated into the outer, rather than the inner, core barrel sections.

Acknowledgements

The work was carried out to an agreed plan, schedule, and extremely tight timescale. The successful outcome is in

Figure 7. Core bit with core securely held in catcher.

Figure 8. Cores boxed for examination.

large measure due to the thought put into the core barrel and bit designs by ROK-MAX and GEOGEM companies and to the efforts of H. Guðmundsson regarding site management of equipment, logistics, and suitable location to set up and service the core barrels. The IDDP committees of Sciences Application Group of Advisors (SAGA) and Deep Vision approved the case for core barrel procurement, which was overseen by Bjarni Palsson of Landsvirkjun. At site the drilling manager, Steinar Már Þórisson, worked with us and both shift crews of the Jardboranir Ltd. drill rig Tyr to allow an excellent test with a good outcome.

We also acknowledge grants from ICDP to G.Ó. Friðleifsson and W.A. Elders, for having the core barrel designed and built and additionally the U.S. NSF (award number EAR-0507625 to Elders) for financing the test coring at Reykjanes. And last but not the least, we thank the IDDP consortium for accepting and supporting the implementation of the core barrel manufacture for scientific purposes.

References

Elders, W.A., and Fridleifsson, G.O., 2005. The Iceland Deep Drilling Project - scientific opportunities. *Proc. World Geothermal Congr.*, Antalya, Turkey, 24–29 April 2005, paper 0626, 6 pp.

Friðleifsson, G.Ó., and Richter, B., 2010. The geological significance of two IDDP-ICDP spot cores from the Reykjanes geothermal field, Iceland. *Proc. World Geothermal Congr.*, Bali, Indonesia, April 25–29 2010, paper 3095, 7 pp.

Authors

Alexander C. Skinner, ACS Coring Services, 13 Riccarton Drive, Currie, Edinburgh, EH14 5PN, Scotland, U.K., e-mail: acscs@blueyonder.co.uk.

Paul Bowers, Rok-Max Drilling Tools Ltd., P.O. Box 87, Truro, Cornwall, TR3 7ZQ, U.K., e-mail: paulbowers@rokmax.com.

Sverrir Þórhallsson, Iceland Geosurvey (ISOR), Grensasvegur 9, Reykjavik, IS-108, Iceland, e-mail: s@isor.is.

Guðmundur Ómar Friðleifsson, HS Orka hf, Brekkustígur 36, 260 Reykjanesbær, Iceland, e-mail: gof@hs.is.

Hermann Guðmundsson, Iceland Geosurvey (ISOR), Grensasvegur 9, Reykjavik, IS-108, Iceland, e-mail: hg@isor.is.

Related Web Links

http://www.icdp-online.org/
http://iddp.is/iddp-papers-at-wgc-2010/

Photo Credits

Figs. 2, 3, 7, and 8 by G.Ó. Friðleifsson, H.S. Orka, and IDDP-PI

30

Integration of Deep Biosphere Research into the International Continental Scientific Drilling Program

by Kai Mangelsdorf and Jens Kallmeyer

Introduction and Workshop Goals

An international workshop on the Integration of Deep Biosphere Research into the International Continental Scientific Drilling Program (ICDP) was held on 27–29 September 2009 in Potsdam. It was organized by the Helmholtz Centre Potsdam GFZ German Research Centre for Geosciences and the University of Potsdam (Germany). Financial support was provided by ICDP. This workshop brought together the expertise of thirty-three microbiologists, biogeochemists, and geologists from seven countries (Finland, Germany, Japan, New Zealand, Sweden, U.K., U.S.A.).

Over the last two decades, microbiological and biogeochemical investigations have demonstrated the occurrence of microbial life widely disseminated within the deep subsurface of the Earth (Fredrickson and Onstott, 1996; Parkes et al., 2000; Pedersen, 2000; Sherwood Lollar et al., 2006).

Considering the large subsurface pore space available as a life habitat, it has been estimated that the biomass of the so-called deep biosphere might be equal to or even larger than that of the surface biosphere (Whitman et al., 1998).

Figure 1. The Mallik Gas Hydrate Research project drill site at the northern edge of the Mackenzie River Delta, Northern Territories Canada. The Mallik project was one of the first ICDP projects containing a deep biosphere component.

Thus, the deep biosphere must play a fundamental role in global biogeochemical cycles over short and long time scales. Its huge size, as well as the largely unexplored biogeochemical processes driving the deep biosphere, makes the investigation of the extent and dynamics of subsurface microbial ecosystems an intriguing and relatively new topic in today's geoscience research. Our knowledge of the deep biosphere is still fragmentary especially in terrestrial environments. While geobiological research is already an integral part in many Integrated Ocean Drilling Program (IODP) sampling missions (Lipp et al., 2008; Roussel et al., 2008; Zink et al., 2003), only a few recent projects within the ICDP have had a geobiological component (Colwell et al., 2005; Gohn et al., 2008; Mangelsdorf et al., 2005). In the recently published book *Continental Scientific Drilling – A Decade of Progress and Challenges for the Future*, Horsfield et al. (2007) argued that exploration of the "GeoBiosphere" should be an integrated component of the activities of ICDP to correct this imbalance.

Thus, the aim of the workshop was to integrate deep biosphere research into ICDP by 1) defining scientific questions and targets for future drilling projects in terrestrial environments and 2) addressing the technical, administrative and logistical prerequisites for these investigations. According to these goals the workshop was segmented into two parts.

Key Scientific Questions and Identification of New Targets

Topics for deep biosphere research in terrestrial systems. With the discovery of deep microbial ecosystems in sedimentary basins–as well as microbial life in granites, deep gold mines, and oil reservoirs—the view of the scientific community was opened to a hidden and largely unexplored inhabited realm on our planet (Fredrickson and Onstott, 1996; Parkes et al., 2000; Pedersen, 2000; Sherwood Lollar et al., 2006). From a surface point of view the deep subsurface is an extreme environment. With increasing burial depth, microbial communities have to cope with increasing temperature and pressure, nutrient limitation, limited porosity and permeability as well as a decrease in the available carbon and energy sources, essentially affecting the composition, extent, life habitats, and the living conditions (Horsfield et al., 2007). The rates of microbial cellular activity in the deep biosphere are estimated to be orders of magnitude lower than those in surface environments. Contrary to surface microbes, which generally have doubling times of minutes to months, the average frequency of cell division of deep subsurface microorganisms is within the range of a century (Fredrickson and Onstott, 1996) and defies our current understanding of the limits of life.

In marine and terrestrial sedimentary basins, buried organic matter is the obvious carbon and energy source for deep microbial life. In the upper part of the sediment column it is thought that intense degradation of organic matter initially increases the recalcitrant proportion of the organic material. However, in deeper successions the bioavailability of organic matter might increase again due to the rising temperature with increasing burial depth, affecting the bond stability of potential substrates in the organic matrix (Parkes et al., 2007; Wellsbury et al., 1997; Glombitza et al., 2009). The organic matter in these ecosystems was initially produced by photosynthesis. Therefore, despite the long delay between production and consumption, these systems are ultimately depending on surface processes.

In environments with little buried organic matter—for instance, in igneous rocks—lithoautotrophic microbial communities are able to synthesize small organic compounds from inorganic sources like hydrogen gas and carbon dioxide, and these simple organic compounds can then be utilized by other heterotrophic microorganisms. Such microbial communities, also called Subsurface Lithoautotrophic Microbial Ecosystems (SLiMEs), form entire ecosystems in the deep subsurface, which are completely independent from photosynthetically produced substrates (Fredrickson and Onstott, 1996; Lin et al., 2005; Lin et al., 2006; Sherwood Lollar et al., 2006; Stevens and McKinley, 1995).

The widely disseminated deep biosphere poses fundamental questions such as the following. What kind of microorganisms populate the deep subsurface? What is their extension and where are their limits? How is their life habitat shaped? What metabolic processes do they perform? What carbon and energy sources do they use? What survival strategies do these microorganisms apply? Does microbial life in the subsurface represent early life on Earth? What is their impact on the global carbon cycle and linked to that on the global climate?

Deep microbes not only create (biogenic gas) but also have the potential to destroy fossil energy resources (biodegradation of oils). Also, research on potential life in the subsurface of other planets (e.g., Mars), studies about the safety of deep nuclear waste disposal sites, bioremediation of polluted sites, deep aquifer exploration, and the search for new biomedicals are drivers of deep biosphere research (Rothschild and Mancinelli, 2001).

Thus, the workshop participants defined general key topics for terrestrial deep biosphere drilling.

- Extent and diversity of deep microbial life and the limits of life
- Subsurface activity and metabolism as well as carbon and energy sources for deep microbial life
- Evolution, survival, and adaptation of deep microbial life
- Resources and applications: natural resources provided or degraded by deep microbial communities and the implication on biotechnology applications
- Interaction of the deep biosphere with the geosphere and implication of deep microbial activity on Earth's climate
- Deep biosphere as a model for early life on Earth and life on other planets

The workshop also addressed more specific topics including a potential target for already scheduled ICDP drilling campaigns of forming a base for a future dedicated ICDP deep biosphere project, and sampling and curation standards to support interoperability between different drilling operations.

Diversity and extent of the deep subsurface biosphere. One exciting question is whether deep terrestrial and lacustrine subsurface communities differ from the subseafloor biosphere (and if yes, how)? There are many structural corresponding subterrestrial and subseafloor environments (e.g., subsurface ecosystems in pressure enhanced hyperthermophilic systems and subseafloor hydrothermal vent systems) that allow a comparison of the indigenous subsurface and subseafloor microbial communities.

In addition to the spatial distribution and diversity of deep microbial life, the question on the temporal diversity in a given subsurface location due to changes in the environmental conditions was addressed. To monitor such changes, the installation of monitoring devices (e.g., osmo samplers), measuring the chemistry and flow rates of fluids, and the use of cartridges with different substrate media were suggested as part of the formation of a natural laboratory under *in situ* conditions.

While investigating the deep biosphere, participants focused on microbial communities. Other components such as phages were mainly overlooked. In the last few years the investigation of phages has become a new topic in deep biosphere research (Engelen et al., 2009). Viral infection of deep microbial communities was discussed as a controlling factor for the deep biosphere through exchange of DNA, killing microorganisms, and also providing essential substrates for non-infected microorganisms due to the viral-induced release of cell components from infected cells (the so-called viral shunt). Thus, this intriguing new scientific field should also form an integrated part of future deep biosphere research in terrestrial systems.

Limits of subsurface life. What determines the biogeography of microorganisms in the deep subsurface? Conceivable factors are the grain size, pore space, and permeability of the sediments and rocks. Furthermore, the content, distribution, and kind of organic matter—and, therefore, the availability of carbon and energy sources—as well as deep fluids and ambient temperature play a significant role with all having strong impact on the habitability in the deep realm. Knowing these factors, can we predict the distribution, extent, and composition of the deep microbial communities? Are deep biosphere communities exploiting all low temperature (<150°C) parts of the rock cycle, and how do the physical and geochemical characteristics of the habitat determine the abundance and distribution of the microbial communities?

In this context it is of interest whether paleopasteurization (Wilhelms et al., 2001) exists in, for instance, organic-rich sediments initially subsided to greater depth with corresponding high ambient temperatures and subsequent uplift into temperature regimes that are usually compatible with deep microbial life. Another aspect is how the biogeography of microbial communities of isolated (closed) terrestrial systems (e.g., no contact to meteoric fluid flows) differs from open systems in time and space, especially with respect to speciation and survival strategies (adaptation/repair mechanisms) in the deep biosphere. Are there ecological

niches, and is there competition between the deep microorganisms affecting the community structure and evolutionary processes?

Processes and interactions. In sediments the deposited and subsided organic matter forms the carbon and energy sources for the indigenous microbial life. Initially, the recalcitrant proportion of the organic matter increases with ongoing subsidence and maturation. On the other hand, early geothermally driven degradation processes already start at comparable low temperatures (>50°C) gradually providing again potential substrates for the deep biosphere (Horsfield et al., 2006). Is there an overlap between biogenic and thermogenic processes being a feedstock for deep terrestrial microbial ecosystems, and what is the role and importance of geological processes to sustain biological systems in the deep biosphere?

As we examine the potential feedstock sources for deep microbial life, it is also of interest to investigate the impact of CO_2, N_2, and radioloytic H_2 production and/or O_2 generation on the metabolic processes and the composition of deep microbial ecosystems, such as in lithoautotrophic communities in a range of different subsurface systems (e.g., U-rich systems, seismically active zones, and high temperature (energy) regimes). What are the rates of formation of these "geogases", and how do the rates change with different settings, depth, and formation ages?

Oil, gas, and coal reservoirs form potential carbon and energy sources for deep microbial life. Thus, what is the relative importance of natural but also abiotic hydrocarbons for

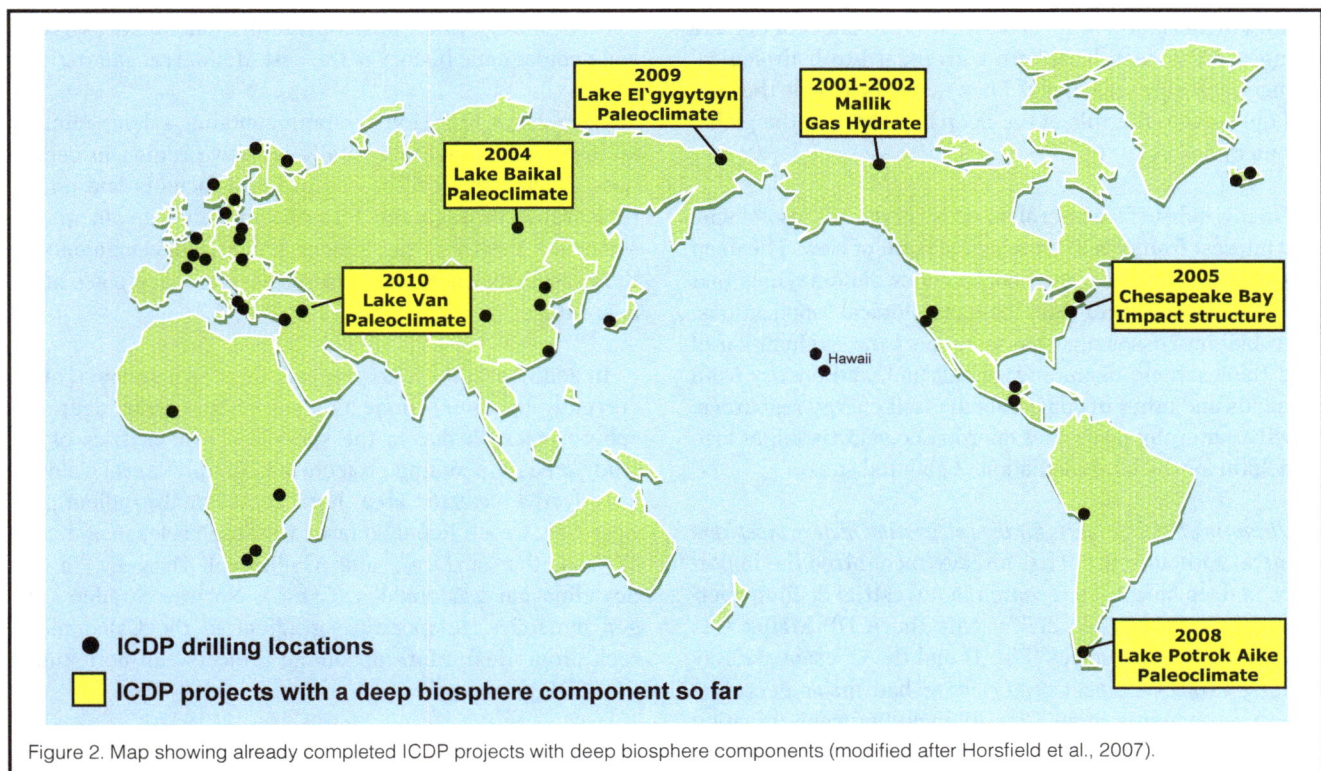

Figure 2. Map showing already completed ICDP projects with deep biosphere components (modified after Horsfield et al., 2007).

supporting deep microbial communities? What role do deep microorganisms play in the formation and destruction of natural resources (oil, gas, coal), and how do their metabolisms influence the geochemistry and mineralogy of the subsurface (i.e., the deposition of ores)?

Also of interest is how natural and human perturbation affect subsurface ecosystems in time and space. For instance, how do deep microbial communities respond to earthquakes, mobilizing fluids (geogases), the disposal of waste, or the sequestration of CO_2 (e.g., Ketzin, Germany or Columbia River basalts, U.S.A.)?

Microorganism populations interact with each other, and this raises further questions. What are the characteristics of such community interactions? What is the genomic inventory of deep microbial organisms? What are the effects of lateral gene transfer and the impact of mutations (incurred by cells that grow so slowly) on the evolution of cells in the deep subsurface? What is the role of phages in syntrophic interactions? Which genomic features support the adaptation and survival of microorganisms in the deep subsurface? Which DNA repair mechanisms do they apply? Why do cells in deep sediments (isolated) seem to be as little diverged as they are in the context of their 16S ribosomal RNA gene or other genes?

Another important aspect is how subsurface microbial communities affect processes in the surface systems and *vice versa*. For the future climate development it is crucial to know how deep microorganisms affect Earth's climate, considering the production of the greenhouse gas methane, especially when permafrost areas are thawing (climate feedback). Furthermore, in reverse, what is the effect of climate change on the deep biosphere with regard to hydrological changes and nutrient supply? These aspects include the general question to the role of the deep biosphere in the global carbon cycle.

Finally, subsurface microbial ecosystems are also of specific interest from a biotechnological point of view. The deep biosphere contains a large pool of genes and enzymes that are potentially useful for biotechnological applications. Microbial processes are able to support the exploitation of ores (bioleaching, biomining), clean fuel, and energy from tar sands and other unconventional fossil energy resources. Specific naturally occurring microbial consortia might also be helpful for the bioremediation of polluted sites.

Upcoming ICDP projects for the integration of deep biosphere research. Although ICDP has already recognized the importance of deep biosphere research in terrestrial drilling operations (Horsfield et al., 2007), only the ICDP Mallik Gas Hydrate Research Project (Fig. 1) and the Chesapeake Bay Drilling Project (impact crater) have had major deep biosphere components so far (Fig. 2), including contamination controls during core retrieval. There was some limited bio-

geochemical research in some lake sediment drilling projects as well: Lake Baikal, Potrok Aike Drilling (see Ariztegui et al., 2010 in this issue), and Lake El`gygytgyn. Although without contamination control, the paleoclimate drilling campaign in Lake Van, Turkey in 2010 also had a deep biosphere component.

In order to extend the number of projects with a deep biosphere component, the workshop identified a list of interesting upcoming ICDP projects where deep biosphere research might be added to already initiated projects. In particular, the ongoing ICDP lake drilling program appears to be an appropriate start to establish deep biosphere research in ICDP (for drilling locations see ICDP homepage, http://www.icdp-online.org). The sediments of lakes provide a unique opportunity to characterize and investigate the subsurface microbial communities in many different climatic zones of the Earth and, therefore, under different environmental conditions. There are three interesting upcoming lake projects in the near future (discussed below).

Lake Ohrid (Macedonia/Albania) is considered to be the oldest continuously existing lake in Europe (assessed age 1–10 Ma). It has a unique aquatic ecosystem with more than 200 described endemic species. The sedimentary successions in the central basin seem to reflect the complete history of the lake being an excellent archive for climate and volcanic activity in the central northern Mediterranean region.

The Dead Sea (Israel, West Bank, and Jordan) is the deepest hypersaline lake in the world. The deep basin of the Dead Sea contains a continuous sedimentary record of Pleistocene to Holocene age which forms an archive for climatic, seismic, and geomagnetic history of the east Mediterranean region.

Lake Issyk-Kul (Kyrgyzstan) contains a long climatic archive of this environmentally sensitive region in central Asia. During this drilling campaign a focus is laid on the time frame through the Pliocene into the late to middle Miocene. However, the recent political developments in Kyrgyzstan may prevent this project from taking place in the near future.

In addition to the lake program there is a series of other currently upcoming projects also of interest for deep biosphere research due to the specific characteristics of the study areas. Upcoming projects are Campi Flegrei Caldera, Italy (active volcanic area, high geothermal gradient), the Eger Rift, Czech Republic (area with high release of mantle CO_2 at the surface), and Collisional Orogeny in the Scandinavian Caledonides (COSC), Norway/Sweden (orogen dynamics, temperature gradient in the Caledonides, rock properties). More upcoming projects can be found at the ICDP homepage.

Potential targets for a dedicated deep biosphere project within ICDP. Another aim of the workshop was to identify potential targets for a dedicated terrestrial deep biosphere drilling campaign, forming the core of a future science proposal within the scope of ICDP. Terrestrial environments provide a broad range of different geological settings with their various associated deep microbial communities. Thus, only a selection of some potential targets for scientific deep biosphere drilling campaigns is presented here (see text box).

- Drilling outward and deeper from within a deep mine (e.g., at the Deep Underground Science and Engineering Laboratory (DUSEL)) for sampling and establishing of a natural laboratory to conduct *in situ* experiments
- Drilling at sites of active serpentinization and other locations where lithoautotrophic communities might occur or play a significant role
- Coring and monitoring of an active fault zone (Eger Rift) and establishment of a natural laboratory to investigate the role seismicity plays in subsurface life
- Drilling along a sequence of oil-bearing rocks to examine biodegradation gradients (e.g., Western Canada and Paris Basin) and with a possible industry connection
- Coring different sedimentary basins with different thermal gradients, different depositional conditions, and organic carbon composition and concentrations to investigate questions on the biogeography of the deep biosphere and the fate of the organic matter as a substrate for deep microbial life
- Drilling at locations where permafrost may be actively releasing biogenic methane (e.g., the high arctic of Canada and Siberia), with a possible link to IODP drilling campaigns, offshore permafrost
- Drilling coal seams to investigate the role of the deep biosphere in coal bed methane generation
- Drilling into an active mud volcano to investigate the role of microbial activity in the generation and degradation of hydrocarbons
- Examining the types of microbes, processes, and activities associated with locations where carbon capture and storage is being attempted (e.g., Columbia River basalts, U.S.A.)
- Drilling a borehole in the backyard of a science institute to establish an easily accessible natural laboratory to perform *in situ* experiments

Furthermore, it was highlighted that each project needs to consider and implement the best methods for collecting high quality samples for microbiological, biogeochemical, and geochemical analysis. This would also entail augmenting the methods book that will be used by future ICDP microbiology efforts. The projects should generally include basic measurements of the geology, geochemistry, mineralogy,

etc. so that the environmental context can be understood for interpreting the microbiological ecosystems. Finally, the need for a central data base for the results of the ICDP projects was emphasized.

Technical, Administrative, and Logistical Prerequisites

The fact that microbial cell abundance in subsurface environments is two to six orders of magnitude lower than at the surface makes the task of recovering uncontaminated supplies extremely difficult. Contamination control requires certain changes to standard drilling protocols, which can be achieved for a relatively small increase in cost when implemented already at an early planning stage. Also, handling and sampling procedures have to accommodate the special requirements to avoid alteration of the samples. Protocols like in IODP (Expedition 311 Scientists, 2006) have to be developed for ICDP drilling operations.

Preventing and assessing contamination. Due to the fact that basically all drilling operations use drilling fluids, contamination of cores through infiltration of drill fluid can only be minimized and not completely avoided. Uncontaminated samples are an absolute necessity for any subsequent analysis; therefore, contamination assessment is a crucial issue for geomicrobiological research in general.

In order to minimize contamination it is advisable to get involved in the planning of the drilling as early as possible, preferably as a Principal Investigator (PI), in order to have full control on the design of the operation. To achieve the best results in a cost effective way, it is also paramount to involve the drilling organization as early as possible. Any changes from standard drilling operations have to be taken into consideration early for the cost estimates; later changes can have dramatic effects on the overall budget. During the initial planning stage it is relatively easy to lay out the drilling operations according to geomicrobiology needs, even without compromising other research areas. Experience has shown that relatively few changes are necessary to minimize contamination of the samples. Most of these changes can be achieved relatively easily and for little extra cost if they are included in the planning at an early stage. Due to the great diversity of sediment/rock types to be drilled, the type of drilling equipment to be used, and other variables, there is no general rule on how to avoid contamination. Such an issue can only be addressed individually, given the specific circumstances.

During the drilling operation, standard microbiological procedures should be followed. While this is nothing new for a knowledgeable scientist, it may very well be so for the drilling staff. In order to ensure smooth drilling and sample handling, the drilling staff needs to be trained, and all procedures have to be discussed with and understood by them.

Still, there are some general issues that help to minimize contamination. Steam cleaning of all drilling equipment has proven to be a relatively cheap but effective way to reduce contamination by removing any foreign rock fragments and hydrocarbons from old pipe grease. A complete sterilization of the entire equipment is usually not necessary because the material will be contaminated again by the time it has traveled down the borehole. When sterilizing equipment, one should always ask the question whether it will be possible to get the sample into the sampler (core barrel, water sampler) without anything non-sterile getting in contact with the sample. Only in such cases does sterilization of equipment really make sense. There are situations, such as water sampling in deep aquifers, where sterilization of samplers may be useful and necessary, but such decisions have to be made on a case-by-case basis.

The choice of drilling technique is of major importance. In few cases (short holes, hard rocks), it is possible to drill without any drilling fluid and use air- or gas-lift techniques instead. High volumes of pressurized air or nitrogen are used in place of conventional drilling fluids to circulate the well bore clean of cuttings and to cool the drill bit. Air drilling can be used where formations are dry, i.e., when there is no influx of water into the hole. Also, normally the specific gravity of the drill mud prevents the hole from closing around the drill string. Boreholes have to be stable to use this technique due to the low density of the gas. So far, there is little experience with this technique for geomicrobiological purposes.

For softer sediments, hydraulic piston coring is the method of choice because this technique has shown to provide the least contaminated samples with regard to penetration of drilling fluid into the core. This is not surprising as the coring itself does not require any drilling fluids and relies solely on the force with which the cutting shoe is driven into the sediment, followed by rotary drilling around the core to extend the diameter of the hole and to push the bottom hole assembly further down. In IODP operations, the advanced piston coring tool (APC) is the prime tool for recovering soft sediments. Although this technique provides the highest quality cores and should, therefore, be carried out as deep as possible, it reaches its limitations in consolidated sediments.

The biggest problems are still the sediments of intermediate stiffness that are too hard for piston and too soft for rotary coring. IODP uses the extended core barrel tool (XCB), but the retrieved samples are often unsuitable for geomicrobiology research because of the high level of contamination and drilling induced disturbances in the core. In many cases the recovered cores consisted of pieces of sediment floating in a solidified mixture of drill mud and cuttings. Rotary drilling becomes the method of choice in consolidated sediments and hard rocks. When using rotary coring, contamination control becomes absolutely crucial, because this technique requires drill fluids and, unlike situations with the APC tool, the core is in direct contact with the rotating drill head. Due to the high pressure of the drill fluid coming out of the drill bit, the rock can be saturated with drill mud several centimeters ahead of the bit.

Drilling should always be carried out with liners to protect the drilled core from further contamination and to ease later handling. Still there will always be some drill mud in the gap between the liner and the drill core, and this mud can seep inwards and contaminate the interior of the core. Additionally, natural or drilling-induced fractures provide pathways through which the mud may enter the core.

Careful adjustment of the drilling conditions, rapid evaluation of the quality of the recovered core material, and, if necessary, changes to the drilling protocol can help to minimize contamination, but some degree of contamination may be expected for all cores collected by rotary drilling. Still, cores suitable for geomicrobiological research can be obtained by this technique.

The composition of the drill fluid has to be monitored carefully; all components of the drill fluids, including the water, should be checked for possible contaminants prior to drilling. The high density of the drill mud is achieved through the addition of clays, which have considerable differences in the microbial load. There is anecdotal evidence that synthetic clays usually contain fewer microbial cells than natural ones, but so far there is no systematic study about the microbial load of different clays. If possible, organic additives (thickeners, emulsifiers, stabilizers, etc.) should be limited to an absolute minimum because they represent a nutrient source for the microbes and can thereby enhance microbial activity. Hydrocarbon-based additives should be avoided at any cost as they interfere with most organic geochemical analyses.

A careful evaluation of all materials during the early planning stage can, therefore, significantly reduce the potential for contamination. Independent of the drilling technique and the composition of the drill mud, some contamination will always occur; therefore, contamination has to be assessed, preferably by multiple techniques.

The most common technique for contamination assessment is the use of fluorescent microspheres. These particles are available in a wide range of sizes (0.5 µm diameter is most common for contamination assessment). Microspheres have the advantage that they can be easily detected by fluorescence microscopy. Having a density very close to 1 g cc[-1], microspheres can be easily separated from the sample by density centrifugation on a cushion of sodium chloride solution. The disadvantage of microspheres is their price, which can add significantly to the total cost of a project.

There are two different ways to apply the microspheres. For hydraulic piston coring the easiest way is to attach small bags filled with spheres to the inner front of the core. As soon as the sediment enters the core barrel, the bag is ripped and the microspheres mix with the drilling fluid and eventually infiltrate the core. Using this technique the concentration of spheres in the mud is not constant and can only roughly be estimated; therefore, it is difficult to assess precisely how much drill fluid per volume of sediment has to enter the core in order to be detected by this technique. Still, this technique has been used for many years in IODP operations with very reliable results. Another way to apply microspheres is to add them directly to the drill mud. In cases where only a few depth intervals are being cored for geomicrobiological analysis, they can be added with a peristaltic pump into the intake of the mud pump. This way, the amount of microspheres can be limited to an absolute minimum. In cases where many depths are being cored, the entire volume of drill mud has to be amended with microspheres. Depending on the well depth and the required volume of drill mud, this approach may quickly reach cost limits. In order to detect sufficiently small concentrations of drill mud in the core, the concentration of microspheres should be at least around 1000 µL⁻¹. Depending on the diameter of the hole and the target depth, the volume of drill mud can vary between single and tens of cubic meters or even more. Microspheres are removed from the mud by various processes (Kallmeyer et al., 2006) and have to be added in regular intervals. Also, large volumes of drill mud can get lost in fractures and have to be replaced by fresh mud, requiring additional addition of microspheres. All these processes have to be taken into account when calculating the number of necessary microspheres.

The addition of known and easily identifiable microorganisms would be an alternative type of particulate tracers. However, this approach will possibly cause major legal problems in many if not most areas.

Solute tracers may offer a viable alternative to added allochthonous microbes and microspheres. In IODP operations perfluorocarbon tracer (perfluoromethylcyclohexane, PFT) has been used on several occasions with good results, although the data may differ somewhat from those obtained with microspheres (Smith et al., 2000a, 2000b). PFT is much cheaper than microspheres, but its detection requires a gas chromatograph, which may cause logistical problems on the drill site. Although this technique is very sensitive, possible incompatibilities with the drill mud matrix have to be evaluated prior to drilling. Also, the samples have to be taken quickly after retrieval of the core, due to the high volatility of PFT.

During the ICDP Chesapeake Bay impact drilling, halon was used as a tracer. Although the results were satisfactory, this technique will most probably not be used in future drilling operations because of the decreasing availability of Halon

as it becomes banned in many countries due to its deleterious effects on the ozone layer.

Fluorescent dyes can be detected by fluorometry, which is a relatively easy and robust technique, allowing for analysis right at the drill site. This may be very helpful in cases where unforeseen changes in drilling operations become necessary, and possible influences on the quality of the recovered material need to be evaluated on the spot. A variety of fluorescent dyes have been used successfully in various operations: rhodamine WT, uranine, lissamine FF, fluorescine, amino G acid (7-Amino-1,3-naphthalenedisulfonic acid). However, quenching of the fluorescence signal due to coloration of the sample may complicate the exact quantification of the infiltration of drill mud into the core.

The level of contamination can vary on a small scale, not just in terms of distance from the outside of the core but also between different sample depths. The drill mud may have infiltrated the core along small cracks, which are not visible upon manual inspection of the core; therefore, even if adjacent contamination controls are "clean", that may not guarantee an uncontaminated sample. Ideally, contamination should be assessed on the sample being analyzed. Redundancy of contamination control is important; at least two different methods should be used in order to ensure good contamination control under all circumstances.

Sampling and sample storage. After the cores are retrieved, sample collection is the next major step. The contaminated outer part of the core has to be removed. Only the uncontaminated inner part can be used for geomicrobiological research (inner coring technique). How much of the outer part needs to be removed and how much uncontaminated material actually becomes available for analysis have to be determined individually. If anaerobic conditions are required for the sample material, sampling has to be conducted in an anaerobic glove box.

Sampling techniques have to be adjusted according to lithology. In soft sediments subsampling can be done with cut-off syringes, whereas in hard sediments the center subcore has to be retrieved by drilling. Cut-off plastic syringes are cheap and can be prepared in large quantities prior to drilling. Metal core drills for hard sediments are much more expensive and usually not available in the same quantities as syringes. The effort in time and manpower of recycling these drills (retrieval of sample, cleaning, sterilization) has to be taken into account when planning the amount of samples that can be processed in a given time.

Sample preservation is another important issue. There are different approaches currently being used, depending on the parameter to be preserved. The standard technique for storage of cell count samples from soft sediment is a solution of similar salinity with formalin. A common method for general geochemical sampling is to put the whole sediment into gas-tight bags flushed with nitrogen and/or containing an oxygen scrubber. Such samples have proven to be a good option for further subsampling in the laboratory.

Another option would be gel preservation. The sample is coated with an antimicrobial gel that prevents surface growth and limits gas exchange. In some cases special waxes with a low melting point have been used to coat samples. However, the application of this technique has so far been limited to samples for physical and chemical analysis, not microbiology. There are no data available whether these waxes give off any volatile compounds that could potentially be used as a carbon source by microbes.

Pore water should be extracted as quickly as possible to avoid alteration during storage. Squeezing yields the highest amounts of pore water but destroys the sediment structure. Rhizon samplers are not as effective but leave the sediment structure intact, thereby allowing the use of the core for other purposes. However, Rhizon samplers apply a vacuum to the retrieved pore water, thus causing the loss of gases, especially CO_2. This loss in CO_2 will inevitably alter the pH of the sample. Although not very efficient, centrifugation can be the method of choice for highly porous and soft sediments.

Samples should be stored according to the parameter to be analyzed. Storage at $4°C$ is preferred for turnover rate measurements and cell counting. For molecular analysis, storage at $-20°C$ may not be cold enough to stop all degradation processes. The best method is still storage in liquid nitrogen, because at that temperature all degradation processes are stopped and oxidation is completely avoided. However, liquid nitrogen may not be available in remote locations, and transport may also be an issue, although with special containers samples can even be sent by airfreight in liquid nitrogen.

Integration of a standard minimum sampling scheme. Compared to the marine realm, terrestrial subsurface microbiology is lagging behind by many years. One of the main reasons for this is the lack of available samples. A minimum sampling scheme that could become a compulsory standard component of all ICDP drilling operations could help to overcome this lack of material. Such a minimum sampling scheme would not interfere with other analyses and could be done with relatively little additional effort. Like in IODP, certain physical parameters should also be measured routinely in order to advance our understanding of subsurface biomass, activities, and habitability. These data would be extremely helpful not just for geomicrobiological research but for other fields as well. Routine measurements should include formation factor as well as downhole temperature and pressure. There have to be at least two different minimum sampling schemes, one for soft and one for hard sediments.

For soft sediments the minimum requirements would be:

- Pore water via Rhizon sampler, split into acidified and chilled aliquots. Such a sample cannot be used for quantification of dissolved inorganic carbon (DIC) or alkalinity due to loss of CO_2. For such measurements a squeezed sample would be preferable. In cases where only one technique can be applied, the Rhizon samplers are still preferable, unless the CO_2-sensitive parameters are of major interest.
- Cell count sample (2-cc syringe sample, stored in 2% formalin), in cases where no contamination control can be made, the sample should be taken from the absolute center of the core and as far away as possible from any visible cracks.
- Dissolved gases sample (2-cc syringe sample, stored in 10% NaOH)
- Elemental parameters (CHN sample): 2-cc sample, stored at $-20°C$ or colder.
- A short whole round core or a large (60 cc) syringe frozen at $-80°C$ for future molecular studies. The cost of such studies is declining rapidly, and the samples will be invaluable.

For hard material the minimum requirement would be a drilled subcore sample, stored in a gas-tight bag, flushed with nitrogen and/or equipped with an oxygen scrubber, stored at $4°C$.

So far, subsurface microbiology in ICDP projects has mainly been done as "one-off" operations, with contamination control and sample handling protocols being developed individually for the specific projects. These individual approaches make it rather difficult to compare data from different projects. By making the minimum sampling schemes a standard part of ICDP operations, a much wider community could use these data.

Data storage. Storage of the logging data can be managed rather easily through the ICDP Operations Support Group (OSG), whereas storage of legacy samples is a much more complicated issue because there are no central storage facilities for ICDP cores. Although legacy samples form a valuable resource for future research, they also represent a great burden and cost factor. As this is an issue that is not just affecting geomicrobiology, it should be addressed on a larger scale.

Drilling and mobile laboratory facilities. ICDP drilling operations are much more diverse with regard to drilling equipment and work environment than at IODP, where the drill ships operate according to well-known standard procedures and provide a good working environment. Still, a large fraction of ICDP operations employs the same drilling equipment, namely the DOSECC rigs and, in the future, the INNOVA Rig as well. A test drilling operation was carried out with DOSECC's Glad 800 Rig at Great Salt Lake, which allowed for equipment testing and, due to the vicinity of DOSECC headquarters, immediate refinement and modification of equipment in the workshop when necessary. There should also be the opportunity for a geomicrobiology test drill to develop and refine the required drilling protocols for biogeochemical and microbiological research. Such an operation should be science driven as opposed to being just a technical exercise, but the geologic setting has to be well known in order to avoid any problems due to unforeseen lithological changes. It would, therefore, be a good option to add such a test drill onto an already scheduled drilling operation. Whereas normal drilling operations are usually run to a rather tight schedule, it is important to allocate sufficient additional time and resources for the testing and not to squeeze this into the already tight schedule of the general project.

For geomicrobiological research, sample processing immediately after retrieval is important to avoid alteration of the samples. Sufficiently equipped laboratories are readily available on the IODP drill ships. This is much different for ICDP operations, where quite often they are located in remote areas with more complicated or impossible access to a suitable laboratory. One solution to this problem is the new BUGLab facility of the Helmholtz Centre Potsdam (GFZ) German Research Centre for Geosciences (Fig. 3). The laboratory is composed of two portable standard 20-ft containers, which can be combined if necessary. Due to their modular structure, the BUGLab containers can be equipped according to the specific requirements of the planned work, allowing the processing of microbiological and biogeochemical samples, on-site analysis of biologically significant transient properties, and on-site analysis of chemical and physical properties that are being useful to guide microbiological and biogeochemical sampling strategies.

Workshop Participants

Lorenz Adrian, UFZ Leipzig, Germany; Rick Colwell, Oregon State University, U.S.A.; Steve D'Hondt, University of Rhode Island, U.S.A.; Clemens Glombitza, GFZ Potsdam, Germany; Ulrich Harms, ICDP, Germany; Ian Head, Newcastle University, UK; Kai-Uwe Hinrichs, MARUM-University of Bremen, Germany; Nils Holm, Stockholm University, Sweden; Brian Horsfield, GFZ Potsdam, Germany; Merja Itävaara, VTT Technical Research Centre of Finland; Jens Kallmeyer, University of Potsdam, Germany; Thomas L. Kieft, New Mexico Institute of Mining and Technology, U.S.A.; Kirsten Küsel, Friedrich Schiller University Jena, Germany; Kai Mangelsdorf, GFZ Potsdam, Germany; Martin Mühling, TU Bergakademie Freiberg, Germany; Richard W. Murray, Boston University, Earth Sciences, U.S.A.; Dennis Nielson, DOSECC, U.S.A.; T.C. Onstott, Princeton University, U.S.A.; R. John Parkes, Cardiff University, U.K.; Karsten Pedersen, University of Gothenburg, Sweden; Matxalen Rey Abasolo, OSG at ICDP, Germany; Axel Schippers, BGR Hannover, Germany; Michael Schlömann, TU Bergakademie

Figure 3. The GFZ BUGLab container is a mobile field laboratory for geomicrobiological and biogeochemical research. The container can be deployed in a wide range of both marine and terrestrial settings, from polar to tropical environments.

Freiberg, Germany; David Smith, University of Rhode Island, U.S.A.; Yohey Suzuki, National Institute of Advanced Industrial Science & Technology, Japan; Volker Thiel, University of Göttingen, Germany; Andrea Vieth, GFZ Potsdam, Germany; Mary Voytek, U.S. Geological Survey, U.S.A.; Maren Wandrey, GFZ Potsdam, Germany; Claudia Wiacek, TU Bergakademie Freiberg, Germany; Heinz Wilkes, GFZ Potsdam, Germany; Matthias Zabel, MARUM-University of Bremen, Germany; Klaus-Gerhard Zink, GNS Science, New Zealand

References

Colwell, F.S., Nunoura, T., Delwiche, M.E., Boyd, S., Bolton, R., Reed, D.W., Takai, K., Lehman, R.M., Horikoshi, K., Elias, D.A., and Phelps, T.J., 2005. Evidence of minimal methanogenic numbers and activities in sediments collected from JAPEX/ JNOC/GSC et al. Mallik 5L-38 gas hydrate production research well. *In* Dallimore, S.R., and Collett, T.S. (Eds.), *Scientific Results from the Mallik 2002 Gas Hydrate Production Research Well Program, Mackenzie Delta, Northwest Territories, Canada.* Geological Survey of Canada, Bulletin 585:1–11.

Engelen, B., Engelhardt, T., Sahlberg, M., and Cypionka, H., 2009. Viral infections as controlling factors of the deep biosphere. National IODP-ICDP meeting in Potsdam, Germany. 16–18 March 2009:52 pp.

Expedition 311 Scientists, 2006. Methods. *In* Riedel, M., Collett, T.S., Malone, M.J., and the Expedition 311 Scientists. *Proc. IODP,* 311: Washington, DC (Integrated Ocean Drilling Program Management International, Inc.). doi:10.2204/iodp.proc. 311.102.2006.

Fredrickson, J.K., and Onstott, T.C., 1996. Microbes deep inside the Earth. *Sci. Am.*, 275:42–47, doi:10.1038/scientificamerican 1096-68.

Glombitza, C., Mangelsdorf, K., and Horsfield, B., 2009. A novel procedure to detect low molecular weight compounds released by alkaline ester cleavage from low maturity coals to assess its feedstock potential for deep microbial life. *Organ.*

Geochem., 40:175–183, doi:10.1016/j.orggeochem.2008.11.003.

Gohn, G.S., Koeberl, C., Miller, K.G., Reimold, W.U., Browning, J.V., Cockell, C.S., Horton, J.W., Kenkmann, T., Kulpecz, A.A., Powars, D.S., Sanford, W.E., and Voytek, M.A., 2008. Deep drilling into the Chesapeake Bay impact structure. *Science*, 320:1740–1745, doi:10.1126/science.1158708.

Horsfield, B., Kieft, T., Amann, H., Franks, S., Kallmeyer, S., Mangelsdorf, K., Parkes, J., Wagner, W., Wilkes, H., and Zink, K.-G., 2007. The GeoBiosphere. *In* Harms, U., Koeberl, C., and Zoback, M.D. (Eds.), *Continental Scientific Drilling: A Decade of Progress and Challenges for the Future.* Berlin-Heidelberg (Springer), 163–211.

Horsfield, B., Schenk, H.J., Zink, K.-G., Ondrak, R., Dieckmann, V., Kallmeyer, J., Mangelsdorf, K., di Primio, R., Wilkes, H., Parker, J., Fry, J.C., and Cragg, B., 2006. Living microbial ecosystems within the active zone of catagenesis: implications for feeding the deep biosphere. *Earth Planet. Sci. Lett.*, 246:55–69.

Kallmeyer, J., Mangelsdorf, K., Cragg, B.A., Parkes, R.J., and Horsfield, B., 2006. Techniques for contamination assessment during drilling for terrestrial subsurface sediments. *Geomicrobiol. J.*, 23:227–239, doi:10.1080/01490450600724258.

Lin, L.-H., Hall, J., Lippmann-Pipke, J., Ward, J.A., Sherwood Lollar, B., DeFlaun, M., Rothmel, R., Moser, D., Gihring, T.M., Mislowack, B., and Onstott, T.C., 2005. Radiolytic H_2 in continental crust: nuclear power for deep subsurface microbial communities. *Geochem. Geophys. Geosyst.*, 6:Q07003, doi: 10.1029/2004GC000907.

Lin, L.-H., Wang, P.-L., Rumble, D., Lippmann-Pipke, J., Boice, E., Pratt, L.M., Sherwood Lollar, B., Brodie, E.L., Hazen, T.C., Andersen, G.L., DeSantis, T.Z., Moser, D.P., Kershaw, D., and Onstott, T.C., 2006. Long-term sustainability of a high-energy, low diversity crustal biome. *Science*, 314:479–482, doi:10.1126/science.1127376.

Lipp, J.S., Morono, Y., Inagaki, F., Hinrichs, K.-U., 2008. Significant contribution of Archaea to extant biomass in marine subsurface sediments. *Nature* 454:991–994. doi:10.1038/nature07174.

Mangelsdorf, K., Haberer, R.M., Zink, K.-G., Dieckmann, V., Wilkes, H., and Horsfield, B., 2005. Molecular indicators for the occurrence of deep microbial communities at the Mallik 5L-38 gas Hydrate Research Well. *In* Dallimore, S.R. and Collett, T.S. (Eds.), *Scientific Results from the Mallik 2002 Gas Hydrate Production Research Well Program, Mackenzie Delta, Northwest Territories, Canada.* Geological Survey of Canada, Bulletin 585:1–11.

Parkes, R.J., Cragg, B.A., and Wellsbury, P., 2000. Recent studies on bacterial populations and processes in subseafloor sediments: a review. *Hydrogeol. J.*, 8:11–28, doi:10.1007/PL00010971.

Parkes, R.J., Wellsbury, P., Mather, I.D., Cobb, S.J., Cragg, B.A., Hornibrook, E.R.C., and Horsfield, B., 2007. Temperature activation of organic matter and minerals during burial has the potential to sustain the deep biosphere over geological timescales. *Organ. Geochem.*, 38:845–852, doi:10.1016/j.orggeochem.2006.12.011.

Pedersen, K., 2000. Exploration of deep intraterrestrial microbial life:

current perspectives. *FEMS Microbiol. Lett.*, 185:9–16, doi:10.1111/j.1574-6968.2000.tb09033.x.

Rothschild, L.J., and Mancinelli, R.L., 2001. Life in extreme environments. *Nature*, 409:1092–1101, doi:10.1038/35059215.

Roussel, E.G., Cambon Bonavita, M.-A., Querellou, J., Cragg, B.A., Webster, G., Prieur, D., and Parkes, R.J., 2008. Extending the sub-sea-floor biosphere. *Science*, 320:1046, doi:10.1126/science.1154545.

Sherwood Lollar, B., Lacrampe-Couloume, G., Slater, G.F., Ward, J.A., Moser, D.P., Gihring, T.M., Lin, L.-H., and Onstott, T.C., 2006. Unravelling abiogenic and biogenic sources of methane in the Earth`s deep subsurface. *Chem. Geol.*, 226:328–339, doi:10.1016/j.chemgeo.2005.09.027.

Smith, D.C., Spivack, A.J., Fisk, M.R., Haveman, S.A., and Staudigel, H., 2000a. Tracer-based estimates of drilling-induced microbial contamination of deep-sea crust. *Geomicrobiol. J.*, 17:207–219, doi:10.1080/01490450050121170.

Smith, D.C., Spivack, A.J., Fisk, M.R., Haveman, S.A., Staudigel, H., and Party, O.L.S., 2000b. Methods for quantifying potential microbial contamination during deep ocean drilling. *ODP Tech. Note*, 28.

Stevens, T.O., and McKinley, J.P., 1995. Lithoautotrophic microbial ecosystems in deep basalt aquifers. *Science*, 270:450–454, doi:10.1126/science.270.5235.450.

Vuillemin, A., Ariztegui, D., Vasconcelos, C., and the PASADO Scientific Drilling Party, 2010. Establishing sampling procedures in lake cores for subsurface biosphere studies: assessing *in situ* microbial activity, *Sci. Drill.*, 10:35–39, doi:10.2204/iodp.sd.10.04.2010.

Wellsbury, P., Goodman, K., Barth, T., Cragg, B.A., Barnes, S.P., and Parkes, R.J., 1997. Deep marine biosphere fuelled by increasing organic matter availability during burial and heating. *Nature*, 388:573–576, doi:10.1038/41544.

Whitman, W.B., Coleman, D.C., and Wiebe, W.J., 1998. Prokaryotes: the unseen majority. *Proc. Natl. Acad. Sci. USA*, 95:6578–6583, doi:10.1073/pnas.95.12.6578.

Wilhelms, A., Larter, S.R., Head, I., Farrimond, P., di-Primio, R., and Zwach, C., 2001. Biodegradation of oil in uplifted basins prevented by deep-burial sterilization. *Nature*, 411:1034–1037, doi:10.1038/35082535.

Zink, K.-G., Wilkes, H., Disko, U., Elvert, M., and Horsfield, B., 2003. Intact phospholipids - microbial "life markers" in marine deep subsurface sediments. *Organ. Geochem.*, 34:755–769, doi:10.1016/S0146-6380(03)00041-X.

Authors

Kai Mangelsdorf, Helmholtz Centre Potsdam, GFZ German Research Centre for Geosciences, Telegrafenberg, 14473 Potsdam, e-mail: K.Mangelsdorf@gfz-potsdam.de.

Jens Kallmeyer, University of Potsdam, Earth and Environmental Sciences, Karl-Liebknecht Str. 25, 14476 Potsdam, e-mail: kallm@geo.uni-potsdam.de.

Photo Credits

Fig. 1 – ICDP

Fig. 3 – Kai Mangelsdorf, GFZ Potsdam

31

The MoHole: A Crustal Journey and Mantle Quest, Workshop in Kanazawa, Japan, 3–5 June 2010

by Benoît Ildefonse, Natsue Abe, Donna K. Blackman, J. Pablo Canales, Yoshio Isozaki, Shuichi Kodaira, Greg Myers, Kentaro Nakamura, Mladen Nedimovic, Alexander C. Skinner, Nobukazu Seama, Eiichi Takazawa, Damon A.H. Teagle, Masako Tominaga, Susumu Umino, Douglas S. Wilson, and Masaoki Yamao

Introduction

Drilling an ultra-deep hole in an intact portion of oceanic lithosphere, through the crust to the Mohorovičić discontinuity (the 'Moho'), and into the uppermost mantle is a long-standing ambition of scientific ocean drilling (Bascom, 1961; Shor, 1985; Ildefonse et al., 2007). It remains essential to answer fundamental questions about the dynamics of the Earth and global elemental cycles. The global system of mid-ocean ridges and the new oceanic lithosphere formed at these spreading centers are the principal pathways for energy and mass exchange between the Earth's interior, hydrosphere, and biosphere. Bio-geochemical reactions between the oceans and oceanic crust continue from ridge to subduction zone, and the physical and chemical changes to the ocean lithosphere provide inventories of these thermal, chemical, and biological exchanges.

The 2010 MoHole workshop in Kanazawa, Japan followed from several recent scientific planning meetings on ocean lithosphere drilling, in particular the Mission Moho Workshop in 2006 (Christie et al., 2006; Ildefonse et al., 2007) and the "Melting, Magma, Fluids and Life" meeting in 2009 (Teagle et al., 2009). Those previous meetings reached a consensus that a deep hole through a complete section of fast-spread ocean crust is a renewed priority for the ocean lithosphere community. The scientific rationale for drilling a MoHole in fast-spread crust was developed in the workshop reports (available online) and most thoroughly articulated in the 2007 IODP Mission Moho drilling proposal (IODP Prop 719MP; www.missionmoho.org).

The 2010 MoHole workshop had two interconnected objectives, which have been discussed jointly between ocean lithosphere specialists, marine geophysicists, and engineers:

- to initiate a roadmap for technology development and the project implementation plan that are necessary to achieve the deep drilling objectives of the MoHole project,
- to identify potential MoHole sites in the Pacific (i.e., in fast-spread crust), where the scientific community will focus geophysical site survey efforts over the next few years.

Selecting drilling sites is essential to identify the range of water depths, drilling target depths, and temperatures that we anticipate, and to better define the technology required to be developed and implemented to drill, sample, and geophysically log the MoHole.

A Brief Summary of the Scientific Rationale for the MoHole

The Moho is the fundamental seismic boundary within the upper part of our planet, yet we have little knowledge of its geological meaning. New deep drilling technology now make it possible to fulfill scientists' long-term aspirations to drill completely through intact oceanic crust, through the seismically defined Moho and then a significant distance (~500 m) into the upper mantle. Our scientific goals (Fig. 1; Christie et al., 2006; Ildefonse et al., 2007; Teagle et al., 2009) can be divided into the following principal tightly interconnected threads.

- What physical properties cause the Mohorovičić discontinuity, and what is the geological nature of this boundary zone?
- How is the (lower) oceanic crust formed at the mid-ocean ridges, and what processes influence its subsequent evolution? What are the geophysical signatures of these magmatic, tectonic, hydrothermal, biogeochemical, and chemical processes?
- What can we infer about the global composition of the oceanic crust, and what are the magnitudes of interactions with the oceans and biology and their influence on global chemical cycles?
- What are the limits of life, and the factors controlling these limits? How do the biological community compositions change with depth and the evolving physical and chemical environments through the oceanic crust?
- What is the physical and chemical nature of the uppermost mantle, and how does it relate to the overlying magmatic crust?

The Mohorovičić Discontinuity

In the oceans, the Moho is commonly a bright seismic reflector at 5–8 km depth, marking a step change to seismic

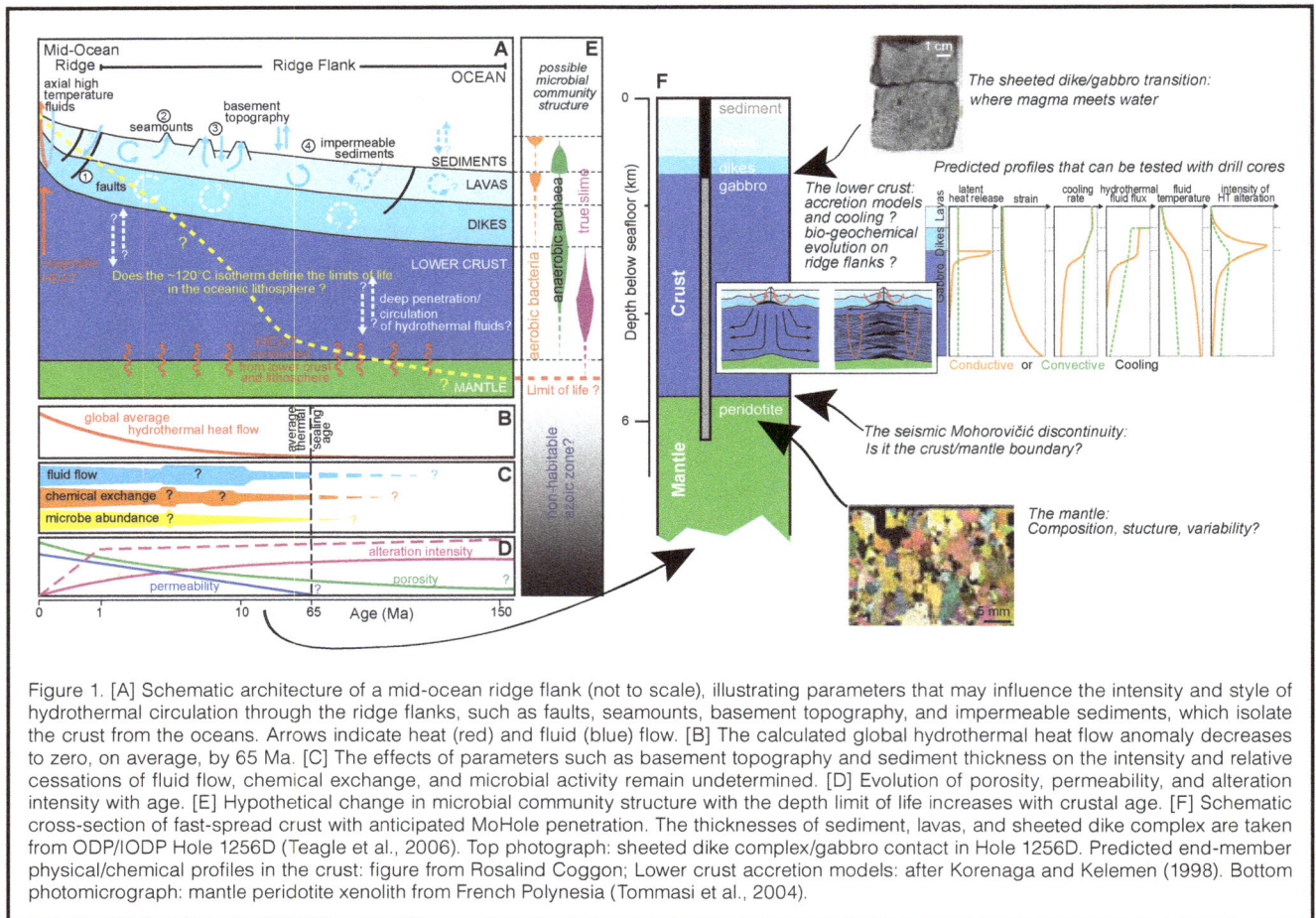

Figure 1. [A] Schematic architecture of a mid-ocean ridge flank (not to scale), illustrating parameters that may influence the intensity and style of hydrothermal circulation through the ridge flanks, such as faults, seamounts, basement topography, and impermeable sediments, which isolate the crust from the oceans. Arrows indicate heat (red) and fluid (blue) flow. [B] The calculated global hydrothermal heat flow anomaly decreases to zero, on average, by 65 Ma. [C] The effects of parameters such as basement topography and sediment thickness on the intensity and relative cessations of fluid flow, chemical exchange, and microbial activity remain undetermined. [D] Evolution of porosity, permeability, and alteration intensity with age. [E] Hypothetical change in microbial community structure with the depth limit of life increases with crustal age. [F] Schematic cross-section of fast-spread crust with anticipated MoHole penetration. The thicknesses of sediment, lavas, and sheeted dike complex are taken from ODP/IODP Hole 1256D (Teagle et al., 2006). Top photograph: sheeted dike complex/gabbro contact in Hole 1256D. Predicted end-member physical/chemical profiles in the crust: figure from Rosalind Coggon; Lower crust accretion models: after Korenaga and Kelemen (1998). Bottom photomicrograph: mantle peridotite xenolith from French Polynesia (Tommasi et al., 2004).

P-wave velocities (Vp) in excess of 8 km s^{-1}. It is generally assumed that the Moho also represents the boundary between mafic igneous rocks (crystallized from magmas that form the crust) and residual peridotites of the upper mantle. However, this interpretation has never been tested, and there are geologically valid scenarios where the Moho might delineate the boundary between mafic and ultramafic cumulate rocks within the crust, or exist below serpentinized peridotites that were previously part of the mantle. Observations and sampling of the Moho, the petrological crust-mantle boundary, and the rocks of the upper mantle are fundamental to understanding the geodynamics and chemical differentiation of our planet. A foremost goal is to reconcile geophysical imaging of the Moho with direct geological observations of cores and downhole measurements (e.g., is the Moho in our study region a sharp compositional boundary or a transition zone of significant thickness?).

Formation of the Lower Crust

On the road to the Moho, we will make paradigm-testing observations of the lower oceanic crust and the deep magmatic, tectonic, and hydrothermal processes that occur at the mid-ocean ridges (Fig. 1). Our principal target will be intact ocean crust formed at a fast-spreading ridge, which should be relatively laterally uniform, and where we have

well-developed theoretical models of crustal accretion that can be tested by drilling. Is the lower oceanic crust formed from the subsidence of a high-level magma chamber, or are there multiple melt bodies at different levels within the oceanic crust (or upper mantle) at fast-spreading ridges?

Magnetic stripes document the history of ocean crust formation and are the very basis of plate tectonic theory, yet we have little information on what contribution the lower crust has to this fundamental signature. Similarly, seismic profiling remains the key tool for investigating the deep crust, but these regional scale measurements have never been calibrated against core or *in situ* measurements. It remains challenging to confidently develop geological interpretations from geophysical measurement of the oceanic crust.

Composition and Hydration of the Ocean Crust

A full penetration will provide the first direct estimate of the bulk composition of ocean crust critical for Earth differentiation models. How deeply do seawater-derived hydrothermal fluids penetrate, and how efficient is hydrothermal circulation at heat extraction and chemical alteration (Fig. 1)? Is fluid flow channeled by major faults, or is it more pervasive? The knowledge of modes of penetration of the hydro-thermal fluids, and of the extent of their interactions with the lithosphere, is required to estimate chemical exchanges with the oceans, as well as to assess the volume and composition of materials transferred to the mantle via subduction.

Limits and Controlling Factors of Life

Understanding the limits of life (Fig. 1), and the factors controlling these limits, is one of the most fundamental goals of geo- and biosciences essential for understanding the origin, evolution, distribution, and future of life on Earth as well as celestial bodies. To date, the limits of life even on our own planet remain poorly defined. The MoHole project provides a unique opportunity to address these limits in the oceanic lithosphere that covers ~60% of our planet. Numerous factors may control the limits of life, such as temperature, water activity, salinity, pH, and energy and carbon sources. Among these, temperature plays a key role, because organisms cannot survive beyond as yet a poorly known temperature threshold (~110–120°C?). The ability of seawater to penetrate into the deep crust or mantle and be available for microorganisms (e.g., minimum pore space) will also have a strong impact on the distribution of living organisms.

Physical and Chemical Nature of the Upper Mantle

Direct observations of the mantle will document how magmas are focused from a broad melting region to a narrow zone of crustal accretion beneath mid-ocean ridges. Measurements across the Moho will quantify the tectonic coupling between the crust and mantle. We presently have little knowledge of the composition and physical state of *in situ* convecting mantle at the ridge axis. A few kilograms of fresh residual peridotite from beneath intact oceanic crust would provide a wealth of new information comparable to the treasure trove obtained from the Apollo lunar samples.

Geophysical Characteristics of the Mohole Project Area

The criteria for best possible deep crustal penetration sites were reformulated during this workshop. The selected target would ideally meet all of the following scientific requirements:

a) Crust formed at fast-spreading rate (>40 mm yr⁻¹ half rate).

b) Simple tectonic setting with very low-relief seafloor and smooth basement relief; away from fracture zones, propagator pseudo-faults, relict overlapping spreading basins, seamounts, or other indicators of late-stage intraplate volcanism. Connection to the host plate active constructive and destructive boundaries would provide important scientific information.

c) Crustal seismic velocity structure should not be anomalous relative to current understanding of "normal" fast-spread Pacific crust, indicative of layered structure.

d) A sharp, strong, single-reflection Moho imaged with Multi-Channel Seismic (MCS) techniques.

e) A strong wide-angle Moho reflection (PmP), as observed in seismic refraction data, with distinct and clearly identifiable sub-Moho refractions (Pn).

f) A clear upper mantle seismic anisotropy.

g) A crust formed at an original latitude greater than ±15°.

h) A location with relatively high upper crustal seismic velocities indicative of massive volcanic formations to enable the initiation of a deep drill hole.

Satisfying requirements for points a–e is essential for success. More flexibility is allowed in meeting points f–h, which are highly desirable but not essential. Several technological constraints limit the range of potential sites:

- Technology for re-circulating drilling mud (riser or alternative; see next section) is currently untested at water depths greater than 3000 m.

- Prior scientific ocean drilling experience is mostly limited to temperatures less than 200°C. Temperatures higher than ~250°C will may limit choices of drill bits and logging tools, may decrease core recovery, and may increase risk of hole failure, or require substantial re-design of drilling equipment. Based on plate cooling models, crust older than ~15–20 Ma should meet this requirement at Moho depths (Fig. 2).

- Thickness of the crustal section above Moho must be at least a few hundred meters less than the maximum penetration/logging/recovery depth of the drilling system to allow significant penetration in mantle peridotites.

- Target area should be in a region with good weather conditions at least eight months out of the year, with calm seas and gentle ocean bottom currents.

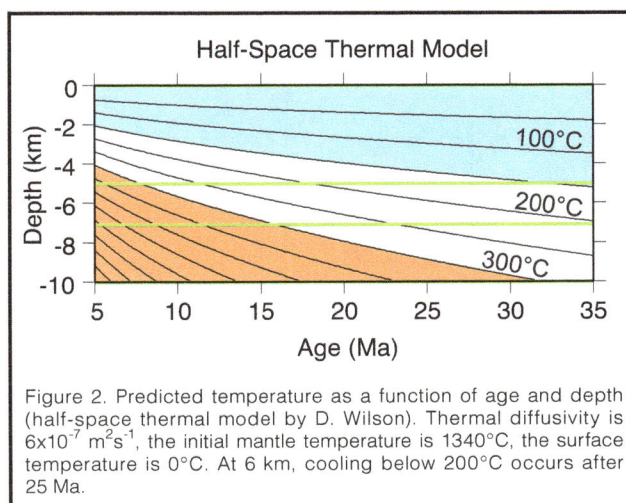

Figure 2. Predicted temperature as a function of age and depth (half-space thermal model by D. Wilson). Thermal diffusivity is 6×10^{-7} m²s⁻¹, the initial mantle temperature is 1340°C, the surface temperature is 0°C. At 6 km, cooling below 200°C occurs after 25 Ma.

- Sediment thickness should be greater than 50 m to support possible riser hardware and other seafloor infrastructure (re-entry cones/uppermost casing strings).
- Targeted area should be close (less than ~1000 km) to major port facilities for logistical practicalities.

Potential Sites

Based on the scientific requirements and technological constraints described above, the workshop participants focused discussions on three areas in the Pacific Basin: Cocos Plate, off Southern and Baja California, and off Hawaii (Table 1; Fig. 3). One of the most important issues to take into consideration is the trade-off between seafloor depth and temperature at Moho depths. Most ocean seafloor subsides below 4000 m by ~25 Ma, whereas at Moho depths of 5–7 km temperatures of 200°C or less are expected for crustal ages of 17–35 Ma (Fig. 2). The respective advantages and disadvantages of the three selected areas are listed in the full workshop report (available online at http://campanian.iodp.org/MoHole/).

The *Cocos Plate region* (Site 1256 area, Fig. 3A) encompasses a section of the Cocos Plate off Central America (from Guatemala to northern Costa Rica) with lithospheric ages between 15 Ma and 25 Ma. At its western limit on 15-Ma crust, this area includes the Ocean Drilling Program (ODP) Hole 1256D (Wilson et al., 2006; Teagle et al., 2006), a site of ongoing Integrated Ocean Drilling Program (IODP) deep

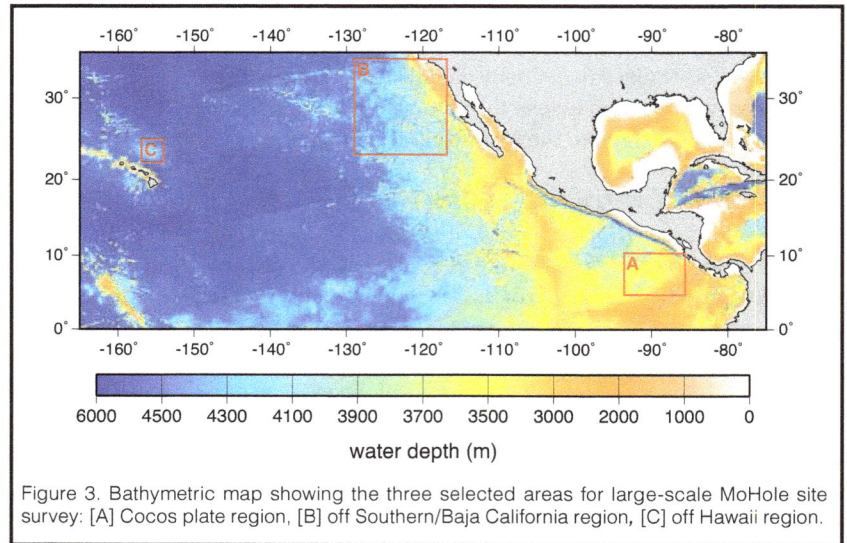

Figure 3. Bathymetric map showing the three selected areas for large-scale MoHole site survey: [A] Cocos plate region, [B] off Southern/Baja California region, [C] off Hawaii region.

drilling into intact ocean crust. MCS (Hallenborg et al., 2003; Wilson et al., 2003) and wide-angle ocean bottom seismometer (OBS) data exist for the 15–17 Ma area in the vicinity of Site 1256. This region sits in superfast crust (half-spreading rate 110 m yr^{-1}), within a corridor that includes a complete tectonic plate life cycle, making it an excellent candidate for understanding ocean crust evolution from a spreading center to subduction. Structure of the crust within this area can be directly related to processes occurring at the modern East Pacific Rise and the Central American subduction zone.

The *off Southern/Baja California region* (Fig. 3B) encompasses a section of the eastern Pacific Plate at ~20°–33°N and ~130°–118°W. Crustal ages are ~20–35 Ma. Very little

Table 1. Principal characteristics of possible candidate sites for the MoHole project.

Candidate Site	Location	Half-Spreading Rate (mm yr^{-1})	Crustal Age (Ma)	Inferred Moho-T (°C)	Water Depth (m)	Sediment Thickness (m)	Crustal Thickness (km)	Total Length to the Moho (km)	Original Latitude
Cocos (Site 1256)	6.7°N, 91.9°W	110	15	>250	3646	250	5,5	8.7–9.2	Near equator
Cocos (Site 844)	8°N, 90.5°W	100	17	>250	3414	290	5,5?	8.7–9.2?	Near equator
Cocos	8.7°N, 89.5°W	100	19	~250	~3400	~300	5,5?	8.7–9.2?	Near equator
off S/Baja California (Deep Tow Site)	31°–33°N, 125°–127°W	60	30–27	<200	4300–4500	~100	~5.5?	~10?	~30–33
off S/Baja California	28°–29°N, 123°–125°W	50	27–22	~200	4200–4400	~100	?	?	~28
off S/Baja California	25°–26°N, 120°–122°W	60	27–22	~200	3900–4300	~80	?	?	~30
off S/Baja California	30.5°–31°N, 121°W	45	20–22	~250	2700–4100	~130	?	?	~25
Hawaii	22.9°–23.7°N, 154.9°–155.8°W	35–40	79–81	~150	4050–4300	~200	~6?	10–10.5?	Near equator
Hawaii	23.5°–23.9°W, 154.5°–154.8°W	35–40	78	~150	4300–4500	~200	~6?	10–10.5?	Near equator

modern geophysical information exists from this region. The best-studied area is in the northernmost part off San Diego, the "Deep Tow" site at 32°25'N, 125°45'W (31–32 Ma; Luyendyk, 1970). Historical data there include deep-tow sidescan and bathymetry, 3.5 kHz profiler, magnetics, and single-channel seismics.

The *off Hawaii region* is located north of Oahu in the flexural arch, where water depths are 4000–4300 m. The crust is ~80 Ma and was formed at an intermediate half-spreading rate of 35–40 mm yr^{-1}. This site offers the lowest temperature at Moho depth (~100°C–150°C), but crustal structure is potentially affected by hotspot volcanism (underplating and/or crustal intrusions), and its significantly older age makes it difficult to relate geochemical changes to modern ocean chemistry or conditions.

Geophysical Surveys: Finding the Right Project Area

The existing geophysical data at all potential sites are not sufficient to identify a clear MoHole Project target area. Consensus at the workshop was that the priority of the community should be directed toward conducting large-scale seismic surveys in the three selected regions, which will lead to the identification of a MoHole target that best satisfies the requirements stated above. These surveys should collect spatially coincident MCS data, wide-angle OBS data, multi-beam bathymetry and gravity. Heat flow and magnetic anomaly data will be useful and should be collected. The characteristics of the required seismic surveys are listed in the full workshop report. JAMSTEC will dedicate three months of science ship time in 2011 for large-scale surveys. There was a consensus that the first survey should be in the off Southern/Baja California region, because so little is known in this area (where depth/age/logistical criteria are viable). Baseline reconnaissance seismic data are urgently required to assess whether this area can possibly meet the scientific requirements. Two additional factors contributed to the choice of this region as short-term priority for initial reconnaissance: crustal ages are greater than near Site 1256 (so temperatures are expected to be cooler), and the existing data suggest that Moho in the Site 1256 area may not be associated with a simple, continuous, strong reflector.

After an appropriate drilling target has been identified, the community should conduct detailed seismic surveys in the vicinity of the specific target, including 3-D multi-streamer MCS and OBS surveys for accurate and geometrically correct imaging of intracrustal reflectors (faults, sills, etc.) and Moho, and to assess crustal structure and thickness variability, and upper mantle velocity structure/anisotropy.

The scope and costs of the surveys required for this project are too large to be undertaken by a single nation or funding agency, hence international collaboration is essential.

Technology Development and Operations

The MoHole initiative is arguably at the point where the framework for the operations can be constructed, since the technology to drill such a hole exists or at least has been shown to be feasible. The technology selection and required engineering development will be key components for the success of the MoHole project. It is important to identify potential issues in drilling and coring engineering from the past and ongoing ocean drilling expeditions, and to find solutions to overcome the problems encountered. The engineering efforts must be directed to ensure that the scientific goals of the MoHole project are achieved. Technology selection process and planning for the key engineering developments should be launched as soon as possible in conjunction with site-survey efforts. To do so, establishing a realistic roadmap, which includes project scoping, development and testing elements all controlled by proper project management, is imperative. All MoHole target sites are located in ultra-deep water of ~4000-m water depth or beyond, and the drilling depth to achieve the MoHole objectives is estimated to extend more than 6000 m below the seafloor. To drill such an ultra-deep borehole, the provision for continuous mud circulation is a top priority technology requirement. Other major areas requiring engineering consideration include logging and coring in high temperature environment, drill bits (specifically designed for abrasive, hard rocks) and drill string (high tensile strength), drilling mud (developed for high temperature environment), and casing/cementing materials and strategies (specifically designed, ideally to the bottom of the hole).

A promising candidate technology for drilling the MoHole is riser drilling, which provides a conduit for the mud to be returned to the vessel for cleaning, evaluation, and recirculation. The D/V *Chikyu* is currently equipped with a deep riser system with a maximum rated water depth of 2500 m. Significant engineering development is required to prepare the D/V *Chikyu* for riser service in water depths ≥4000 m. In addition to riser drilling, several other technologies are being considered to drill safely and efficiently to the target depth, including Surface Blow Out Preventer (BOP) with slim riser pipe (casing pipe) and Shut In Device (SID), or Riserless Mud Recovery (RMR; Myers, 2008) with mud circulation pump and mud return line. The lithologies intersected by the borehole drilled to Moho depths are unequivocally expected to be free from overpressures, hydrocarbons, or other geohazards. However, future regulatory changes may require the use of blow-out prevention in mud circulation systems. Hence, although a BOP will likely not be needed for well control, the use of a BOP will be considered by the MoHole planning group.

Drilling the MoHole will be a challenging enterprise requiring years of detailed preparation, planning, and engineering. Operationally, major challenges will be associated with collecting the cored material, making the *in situ* measurements, installing casing, and keeping the borehole open for successive episodes of deepening in a multiyear, multiphase operation. To gear up for the operations, all issues related to drilling, casing, coring, and logging must be adequately explored and included in a comprehensive and complete operation plan, as soon as the site characteristics are known. Key elements of the drilling/coring/downhole measurements were listed during the workshop (see full workshop report). The well design of the primary site may

require data from a pilot hole, to properly evaluate parameters such as mud weights and casing set points. A pilot hole may be either a separate hole or simply a pilot section of the main hole. Drilling engineering data from a pilot section will be critical in managing the pressure, temperatures, and stress within the borehole.

After completion of drilling, coring, and borehole logging, the MoHole should be used for further experiments, including vertical seismic profiles (VSP), and long-term monitoring. Given the extreme borehole depth to be drilled, at least two offset VSPs are ideally required, one at the estimated halfway point of the well and perhaps one at final depth. Instrumenting the MoHole will eventually become a key, second-stage goal. Hence, the sub-sea equipment and borehole must be constructed to accommodate observatory science (e.g., fluid monitoring, microbiology incubation experiments). This implies ROV access to the wellhead and the ability to access the borehole through a BOP or SID.

Keys for Success

The keys for a successful MoHole project, as identified during the workshop, include scientific (essentially sampling strategy) considerations, as well as technology development, industry engagement, and public engagement through outreach activities and education. The MoHole project will be one of the largest scientific endeavors in Earth science history, and this challenge should provide precious opportunities to a diversity of scientists, engineers and technologists. One of the keys to success will be the sharing of the opportunities and achievements across a broad spectrum of Earth and life scientists.

The size and duration (ten years or more) of the MoHole project will require an appropriately supported, centralized science operations and engineering management group to oversee the successful initiation and completion of the project. This international group will be key to success and should be created as early as feasible. The envisioned, ideal timeline for the MoHole project is to complete prospective 2-D site survey (including data analysis) in ~2014, choose the MoHole site and conduct 3-D site survey in ~2015, start preparing operations in ~2015-2016, start drilling in ~2018, and reach the mantle in ~2022.

Scientific Coring, Sampling, and Measurements

Many of our primary scientific goals will require continuous core samples. To be regarded as successful, the MoHole project must at least return the following (see also Fig. 4):

- Continuous core, including samples of all boundaries, across the region identified by seismic imaging as the Moho, and the lithologic transition from cumulate

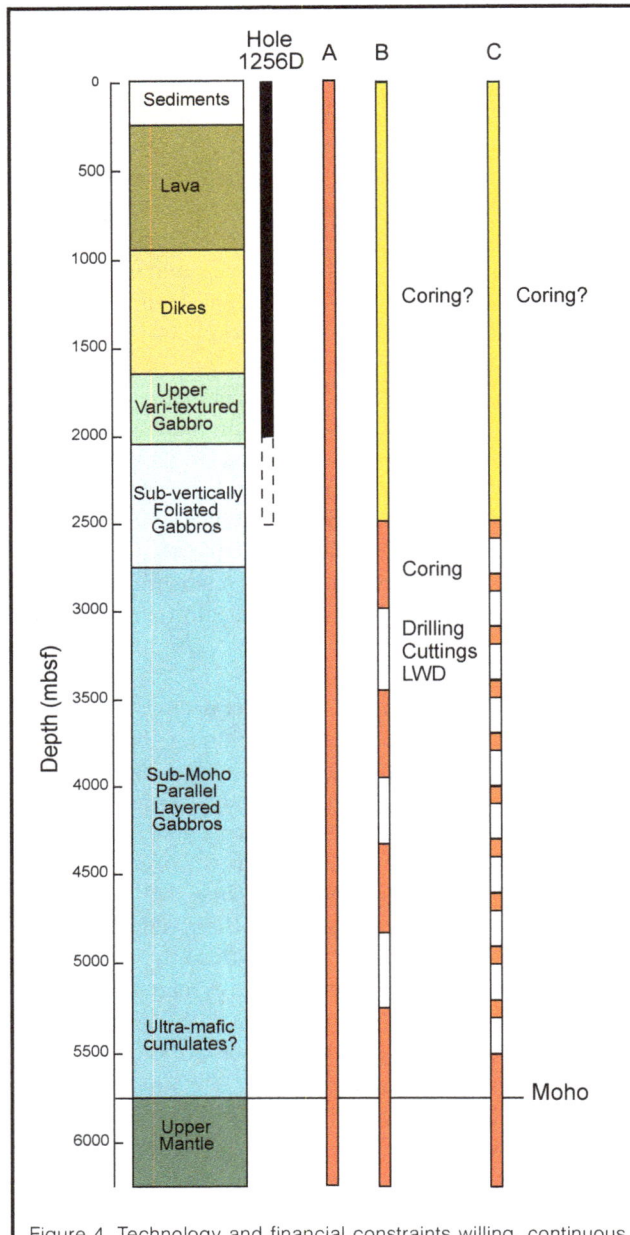

Figure 4. Technology and financial constraints willing, continuous coring all the way to the Moho and then a significant distance (~500 m) into the uppermost mantle [A] would be the best approach to achieve the scientific goals of this project. However, approaches that mix spot coring ([B] long coring of key sections or [C] 10-m coring before bit change every 50 m) with continuous wireline coring may need to be considered. Significant lengths of continuous cores across major lithologic and geophysical transitions are mandatory to answer the fundamental scientific questions.

magmatic rocks to residual peridotites (these may or may not be the same target)

- Continuous coring of the lower 500 m of the mafic and ultramafic cumulate rocks in the oceanic crust
- Continuous coring of 500 m of peridotites and associated lithologies in the uppermost mantle below the Moho
- Sufficient cores from intervals of the lower oceanic crust to test models of crustal accretion and melt movement, to resolve the geometry and intensity of hydrothermal circulation, and to document the limits and activity of the deep microbial biosphere
- A continuous, comprehensive suite of geophysical logs (wireline, Logging While Drilling/Coring) and borehole experiments to measure *in situ* physical properties, to acquire borehole images, and to identify key geophysical and lithologic regions and transitions (e.g., Layer 2-3 boundary, the Moho) throughout the ocean crust and into the upper mantle.

Due to the expected relatively coarse grain size of the rocks to be encountered in the MoHole and the fine scales of expected lithologic/geochemical variation, it is anticipated that lithological records provided by mud/chip logging will be insufficient to address the scientific questions posed. However, a continuous series of mud, cuttings, and gas logs will provide useful supplementary information in areas of poor or no core recovery, and should be routine throughout the experiment. In addition to sampling and analyzing rocks, measurements of temperature and chemical compositions of fluids are required, together with biological analyses such as cell counting and DNA/RNA analyses.

Drilling Technology and Industry Engagement

Technologies that are applicable to the MoHole project are now being developed within the oil and gas industry. These were presented at the workshop by industry representatives. Conversely, some of the required technologies are very specific to scientific drilling, such as logging and coring at high temperatures (e.g., the IDDP project, Skinner et al., 2010) or drilling the hard crustal and mantle rocks. Development of such technologies will be a primary key for success, but to achieve this it may be necessary to complement IODP financial support with external funding.

As the oil and gas industry conducts operations in increasingly deep water, important keys for success will be continuous collaboration with industry and introduction of new technologies to the MoHole project where applicable. It will be necessary to establish a strategy to engage the industry in the project, exchange personnel, and plan joint development work. This can occur at several levels, including i) continuing *ad hoc* collaboration, through inviting oil and gas industry representatives to participate in planning activities and community workshops, ii) contracting services

from planning to execution, and iii) participation of engineers and scientists engaged in the MoHole project to industry workshops, symposiums, and technology development forums.

Public Engagement, Outreach, and Education

Another key component of the success of a Mohole project will be to improve public support and understanding of the scientific goals and excitement of the project. Engaging the public through outreach and education activities, as well as being pro-active in advertising the project to the wider scientific community and engaging new groups of scientists, should be integral parts of the activities carried out by the MoHole project scoping group, under the umbrella of IODP and future international collaboration for scientific ocean drilling. One tool to be implemented rapidly, as soon as scoping activities commence, is a dedicated, dynamic and engaging MoHole web page.

References

Bascom, W.N., 1961. *A Hole in the Bottom of the Sea: The Story of the Mohole Project.* Garden City, New York (Doubleday and Company, Inc.): 352 pp.

Christie, D.M., Ildefonse, B., et al., 2006. *Mission Moho - Formation and Evolution of Oceanic Lithosphere.* Full workshop report. Portland, OR, U.S.A., 7–9 September 2006. Available online at http://www.iodp.org/mission-moho-workshop.

Hallenborg, E., Harding, A.J., Kent, G.M., and Wilson, D.S., 2003. Seismic structure of 15 Ma oceanic crust formed at an ultrafast spreading East Pacific Rise: evidence for kilometer-scale fracturing from dipping reflectors. *J. Geophys. Res.*, 108(B11):2532, doi: 10.1029/2003JB002400.

Ildefonse, B., Christie, D.M., and Mission Moho Workshop Steering Committee, 2007. Mission Moho workshop: drilling through the oceanic crust to the mantle. *Sci. Drill.*, 4:11–18. doi:10.2204/iodp.sd.4.02.2007.

Korenaga, J., and Kelemen, P.B., 1998. Melt migration through the oceanic lower crust: a constraint from melt percolation modeling with finite solid diffusion. *Earth Planet. Sci. Lett.*, 156(1–2):1–11, doi:10.1016/S0012-821X(98)00004-1.

Luyendyk, B.P., 1970. Origin and history of abyssal hills in the northeast Pacific Ocean. *GSA Bull.*, 81(8):2237–2260, doi:10.1130/0016-7606(1970)81[2237:OAHOAH]2.0.CO;2.

Myers, G., 2008. Ultra-deepwater riserless mud circulation with dual gradient drilling. *Sci. Drill.*, 6:48–51. doi:10.22 04/iodp. sd.6.07.2008.

Shor, E.N., 1985. A chronology from Mohole to JOIDES. *In* Drake, E.T., and Jordan, W.M. (Eds.), *Geologists and Ideas; A History of North American Geology.* Geol. Soc. Am. Spec. Publ. 4. Boulder, CO (Geological Society of America): 391–399.

Skinner, A., Bowers, P., Þórhallsson, S., Friðleifsson, G.O., and Guðmundsson, H., 2010. Drilling and operating a core barrel for the Iceland Deep Drilling Project. *GeoDrilling*

International, May 2010:18–21.

Teagle, D.A.H., Alt, J.C., Umino, S., Miyashita, S., Banerjee, N.R., Wilson, D.S., and the Expedition 309/312 Scientists, 2006. *Proc. IODP*, 309/312: Washington, DC (Integrated Ocean Drilling Program Management International, Inc.). doi:10.2204/iodp.proc.309312.2006

Teagle, D., Ildefonse, B., Blackman, D.K., Edwards, K., Bach, W., Abe, N., Coggon, R., and Dick, H., 2009. Melting, magma, fluids and life: challenges for the next generation of scientific ocean drilling into the oceanic lithosphere. Full workshop report. National Oceanography Centre, University of Southampton, 27–29 July 2009, available online at http://www.interridge.org/WG/DeepEarthSampling/workshop2009.

Tommasi, A., Godard, M., Coromina, G., Dautria, J.M., and Barsczus, H., 2004. Seismic anisotropy and compositionally induced velocity anomalies in the lithosphere above mantle plumes: a petrological and microstructural study of mantle xenoliths from French Polynesia. *Earth Planet. Sci. Lett.*, 227:539–556. doi:10.1016/j.epsl.2004.09.019

Wilson, D.S., Hallenborg, E., Harding, A.J., and Kent, G.M., 2003. Data report: site survey results from cruise EW9903. *In* Wilson, D.S., Teagle, D.A.H., Acton, G.D., et al., *Proc. ODP, Init. Repts.*, 206: College Station, TX (Ocean Drilling Program):1–49, doi:10.2973/odp.proc.ir.206.104.2003

Wilson, D.S., Teagle, D.A.H., Alt, J.C., Banerjee, N.R., Umino, S., Miyashita, S., Acton, G.D., Anma, R., Barr, S.R., Belghoul, A., Carlut, J., Christie, D.M., Coggon, R.M., Cooper, K.M., Cordier, C., Crispini, L., Durand, S.R., Einaudi, F., Galli, L., Gao, Y.J., Geldmacher, J., Gilbert, L.A., Hayman, N.W., Herrero-Bervera, E., Hirano, N., Holter, S., Ingle, S., Jiang, S.J., Kalberkamp, U., Kerneklian, M., Koepke, J., Laverne, C., Vasquez, H.L.L., Maclennan, J., Morgan, S., Neo, N., Nichols, H.J., Park, S.H., Reichow, M.K., Sakuyama, T., Sano, T., Sandwell, R., Scheibner, B., Smith-Duque, C.E., Swift, S.A., Tartarotti, P., Tikku, A.A., Tominaga, M., Veloso, E.A., Yamasaki, T., Yamazaki, S. and Ziegler, C., 2006. Drilling to gabbro in intact ocean crust. *Science*, 312(5776):1016–1020. doi: 10.1126/science.1126090.

Authors

Benoît Ildefonse, CNRS, Université Montpellier 2, CC60, 34095 Montpellier cedex 5, France, e-mail: benoit.ildefonse@um2.fr.

Natsue Abe, Institute for Research on Earth Evolution (IFREE), Japan Agency for Marine-Earth Science and Technology (JAMSTEC), 2-15 Natsushima-cho, Yokosuka 237-0061, Japan.

Donna K. Blackman, Scripps Institution of Oceanography, UCSD, La Jolla, CA 92093-0225, U.S.A.

J. Pablo Canales, Woods Hole Oceanographic Institution, Department of Geology and Geophysics, Woods Hole, MA 02543, U.S.A.

Yoshio Isozaki, Marine Technology Center (MARITEC), JAMSTEC, 2-15 Natsushima-cho, Yokosuka 237-0061, Japan.

Shuichi Kodaira, Institute for Research on Earth Evolution (IFREE), Japan Agency for Marine-Earth Science and Technology (JAMSTEC), 3173-25 Showa-machi, Kanazawa-ku, Yokohama 236-0001, Japan.

Greg Myers, Consortium for Ocean Leadership, 1201 New York Avenue NW, Suite 400, Washington, DC 20005, U.S.A.

Kentaro Nakamura, Precambrian Ecosystem Laboratory, JAMSTEC, 2-15 Natsushima-cho, Yokosuka 237-0061, Japan.

Mladen Nedimovic, Department of Earth Sciences, Dalhousie University, Edzell Castle Circle, Halifax, NS B3H 4J1, Canada.

Alexander C. Skinner, ACS Coring Services, 13 Riccarton Drive, Currie, Edingburgh, EH14 5PN, Scotland, U.K.

Nobukazu Seama, Research Center for Inland Seas, Kobe University, 1-1 Rokkodai, Nada, Kobe 657-8501, Japan

Eiichi Takazawa, Department of Geology, Faculty of Science, Niigata University, 2-8050, Niigata, 950-2181, Japan.

Damon A.H. Teagle, National Oceanography Centre, Southampton, University of Southampton, U.K.

Masako Tominaga, Department of Geology and Geophysics, Woods Hole Oceanographic Institution, Woods Hole, MA 02543, U.S.A.

Susumu Umino, Department of Earth Sciences, Kanazawa University, Kakuma-Machi, Kanazawa-Shi, Ishikawa 920-1192, Japan.

Douglas S. Wilson, Department of Earth Science, UCSB, Santa Barbara, CA 93106-9630, U.S.A.

Masaoki Yamao, Center for Deep Earth Exploration (CDEX), JAMSTEC, 2-15 Natsushima-cho, Yokosuka 237-0061, Japan.

Related Web Link

http://campanian.iodp.org/MoHole/

Australian Earth Sciences Convention Features Ocean Drilling

In July 2010, a selection of national and international speakers, industry leaders, key decision makers, and assorted geoscientists (650 in all) met in Canberra for the Australian Earth Sciences Convention. Australia and New Zealand (ANZIC) took this opportunity to showcase our past, current, and future IODP expeditions. The following papers, nearly all on preliminary expedition results, were given at an IODP symposium held on 7 July:

- **Kevin Welsh** (keynote speaker) – IODP 318 Cenozoic East Antarctic ice sheet evolution from Wilkes Land Margin sediments: preliminary results.
- **Neville Exon** – Australia's involvement in IODP: what it means for our scientists?
- **Bob Carter** – Preliminary results from IODP Expedition 317 (Canterbury Basin, New Zealand).
- **John Moreau** – Biogeochemical and geomicrobiological evidence for an ultra-deep anaerobic methane oxidation zone in the Nankai Trough subseafloor.
- **Kelsie Dadd** – IODP Expedition 323 in the Bering Sea: environmental change over 5 million years recorded in deep-sea sediment.
- **Christian Ohneiser** – Magnetostratigraphic records from Eocene-Miocene sediments cored in the equatorial Pacific: initial results from the Pacific Equatorial Age Transect (PEAT), IODP Expeditions 320/321.
- **David Murphy** – Sr, Nd, and Pb isotope data from the Shirshov Massif of the Shatsky Rise, northwest Pacific.
- **Simon George** – Geochemistry from the Cante

Many scientists (average of 40–50) attended these talks, which were followed by a number of visits to the IODP exhibition booth. There they asked a range of questions regarding applications for shipboard positions, the science, the life on board ship, and general information about the membership and organization of IODP. ANZIC brochures, *Scientific Drilling* journal, and expedition fact sheets were displayed and distributed. Video presentations, especially those on the Wilkes Land expedition, generated interest.

Please note that there will be an IODP symposium at the 34th International Geological Congress in Brisbane, which is scheduled for 2–10 August 2012. The emphasis is likely to be on the Southern Hemisphere Oceans.

New Drilling Platform Expands Climate Studies in Lakes

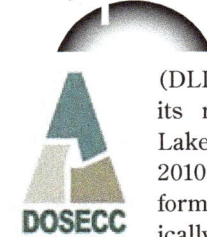

DOSECC's Deep Lake Drilling System (DLDS) platform made its maiden voyage on Lake Van, Turkey in July 2010. The DLDS platform is designed specifically for deep lake drilling and will enable researchers to sample previously inaccessible deep lake sediments. The DLDS platform consists of two main parts: the drilling rig and associated equipment, and the barge itself. The drilling rig is an Atlas Copco T3WDH rig, a top-head-drive rotary rig designed for water well and oil and gas drilling. DOSECC has made extensive modifications to the rig to turn it into a deep coring rig. The DLDS is designed to drill to 1400 m.

The drilling barge, made by Damen in the Netherlands, is a modular system which enables easy shipping anywhere in the world. It is constructed with six separate containers connected in a two-by-three configuration with a moon pool built into one of the modules. The barge is 24.4 m long by 7.3 m wide. Along with the drilling rig, pipe, mud tanks, and associated supplies, the platform also accommodates a science lab and a driller's shack. The science lab is used for on-board sampling from the core catcher and for labeling and orienting the core samples. During the drilling crew shift change, the cores are transported to shore for additional sampling and testing. Four separate winches and anchors with 2 km of cable each keep the barge on station.

In early August, DOSECC operated the barge in Lake Van at a water depth of 360 m. The deepest hole planned for the Lake Van project is 250 m below the lake floor. Upcoming projects for the DLDS include the Dead Sea in Israel and Lake Ohrid in Macedonia. The system will then return to the U.S. following the Lake Ohrid project.

IODP/ECORD - Canada 2010 Summer School: A Great Success

The summer school by ECORD/ IODP-Canada was a great success from 27 June to 12 July. Nineteen students and postdoctoral fellows from Canada, France, Germany, Serbia, Portugal, the U.K, and the U.S.A. participated in a two-week intensive training in marine geology and paleoceanography. Participants had sailing and sampling experience on board the R/V *Coriolis II* in the St. Lawrence Estuary and Saguenay Fjord; on board they acquired theoretical and practical knowledge on cutting-edge techniques for sampling and analyzing geological and geophysical data. Courses, lectures, practical exercises, and laboratory visits were offered at Université du Québec à Rimouski (UQAR), the Institut national de la recherche scientifique - Centre – Eau Terre

The participants during a field trip at the Parc national du Bic. Photo by H. Gaonac'h

Environnement (INRS-ETE), and the Université du Québec à Montréal (UQAM), in addition to field trips in Gaspesia and St. Lawrence Lowlands giving students an extensive scientific portrait of paleoceanography and paleoclimatology in polar and sub-polar environments. This summer school was possible thanks to an impressive group of scientists sharing their knowledge with participants. More than a dozen researchers from ECORD countries actively involved in IODP activities presented the most recent state-of-the-art theories and practices in high latitude geophysics, geochemistry, paleontology, geomorphology, oceanography, sedimentology, sea-ice modeling, and gas hydrates. Speakers included: Hans Asnong, Anne de Vernal, Claude Hillaire-Marcel, and Taoufik Radi (UQAM, Canada), Gilles Bellefleur (GSC-Ottawa, Canada), Xavier Crosta, and Frédérique Eynaud (Université Bordeaux I, France), Mathieu Duchesne (GSC-Québec, Canada), Pierre Francus (INRS-ETE, Canada), Martin Frank (IFM-GEOMAR, Germany), Yves Gélinas (Concordia University, Canada), Joël Guiot and Guillaume Massé (CNRS, France), Patrick Lajeunesse (Université Laval, Canada), Jean-François Lemieux (New York University, U.S.A.), Matt O'Regan (Cardiff University, U.K.), Joseph Ortiz (Kent State University, U.S.A.), Frank Rack (University of Nebraska-Lincoln, U.S.A.), André Rochon, Guillaume St-Onge, and Bjorn Sundby (UQAR, Canada), Ruediger Stein (AWI, Germany).

IODP-Canada is grateful to the many institutions which sponsored this summer school: the GEOTOP Research Center, the Institut des sciences de la mer de Rimouski (ISMER), INRS-ETE, UQAM, the Canadian Consortium for Ocean Drilling (CCOD), the European Consortium for Ocean Research Drilling (ECORD), and the MobilUQ program of the Université du Québec. For more details on the 2010 IODP/ECORD-Canada Summer School, please contact the IODP-Canada office, e-mail: coordinator@mail.iodpcanada.ca.

IODP/ICDP Canada Booth at GeoCanada 2010

IODP-Canada and ICDP-Canada organized a booth at the most recent Geocanada 2010 meeting in Calgary, Alberta. Many visitors came to the booth open on 10–12 May 2010. GeoCanada 2010 was sponsored by several Canadian associations, notably the Geological Association of Canada (GAC) and the Mineralogical Association of Canada (MAC). The information provided at the booth included the list of recent and current drilling expeditions coordinated by IODP in oceanic environments and by ICDP on continents; it also included the various targeted scientific issues such as past climate change, the geology and geophysics of oceanic and continental crusts, the bottom of the ocean biosphere, available resources such as gas hydrates, triggers in active seismic zones, etc. IODP summer schools organized for summer 2010 were also announced. Students and researchers present at the booth came from the University of Alberta in Edmonton, the Université du Québec à Montréal, and the University of Toronto. This successful activity was part of the IODP and ICDP Canada programs to better promote IODP and ICDP activities in Canada. More information on their homepage: http://www.iodpcanada.ca/, for questions contact: coordinator@mail.iodpcanada.ca.

Solution for Riser Drilling in Strong Current

In the NantroSEIZE drilling area, the very strong "Kuroshio Current" exists. When a riser is deployed in such a current, Vortex Induced Vibration (VIV) of the riser occurs. VIV is a cross-flow vibration of a circular cylinder placed in a current. Long term VIV will cause fatigue fractures of the riser.

In order to suppress VIV of the riser, JAMSTEC decided to apply a riser fairing device, which is one possible way to suppress VIV. The shape of the fairing is like a fin, and a current against the riser is redirected so that vortex shedding behind the riser is reduced. The fairing also rotates freely about the riser and aligns with direction of the current passively. In this way, VIV is suppressed.

In IODP Expedition 319, the D/V *Chikyu*'s first riser drilling in NantroSEIZE was carried out. Attached to upper riser joints were 132 sets of fairings. Though the duration of strong currents over 2.5 knots was only about 20 hours in this operation, it was confirmed that VIV of the riser was suppressed sufficiently.

In future NantroSEIZE operations, the riser will be exposed to stronger currents for longer periods. The fairing will be essential for the suppression of VIV and the success of NantroSEIZE.

Permissions

List of Contributors

Gilbert F. Camoin
Centre Européen de Recherche et d'Enseignement des Géosciences de l'Environnement (CEREGE) UMR 6635 CNRS, Europôle Méditerranéen de l'Arbois, BP 80, F-13545 Aix-en- Provence cedex 4, France

Yasufumi Iryu
Institute of Geology and Paleontology, Graduate School of Science, Tohoku University, Aobayama, Sendai 980-8578, Japan

Dave B. McInroy
British Geological Survey, Murchison House, West Mains Road, Edinburgh EH9 3LA, U.K

Dale S. Sawyer
Department of Earth Science, Rice University, MS 126, P.O. Box 1892, Houston, Texas 77251- 1892, U.S.A

Millard F. Coffin
Ocean Research Institute, University of Tokyo, 1-15-1 Minamidai, Nakano-ku, Tokyo 164-8639,Japan

Timothy J. Reston
Subsurface Group, Earth Sciences, School of Geography, Earth and Environmental Sciences, University of Birmingham, B15 2TT, U.K

Joann M. Stock
Seismological Laboratory 252-21, California Institute of Technology, 1200 East California Boulevard, Pasadena, Calif. 91125, U.S.A

John R. Hopper
Department of Geology and Geophysics, Texas A&M University, College Station, Texas 77843-3115, U.S.A

Steven D'Hondt
Graduate School of Oceanography, University of Rhode Island, Narragansett Bay Campus, South Ferry Road, Narragansett, R.I. 02882, U.S.A

Fumio Inagaki
Kochi Institute for Core Sample Research, Japan Agency for Marine-Earth Science and Technology (JAMSTEC), B200 Monobe, Nankoku, Kochi, 783-8502, Japan

Timothy Ferdelman
Max Planck Institute (MPI) for Marine Microbiology, Celsiusstr. 1, D-28359, Bremen, Germany

Bo Barker Jørgensen
Max Planck Institute (MPI) for Marine Microbiology, Celsiusstr. 1, D-28359, Bremen, Germany

Kenji Kato
Department of Geosciences, Faculty of Science, Shizuoka University, Shizuoka, 422-8529, Japan

Paul Kemp
Center for Microbial Oceanography: Research and Education, 1000 Pope Road, Marine Sciences Builiding, Honolulu, Hawaii 96822, U.S.A

Patricia Sobecky
School of Biology, Georgia Institute of Technology, Atlanta, Ga. 30332, U.S.A

Mitchell Sogin
Marine Biological Laboratory, 7 MBL Street, Woods Hole, Mass., 02513-1015, U.S.A

Ken Takai
Subground Animalcule Retrieval Program, Extremobiosphere Research Center, Japan Agency for Marine-Earth Science and Technology (JAMSTEC), 2-15 Natsushima-cho, Yokosuka, Kanagawa, 237-0061, Japan

M. Teresa Mariucci
Simona Pierdominici, Paola Montone, Istituto Nazionale di Geofisica e Vulcanologia sezione di Sismologia e Tettonofisica, Via di Vigna Murata 605, 00143, Rome, Italy

Hideaki Motoyama
National Institute of Polar Research, Kaga 1-9-10, Itabashi-ku, Tokyo 173-8515, Japan

Jörg Pross
Institute of Geosciences, University of Frankfurt, Altenhöfer Allee 1, D-60438 Frankfurt, Germany

Ulrich C. Müller
Institute of Geosciences, University of Frankfurt, Altenhöfer Allee 1, D-60438 Frankfurt, Germany

Ulrich Kotthoff
Institute of Geosciences, University of Frankfurt, Altenhöfer Allee 1, D-60438 Frankfurt, Germany

Polychronis Tzedakis
School of Geography, University of Leeds, West Yorkshire LS2 9JT, U.K

Alice Milner
School of Geography, University of Leeds, West Yorkshire
LS2 9JT, U.K

Gerhard Schmiedl
Geological-Paleontological Institute, University of Hamburg, Bundesstraße 55, D-20146 Hamburg, Germany

Kimon Christanis
Department of Geology, University of Patras, GR-265.00 Rio-Patras, Greece

Stavros Kalaitzidis
Department of Geology, University of Patras, GR-265.00 Rio-Patras, Greece

Henry Hooghiemstra
Institute for Biodiversity and Ecosystem Dynamics, University of Amsterdam, Kruislaan 318, NL-1098 SM Amsterdam, The Netherlands

Ernst Huenges
GFZ Potsdam, Telegrafenberg, D- 14473, Potsdam, Germany

Inga Moeck
GFZ Potsdam, Telegrafenberg, D- 14473, Potsdam, Germany

Philippe Gaillot
CDEX-IFREE, Japan Agency for Marine- Earth Science and Technology, Yokohama Institute for Earth Science, 3173-25 Showa-machi, Kanazawa-ku, Yokohama, Kanagawa, 236-0001 Japan

Tim Brewer
Department of Geology – Geophysics and Borehole Research, University of Leicester, University Road, Leicester, LE1 7RH, U.K

Philippe Pezard
Laboratoire de Géophysique et d'Hydrodynamique en Forage, Geosciences Montpellier, University of Montpellier 2, France

En-Chao Yeh
Department of Geosciences, National Taiwan University, No.1, Sec. 4, Roosevelt Road, Taipei 106, Taiwan

Earl Davis
PGC, Geological Survey of Canada, 9860 West Saanich Road, North Saanich, Sidney, British Columbia V8L 4B2, Canada

Keir Becker
RSMAS/MGG, University of Miami, 4600 Rickenbacker Causeway, Miami, Fla. 33149-1098, U.S.A

Nicole Biebow
Alfred Wegener Institute (AWI), Am Handelshafen 12, D-27570, Bremerhaven, (Building E-3335), Germany

Jörn Thiede
Alfred Wegener Institute (AWI), Am Handelshafen 12, D-27570, Bremerhaven, (Building E-3221), Germany

Tim Freudenthal
Marum Center for Marine Environmental Sciences, University of Bremen, Leobener Str. D-28359 Bremen, Germany

Gerold Wefer
Marum Center for Marine Environmental Sciences, University of Bremen, Leobener Str. D-28359 Bremen, Germany

Lothar Wohlgemuth
Operational Support Group ICDP, GFZ Potsdam, Telegrafenberg A34, 14473 Potsdam, Germany

Ulrich Harms
Operational Support Group ICDP, GFZ Potsdam, Telegrafenberg A34, 14473 Potsdam, Germany

Jürgen Binder
Herrenknecht Vertical GmbH, 77963 Schwanau, Germany

Yasufumi Iryu
Institute of Geology and Paleontology, Graduate School of Science, Tohoku University, Aobayama, Sendai 980-8578, Japan

Hiroki Matsuda
Department of Earth Sciences, Faculty of Science, Kumamoto University, Kurokami 2-39-1, Kumamoto 860-8555, Japan

Hideaki Machiyama
Kochi Institute for Core Sample Research, Japan Agency for Marine-Earth Science and Technology (JAMSTEC) Monobe-otsu 200, Nangoku, Kochi 783-8502, Japan

Werner E. Piller
Institute of Earth Sciences (Geology and Palaeontology), University of Graz, Heinrichstrasse 26, A- 8010 Graz, Austria

Terrence M. Quinn
John A. and Katherine G. Jackson School of Geosciences, Department of Geological Sciences, The University of Texas at Austin, 1 University Station C1100, Austin, Texas 78712-0254, U.S.A

Maria Mutti
Institut für Geowissenschaften, Universität Potsdam-Postfach 60 15 53, D-14415 Potsdam, Germany

Kozo Takahashi
Department of Earth & Planetary Sciences, Graduate School of Sciences, Kyushu University, Hakozaki 6-10-1, Higashi-ku, Fukuoka 812-8581, Japan

A. Christina Ravelo
Ocean Sciences Department, University of California, 1156 High Street, Santa Cruz, CA 95064, U.S.A

Carlos Alvarez Zarikian
Integrated Ocean Drilling Program & Department of Oceanography, Texas A&M University, 1000 Discovery Drive, College Station, TX 77845-9547, U.S.A

Mark Zoback
Department of Geophysics, Stanford University, Stanford, CA 94305-2215, U.S.A

Stephen Hickman
U.S. Geological Survery, 345 Middlefield Road MS 977, Menlo Park, CA 94025-3591, U.S.A

William Ellsworth
U.S. Geological Survery, 345 Middlefield Road MS 977, Menlo Park, CA 94025-3591, U.S.A

Martin Melles
Institute of Geology and Mineralogy, University of Cologne, Zuelpicher Str. 49a, D-50674 Cologne, Germany

Julie Brigham-Grette
Department of Geosciences, University of Massachusetts, 611 North Pleasant Street, Amherst, MA 01003, U.S.A

Pavel Minyuk
North-East Interdisciplinary Scientific Research Institute, FEB RAS, 16 Portovaya St., 685000, Magadan, Russia

Christian Koeberl
Department of Lithospheric Research, University of Vienna, Althanstrasse 14, A-1090 Vienna, Austria (and: Natural History Museum, A-1010 Vienna, Austria)

Andrei Andreev
Institute of Geology and Mineralogy, University of Cologne, Zuelpicher Str. 49a, D-50674 Cologne, Germany

Timothy Cook
Department of Geosciences, University of Massachusetts, 611 North Pleasant Street, Amherst, MA 01003, U.S.A

Grigory Fedorov
Arctic and Antarctic Research Institute, Bering Street, 199397 St. Petersburg, Russia

Catalina Gebhardt
Alfred Wegener Institute for Polar and Marine Research, Am Alten Hafen 26, D-27568 Bremerhaven, Germany

Eeva Haltia-Hovi
GFZ German Research Centre for Geosciences, Potsdam, Telegrafenberg C321, D-14473 Potsdam, Germany

Maaret Kukkonen
Institute of Geology and Mineralogy, University of Cologne, Zuelpicher Str. 49a, D-50674 Cologne, Germany

Norbert Nowaczyk
GFZ German Research Centre for Geosciences, Potsdam, Telegrafenberg C321, D-14473 Potsdam, Germany

Georg Schwamborn
Alfred Wegener Institute for Polar and Marine Research, Telegrafenberg A43, D-14473 Potsdam, Germany

Volker Wennrich
Institute of Geology and Mineralogy, University of Cologne, Zuelpicher Str. 49a, D-50674 Cologne, Germany

Nikolay I. Vasiliev
Drilling Department, St. Petersburg State Mining Institute, 2, 21 Line, St. Petersburg 199106, Russia

Pavel G. Talalay
Polar Research Center, Jilin University, No. 6 Ximinzhu Street, Changchun City, Jilin Province 130026, China

Mordechai Stein
Geological Survey of Israel, 30 Malkhe Israel St. Jerusalem, 95501, Israel

Zvi Ben-Avraham
Department of Geological Sciences, Louis Ahrens Building, Library Road, University of Cape Town, Rondebosch, 7700, Republic of South Africa

Steve Goldstein
Lamont-Doherty Earth Observatory, 213 Comer, 61 Route 9W - P.O. Box, 1000 Palisades, NY 10964- 8000, U.S.A

Amotz Agnon
Institute of Earth Sciences, Hebrew University of Jerusalem, Edmond J. Safra campus, Givat Ram, 91904, Jerusalem

Daniel Ariztegui
Department of Geology and Paleontology, University of Geneva, 13, Rue des Maraîchers, CH-1205 Genève, Switzerland

Achim Brauer
Helmholtz Centre Potsdam, GFZ German Research Centre for Geosciences, Section 5.2, Climate Dynamics and Landscape Evolution, Telegrafenberg, C 323 D-14473 Potsdam, Germany

Gerald Haug
ETH Zürich, Geologisches Institut NO G 51.1, Sonneggstrasse 5, 8092 Zürich, Switzerland

Emi Ito
Geology and Geophysics, Room 108, PillsH 0211, 310 Pillsbury Drive SE, Minneapolis, MN 55455, U.S.A

Yoshinori Yasuda
International Research Center for Japanese Studies, 3-2 Oeyama-cho, Goryo, Nishikyo-ku, Kyoto 610-1192, Japan

Bradley Weymer
SAFOD Curatorial Specialist and Graduate Assistant Researcher, Integrated Ocean Drilling Program and SAFOD, Texas A&M University, 1000 Discovery Drive, College Station, TX 77845-9547, U.S.A

John Firth
Curator GCR Superintendent, Integrated Ocean Drilling Program and SAFOD, Texas A&M University, 1000 Discovery Drive, College Station, TX 77845-9547, U.S.A

Phil Rumford
GCR Superintendent, Integrated Ocean Drilling Program and SAFOD, Texas A&M University, 1000 Discovery Drive, College Station, TX 77845-9547, U.S.A

Frederick M. Chester
Professor Associate Professor, Department of Geology and Geophysics, TAMU, Center for Tectonophysics and Department of Geology & Geophysics, Texas A&M University, College Station, TX 77843-3115, U.S.A

Judith S. Chester
Associate Professor, Department of Geology and Geophysics, TAMU, Center for Tectonophysics and Department of Geology & Geophysics, Texas A&M University, College Station, TX 77843-3115, U.S.A

David Lockner
U.S. Geological Survey, Menlo Park, U.S. Geological Survey Earthquake Science Center, 345 Middlefield Road, MS/977Menlo Park, CA 94025, U.S.A

Constance Bertka
Deep Carbon Observatory, Senior Consultant, Geophysical Laboratory, 5251 Broad Branch Road NW, Washington, DC 20015, U.S.A

Donna K. Blackman
IGPP, Scripps Institution of Oceanography, University of California San Diego, 9500 Gilman Drive, La Jolla, CA 92093-0225, U.S.A

Benoit Ildefonse
CNRS, Géosciences Montpellier, Université Montpellier 2, CC 60, 34095 Montpellier Cédex 05, France

Peter B. Kelemen
211 Comer, 61 Route 9W – P.O. Box 1000, Palisades, NY 10964-8000, U.S.A

Andrea Johnson Mangum
Deep Carbon Observatory, Program Associate, Geophysical Laboratory, 5251 Broad Branch Road NW, Washington, DC 20015, U.S.A

Greg Myers
Senior Technical Expert: Engineering and Technology, Department: Ocean Drilling, Consortium for Ocean Leadership, 1201 New York Avenue NW, 4th Floor, Washington, DC 20005, U.S.A

Jason Phipps-Morgan
Professor, Earth and Atmospheric Sciences, Cornell University, Snee Hall, Room 2122, Ithaca, NY 14853, U.S.A

Matthew Schrenk
Assistant Professor, Howell Science S301B, Department of Biology, East Carolina University, Greenville, NC 27858, U.S.A

Kiyoshi Suyehiro
(corresponding author), President & CEO, IODP-MI, Tokyo University of Marine Science and Technology, Office of Liaison and Cooperative Research 3rd Floor, 2-1-6, Etchujma, Koto-ku, Tokyo 135-8533, Japan

Yoshiyuki Tatsumi
Institute for Research on Earth Evolution (IFREE), Japan Agency for Marine-Earth Science and Technology, 2-15 Natsushima-cho, Yokosuka-city, Kanagawa 237-0061, Japan

Jessica Warren
Assistant Professor, School of Earth Sciences, Department of Geological and Environmental Sciences, Stanford University, 450 Serra Mall, Stanford, CA 94305, U.S.A

Ilmo T. Kukkonen
Geological Survey of Finland, Espoo, P.O. Box 96, FI-02151 Espoo, Finland

Maria V.S. Ask
Luleå University of Technology, SE-971 87 Luleå, Sweden

Odleiv Olesen
Geological Survey of Norway, NO-7491 Trondheim, Norway

Henning Lorenz
Uppsala University, Department of Earth Sciences, Villavägen 16, 752 36 Uppsala, Sweden

David Gee
Uppsala University, Department of Earth Sciences, Villavägen 16, 752 36 Uppsala, Sweden

Christopher Juhlin
Uppsala University, Department of Earth Sciences, Villavägen 16, 752 36 Uppsala, Sweden

Anthony W. Walton
Chair, Board of Directors, DOSECC Inc., Department of Geology, The University of Kansas, Lawrence, KS 66045, U.S.A

Maarten de Wit
Africa Earth Observatory Network, University of Cape Town, Rondebosch 7701, South Africa

Lisa McNeill
School of Ocean and Earth Science, National Oceanography Centre, Southampton, University of Southampton, Southampton, SO14 3ZH, UK

Demian Saffer
The Pennsylvania State University, University Park, PA 16802, U.S.A

Tim Byrne
Center for Integrative Geosciences, University of Connecticut, Storrs, CT 06269, U.S.A

Eiichiro Araki
Earthquake and Tsunami Research Project for Disaster Prevention, Japan Agency for Marine-Earth Science and Technology (JAMSTEC), Kanagawa 237-0061, Japan

Sean Toczko
Center for Deep Earth Exploration (CDEX), Japan Agency for Marine-Earth Science and Technology (JAMSTEC), Kanagawa 237-0061, Japan

Nobu Eguchi
Center for Deep Earth Exploration (CDEX), Japan Agency for Marine-Earth Science and Technology (JAMSTEC), Kanagawa 237-0061, Japan

Kyoma Takahashi
Center for Deep Earth Exploration (CDEX), Japan Agency for Marine-Earth Science and Technology (JAMSTEC), Kanagawa 237-0061, Japan

Michael B. Underwood
Department of Geological Sciences, University of Missouri, Columbia, MO 65203, U.S.A

Saneatsu Saito
Institute for Research on Earth Evolution, Japan Agency for Marine-Earth Science and Technology, 2-15 Natsushima-cho, Yokosuka 237-0061, Japan

Yu'suke Kubo
Center for Deep Earth Exploration, Japan Agency for Marine-Earth Science and Technology, 3173-25 Showa-machi, Kanazawa-ku, Yokohama 236-0001, Japan

Gregory Mountain
(Co-chief Scientist), Department of Earth and Planetary Sciences, Rutgers University, 610 Taylor Road, Piscataway, NJ 08854, U.S.A

Dr. Jean-Noël Proust
(Co-chief Scientist), Géosciences, CNRS, Université Rennes1, Campus de Beaulieu, 35042 Rennes, France

Aurèle Vuillemin
Section of Earth & Environmental Sciences, University of Geneva, Rue des Maraîchers 13, CH-1205 Geneva, Switzerland

Daniel Ariztegui
Section of Earth & Environmental Sciences, University of Geneva, Rue des Maraîchers 13, CH-1205 Geneva, Switzerland

Crisogono Vasconcelos
Geological Institute, ETH Zürich, Sonneggstr. 5, 8092 Zürich, Switzerland

Alexander C. Skinner
ACS Coring Services, 13 Riccarton Drive, Currie, Edinburgh, EH14 5PN, Scotland, U.K

Paul Bowers
Rok-Max Drilling Tools Ltd., P.O. Box 87, Truro, Cornwall, TR3 7ZQ, U.K

Sverrir Þórhallsson
Iceland Geosurvey (ISOR), Grensasvegur 9, Reykjavik, IS-108, Iceland

Guðmundur Ómar Friðleifsson
HS Orka hf, Brekkustígur 36, 260 Reykjanesbær, Iceland

Hermann Guðmundsson
Iceland Geosurvey (ISOR), Grensasvegur 9, Reykjavik, IS-108, Iceland

Kai Mangelsdorf
Helmholtz Centre Potsdam, GFZ German Research Centre for Geosciences, Telegrafenberg, 14473 Potsdam

Jens Kallmeyer
University of Potsdam, Earth and Environmental Sciences, Karl-Liebknecht Str. 25, 14476 Potsdam

Benoît Ildefonse
CNRS, Université Montpellier 2, CC60, 34095 Montpellier cedex 5, France

Natsue Abe
Institute for Research on Earth Evolution (IFREE), Japan Agency for Marine-Earth Science and Technology (JAMSTEC), 2-15 Natsushima-cho, Yokosuka 237-0061, Japan

Donna K. Blackman
Scripps Institution of Oceanography, UCSD, La Jolla, CA 92093-0225, U.S.A

J. Pablo Canales
Woods Hole Oceanographic Institution, Department of Geology and Geophysics, Woods Hole, MA 02543, U.S.A

Yoshio Isozaki
Marine Technology Center (MARITEC), JAMSTEC, 2-15 Natsushima-cho, Yokosuka 237-0061, Japan

Shuichi Kodaira
Institute for Research on Earth Evolution (IFREE), Japan Agency for Marine-Earth Science and Technology (JAMSTEC), 3173-25 Showa-machi, Kanazawa-ku, Yokohama 236-0001, Japan

Greg Myers
Consortium for Ocean Leadership, 1201 New York Avenue NW, Suite 400, Washington, DC 20005, U.S.A

Kentaro Nakamura
Precambrian Ecosystem Laboratory, JAMSTEC, 2-15 Natsushima-cho, Yokosuka 237-0061, Japan

Mladen Nedimovic
Department of Earth Sciences, Dalhousie University, Edzell Castle Circle, Halifax, NS B3H 4J1, Canada

Alexander C. Skinner
ACS Coring Services, 13 Riccarton Drive, Currie, Edingburgh, EH14 5PN, Scotland, U.K

Nobukazu Seama
Research Center for Inland Seas, Kobe University, 1-1 Rokkodai, Nada, Kobe 657-8501, Japan

Eiichi Takazawa
Department of Geology, Faculty of Science, Niigata University, 2-8050, Niigata, 950-2181, Japan

Damon A. H. Teagle
National Oceanography Centre, Southampton, University of Southampton, U.K

Masako Tominaga
Department of Geology and Geophysics, Woods Hole Oceanographic Institution, Woods Hole, MA 2543, U.S.A

Susumu Umino
Department of Earth Sciences, Kanazawa University, Kakuma-Machi, Kanazawa-Shi, Ishikawa 920- 1192, Japan

Douglas S. Wilson
Department of Earth Science, UCSB, Santa Barbara, CA 93106-9630, U.S.A

Masaoki Yamao
Center for Deep Earth Exploration (CDEX), JAMSTEC, 2-15 Natsushima-cho, Yokosuka 237- 0061, Japan